Introduction and Application of Organic Fertilizers as Protectors of Our Environment

Introduction and Application of Organic Fertilizers as Protectors of Our Environment

Edited by

Munir Ozturk,
Nudrat Aisha Akram,
Bengu Turkyilmaz Unal
and Muhammad Ashraf

Cambridge
Scholars
Publishing

Introduction and Application of Organic Fertilizers as Protectors of Our Environment

Edited by Munir Ozturk, Nudrat Aisha Akram, Bengu Turkyilmaz Unal and Muhammad Ashraf

This book first published 2022

Cambridge Scholars Publishing

Lady Stephenson Library, Newcastle upon Tyne, NE6 2PA, UK

British Library Cataloguing in Publication Data
A catalogue record for this book is available from the British Library

ISBN (10): 1-5275-8098-9
ISBN (13): 978-1-5275-8098-5

CONTENTS

FOREWORD

PROF. NAFEES A. KHAN

DEPARTMENT OF BOTANY
PH. D. (ALIG.), D. SC. ALIGARH MUSLIM UNIVERSITY
PROFESSOR OF PLANT PHYSIOLOGY ALIGARH, INDIA

The world population is likely to reach around 9 billion by the year 2050. Reports from different agencies including United Nations suggested that the food crisis will be the mainstream concern for all humankind. In order to maintain the demand and supply of the food for burgeoning population along with economic growth, the production of food must be increased up to 50-70% by the year 2050. The agricultural system in the current scenario is facing several challenges such as climate change, biotic stresses, poor management in the field practices, irrigation mismanagement, extensive use of chemicals-based fertilizers, and the limited promotion of the organic agriculture around the world. The use of fertilizers was responsible for increased grain production during 1960's revolution. The fertilizers contribute about 40-60% of the food production worldwide. Developing best fertilizer management practices should include the use of fertilizers in an effective, efficient and safe manner to ensure sufficient crop production especially in developing countries. Additionally, the agricultural practices are continuously facing changes in nature and fertilizer practices, varying with crop production system, time and location, in turn significantly affect the final produce under different crop production system. Accordingly, more convincing research providing double benefits of improving fertilizer efficiency and also minimizing losses from organic/natural resources is needed to determine benefits, costs and adequate agricultural practices in order finally to avoid the necessity of using synthetic inorganic fertilizers.

This book entitled 'Introduction and Application of Organic Fertilizers as Protectors of Our Environment' is anticipated to offer an overview of developing scientific issues related with the use of organic fertilizers that have potentiality to protect lithosphere. It comprises of 22 chapters highlighting some key research activities towards developing a sustainable agriculture and environment. The first chapter explains the potential use of plant growth promoting rhizobacteria (PGPR) and suggests how it can be considered as an alternative to inorganic fertilizers to advance crop/plant yield and quality. The second chapter provides information on organic agriculture and its role in providing its supports towards sustainable ecosystem and human health; whereas the third chapter discusses the plant-microbe associations and their interface between unavailability and availability of soil nutrients. The fourth chapter highlights the importance of managing the indigenous mycorrhizae when soil borne diseases increase and the soil nutrients deplete under field conditions. Chapter five summarizes the latest information on mutations leading to a better tomorrow in the production of improved crop varieties, because crop improvement through induced mutation adds credence. The sixth chapter describes the potential benefits, tolerance and implementation of arbuscular mycorrhiza fungi and biochar interactive use in the remediation of polluted soils and also plant nutrition in sustainable agriculture. The sixth chapter aims to highlight useful information on algal bio-fertilizer for developing an ecofriendly and non-polluting approach in improving the crop productivity. The seventh chapter discusses how vermicomposting could enhance the degradation of organic wastes and enable reduction of environmental pollution. It also highlights how vermi-compost method can provide good economic benefit compared to other methods to support sustainability in agriculture. The chapter eight enlightens different aspects of melatonin. The authors highlight the potential mechanisms underlying melatonin-mediated reduction in abiotic stress-impacts on crops, reactive oxygen species-detoxification and melatonin-mediated improved crop production. In Chapter nine authors have presented a detailed overview on the topic of plant metabolomics highlighting the crucial role of metabolomics under plant growth/development and response of plants to different constraints. The chapter ten attempts to enlighten the information on the damages caused by conventional farming as well as correct and conscious use of fertilizers allowed in organic farming. The eleventh chapter covers principles of fertilizer use in organic agriculture. The chapter twelve highlights the organic farming system and organic fertilizers in improving the quality and quantity of vegetables. In chapter thirteen the topic on phenomics has been discussed with a perfect goal to link the gap between plant function, genomics and agronomic characters. The fourteenth chapter takes new direction and discusses the complex interrelation between plant nutrients and abiotic stresses like water scarcity, heat stress and elevated CO_2 concentration. Fifteenth chapter deals with the advanced technology based work and summarizes the details of green nanotechnology together with its applications like nano-fertilizers, nano-pesticides and nanoherbicides. Sixteenth chapter is devoted to the importance of ecological zoning in South Kazakhstan and describes the developmental trends of landscape structure of main territory and the agriculture land, based on local ecological studies and suggests crucial findings. Structural and functional genomics of rice is the topic of chapter seventeen. It discusses the use of rice as a model plant for genomic study of other crops. Green nanotechnology is the subject matter of chapter eighteen. It deals with the goals of green nanotechnology. Potential of organic fertilizers for fruits and vegetable production is an important subject matter in horticulture and discussed in chapter nineteen. The authors suggest that organic fertilizers are potentially better alternatives over traditional fertilizers which pose the risks of ground water contamination via nitrate leaching. In this chapter, insights into various types of organic fertilizer have been presented and interactive effects of these fertilizers on soil characteristics and vegetable production have been discussed. Chapter twenty presents the results expected to be used by the local executive bodies for industrial and agricultural development of the area studied. The chapter twenty-one deals with the topic of organelleomics in particular mitochondria and chloroplast of plants. The chapter brings forth an emerging research area and presents age old goals of crop improvement and yield. The last chapter (chapter twenty-two) is based on the use of green manures that can help to recover degraded soils.

The contributions made by the authors and also the efforts of editors are timely in the changing scenario of agriculture and crop productivity. The publication of this book has very significant importance for scientists, teachers, students, agricultural agronomists and farmers to understand the advantages of organic fertilizers. I sincerely hope that the information presented in this book will be of great value to those directly engaged in the handling and use of organic fertilizers and will be proven a good use of organic fertilizers to accomplish a sustainable agriculture without environmental orientations.

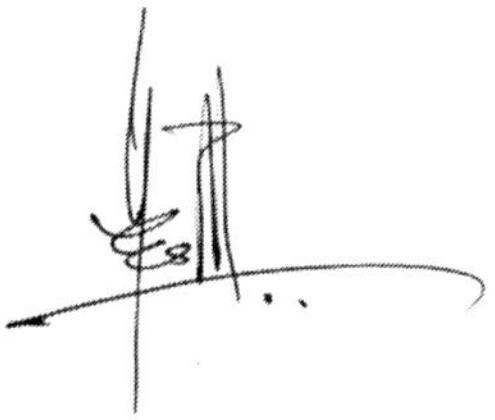

Dated: October 15, 2021
Prof. Nafees A. Khan

PREFACE

This book aims to deal with organic fertilizers, as there is an increasing demand for organic production and protection of environment. From such perspectives, this book is very timely as many countries are now trying to commercially produce organic fertilizers and are practicing organic agriculture not only to improve their soil resources but also produce environmentally sustainable and safe food for healthy human life. The organic fertilizers are commonly produced through different ways using organic substances. Animal manure and agricultural waste are recycled as the main raw materials, this way they are not only creating economic benefits for the producers, but also making a great contribution to soil and natural environments. The rapid development of livestock and poultry farming produces a lot of wastes. The organic fertilizers are commonly produced through different ways using organic substances. Animal manure and agricultural waste are recycled as the main raw materials, this way they are not only creating economic benefits for the producers, but also making a great contribution to soil and natural environments. Biological organic fertilizer is widely used in farmland, fruit trees, flowers, landscaping, high grade lawn and soil improvement. This book contains state-of-the-art information in organic farming for researchers, educators, graduate students and industry. It consists of 22 Chapters.

Chapter 1 provides the reader with information about "Plant Growth Promoting Rhizobacteria" (PGPR) which are the natural soil microorganisms and are generally colonized in plant roots and are called rhizobacteria in the so-called rhizosphere. They can be considered as alternative way for inorganic fertilizer with a specific goal to advance plant development, quality and yield. The aim of *Chapter 2* is to examine the aspects and problems of conventional agricultural production that prevent sustainability and to suggest solutions based on organic farming practices as to overcome this problem. Organic production model supports ecosystem and human health. The *third Chapter* deals with the plant microbial association. In the plant production environment there are completely different kinds of microbial interactions with different microbes. Microbial interactions are ubiquitous, diverse, critically important in the function of any biological community, and are crucial in global biogeochemistry. The organic production model supports ecosystem and human health.

Importance of organic agriculture from the microbiological point of view is discussed in *Chapter 4.* It is known that the soil and water qualities are significantly affected by the use of chemicals in agriculture. For several reasons, microbiologically, using plant rhizosphere mechanisms such as management of indigenous mycorrhizae or/and selected mycorrhizal application may increase the soil fertility and increase the biomass production ultimately contributing to the agricultural sustainability.

The *Chapter 5* summarizes the latest information on mutations leading to a better tomorrow in the production of improved crop varieties, because crop improvement through induced mutation adds credence. The review suggests that induced mutation is an indomitable approach and has contributed enormously towards economic step up of developing as well as developed countries. Similarly, the authors in *Chapter 6* deal also with the importance of exploitation of mycorrhizae and their function as biological and organic amendments, which are the cost effective and environment friendly approaches for plant growth and phytoremediation in polluted soils. This chapter also points out the potential benefits, tolerance and implementation of mycorrhizae use in the remediation of polluted soils and plant nutrition in sustainable agriculture.

Green fertilization is the subject matter of *Chapter 7* which is considered for agriculture in a long time. Different plants are used for green fertilization; leguminous plants are generally preferred and are accepted as the best green fertilization plants, due to their content of nutrients. The green fertilization technique involves burying green plant materials in the ground to increase the quality of the soil for plant production. Vermicompost is the subject matter of *Chapter 8.* They enhance degradation of organic wastes and enable the reduction of environmental pollution, consequently increase the sustainability of the soil fertility and protect environmental and human health.

In *Chapter 9* authors have presented a detailed overview on the topic of plant metabolomics and their role in crop improvement. This topic reflects an integration of gene expression, protein interaction and other different regulatory processes being closer to the phenotype than mRNA transcripts or proteins alone. It summarizes the latest data on circadian rhythms, crop improvement, metabolite profiling, crop breeding, environmental stresses, spatial and temporal characterization of metabolome with examples from the studies on *Arabidopsis thaliana*. It discusses the crucial role of metabolomics under plant growth, development and response of plants to different biotic and abiotic environmental constraints.

Production of indoleamine in plants has emerged as an interest in plant sciences and this matter is discussed in *Chapter 10.* The indoleamine melatonin as a free radical scavenger, electron donor, and antioxidant used largely in vitro and in vivo studies. Melatonin has been confirmed to aid plants to grow in the harsh environment. This chemical is recognized to have great potential to promote crop production especially under environmentally stress conditions.

The Authors of *Chapter 11* discusses organic fertilizers such as compost, humic acids, fulvic acids, leonardit, farmyard manure, green manures, seaweeds, mycorrhizas. Some fertilizers include enzymes, amino acids, seaweed extracts and microorganisms. They suggest that excessive use of fertilizers should not be permitted in organic production, as in the use of commercial fertilization. It is suggested and explained here the damage caused by conventional farming as well as correct and conscious use of fertilizers that are allowed in organic farming. The principles of fertilizer use in organic agriculture is explained in *Chapter 12*. It is known that organic agriculture systems usually don't use chemicals as fertilizers. The chapter indicates that organics that are used are naturally derived materials of biological or mineral origin and have low nutrient concentrations or solubility.

The *chapter 13* discusses the topic on phenomics with a perfect goal to link the gap between plant function, genomics and agronomic characters. The aim of authors here has been to enlighten the facts related to the field based phenomics, a simultaneous proximal feeling for spectral reflectance, canopy warmth, and vegetation architecture. *Chapter 14* aimed to explain the application of *pranic* agriculture protocol on tomatoes, pole beans, cucumber and European cucumber plants. Prana is the energy needed by the plants for its growth. Authors study shows that pranic agriculture could play a significant role in improving the quality and quantity of vegetable production.

Positive affect of organic fertilizers to improve the physical and chemical productivity characteristics and microbial properties of the soil is the subject matter of *Chapter 15*. Organic fertilizers all add important plant nutrients as well as significant amount of carbon to the soil. It is well known that unconscious and excessive use of fertilizers in organic farming also can lead to a damage of the environment. Here authors provide information about the damage caused by conventional farming as well as correct and conscious use of fertilizers in organic farming.

Chapter 16 is a review paper aimed to discuss the complex interrelation between plant nutrients (nitrogen, phosphorus, potassium and magnesium), water scarcity, heat stress and elevated CO_2 concentration. Many climate change models suggest that the effects of the occurrence and duration of heat and drought stress on major crops will increase in many regions, negatively affecting the food security. In addition to soil scientists plant breeders, botanists and molecular geneticists are trying to identify genes involved in drought tolerance under water scarce conditions.

Structural and functional genomics of rice is the topic of *chapter 17*. It discusses the use of rice; with a genome size of 430 Mb having 12 chromosomes and >50,000 genes; as a model plant for genomic study of other crops. Different techniques including cytogenetic, molecular genetics and physical mapping have been discussed for constructing rice genome map thus providing information regarding general features of the rice genome. Green nanotechnology is the subject matter of *Chapter 18*. It deals with the goals of green nanotechnology and includes its applications in nano fertilizers, pesticides and herbicides in the soil and plant diseases management.

Potential of organic fertilizers for fruits and vegetable production is an important subject matter in horticulture and discussed in *Chapter 19*. The authors suggest that organic fertilizers are potentially better alternatives over traditional fertilizers which pose the risks of ground water contamination via nitrate leaching. In this chapter, insights into various types of organic fertilizer have been presented and interactive effects of these fertilizers on soil characteristics and vegetable production have been discussed.

Chapter 20 was devoted to the importance of ecological zoning in South Kazakhstan in agricultural production and describe the developmental trends of landscape structure of main territory and the agriculture land. The *chapter 21* deals with the topic of organellomics in particular mitochondria and chloroplast of plants. The chapter brings forth an emerging research area.

Several theories propounded to elucidate the evolution of both mitochondria and chloroplast, endosymbiont theory of evolution, lateral gene transfer theory and nucleosymbiosis or fusion theory have been summarized together with the mechanism of RNA splicing and different proteins involved in the process. The chapter therefore presents age old goals of crop improvement and yield. The results expected to be used by the local executive bodies for industrial and agricultural development of the area studied. Green manure use in organic agriculture is the subject matter of the last *Chapter* (*22*) of the book. The use of green manures can help to recover degraded soils and improve crop yields with low cost with easy application techniques.

Prof. Dr. Ahmet Ruhi Mermut
University of Saskatchewan, Canada
Harran University, Şanlıurfa, Turkey

CHAPTER 1

PLANT GROWTH-PROMOTING RHIZOBACTERIA AS AN ORGANIC FERTILISER SOURCE

METIN TURAN,[1*] EKREM OZLU,[2] ERTAN YILDIRIM,[3] NURGUL KITIR,[4] ADEM GUNES,[1] HIKMET KATIRCIOĞLU,[5] SANEM ARGIN,[6] ILGIN KARACAN,[1] CANSU COBAN,[1] DILEK UCAR,[6] DENIZ OLGUN,[1] HAZAL ELITAS,[1] NUKHET HALILOĞLU,[1] DILA OMACER,[1] SHEYDA SHAKOORY[1]

[1] Department of Genetics and Bioengineering, Faculty of Engineering, Yeditepe University, Istanbul, Turkey.
[2] Department of Soil Science, Wisconsin-Madison University, Madison, Wisconsin, USA
[3] Department of Horticulture, Faculty of Agriculture, Ataturk University, Erzurum, Turkey
[4] Department of Plant Production and Technologies Faculty of Agriculture and Natural Sciences Konya Food and Agriculture University, Konya, Turkey
[5] Secondary School Science and Mathematics Education, Gazi University, Ankara, Turkey
[6] Department of Agricultural Trade and Management Yeditepe University, Istanbul, Turkey. m_turan@hotmail.com

Abstract

In order to meet the increased demand for food resulting from population growth, environmental damage is increased owing to the intensive use of chemicals in agriculture. The high usage of chemicals and inorganic fertilizers result in the degradation of arable land and an increase in the negative effects of plant pathogens present in the soil. In addition, plants face other stress conditions such as salinity, drought, temperature changes, frost, and heavy metals, which negatively affect their growth, especially the growth of roots. Plant growth-promoting Rhizobacteria (PGPR) can be considered an alternative to the use of inorganic fertilisers with the specific goals of advancing plant development, quality and yield. PGPR are natural soil microorganisms, which are generally colonised in plant roots and grow in a variety of ways in the rhizosphere and in the phytospheres. PGPR generally play various roles in increasing the nutrient ratio of the plant, encourage plant growth, and control organic pollutants and diseases by colonising the plant roots and leaves. They can act as biofertilisers or biocontrol agents due to their inherent capabilities, and as phytostimulators when the PGPR are applied to the plant roots. Furthermore, PGPR have agronomic significance on account of plant growing cycles on the planet in differing spells of cold conditions. PGPR can be examined in two subsections of the mechanism of actions, direct and indirect. PGPR might enhance the accessibility of the plant to nutrients and water by providing for the integrity of the plant root under stress conditions and, indirectly, affecting the repair of damaged areas. In addition, PGPR found in the soil are capable of increasing the availability of phosphorus to the plant by producing organic acids and enzymes. Moreover, PGPR have potential in protecting the plant against cold stress by producing antifreeze proteins, and in being able to produce metabolites like plant development controllers, which can specifically advance growth and plant nutrient take-up. In summary, PGPR can find applications in bioengineering, phytostimulation, and rhizoremediation.

Key words: PGPR, soil enzymes, mitigation of cold stress, nutrient uptake

Introduction

Plant growth-promoting rhizobacteria (PGPR) are potential alternatives to inorganic fertilisers in terms of promoting crop growth, yield, and quality. The bacterial characteristics they possess, such as biological nitrogen fixation, the solubilisation of inorganic phosphate, the mineralisation of organic phosphate, nutrient uptake, 1-aminocydopropane-1-carboxylic acid (ACC) deaminase activity, and the production of siderophores and phytohormones, can be assessed as plant growth-promotion traits. PGPR, which increase plant tolerance against biotic and abiotic stresses, are also capable of promoting plant growth by colonising the plant root. An increase in plant growth due to the addition of PGPR occurs through increases in germination rates, root growth, yield, leaf area, chlorophyll content, nitrogen content, protein content, shoot and root weight, tolerance to drought, salinity, heavy metal stress and delayed leaf senescence. The addition of PGPR can increase agronomic efficiency by reducing costs and environmental pollution. If the selected organisms and inoculation are efficient, the use of harmful materials can be reduced or eliminated. This chapter deals with various aspects: (i) fixation of nitrogen, solubilisation/mineralisation of mineral and organic phosphates and other nutrients; (ii) production of plant hormones, ACC deaminase and siderophore; (iii) promotion of soil health, plant physiology, vigour, yield and quality; and (iv) alleviation and mitigation of abiotic stress on wheat by PGPR. Further elucidation of the different mechanisms may help to make microorganisms a valuable asset in developing future agriculture.

PGPR and Nutrient Uptake

PGPR are natural soil microorganisms that grow in free soil in the rhizosphere or in phytospheres, to promote plant growth and yield. They generally colonise plant roots (root surface or root redicardial tissue) and are called rhizobacteria that support plant growth.

PGPR can be examined in two sub-categories in terms of the mechanism of action. Direct-acting PGPR are effective at each step, from sowing to production, whereas indirect PGPRs affect plant development by bio-control. Depending on the category, PGPR generally play various roles—to increase the nutrient ratio of the plant, encourage plant growth, and control organic pollutants and diseases. Thus, PGPR can find applications in bioengineering, phytostimulation, and rhizoremediation.

Mechanism of direct action: Fig. 1.

- Asymptomatic nitrogen fixation.
- Increase inorganic phosphate solubility.
- Mineralisation of organic phosphorus compounds
- Beneficial bacterium maintains alkaline environment sustainability by improving mineral intake of plants
- Siderophore (iron and organic acid) production
- Facilitating the uptake of a plant-derived compound (phytohormones, etc.) by bacteria
- Production of plant hormones, such asoxine, gibberellin, cytokinin
- Inhibition of ethylene synthesis by ACC deaminase activity
- Reducing the environmental stress to which the plant is exposed by moderating the environment
- Vitamin synthesis
- Increase in root permeability

The induction of plant growth with PGPR depends on all these effects: germination rate, chlorophyll content responsible for photosynthesis, Mg-N content, and delay formation of the absent layer present in the leaf to help in plant growth.

On the other hand, PGPR affect plant growth directly or indirectly (PGPR soil bioremediation) as follows:

- Encouraging plant growth
- Increasing plant tolerance
- Reducing diseases by antibiotic production
- Protecting the plant by removing the xenobiotics.

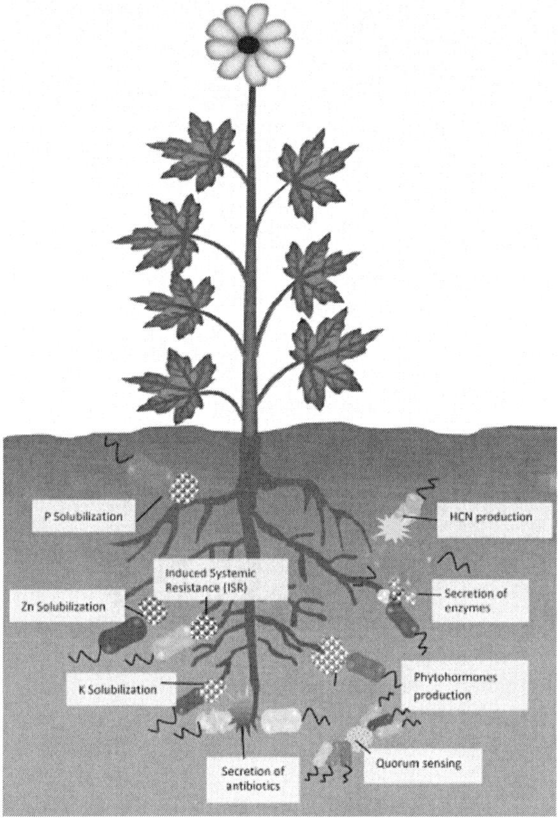

Fig. 1. Key influences of PGPR (Mishra, Prakash, and Arora 2016)

PGPR Bacteria and Made-Up Studies

Acinetobacter, *Alcaligenes*, *Arthrobacter*, *Azospirillium*, *Azotobacter*, *Agrobacterium*, *Bacillus*, *Beijerinckia*, *Burkholdria*, *Clostridium*, *Enterobacter*, *Erwinia*, *Flavobacterium*, *Klebsiella*, *Micrococcus*, *Pseudomonas*, *Rhizobium* *Rhodobacter*, *Rhodospirrilum*, *Serratia* and *Xanthomonas* belong to the PGPR family (Bloemberg and Lugtenberg 2001, Rodriiguez and Fraga 1999, Sturz and Nowak 2000, Sudhakar et al. 2000, Vessey 2003, Esitken 2011, Niranjan Raj, Shetty, and Reddy 2006).

The stimulation of different products with PGPR is demonstrated by laboratory and field experiments. It was determined that *Pseudomonas putida* and *Pseudomonas fluorescens* bacterial strains enhanced the root and shoot length in canola, lettuce, potato, radish, bean, rice, sugar beet, wheat, tomato, apple, citrus and subtropical plants. Wheat yield was reported to be increased by 30% with inoculation of *Azotobacter*, and by 43% with *Bacillus* inoculation; and *Bacillus megaterium* and *Azotobacterchroococcum* increased yields by 10–20% in some crops. *Azospirillum* has been found to increase efficiency in corn, sorghum and wheat, and *Bacillus* has been found to do the same in peanuts, potatoes, sorghum and wheat (Rodrıiguez and Fraga 1999). Moreover, it has been determined that different PGPR strain treatments increased the dry weight of root and shoot for bean (Vedder-Weiss et al. 1999), lettuce, cucumber and tomato; dry weight of root and leaf for onion (Van Peer and Schippers 1988); fruit number and yield of cucumber and tomato (Gagne et al. 1993, McCullagh et al. 1996); fresh and dry weight of basil (Golpayegani and Tilebeni 2011); and plant growth and plant yield of cabbage (Yildirim et al. 2016).

The Role of PGPRs in Food Intake

Together with environmental damage and an increasing population, enabling people to consume good quality food has become much harder. It is expected that today's population of around 7.6 billion will reach 10 billion in the next 50 years (Glick 2014). The intensive use of chemicals and excessive tillage in agriculture increase the necessity of enhancing plant yield per unit area due to increases in the global population. This causes degradation in agricultural lands and increases the negative effectiveness of plant pathogens present in the soil. In addition, unbalanced and unplanned agriculture has destroyed the soil integrity of fertile soils and caused high levels of degradation of soil chemistry. In addition, global warming, which affects day-to-day wellbeing, causes an increase in biological stress in agricultural areas (Carmen and Roberto 2011).

One example is soil salinity, one of the most important issues in agricultural land; and frost damage owing to seasonal imbalances, turns 1–2% of fertile land every year into inefficient areas. Soil salinization depredates 7% of the soil and 20% of the total arable area (Rasool et al. 2013).

Most of the salt is found in areas where the plant is growing and limits its development, thus reducing plant yield. Furthermore, plant growth and metabolism have seriously changed in terms of physiological, morphological, and biochemical properties because of salinity (Gupta and Huang 2014). On the other hand, drought is one of the significant issues influencing fertility in agricultural lands. Critical issues in plant growth for above 50%

of arable land may be due to more severe and frequent drought-induced climate change by 2050 (Vinocur and Altman 2005). Drought stress may be a reason for the destructive consequences for plant development and metabolic processes in essential field plants by influencing water connections, photosynthetic digestion and nutrient uptake.

Under all these circumstances, PGPR play an essential role in expanding protection from both biotic and abiotic stresses, as sustainable agriculture is at stake (Etesami and Beattie 2017, Meena et al. 2017).

The availability of soluble salts and high contents of soil Na^{+} and Cl^{-} ions reduce nutrient uptake by plants. For example, high osmotic pressures, like N, P, K, soluble salts, and high ion contents, cause water to translocate in plants (Feigin 1985, Sharpley et al. 1992). In addition, some elements in the soil negatively affect the nutrient intake of the plant. For example, the high levels of Cd^{+2} in the soil, decreases the intake of Ca^{+2}, Mg^{+2}, Fe^{+2} and Zn^{+2}, by the plant.

1. The Direct Effect of PGPR in Nutrient Uptake Efficiency

1.1. Nitrogen Fixation

Molecular nitrogen (N_2) is one of the most significant components of both the atmosphere and the earth. One of the factors that makes this component so important is its involvement in vital structures such as nucleic acids and proteins. In order to participate in biological systems, N_2 must combine with hydrogen to form a complex that can be reduced. Nitrogen can be converted to nitrate by nitrite bacteria, and can be used by plants. This binding occurs both by symbiotic and non-symbiotic nitrogen binding bacteria via the action of nitrogenase enzyme. The nif gene, which has a part in the synthesis and purpose of the nitrogenase complex, namely the nitrogenase enzyme, is reported to be in a cluster of 7 operons, 20–24 kb.

For plants, this reduction process occurs with legumes and rhizobia in their roots. The reduction of nitrogen with *Rhizobium* accounts for about 20% of all global nitrogen reductions and occupies a large space. It is reduced to 75–300 kg N per hectare, per year.

PGPR can upgrade plant N uptake because the cooperative and non-harmonious N_2 fixation improves different situations, for example, the mineralisation of organic forms of soil nitrogen and the expansion of plant root framework by the generation of phytohormone IAA and ACC deaminase. These bacteria, which are essential for soil micro-ecological dynamics and able to bind nitrogen, are collected into two groups: symbiotic and nonsymbiotic. They are known as symbiotic bacteria that bind *Mesorhizobium*, *Sinorhizobium*, *Azorhizobium* and *Bradyrhizobium* nitrogen to 16S rRNA sequence homology, or nod genes (which promote the formation of nodules in plant roots). The non-nociceptive bacteria binding to nitrogen can be classified as *Azoarcus*, *Gluconacetobacter*, *Diazotrophicus*, *Azotobacter*, *Herbaspirillium*, *Archomobacter*, *Acetobacter*, *Arthrobacter*, *Azospirillum*, *Azomonas*, *Bacillus*, *Beijerinckkia*, *Clostridium*, *Corynebacterium*, *Derxia*, *Enterobacter*, *Klebsiella*, *Pseudomonas*, *Rhodopseudomonas* and *Xanthobacter*.

The expansion in nutrient uptake due toPGPR has been noted in numerous crops, namely, rocket (Dursun, Ekinci, and Dönmez 2008), chickpea (Elkoca, Kantar, and Sahin 2007); barley (Cakmakci, Dönmez, and Erdoğan 2007), and cabbage (Turan et al. 2014, Yildirim et al. 2016). This expansion in plant nutrient uptake might be ascribed to the mechanisms of non-symbiotic N_2-fixation.

1.2. Solubility of Complex Phosphate (P)

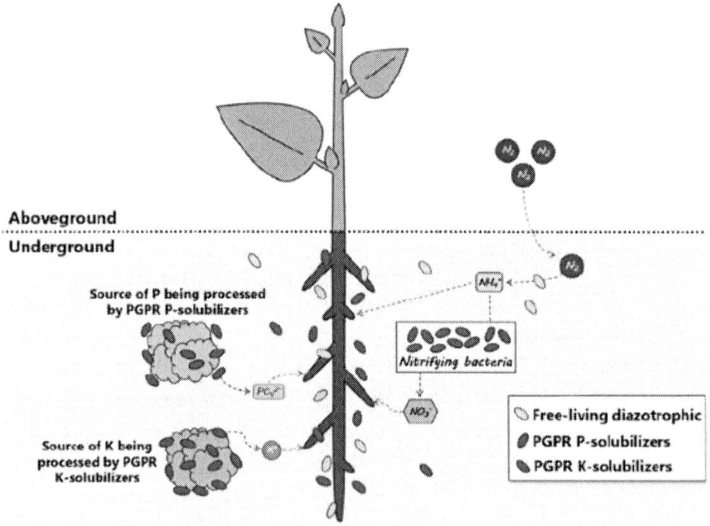

Fig. 2.a. Mechanism of phosphate (P) and potassium (K) uptake in plants (Figueiredo et al. 2016)

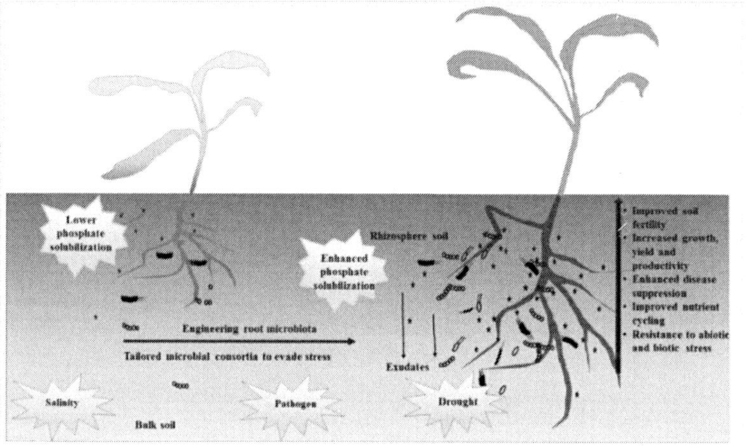

Fig. 2.b. Phosphate solubilisation (Figueiredo et al. 2016)

After nitrogen, the fundamental mineral component that restrains the development of plants is P, which is taken with soluble forms ($H_2PO_4^-$) or (HPO_4^{-2}). $H_2PO_4^-$ and HPO_4^{-2} anions are highly restricted due to their co-precipitation with Ca^+, Mg^{+2}, Fe^{+3}, Mn^{+2}, Zn^{+2} and Al^{+3} cations, and P, in the form that can be taken, is rather rare. It is found especially as organic precipitates on the mineral surfaces of the earth (Richardson and Simpson 2011).

PGPRs have a leadership role in this P cycle from soil to vegetation at different stages. These are: PGPRs in soil and insoluble phosphates (Fig.2. a., b.) (Fig. 3.)

i) Acidification
ii) Chelates
iii) Direct reaction
iv) Secretion of siderophores
v) Secretion of mineral solvent components
vi) IAA production
vii) ACC deaminase activity
viii) Production of extracellular polysaccharides
ix) Increase in P intake of plants by inducing mechanisms of conversion to Fe and Al.

Organic acids and phosphoric acids play a critical part in the mineralisation of soil organic phosphorus and, therefore, in making it ready for plant use. The PGPR bacteria found in the soil are capable of dissolving phosphorus by producing organic acid. These are: *Bacillus polymyxa*, *B. megatarium*, *B. circulans*, *B. subtilis*, *B. firmus*, *Pseudomonas striata*, *P. rathonia*, *Rhizobium leguminosarum* and *R. meliloti*. Even some *Bacillus* species are able to produce a mixture of lactic, isovaleric, isobutyric and acetic acid groups. The best phosphate solvent root bacteria include *Pseudomonas*, *Bacillus* and *Rhizobium* (Antoun 2003).

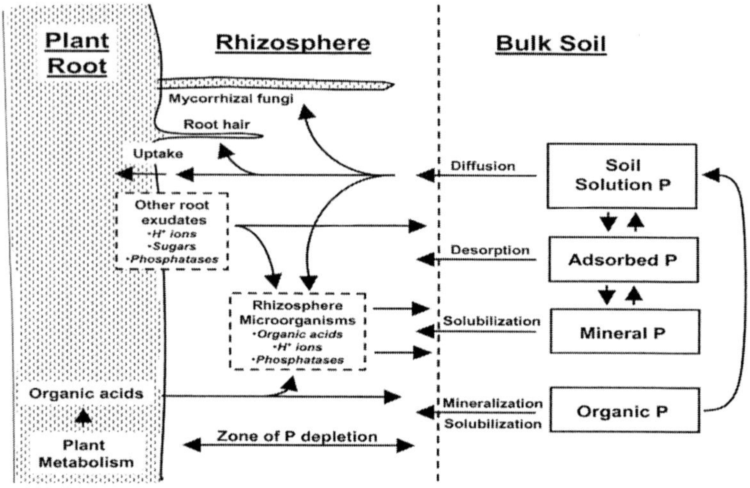

Fig. 3. Solubilisation mechanism of phosphate.

1.3. Uptake of Potassium (K)

Potassium (K) is one of the macro-elements which play an essential role in plant development. The maximum amount of K a plant needs to be able to grow adequately is 50–300 kg ha^{-1} (Zörb, Senbayram, and Peiter 2014).

The K resolution, where PGPR predominate, is almost identical to the P dissolution stages of the action mechanisms. Generally, potassium-solubilizing bacteria (KSB) dissolve K-bearing minerals, and release K by creating capsular polysaccharides, hydroxyl anions, exopolysaccharides, siderophores, organic ligands, extracellular catalysts, synthesis of organic and inorganic acids and biofilms on rhizospheric mineral surfaces (Etesami and Beattie 2017, Meena, Maurya, and Verma 2014, Uroz et al. 2009).

Besides minerals bearing potassium, some PGPR (i.e. *Bacillus*, *Burkholderia* and *Pseudomonas*), which are also known as silicate solvent bacteria, can break down silicon from the silicon minerals. Organic acids are, in particular, the most active substances for dissolving silicate minerals, for example, gluconic acid, feldspar, muscovite and biotite (Fig. 2a.).

1.4. Uptake of Micronutrients

PGPR increase the uptake of plant micronutrients (Fe, Mn, Zn, Cu, etc.). This increase is achieved in different ways. As an example, it can be done by lowering the soil pH to make it alkaline and increasing the air gaps between the soil particles.

1.4.1. Uptake of Iron (Fe)

Iron (Fe) is an important micronutrient, and is especially lacking in plants grown under dry and calcareous conditions. Siderophore-producing strains of PGPR may enhance iron-feeding in the plant. Iron (Fe) plays an important role in the energy metabolism of plants. Despite this, it is very difficult for the plant to pick up the debris from the soil. Because Fe does not exist as free ions in the soil, it is found as a ferric hydroxide polymer. In order to be able to obtain the iron found in this limited amount of soil, the bacteria living in the rhizosphere produce a water-soluble molecule, siderophore. Siderophores are organic compounds with low molecular weight and high affinity for binding some elements, such as Fe^{3+} and other metal ions, and for increasing plant availability. Siderophores are synthesised both by the PGPR bacteria found in the root region and by the plant itself (Altin and Tayyar 2005).

Fig. 4 shows intracellular transport of iron by siderophores carried out by microorganisms. This transport actually occurs in two ways: low coherent iron transport and high coherent iron transport. In low compatible iron transport, Fe^{+3} is transported into the cell through the cell membrane-associated receptor proteins.

In high compatible iron transport, siderophore forms a complex with Fe^{+3}, and the cell is trapped by zygotic receptor proteins. Iron is released from the Fe^{+3} + siderophore complex entering the cell, and this separation is by way of reduction. The iron-releasing siderophores re-enter the cell and form the Fe^{+3} + siderophore complex again.

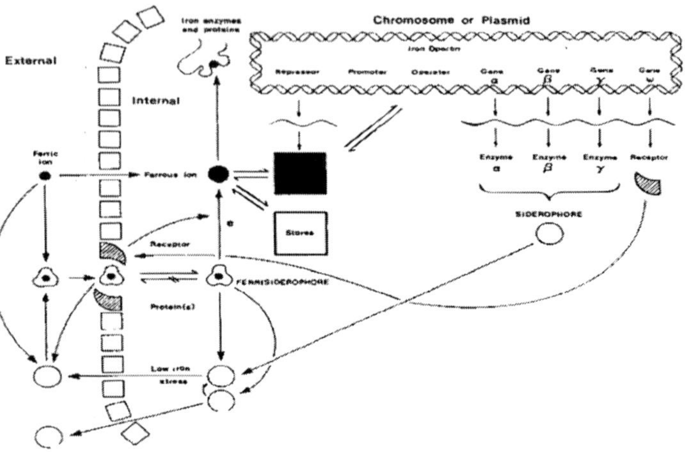

Fig.4. Low and high compatible iron transport in microorganisms.

1.4.2. Uptake of Manganase (Mn)

Some PGPR bacteria (*Bacillus*, *Pseudomonas and Geobacter*) reduce Mn^{+4} to Mn^{+2}. These bacteria increase the availability of Mn to the plant, while Mn^{+2} in the soil affects plant root growth. When the increase in the root secretion is carried out entirely by PGPR bacteria, the electron necessary for the reduction of MnO_2 to Mn^{+2} is obtained by decomposition of the carbonaceous compounds in the root secretion, and the proton required is obtained by proton elimination from the stem cells. Moreover, roots and PGPRs create chelating compounds (phenolic compounds and organic acids) that shape soluble complexes with Mn, Fe, and different components that keep their re-precipitation.

1.4.3. Uptake of Selenium (Se)

Selenium is the main component of selenoproteins, and plays an important role for plants. In addition to providing the necessary nutrients for plants, some PGPR (e.g. *Paenibacillus sp.*, *Bacillus sp.*, *Klebsiella*, *Acinetobacter sp.*, *Stenotrophomonas sp.*, *Enterobacter sp.*, *and Pseudomonas sp.*) are known as selenobacteria, which improve selenium accumulation in plants.

Selenobacteria are bacteria with a high selenium tolerance. These bacteria increase the transfer of selenium to shoots. In addition, they not only cause the amount of selenium in the crop to increase, but also remove the selenium stress that is caused by the excess of selenium present in the soil and seen in plants.

2. Indirect Effect of PGPR in Nutrient Uptake Efficiency

Plants face stress conditions, such as salinity, drought, temperature, frost and heavy metals. These stress conditions negatively affect plant growth, especially the growth of roots. This effect on roots causes a decrease in nutrient uptake.

These aforementioned drawbacks, stemming from stress in plant roots, can be solved by the PGPR producing IAA (indole-3-acetic acid), and PGPR containing the enzyme ACC (1-aminocyclopropane-1-carboxylate) deaminase. These PGPR might enhance the accessibility of the plant to the nutrients and water by providing the integrity of the plant root under stress conditions and, indirectly, affecting the repair of damaged areas. (Fig. 5)

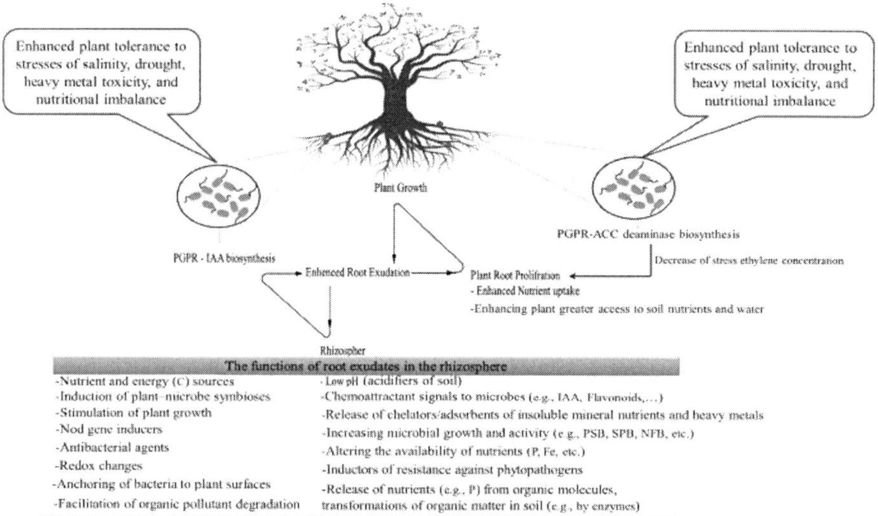

Fig. 5. The functions of root exudates in the rhizosphere

Root secretions usually contain substances essential to bacteria in the rhizosphere. These are saccharide, organic acid, L-Trp, phenols, vitamins, nucleotides and enzymes.

PGPR in the rhizosphere synthesise and secrete IAA. This IAA can build the plant cells and/or develop plant tissues (expanded root development and root length), thus together with the plant-determined incorporated IAA, a bigger root surface zone is achieved. The plant is therefore provided with more supplements and water, and discharges more leachate. IAA, likewise, gives extra supplements to help the rhizosphere bacteria develop by loosening the cell walls of the plant and increasing the root exudation. The release of more supplements increases microbial dynamics as well as IAA, and this procedure proceeds in a cycle. PGPRs producing ACC deaminase might expand root development and root length by diminishing ethylene stress, thus increasing plant development and root exudates.

Hormones, like IAA (indole-3-acetic acid), cytokinin and gibberellin are viable in plant development. IAA (the so-called auxin hormone) and cytokinin are involved in the intake of plant nutrients, and indirectly affect the nutrient intake of plants by inducing events, for instance, cell division, seed germination, root improvement, chlorophyll sequestration and leaf advancement. While the major hormone produced by PGPR is IAA (Fallik et al. 1989), the hormones such as indole lactic acid, indole-3-butyric acid, indole-3-ethanol, indol-3-methanol, some gibberellin and some cytokinin are also produced by PGPR (Bashan and Levanony 1990, Crozier et al. 1988, Tien, Gaskins, and Hubbell 1979). For instance, IAA synthesis, which was induced by indole-pyruvic acid and indole-3-acetic aldehyde, was detected in *Erwiniaherbicola*, some saprophytic *Pseudomonas* and *Agrobacterium* species, *Bardyrhizobium*, *Rhizobium*, *Azospirillum*, *Klebsiella* and *Enterobacter*. IAA, which is derived from Tryptophan, is found in *Cyanobacterium* bacteria, whereas IAA production, which is independent from Tryptophan, was detected in *Azospirila* and *Cyanobacteria* (Imriz 2014).

Ethylene is a very important hormone for plants, but its over-expression in stress conditions damages the plant. This hormone affects seed germination, root growth, root nodulation, flowering and fruit formation in plants. The 1-aminocyclopropane-1-carboxylate (ACC) deaminase, one of the previously mentioned enzymes stabilises the ethylene hormone by inhibiting the excessive secretion of ethylene hormone into this sheath (Glick 1995, Glick, Penrose, and Li 1998).

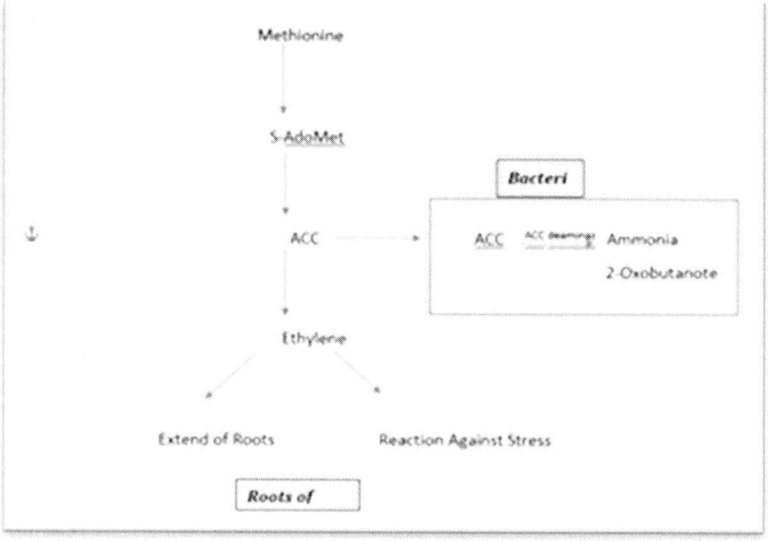

Fig. 6. ACC Mechanism of bacteria and plants

After the ethylene signal is synthesised in the ACC plant it is taken up by the bacteria from the plant roots and hydrolysed to the enzyme and 2-oxobutanoate by the enzyme produced by the bacterium. Thus, the level of ACC concentration in the plant is reduced, and excessive ethylene formation is prevented. At high concentration, accumulation of ethylene hormone inhibits plant growth. Ethylene production is catalysed by the conversion of S-adenosyl methionine (SAM) to aminocyclopropane-1-carboxylate (ACC), followed by a two-step process catalysed by the enzyme ACC-oxidase, which contains ethylene conversion of ACC. In Figure 6, the possible mechanism of ACC-deaminase activity is shown schematically (Imriz 2014).

ACC-deaminase action was identified in numerous bacterial strains of the variety *Acinetobacter*, *Achromobacter*, *Arobacterium*, *Alcalienes*, *Azospirillum*, *Bacillus*, *Burkholderia*, *Enterobacter*, *Pseudomonas*, *Ralstonia* and *Rhizobium* (Kang et al. 2010, Wang et al. 2000, Zahir et al. 2009).

3. Plant Morphology and Physiology as Influenced by PGPR

In the past, the internal and external parts of plants were studied collectively in the branch of botany known as plant anatomy. In the mid-twentieth century, the study of plants was split in two: while plant anatomy was about the internal structure of plants, plant morphology focused on their external structure.

It is known that plants consist morphologically of the shoot system and the root system. The shoot system is found above ground and has organs such as flowers, buds, fruits, leaves and stems, whereas the root system is found underground and has organs like tubers, roots, and rhizomes. The root system is important in terms of plant anchorage to the soil, absorption of nutrients, water and ions, and being a storage area for some plants. Moreover, the root system is connected with soil microbial populations (Beneduzi, Ambrosini, and Passaglia 2012). The microbial community, which is associated with plant roots, is known as the rhizo-microbiome. PGPR are a bacteria community that colonises the plant root and advances plant development; hence, the root structure must be known in detail for understanding the effect of PGPR on plant morphology (Claben-Bockhoff 2001).

As mentioned above, the root is the underground part of the plant structure. The initial part of the root that directly grows from the radicle is known as the primary root. The primary root branches can develop into secondary and tertiary roots. Whilst some plants have strong and long primary roots (taproot), some of them have short and very thin roots (fibrous root). Roots have different sections. Apical meristematic tissues are found at the growing part of the root, and are protected by the root cap. PGPR colonise at lateral roots and root hairs where they can show their beneficial properties to plants.

Roots have their own shape and characteristics, known as root system architecture; this includes the distribution of the lateral and primary roots, the length, and the number of the roots. The root system architecture can be changed by some factors, and PGPR are one of them. PGPR change the root system architecture by changing hormonal balance in the root system. Cytokinin, auxin, ethylene, abscisic acid (ABA) and gibberellin are hormones which play a role in root development. Auxin and cytokinin balance is important for plant organogenesis and root architecture. The balance can be damaged by PGPR. They can create a wide assortment of phytohormones like auxins and cytokinin.

Indole-3-acetic acid (IAA) is the best-documented auxin created by PGPR. IAA produced by PGPR has many roles in plant growth. Low centralisation of IAA prompts essential root extension, yet the high grouping of IAA prompts root hair formation, the shortening of essential root length and an increment in lateral root arrangement. PGPR also have an indirect effect on the plant auxin pathway. Some PGPR types, like *Azospirillumbrasilense*, can produce NO (nitric oxide) during root colonisation. NO has a role in the auxin-signalling pathway in plants: it controls lateral root formation.

Ethylene is another type of phytohormone, and it can be produced by some PGPR, such as *Azospirillumbrasilense*. Its major roles are: hindrance of root lengthening and auxin transport, advancement of abscission, senescence of different organs, and it prompts natural product maturing (Fig. 7).

Some studies show that PGPR can produce ABA or gibberellic acid. ABA is a hormone that has a role in the protection of the plants against drought. When plants undergo water stress, the ABA level will increase and stomata become smaller than before in order to limit water loss. It is also known that this hormone plays important roles in lateral root development (Vacheron et al. 2013).

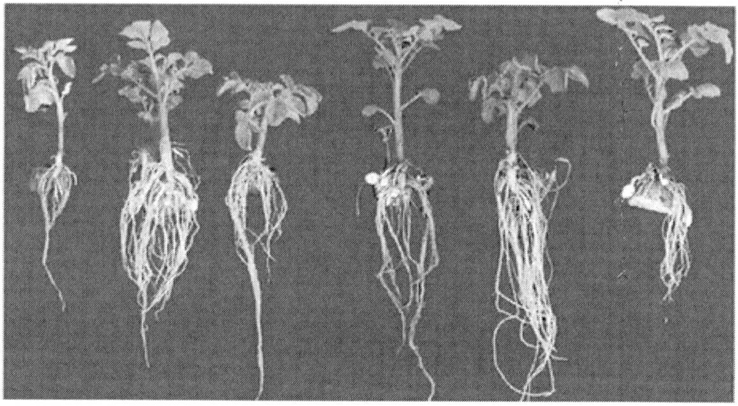

Fig. 7. The figure shows the root structures of plants. The first plant is chosen as a control group, so it is not in contact with any PGPR. Other plants undergo different types of PGPR and they show different root characteristics (Anonymous 2016).

Besides the morphological effects, PGPR can affect plants physiologically. PGPR can increase the nutrient uptake or availability from the root by changing the root structure, and increasing plant nutrition will stimulate the primary metabolism and enhance growth.

PGPR can also increase the nutritional supply that is found in soil. Phosphate solubilisation is the key process for increasing the nutritional supply of the soil. Plants can take phosphorus from the soil but sometimes there is an insufficient amount of available phosphorus in soil. In that case, the PGPR can dissolve insoluble forms of phosphate and plants can use them. Some other PGPR make N_2 fixation and can supply nitrogen to the plants. With the nutritional supplies, plant growth will increase (Bharti et al. 2016).

4. Biological Nitrogen Fixation

Nitrogen has a very important place in the life cycle of human beings and animals as plants grow and develop accordingly. Nitrogen is involved in the protein, amino acid, amide, ATP, and nucleic acid present in the plant, and in chlorophyll, which is a necessary pigment for photosynthesis. Owing to the rapid increase in the global population, the requirement for food is, likewise, expanding. In order to meet nutritional requirements, it is necessary to increment both harvest creation and to expand the yield. Nitrogen is one of the most important nutrients that increase yield; it is among the most important nutrients because it is already present in the organic structure of the plant and makes the plant grow faster. Nitrogen cannot be used directly by organisms: for this reason, productive crops cannot be grown when there is a lack of nitrogen in the soil.

Plants suffer from nitrogen deficiency when they cannot get enough nitrogen. The following effects are observed in plants as a result of insufficient nitrogen:

- Leaves shrink
- There is a noticeable redness in colour in plants
- The plants bloom early, and age quickly
- Leaf and trunk systems become weak

The main element of nitrogen contained in the soil is the organic matter. It is not possible for plants to directly access it due to its being found in organic matter. 78% of the air we breathe is nitrogen, and there are three ways this can be changed into a form that can be taken in by plants:

a. Non-Biological Nitrogen Fixation:

Non-biological nitrogen fixation is the transformation of nitrogen in the atmosphere through weather events (lightning, e.g.) into a form that plants can use. This nitrogen fixation path takes up much less space in nitrogen fixation than other ways.

b. Industrial Nitrogen Fixation:

In industrial fixation, very high temperatures and pressures are obtained, mostly by the use of fossil fuels. One of the main outcomes of this is environmental pollution. As a result of industrial fixation, the use of nitrogenous fertilisers with high nitrogen content damages the ozone layer, causing many problems such as nitrite and nitrate accumulation in food, and pollution in stagnant water.

c. Biological Nitrogen Fixation:

Nitrogen in the atmosphere fixed by microorganisms is called biological nitrogen fixation.

Only microorganisms can reduce the gas phase of molecular nitrogen in living things found in soil. The nitrogenase enzyme, used in the reduction of nitrogen by the genetic structures of these microorganisms can be synthesised by these microorganisms. Within this genetic structure of prokaryotes, 17 nitrogen fixation genes are widely available. These genes allow (1) the electron transfer to bond Mo and Fe to the nitrogenase protein; (2) the preservation of the nitrogenase from O_2; and (3) the regulation of the nitrogenase activity. As indicated by these procedures, we can comprehend that natural nitrogen fixation is an exceptionally delicate and complex structure.

The prokaryotic microorganisms that can bind nitrogen are all bacteria. These bacteria live either in free soil or with another living organism. *Rhizobium* bacteria, which are found in symbiotic relationships in the roots of leguminous plants, are provided for planting. The bacteria live by taking carbohydrates from the plant, and the bacteria also give the plant the nitrogen it takes from the air.

All plants, other than leguminous plants, can be utilised as a part of the type of nitrogen (NH_4^+) and nitrate (NO_3^-).

The bacteria that fix nitrogen by establishing a symbiotic relationship are divided into two groups. The first is the group of *Rhizobium* bacteria that can survive symbiotic life with leguminous plants, as mentioned above, and the second is the group of *Actinomycetes,* which can build up an advantageous association with high plants from tree species other than leguminous plants. Symbiotic nitrogen fixation is mainly performed by *Rhizobium* bacteria.

Reducing atmospheric nitrogen is a complex procedure requiring energy. The nitrogen molecule is created by a triple covalent bonding of two nitrogen atoms. This molecule is exceedingly latent and receptive. This ternary bond, which is difficult to reach, requires about 225 kcal energy to break. Biological nitrogen fixation is required to decrease di-nitrogen (N_2) to ammonia (NH_3)—which is an inert gas—by plants through the organisation of certain microorganisms. Another name for biological nitrogen fixation is diazotroph. These microorganisms obtain their energies by oxidising organic molecules. Free-living microorganisms do not make photosynthesis, but they obtain their energy from different organisms. Photosynthetic microorganisms like cyanobacteria utilise sugar, which is delivered through photosynthesis, to generate energy. Microorganisms that establish a symbiotic relationship with nitrogen fixation obtain these compounds from host plants.

Diazotrophs can be asymbiotic (free-living) or symbiotic:

i. **Free-living Nitrogen-binding Bacteria**: *Azotobacter* (aerobic), *Beijerinckia* (aerobic) and *Clostridium* (anaerobic) bacteria are nitrogen-fixing saprophytic bacteria. *Desulphovibrio* is a chemotrophic nitrogen-restricting bacterium. *Rhodopseudomonas*, *Rhodospirillum* and *Chromatium* are photoautotrophic nitrogen-restricting bacteria. These bacteria are found to contribute 10–25 kg of nitrogen, per hectare, per year.

ii. **Free-living Nitrogen-binding Cyanobacteria:** Many free-living blue-green algae (cyanobacteria) can settle nitrogen, e.g. *Anabaena*, *Nostoc*, *Aulosira*, *Cylindrospermum*, *Trichodesmium*. These bacteria are also very important in an ecological sense because they can live in places where denitrifying bacteria are active. These bacteria are found to contribute 20–30 kg of nitrogen per hectare, per year.

iii. **Symbiotic-living Nitrogen-binding Cyanobacteria**: *Anabaena* and *Nostoc* species are symbiotic with lichens. *Azolla pinna* is a water fern; its leaves have *Anabaena azollae.* For nitrogen fixation, it is mostly grafted onto rice fields.

iv. **Symbiotic Nitrogen-binding Bacteria:** *Rhizobium* aerobic are gram negative, nitrogen-binding bacteria. *Papilionaceous* lives symbiotically in its roots. *Sesbaniarostrata* has *Rhizobium* in root knobs and *Aerorhizobium* in stem nodules. It is a symbiote in the root knobs of numerous leguminous plants, for example, Frankia, Casuarina and Alnus. *Xanthomonas* and *Mycobacterium* bacteria are in a symbiotic relationship in the leaves of a few individuals from the *Rubiaceae* and *Myrsinaceae* (e.g. *Ardisia*) family. Many species of *Rhizobium* bacteria live free in soil. However, *Rhizobium* bacteria cannot bind nitrogen alone.

Frankia and *Rhizobium*, which symbiotically fix nitrogen, live freely in soil and can bind nitrogen in the atmosphere. It is known that the amount of nitrogen bound by biological means on the planet is 175 million tonnes per year; 50% of this nitrogen is provided by the leguminous-*Rhizobium* relationship.

4.1. Stages of Biological Nitrogen Fixation

4.1.1. Nodule Formation Mechanism

Nodule formation in leguminous plants results in interactions between the plant and *Rhizobium* bacteria.

The formation of nodules in leguminous plants results in mutual interactions between the plant and the plant *Rhizobium*. It forms nodules that provide or do not fix nitrogen by bacteria that provide nodule formation on the root of the plant. In other words, there is only one type of bacterium that infects a specific group of legume plants. These "host" plants that accept bacterium have been grouped under cross-inoculation groups. A cross-immunisation assembly alludes to a legume that structures a dynamic nodule when it is associated with bacteria detached from the nodules of any one specific plant gathering. *Rhizobium* bacteria shape nodules in the base of the plant in three stages:

a) Period Before Infection:

There are few connections between the bacteria and the plant during this period. The substance, tryptophan, secreted from the root of the previous plant, stimulates the multiplication of bacteria in the rhizosphere. As a result of this stimulation, the bacteria also form indole acetic acid (IAA) as a root growth hormone. It causes the root of the plant to develop rapidly and, at the same time, the root forms into a special shape. Later on, the bacteria secrete polygalactase (PG), which makes the stem cells flexible. These metabolites provide flexibility to the stem cells. Thus, bacteria can easily enter the stem cells.

b) Occurrence of Infection Lane:

Bacteria that have entered stem cells form a tube that reaches the cortex wall of the plant in capillary roots. This structure is called an infection thread. If infection thread is initiated, the root causes the bacteria in the meristem cells to move to the cortex cells. In the meantime, the bacteria multiply rapidly, and at the same time, the host continues to multiply in the plant cells and initiate nodule formation.

c) Nodule Formation:

The infective strip reaches the cell in the cortex area with tetraploid cells (the cell with twice the number of general chromosomes). These and neighbouring cells around it multiply by a rapid division. Then, the stem cells are dispersed in the cytoplasm. After multiplication, they lose their characteristic rod or short bar shapes. They become coarse X, Y or irregular pleomorphic shapes. The formation of the *Rhizobium* bacteria is called the new bacterium. Bacteroides do not divide within the nodule—they cannot be developed in artificial food environments. Nitrogen fixation takes place only in bacterial forms (Haktanır and Arcak 1997) (Fig. 8, 9).

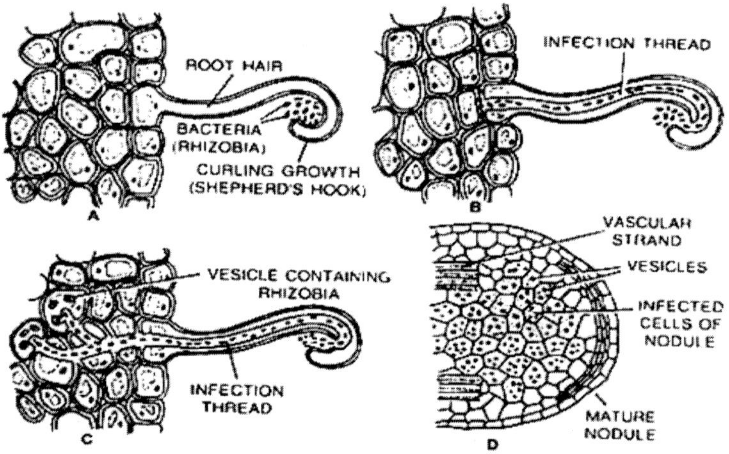

Fig. 8. Development of the Root Nodule (Anonymous)

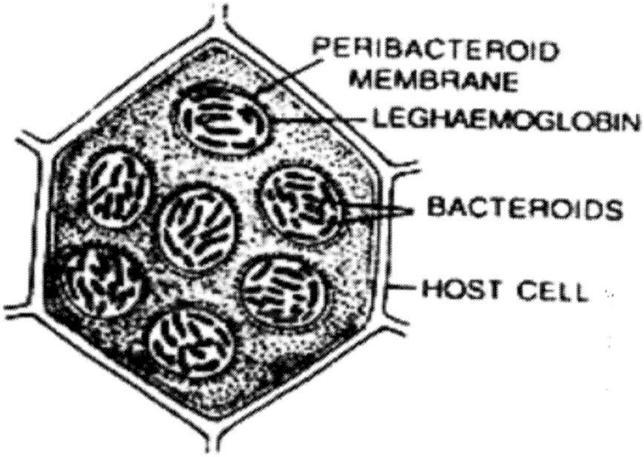

Fig. 9. Appearance of bacteroides in infected cortical cells (Anonymous).

4.1.2. Mechanism of Nitrogen Fixation

Nodules in nitrogen fixation have a task-like nitrogen fixation zone. An enzyme complex called leghemoglobin (leguminous hemoglobin) and nitrogenase contain all the biochemicals necessary for this process.

Nitrogenase contains two components. These are Mo-Fe protein (molybdoferredoxin) and Fe-protein (azoferredoxin). Nitrogenase converts the atmospheric nitrogen (N_2) to ammonia ($2NH_3$). Ammonia (NH_3) is the first stable product of nitrogen fixation.

The Reaction Equation is:

$$N_2 + 8e- + 16\ ATP \xrightarrow[\text{nitrogenase}]{\text{Mg2-}} 2NH_3 + H_2 + 16ADP + 16Pi$$

Nitrogenase is very sensitive to oxygen. It contains a reddish-pink pigment called leghemoglobin (Lb) to protect the enzymes in the nodules. Leghemoglobin (Lb) is an oxygen scavenger.

In the first stage in nitrogen fixation, the free nitrogen in the atmosphere is the first step towards the MoFe protein. In this process, ferrodoxin (nitrogenase reductase) acts as an electron donor to Fe-protein. It hydrolyzes ATP and reduces the MoFe protein. The MoFe protein, in turn, reduces the substrate N_2. Electrons and ATP are obtained by photosynthesis and the respiration of host cells.

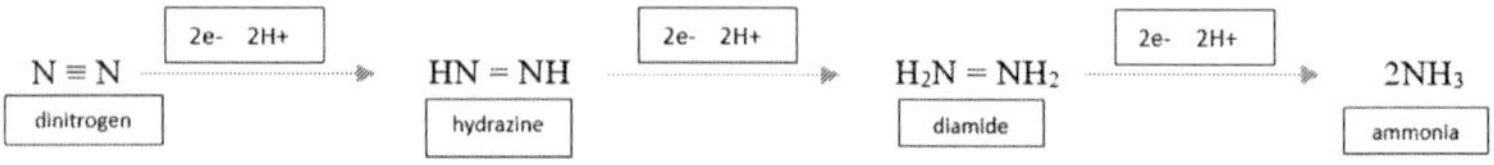

The intermediates of N_2 fixation.

Nitrogen-binding systems can grow in nitrogen-poor soil and are simultaneously a source of protein. Nitrogen is provided for soil fertility. Adenosine Triphosphate (ATP) is the source of energy required for the degradation and reduction of nitrogen N_2 to ammonia.

ATP is obtained by the breakdown of sugars and related molecules. Sugars are produced by the photosynthetic plant and transferred to the nodules. The plant fixes 1–20 grams of carbon (C) with the photosynthesis for each gram of N_2 which is bound by *Rhizobium* bacteria. To explain this, it is an indication that the symbiotic relationship and the N_2 fixation plants require additional energy to produce more products of photosynthesis.

It is accepted that nitrogen-binding systems require more phosphorus (P) than nitrogen-free systems. Phosphorus is required for plant development, nodule formation and advancement, and ATP synthesis. Each step of these processes is very important for nitrogen fixation.

4.2. Factors Affecting Nitrogen Fixation

Microorganisms found in soil are affected by environmental conditions. Microorganisms multiply rapidly and fulfil their natural functions in conditions suitable for them. But in adverse conditions, they struggle to do so. Since the constant change of these conditions affects microorganisms, the amount of nitrogen obtained by biological nitrogen fixation can vary greatly from region to region.

There are many factors that affect biological nitrogen fixation. These factors can be classified under certain groups.

a) Edaphic Factors

Edaphic factors are factors related to soil conditions. The six main edaphic factors that limit biotic nitrogen fixation are:

- Excessive soil nutrition
- Drought
- Soil acidity
- P deficiency
- Excess mineral N

- Missing Ca, Mo and Co

To briefly explain these factors:

• Excessive humidity and water pressure prevent the development of root and nodulation zones. They also affect the O_2 diffusion in the root system of the plant. By contrast, *Sesbaniarostrata* and *Aeschynomene* spp. can fix N_2 even under these conditions because these microorganisms are not on the roots, but on the stems of the plants.

• Drought reduces the number of Rhizobia in the soil. Accordingly, nitrogen fixation and nodulation are inhibited. If the drought lasts for a long time, the nodules undergo degradation. However, rooted legumes that can use nuts in the lower soil layers may continue nitrogen fixation when the soil starts to dry. Mycorrhiza is a symbiotic relationship amongst fungi and plant roots. Some microspheres develop outside the root zone. These are called ectomycorrhiza. An example of an ectomycorrhiza is *Baletussuillus*. Ectomycorrhizae (e.g. *Glamus* sp.) form an inward-growing hyphae network extending to the root, penetrating between the epidermis and cortex of plants. This is called the hartig network. This network is a site of nutrient exchange between the fungus and the host plant and supplies chemical elements required for plant growth, such as potassium and provides compounds, such as nitrate. These are vesicular-arbuscular mycorrhizas, commonly referred to as VAM.

• Problems such as soil acidity, low Ca level, and aluminum and manganese toxicity, negatively affect nitrogen fixation and nodule formation.

• Various micro-elements (Cu, Mo, Co, B) are required for N_2 fixation. Some of them, for example Mo, are components of nitrogenase.

b) Climatic Factors

Two of the most important climatic factors of biological nitrogen fixation are temperature and light. Biological nitrogen fixation is a process involving enzymatic reactions. For this reason, the fixation mechanism is adversely affected in high temperature conditions. However, symbiotic interactions exist between high temperature (>35°C) and low (<25°C) temperatures. For this reason, the harmony between the plant and the microorganism can be changed according to the temperature.

Light is necessary for the mechanism of photosynthesis. In turn, it affects nitrogen fixation. There is a difference in the biological nitrogen fixation when a plant is in shadow and when it receives more light.

c) **Biotic Factors**

The absence of biological nitrogen fixation in biotic factors is most influential. Other limiting factors are:

- Extreme foliage of host plant.
- Product competitiveness.
- Problems caused by insects and nematodes.

4.3. Vaccination of Legumes

If no legume plants have previously been grown in a region, the relevant bacteria are usually absent or inactive. The amount of mineral nitrogen in the soil is constantly being changed by the external actions of farmers. When mineral nitrogen is present in excess in the soil, the nitrogenase enzyme is inhibited. Nitrogenase inhibition also results in decreased nitrogen fixation.

If there is no effective and specific *Rhizobium* bacterium in the soil in a region, or if there are very low amounts, it is necessary to introduce Rhizobia to the soil so that nitrogen fixation can take place and the nodulation process can occur. If there is no definite data on the number and effectiveness of Rhizobia in the soil in agricultural systems, leguminous plants should be inoculated with Rhizobia, at appropriate doses.

5. Mitigation of Cold Stress in Plants by PGPR

PGPR are bacteria that can colonise plant roots and help advance plant development. PGPR can be considered an alternative form of inorganic fertiliser, with specific goals to advance plant development, quality and yield. They can act as biofertilisers or biocontrol agents as a result of their inherent capabilities, and as phytostimulators when applied to plant roots. Environmental stresses result in major problems for agricultural productivity. Studies have shown the major potential of PGPR in decreasing many abiotic stresses in plants, such as salinity, high temperature, drought, cold stress, heavy metal toxicity, etc. Plants can survive either through refraining cooling of tissue water, or through freezing tolerance under extreme cold conditions. Some species of plants have developed traits which enable them to tolerate cooling or freezing temperatures by raising their anti-freezing response: this is known as acclimation.

Cold stress is a difficult issue for the sustainability of plant yields. Moreover, cold stress can be a reason for large-scale crop losses. Plants show different phenotypic indications of reacting to cold stress, for example, hindered seedlings, yellowing of leaves, poor germination, leaf-withering and, on occasion, the death of their tissues. In addition, cold stress prevents reproductive development of the plant. Physiological, metabolic, cellular and molecular dysfunction become extreme under the natural temperature a plant is designed to expect. Temperature conditions can vary from one plant to another. The optimum temperature for one plant may be stressful for another plant. For instance, soybean, tomato and banana suffer damage when the temperature is below 10–15°C. This damage is a factor of a plant's susceptibility to cold stress.

Cold-tolerant PGPR have been used to support plants against cold stress. Certain sorts of bacteria, such as *E. herbicola* and *P. syringae* with ice-nucleating dynamics, go about the ice nuclei, which prevent plants from supercooling, and frost damage at sub-zero temperature. As they are cold-tolerant, PGPRs can be used to inoculate the plant. Thus, the growth of plants and nutrient uptake are provided under lower temperature conditions. The bacterium enhances the dry biomass, and root and shoot length. The required cellular metabolites, such as starch, sugars, proteins and amino acids are also improved with inoculation of a cold-tolerant bacterium. Cold-adapted potential plant growth-promoting bacteria can involve applications, like inoculants, for advancing plant development, and biocontrol agents for plant growth against cold stress. Furthermore, epiphytic or endophytic PGPR increase plant development and yield when their obstruction is improved in response to stress. PGPR can straightforwardly encourage the proliferation of the plant host system through phytohormone generation in the root zone. These hormones increase the length of root hairs. At the point when the plant root surface area builds up, uptake of essential nutrients, such as water and minerals from the large volume of soil, becomes easier.

PGPR have agronomic significance because of the way plant growing cycles on earth function during differing spells of cold. However, they have potential in terms of how they function in cold conditions and enable metabolites, like plant development controllers to be produced, which can specifically advance growth and plant nutrient take-up. PGPR can prompt formative differentiations in host plants and fundamental protection for pathogens, enhance nutrient and water uptake, and affect phytohormone production.

Cold stress is a noteworthy ecological stress that chokes agricultural plant yield since it influences their quality and post-harvest life. Some plant types acquire resilience to chilling and solidifying when they are subject to sub-lethal cold stress, a procedure known as "cold acclimation". Nowadays, most essential plants are not fit for cold acclimation. Cold-tolerant PGPR are ordinarily utilised by the agro-biological community, and assume key roles as biocontrol agents, nitrogen fixatives and mitigators of plant cold stress.

6. Principles, Applications and Future Aspects of PGPR

6.1. Principles of PGPR

Plant development-advancing mechanisms are divided into two mechanisms: roundabout and direct. The roundabout mechanism occurs outside of the plant; direct mechanism occurs within the plant and influences metabolism of plants. Thus, direct mechanisms contain plant development regulators and instigate the metabolism of plants (Goswami, Thakker, and Dhandhukia 2016).

6.2. Applications of PGPR

PGPR propose a contrasting option to the utilisation of chemicals for plant development in various applications. The principle utilisations of bacteria for enhanced plant growth are agriculture, horticulture, forestry and environmental reclamation (Lucy, Reed, and Glick 2004).

6.2.1. Applications of PGPR in Agriculture

Free-living PGPR applications are the most seriously contemplated in agriculture. When PGPR is used to inoculate, then germination rates, root development, leaf area, shoot and root weights are enhanced. Additionally, they increase chlorophyll, magnesium, nitrogen and protein content, and give resistance to drought. Another important use of PGPR is for disease resistance for the plant. This is also known as biocontrol. However, the utilisation of PGPR has been restricted due to irregularity of results from laboratory, greenhouse, and field studies.

Climatic variability also has a great influence on the effectiveness of PGPR, and different types of soil may cause the effectiveness of PGPR to vary. According to a study with wheat and a *Pseudomonas*, the soil became less fertile and the plant growth stimulation from PGPR was greater.

If effective PGPR are chosen for plants at specific sites, the nutritional level of the soil has to be considered. It is also necessary to consider how PGPR will perform at that site. After inoculation, the soil moisture rate is affected by PGPR. In some cases, specific species of bacteria may only support bacterial growth in certain products (Labandera-Gonzalez, 1994). In one example, only one of four *Pseudomonas* strains supporting radish growth was found to be effective in the development of potato reproduction. Moreover, germination and increases in maximum yield are observed in crops that have been inoculated with PGPR strains (Dobbelaere et al. 1999).

6.2.2. Applications of PGPR in Forestry

There is less research on the utilisation of PGPR in forestry than there is in agriculture. In the 1980s, PGPR and their effects on angiosperms were investigated; however, PGPR on gymnosperms have only been investigated since the 1990s. When PGPR are tested in forest environments, the soil type should be considered an important factor. Some woodland soils are acidic, particularly coniferous forests, and some PGPR are impervious to low pH conditions. The survival of PGPR in winter is particularly challenging, especially in trees in the coldest regions. For example, in Canada, Scandinavia and Russia, trees, and the majority of agricultural crops, are perennial.

6.2.3. Applications of PGPR for Environmental Remediation

The emergence of bacteria with plants for environmental applications is an extension of PGPR technology. Recent work in the same field incorporates diverse uses, like helping plant development in acidic conditions, resisting high temperature stress, and using PGPR in phytoremediation technologies. Apart from environmental uses, these studies may also have an effect on agricultural or forestry practices. Phytoremediation is a technology which is used for degrading or stabilising hazardous substances in an environment by using plants. These plants should collect high amounts of contaminant and produce large biomass. Some plants grow in contaminated areas, so toxicity is inherited in this way. In this situation, adding PGPR can help plant growth. Phytoremedation allows plants to grow well and enhances detoxification. Some rhizobacteria types are useful for plants by mobilising soil contaminants and promoting growth (Huang 2003). During phytoremediation, plant properties, like biomass, health, nutrition and contaminant uptake are improved by PGPR. Although PGPR play an important role, studies associated with plants are limited (Huang et al. 2005).

6.2.4. Applications of PGPR as Multifunctional Agents

The influence of PGPR on plant proficiency shifts within the setting of laboratory, greenhouse and field trials. The reasons for these issues are that the soils come from unpredictable environments and some outcomes are not accomplished. Climatic changes have a large effect on PGPR, but negative growth conditions in these fields are accepted as a normal function of agriculture. Global dependence on agricultural chemicals can be reduced with widespread use of PGPR. Today, they are also an easily accessible technology for agriculturists in both developed and developing countries. PGPR ought to be coordinated effectively for the right plant and conditions, so that specific and genuine outcomes can be obtained (Kaymak 2011).

6.3. Future Development of PGPR

Soil fertility is important and needed for high yields and to enhance crop production in an ecological manner. A new engineering branch, called rhizoengineering, has been developed for these purposes. It includes different strategies for improving plant health, yield and manipulating microorganisms related to plant roots. Future developments of PGPR will therefore rely on concepts used within the field of rhizoengineering. Future research in rhizosphere biology will increase knowledge about rhizosphere biology, develop biotechnological approaches and provide coordinated administration of soil microbial populations. New techniques ought to be investigated utilising bioinoculants for high esteem plants, like fruits, vegetables, and flowers (Gupta et al. 2015). Single inoculation could be a proficient strategy for diminishing the unsafe effects of stress on plant development. Furthermore, ice-nucleating PGPR could be effective for plant development at low temperatures. Nitrogen obsession and phosphate solubilisation are preceded by PGPR; however, there are fewer studies about potassium solubilisation, which is the third fundamental macronutrient for plant growth (Shinwari et al. 2015). This would increase the scope of inoculants used and the confidence of farmers in their usage. The marketing of bioinoculant products in the future, and the release into the environment of these transgenic and eco-friendly materials, will depend on biosafety data (Spaepen and Vanderleyden 2011).

References

Altin, N, and B Tayyar. 2005. "General properties and efficiency of root bacteria." *Anadolu Journal of Aegean Agricultural Research Institute* 15 (2):87–103.

Anonymous. 2016. "Rizobatteri per la coltivazione sostenibile della patata." http://www.freshplaza.it/article/85188/Rizobatteri-per-la-coltivazione-sostenibile-della-patata.

Antoun, H. 2003. "Field and greenhouse trials performed with phosphate-solubilizing bacteria and fungi." First International meeting on microbial phosphate solubilization, Salamanca, Spain, 09 December 2002–02 January 2003.

Bashan, Y, and H Levanony. 1990. "Current status of *Azospirillum* inoculation technology: *Azospirillum* as a challenge for agriculture." *Canadian Journal of Microbiology* 36 (9):591–608. 10.1139/m90-105.

Beneduzi, A, A Ambrosini, and LMP Passaglia. 2012. "Plant growth-promoting rhizobacteria (PGPR): their potential as antagonists and biocontrol agents." *Genetics and Molecular Biology* 35 (4):1044–1051.

Bharti, N, S S Pandey, D Barnawal, V K Patel, and A Kalra. 2016. "Plant growth promoting rhizobacteria *Dietzia natronolimnaea* modulates the expression of stress responsive genes providing protection of wheat from salinity stress." *Plant Anatomy and Morphology. JSTOR. Plant Biology - Root Morphology* 6:34768.

Bloemberg, Guido V, and B JJ Lugtenberg. 2001. "Molecular basis of plant growth promotion and biocontrol by rhizobacteria." *Current Opinion in Plant Biology* 4 (4):343–350.

Cakmakci, R, M F Dönmez, and ÜErdoğan. 2007. "The effect of plant growth promoting rhizobacteria on barley seedling growth, nutrient uptake, some soil properties, and bacterial counts." *Turkish Journal of Agriculture and Forestry* 31 (3):189–199.

Carmen, B, and D Roberto. 2011 "Soil bacteria support and protect plants against abiotic stresses." In *Abiotic Stress in Plants-Mechanisms and Adaptations*, edited by A Shanker and B Venkateswarlu. London, UK: InTech.

Claben-Bockhoff, R. 2001. "Plant morphology: the historic concepts of Wilhelm Troll, Walter Zimmermann and Agnes Arber." *Annals of Botany* 88 (6):1153–1172.

Crozier, A, P Arruda, J M Jasmim, A M Monteiro, and G Sandberg. 1988 "Analysis of indole-3-acetic acid and related indoles in culture medium from *Azospirillum lipoferum* and *Azospirillum brasilense*." *Applied and Environmental Microbiology* 54 (11):2833–2837.

Dobbelaere, S, A Croonenborghs, A Thys, A Vande Broek, and J Vanderleyden. 1999. "Phytostimulatory effect of *Azospirillum brasilense* wild type and mutant strains altered in IAA production on wheat." *Plant and Soil* 212 (2):153–162.

Dursun, A, M Ekinci, and M F Dönmez. 2008. "Effects of inoculation bacteria on chemical content, yield and growth in rocket (*Eruca vesicaria* subsp. *sativa*)." *Asian Journal of Chemistry* 20 (4):3197–3202.

Elkoca, E, F Kantar, and F Sahin. 2007. "Influence of nitrogen fixing and phosphorus solubilizing bacteria on the nodulation, plant growth, and yield of chickpea." *Journal of Plant Nutrition* 31 (1):157–171.

Esitken, A. 2011. "Use of plant growth promoting rhizobacteria in horticultural crops." In *Bacteria in Agrobiology: Crop Ecosystems*, edited by D K Maheshwari, 189–235. New York, USA: Springer.

Etesami, H, and G A Beattie. 2017. "Plant-Microbe Interactions in Adaptation of Agricultural Crops to Abiotic Stress Conditions." In *Probiotics and Plant Health*, edited by V Kumar, M Kumar, S Sharma and R Prasad, 163–200. Singapore: Springer.

Fallik, E, Y Okon, E Epstein, A Goldman, and MFischer. 1989. "Identification and quantification of IAA and IBA in *Azospirillum brasilense*-inoculated maize roots." *Soil Biology and Biochemistry* 21 (1):147–153.

Feigin, A. 1985. "Fertilization management of crops irrigated with saline water." In *Biosalinity in Action: Bioproduction with Saline Water*, edited by Dov Pasternak Anthony and San Pietro, 89. 285–299. 10.1007/BF02182248. Springer.

Figueiredo, M V B, A Bonifacio, A Cerqueira Rodrigues, and F Fde Araujo. 2016. "Plant Growth-Promoting Rhizobacteria: Key Mechanisms of Action." In *Microbial-mediated Induced Systemic Resistance in Plants*, edited by Devendra K Choudhary and Ajit Varma, 23–37. Springer.

Gagne, S, L Dehbi, DLe Quere, F Cayer, J-L Morin, R Lemay, and NFournier. 1993. "Increase of greenhouse tomato fruit yields by plant growth-promoting rhizobacteria (PGPR) inoculated into the peat-based growing media." *Soil Biology and Biochemistry* 25 (2):269–272.

Glick, B R. 1995. "The enhancement of plant growth by free-living bacteria." *Canadian Journal of Microbiology* 41 (2):109–117.

Glick, B R. 2014. "Bacteria with ACC deaminase can promote plant growth and help to feed the world." *Microbiological Research* 169 (1):30–39.

Glick, B R, D M Penrose, and J Li. 1998. "A model for the lowering of plant ethylene concentrations by plant growth-promoting bacteria." *Journal of Theoretical Biology* 190 (1):63–68.

Golpayegani, A, and H G Tilebeni. 2011. "Effect of biological fertilizers on biochemical and physiological parameters of basil (*Ociumum basilicm* L.) medicine plant." *American-Eurasian Journal of Agricultural and Environmental Sciences* 11 (3):411–416.

Goswami, D, J N Thakker, and P C Dhandhukia. 2016. "Portraying mechanics of plant growth promoting rhizobacteria (PGPR): A review." *Cogent Food and Agriculture* 2 (1):1127500.

Gupta, B, and B Huang. 2014. "Mechanism of salinity tolerance in plants: physiological, biochemical, and molecular characterization." *International Journal of Genomics* 2014 (1–18).

Gupta, G, S S Parihar, N K Ahirwar, SK Snehi, and V Singh. 2015. "Plant Growth Promoting Rhizobacteria (PGPR): current and future prospects for development of sustainable agriculture." *Journal of Microbial and Biochemical Technology* 7 (2):096–102.

Haktanır, K, and S Arcak. 1997. "Soil Biology (Entrance of Soil Ecosystems)." *Ankara University Faculty of Agriculture, Department of Soil Science* (1486):447.

Huang, X-D. 2003. "Applications of free living plant growth-promoting rhizobacteria." *Antonie Van Leeuwenhoek* 86 (1):1–25.

Huang, X-D, Y El-Alawi, J Gurska, B R Glick, and B M Greenberg. 2005. "A multi-process phytoremediation system for decontamination of persistent total petroleum hydrocarbons (TPHs) from soils." *Microchemical Journal* 81 (1):139–147.

Imriz, G. 2014. "Plant Growth Promotin Rhizobacteria (PGPR) and Efficiency Mechanisms in Plant Production." *Journal of Electronic Microbiology.*

Kang, B G, W T Kim, H S Yun, and S C Chang. 2010. "Use of plant growthpromoting rhizobacteria associated with chickpea (*Cicer arietinum* L.)." *International Journal of Plant Production* 1 (2):141–152.

Kaymak, H C. 2011. "Potential of PGPR in Agricultural Innovations." In *Plant Growth and Health Promoting Bacteria*, edited by D K Maheshwari, 45–79. Berlin: Springer.

Lucy, M, E Reed, and B R Glick. 2004. "Applications of free living plant growth-promoting rhizobacteria." *Antonie Van Leeuwenhoek* 86 (1):1–25.

McCullagh, M, R Utkhede, JG Menzies, ZK Punja, and TC Paulitz. 1996. "Evaluation of plant growth-promoting rhizobacteria for biological control of Pythium root rot of cucumbers grown in rockwool and effects on yield." *European Journal of Plant Pathology* 102 (8):747–755.

Meena, K K, A M Sorty, U M Bitla, K Choudhary, P Gupta, A Pareek, DP Singh, R Prabha, P K Sahu, and V K Gupta. 2017. "Abiotic stress responses and microbe-mediated mitigation in plants: the omics strategies." *Frontiers in Plant Science* 8:172.

Meena, V S, BR Maurya, and J P Verma. 2014 "Does a rhizospheric microorganism enhance K^+ availability in agricultural soils?" *Microbiological Research* 169 (5–6):337–347.

Mishra, J, J Prakash, and N K Arora. 2016. "Role of Beneficial Soil Microbes in Sustainable Agriculture and Environmental Management." *Climate Change and Environmental Sustainability* 4 (2):137–149. 10.5958/2320-642X.2016.00015.6.

Niranjan R, S, H S Shetty, and MS Reddy. 2006. "Plant growth promoting rhizobacteria: potential green alternative for plant productivity." In *PGPR: Biocontrol and Biofertilization*, edited by Zaki A Sıddıqui, 197–216. The Netherlands: Springer.

Rasool, S, A Hameed, M M Azooz, T O Siddiqi, and P Ahmad. 2013. "Salt Ctress: Causes, Types and Responses of Plants." In *Ecophysiology and Responses of Plants under Salt Stress*, edited by P Ahmad, M M Azooz and M N V Prasad, 1–24. New York, USA: Springer.

Richardson, Alan E, and Richard J Simpson. 2011. "Soil microorganisms mediating phosphorus availability update on microbial phosphorus." *Plant Physiology* 156 (3):989–996.

Rodrıiguez, H, and R Fraga. 1999. "Phosphate solubilizing bacteria and their role in plant growth promotion." *Biotechnology Advances* 17 (4–5):319–339.

Sharpley, A N, J J Meisinger, J F Power, and D L Suarez. 1992. "Root Extraction of Nutrients Associated with Long-Term Soil Management." In *Advances in Soil Science-Limitations to Plant Root Growth*, 151–217. Springer.

Shinwari, K I, A U Shah, M Irfan Afridi, MZeeshan, H Hussain, J Hussain, and O Ahmad. 2015. "Application of plant growth promoting rhizobacteria in bioremediation of heavy metal polluted soil." *Asian Journal of Multidisciplinary Studies* 3 (4).

Spaepen, S, and J Vanderleyden. 2011. "Auxin and Plant-Microbe Interactions." *Cold Spring Harbor Perspectives in Biology* 3 (4): a001438.

Sturz, AV, and J Nowak. 2000. "Endophytic communities of rhizobacteria and the strategies required to create yield enhancing associations with crops." *Applied Soil Ecology* 15 (2):183–190.

Sudhakar, P, GN Chattopadhyay, SK Gangwar, and JK Ghosh. 2000. "Effect of foliar application of *Azotobacter*, *Azospirillum* and *Beijerinckia* on leaf yield and quality of mulberry (*Morus alba*)." *The Journal of Agricultural Science* 134 (2):227–234.

Tien, TM, MH Gaskins, and DH Hubbell. 1979. "Plant growth substances produced by *Azospirillum brasilense* and their effect on the growth of pearl millet (*Pennisetum americanum* L.)." *Applied and Environmental Microbiology* 37 (5):1016–1024.

Turan, M, MEkinci, EYildirim, A Güneş, K Karagöz, R Kotan, and ADursun. 2014. "Plant growth-promoting rhizobacteria improved growth, nutrient, and hormone content of cabbage (*Brassica oleracea*) seedlings." *Turkish Journal of Agriculture and Forestry* 38 (3):327–333. doi: 10.3906/tar-1308-62.

Uroz, S, C Calvaruso, M-P Turpault, and P Frey-Klett. 2009. "Mineral weathering by bacteria: ecology, actors and mechanisms." *Trends in Microbiology* 17 (8):378–387.

Vacheron, J, G Desbrosses, M-L Bouffaud, B Touraine, Y Moënne–Loccoz, D Muller, L Legendre, F Wisniewski-Dye, and C Prigent-Combaret. 2013. "Plant growth-promoting rhizobacteria and root system functioning." *Frontiers in Plant Science* 4:356.

VanPeer, R, and B Schippers. 1988. "Plant growth responses to bacterization with selected *Pseudomonas* spp. strains and rhizosphere microbial development in hydroponic cultures." *Canadian Journal of Microbiology* 35 (4):456–463.

Vedder-Weiss, D, E Jurkevitch, S Burdman, D Weiss, and Y Okon. 1999 "Root growth, respiration and beta-glucosidase activity in maize (*Zea mays*) and common bean (*Phaseolus vulgaris*) inoculated with *Azospirillum brasilense*." *Symbiosis* 26: 363–377.

Vessey, J K. 2003. "Plant growth promoting rhizobacteria as biofertilizers." *Plant and soil* 255 (2):571–586.

Vinocur, B, and A Altman. 2005. "Recent advances in engineering plant tolerance to abiotic stress: achievements and limitations." *Current Opinion in Biotechnology* 16 (2):123–132.

Wang, C, E Knill, B R Glick, and G Défago. 2000. "Effect of transferring 1-aminocyclopropane-1-carboxylic acid (ACC) deaminase genes into *Pseudomonas* fluorescens strain CHA0 and its gac A derivative CHA96 on their growth-promoting and disease-suppressive capacities." *Canadian Journal of Microbiology* 46 (10):898–907.

Yildirim, E, M Turan, A Dursun, M Ekinci, R Kul, FParlakova Karagoz, M F Donmez, and N Kitir. 2016. "Integrated Use of Nitrogen Fertilization and Microbial Inoculation: Change in the Growth and Chemical Composition of White Cabbage." *Communications in Soil Science and Plant Analysis* 47 (19):2245–2260.

Zahir, Z A, U Ghani, M Naveed, S M Nadeem, and H N Asghar. 2009. "Comperative effectiveness of rhizobacteria containing ACC-deaminase for growth promotion of pea (*Pisum sativum*) under drought conditions." *Journal Microbiology and Biotechnology* 18 (5):958–963.

Zörb, C, M Senbayram, and E Peiter. 2014. "Potassium in agriculture–status and perspectives." *Journal of Plant Physiology* 171 (9):656–669.

CHAPTER 2

ASPECTS OF ORGANIC FARMING ON SUSTAINABILITY

SEVINC SENER,[1*] DEMIR OZDEMIR[2]

[1]Akdeniz University, Faculty of Agriculture, Department of Horticulture, Antalya-TURKEY
[2]Akdeniz University, Faculty of Agriculture, Department of Agricultural Biotechnology, Antalya-TURKEY
ssener@akdeniz.edu.tr

Abstract

People have always had their dietary needs met, directly or indirectly, by agricultural produce. With the increasing world population, the need for human populations to maximise agricultural productivity models over time is necessary to meet an increasing demand for food. Recently, thousands of different synthetic agro-chemicals have been used to suppress pests and diseases, and to increase the yield in "modernised" agriculture. Nowadays, conventional pest and nutrient management systems have increased dependency on synthetic chemical inputs, as well as making agriculture unsustainable. The problems of conventional agriculture production have led people to seek alternative production systems such as "Organic Production Models". Organic agriculture is a global, fast-growing, sustainable production model that supports the ecosystem and human health. This system improves the degraded ecological balance, reduces agricultural inputs and activities, and uses natural preparations instead of synthetic fertilisers, pesticides and hormones that are harmful to human and environmental health. The aim of this chapter is to examine the aspects of conventional agricultural production that prevent sustainability, and to suggest solutions based on organic farming practices to overcome this problem.

Key words: organic farming, environmental, sustainability, conventional agriculture.

Definition and Scope

With the increasing world population, the need to maximise productivity has grown over time to try to meet an increasing demand for food. People have tried to protect plants and animals from pest and diseases, and to increase their productivity by using different methods, even before the discovery of chemicals. By the twentieth century, a number of changes had occurred in agricultural production systems, such as the increasing use of technology, mechanisation, and synthetic chemicals. These changes, which focused on the maximisation of yields, clogged the agricultural production system in different ways, and became unsustainable over time. Aiming solely at increased efficiency in agricultural production systems will not lead to success as it can only produce short-term solutions. Hence, agriculture cannot be based solely on economic fundamentals: it is an integrated system that includes ecological and social patterns. Yunlong and Smit (1994) suggested that agricultural sustainability should be assessed from the holistic perspective of ecological soundness, social acceptability, and economic viability.

Globally, conventional pest and nutrient management, which is applied in most cultivated areas, increases the dependency on synthetic chemical inputs, as well as making agriculture unsustainable. The problems of conventional agriculture production have led people to seek alternative production models such as "Organic Production Models". Organic agriculture is a production model that supports ecosystems and human health. In this model, traditional and innovative methods are used in an integrated manner: there is minimal use of effective applications on the environment, and fair relations among living beings are supported, thus raising the quality of life. Organic agriculture is based on ecological processes that are compatible with nature and does not use inputs that can have harmful effects. The goal of organic farming is to improve the ecological balance, which has deteriorated due to conventional agricultural practices; to reduce agricultural inputs and activities; and to use natural preparations instead of synthetic chemicals that are harmful to human and environmental health. Organic farming can be defined as a system where the use of synthetic fertilisers, pesticides, antibiotics and hormones is prohibited, and which is aimed at producing food with minimal harm to ecosystems (McIntyre 2009, Swedish Control Association of Ecological Farming 2003).

Why Organic Agriculture?

Plant health can be impaired by biotic and abiotic stress factors. Particularly in developing countries, synthetic chemicals are used to protect plant and soil health, and increase the yield, rather than integrated pest management. Therefore, pesticides are cost-effective, easily available, give fast results, and display a wide spectrum of bioactivity. The most arguable practices in conventional agriculture in many countries are: performing monoculture agricultural systems and ignoring crop rotation, industrialised animal production, and the uncontrolled uses of pesticides. All these misapplications cause an increase in the incidence of pests and diseases. More synthetic chemical inputs have been used to prevent and treat increasing pests and diseases. Problems that arise due to these misapplications are increasing day-by-day and bring further problems. For example, pesticides, which are used to remove pests and disease agents, are also damaging to natural enemies: in fact, those agents that cannot be controlled naturally prompt harmful effects in the following production season. Another problem that is caused by the frequent use of synthetic chemicals for plant protection is the passage of some minor pests into major pests. Furthermore, resistance can occur because of high doses or misapplied pesticides in targeted organisms. Synthetic fertilisers damage the soil flora and fauna and, as a result, more fertilising is needed every year in order to be able to yield. Another important point is that the pesticide residue problems in agricultural products cause deterioration in human health.

The negative effects of the conventional agricultural model are not limited to human and environmental health problems, but also bring economic problems. These problems include: the energy problem, the rapid increase in oil prices, the decrease in the rate of the yield, and price fluctuations due to monoculture production. The food and agriculture policies of the modern system which have failed to achieve sustainable outcomes in developing countries, and the environmental impacts and health problems created by conventional agriculture over time have consequently led to the search for an alternative agricultural model. As a result of this research, the organic farming model emerged, and became a rapidly growing agricultural sector across the globe. Organic farming is practiced in 172 countries on 43.7 million hectares of agricultural land, managed organically by approximately 2.3 million farmers. Thus, global sales of organic food and drink reached 80 billion US dollars in 2014 (FiBL2016).

Principles of Organic Agriculture

Organic agriculture relies on ecological processes, biodiversity and cycles adapted to local conditions. The principle guidelines for organic production are to use materials and practices that integrate the parts of the farming system into an ecological whole. Organic farming methods cannot completely eliminate environmental pollution, but they can minimise it, because the methods applied in this farming method are not completely free of residues (Datta 2012). In organic farming, all production stages are tracked and planned, products are certified and there are control mechanisms between farmers and consumers.

Organic agriculture is a sustainable system which prohibits the use of synthetic molecules or chemicals such as pesticides, herbicides, chemical fertilisers, antibiotics, hormones, growth stimulants, etc. It relies on methods such as crop rotation, natural nitrogen fixation, biologically active soil, recycled farm manure and crop residues, cultural methods, biological or mechanical weed and pest control.

Cultural methods, among which are some of the oldest agricultural techniques, are based on applications and tactics which play critical roles in disrupting pest infestation of crops, including: crop rotation; intercropping; use of resistant or tolerant varieties; farm/field location; soil management (organic fertiliser, composting, conservation tillage); water management; soil tillage in the night (to reduce weeds); use of pest-free seed and planting stock; time of planting/sowing (moving up or delaying the growing period);avoiding high planting densities; flooding of fields (to combat soil-borne harmful organisms); recording the whole process (abiotic factors, irrigation and fertiliser programs); and trap crops.

Other methods besides cultural ones, which are called active methods, are described as physical, mechanical, biological control methods, and biopesticide practices. Physical and mechanical methods are: heat (hot water) treatment; sanitation; soil solarisation; usage of traps (light, colour, water, sticky, pheromone and food); mulching; low temperatures (to inhibit storage pests); microwaves; UV (Ultraviolet waves to eradicate water-borne micro-organisms) and IR (Infrared waves to control weeds); filters (used to remove pest and weed seeds from water). Biological methods regulate the pest population densities by using their native, or introduced, natural enemies. The use of natural enemies of pests, such as parasitoids, predators, and pathogens, and management of the crop variety to enhance the enemies' impact, has an important role in crop protection. Natural enemies may be indigenous or introduced from elsewhere. Biopesticides, obtained from natural plant products, have been found to be advantageous in terms of the features listed below: decomposing quickly and being less toxic (Fig. 1). On the other hand, it should not be overlooked that these products also have disadvantages regarding synthetic chemicals due to various properties (slow or delayed effect, lack of persistence and wide spectrum activity, rapid degradation by UV lights, storage problems, not being easily available, and difficulty of culturing in large quantities).

Fig. 1. Biopesticide application (Original)

According to the International Federation of Organic Agriculture Movements (IFOAM), organic agriculture is based on the principle of health, ecology, fairness and care: "The aim of the principles is both to inspire the organic movement and to describe the purpose of organic agriculture to the wider world".

"Organic agriculture should sustain and enhance the health of soil, plant, animal, human and planet as one and indivisible, and these are the principles of health". The principal emphasis is on the importance of healthy soil because it is associated with plants, animals and humans.

Principles of ecology are defined as: "Organic agriculture should be based on living ecological systems and cycles, work with them, emulate them and help sustain them. This principle roots organic agriculture within living ecological systems. Organic agriculture should attain ecological balance through the design of farming systems, establishment of habitats, and the maintenance of genetic and agricultural diversity" (Luttikholt 2007).

Principle of fairness: "It should build on relationships that ensure fairness with regard to the common environment and life opportunities. Fairness is characterised by equity, respect, justice and stewardship of the shared world, both among people and in their relations to other living beings".

Principle of care: "It should be managed in a precautionary and responsible manner to protect the health and wellbeing of current and future generations and the environment". This principle emphasises responsibility to protect the environment.

The general purposes of these principles require some main applications, including:

- To protect soil health and increase productivity
- To prevent environmental pollution from affecting agricultural production
- To increase the quality of food
- To protect wildlife and biodiversity and maintain the genetic diversity
- To design a collaborative production between crop production and animal husbandry
- To protect natural resources, such as soil, water and air

1. Organic Plant Production and Sustainability

After the Second World War, the use of synthetic chemicals in food processing to meet food needs was allowed, and following this it became impossible to get produce from certain plants without pest control. Since then, the use of pesticides has increased tremendously. Moreover, other highly technological and industrial inputs have become more frequently used in recent years, such as genetically modified organisms and tissue-cultured plants, etc.

Worldwide pesticide production increased at a rate of about 11% per year, from 0.2 million tonnes in the 1950s to more than 5 million tonnes by 2000 (Carvalho 2017). Nowadays, thousands of different synthetic pesticides and fertilisers are widely used to suppress pests and diseases, and to increase the agricultural yield. The worldwide consumption of pesticides is about two million tonnes per year (Dey et al. 2014). In contrast to conventional systems, organic agriculture offers alternatives to agro-chemicals and conventional practices.

In order to achieve sustainability in agriculture, cultural, physical, mechanical, biological and biopesticide control methods are being used, and include many practices and tactics. However, some of them are vital, such as crop rotation, natural plant nutrition, biodiversity, mulching for soil and water management, and crop protection.

1.1. Crop Rotation

Crop rotation is the successive cultivation of different crops in a specified order on the same fields, which mitigates the build-up of pathogens and pests, and also improves soil structure and fertility. Crop rotation creates automatic pest control by eliminating their food source. Different types of crops in the same area in sequenced seasons help to improve soil stability and thus bring down soil erosion, and enhance water infiltration by alternating between crops with deep roots and those with shallow roots. In crop rotation planning, many factors such as soil type, farm size, labour supply, climate, growing practices, and market need to be considered. Effective crop rotation should include a legume crop in the regular sequence because of biological nitrogen fixation and protecting degradation of soil fertility (Fig. 2) (Kumar and Goh 1999). In addition, a legume crop should always precede a nitrogen-depleting one; a low biomass crop should be offset with a high biomass cover crop. Drinkwater, Wagoner, and Sarrantonio (1998) reported that long-term, legume-based and organic production systems have resulted in an increase in soil organic matter and reduced nitrate runoff. With this result it can be stated that crop rotation is the main, important pattern of agriculture. Consequently, enhancement of the soil structure, increasing soil fertility, control of weeds, pests and diseases, producing different types of output, and reducing risks are the primary benefits of crop rotation.

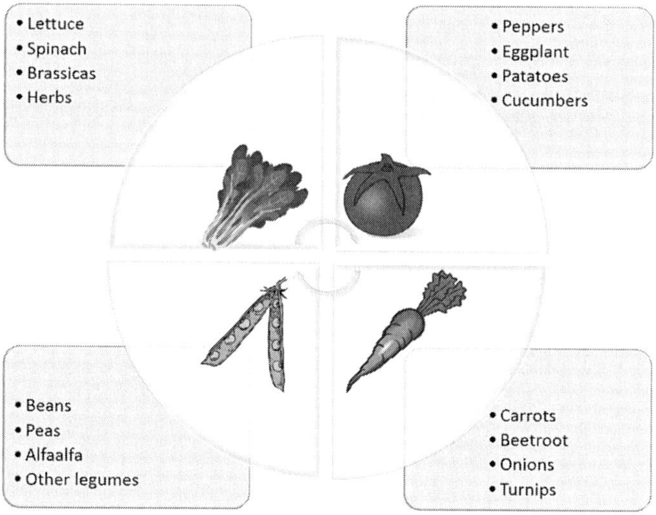

Fig. 2. Crop rotation guide for vegetable garden

1.2. Natural Plant Nutrition

Organic fertilisers such as manure enhance the physical condition of the soil by improving its water-holding capacity and optimising drainage. They support soil microbial activity and replenish essential micronutrients in the soil by providing organic acids that help to dissolve soil nutrients. Gunapala and Scow (1998) have demonstrated that the amount of soil nitrogen in fields under organic production was positively correlated with soil microbial components, but negatively correlated in conventional production systems. Fertilisation with organic fertilisers affects plant diseases and weeds and resistance to erosion. Broadly, organic fertilisers include: composts and urban wastes, animal wastes and manures, human excreta, crop residues, agroindustrial biological wastes, vermicompost, oil cakes, and green manures. Applications for natural plant nutrition are crop rotations, cover crops, green manures, animal manures, application of allowable substances in limited quantities, and composting.

1.3. Biological Diversity

Biodiversity is the variability among living organisms from all sources. It has a significant role in ecosystems, such as the recycling of nutrients, the regulation of microclimate and local hydrological processes, the suppression of undesirable organisms and the detoxification of noxious chemicals (Altieri 1999). Biodiversity plays an important role in the cost reduction of agricultural inputs and preventing environmental pollution. Among the benefits of biological diversity in agricultural production are: more beneficial insects, greater below-ground diversity, disease-resistant local animal breeds and cultivars, better nutrient cycling, disease suppression, tillage, carbon and nitrogen fixation, pollination, waste removal, and aesthetic value. Organic farming has a positive impact on biodiversity compared with monoculture agriculture. Conventional agriculture has destructive effects on natural flora and fauna due to the wide use of a broad spectrum of synthetic chemicals. In contrast, organic systems benefit biodiversity by supporting soil micro-organisms and other organisms, bird life and mammals, and the cultivating of more diverse plants. Some research has shown that the rate of biological diversity in organic farms is higher than conventional farms (Gabriel et al. 2010, Hole et al. 2005, Frieben and Kopke 1995, Kasperczyk and Knickel 2006, Mader et al. 2002). Some of the practices can be applied to agricultural land to increase the biological biodiversity, such as: intercropping, companion planting, establishment of beneficial habitats, crop rotations, cover crops, etc.

1.4. Mulching

Mulching is of great importance, not only in terms of crop protection, but also for making agriculture sustainable and eco-friendly. Deficit plant nutrients and control of weeds are the main problems in organic agriculture. Mulching retains soil moisture, influences the soil temperature, prevents soil erosion, yields clean fruits, and increases the yield (Acharya, Hati, and Bandyopadhyay 2005, Kumar and Dey 2011). A number of materials can be used for mulching, such as: sawdust, grass clippings, hay, paper, compost, textile, stone, etc. (Fig. 3). Organic mulch materials are incorporated naturally into the soil and can improve soil productivity. Decomposition of mulches is dependent on a number of factors, such as temperature, rain, activity of micro-organisms, worms and other organisms. Black plastic mulch material has been widely used in agricultural practices because of its obtainability, low-cost, and low rate of decomposition properties (Anttonen et al. 2006). Organic materials and living mulches are also preferred owing to their air and water permeability properties. Legumes increase the active organic matter of soil and add nitrogen when used as living mulches (cover crop), and also the next crop can use them.

Fig. 3. Sawdust and textile mulch application for strawberries

2. Organic Agriculture and Health

2.1. Organic Agriculture and Human Health

Many nutritionists recommend consuming fresh fruits and vegetables on a daily basis for a healthy life. However, conventionally-cultivated fruits and vegetables can threaten health because of their high concentrations of mineral deposits and pesticide residues. Some studies have highlighted that, like pesticide residues in fresh vegetables and fruits, animal products can also contain pesticide residues (milk, fish, meat, egg and their by-products) (Carvalho 2017, Claeys et al. 2011, Le Doux 2011, EC (European Commission) 2009, Van Boxstael et al.2013). The levels of residue depend on the physical and chemical properties of the pesticide, as well as on the dose, dissolution rate, and environmental factors. A pesticide that is applied on a crop and volatilised into the atmosphere is affected by the volatility of the pesticide, vapour pressure, temperature and wind movement; the synthetic chemicals may be degraded by sunlight, moisture and temperature.

Pesticides are used to control pests that are devastating the crops so that benefits result from increased crop yields. In addition to this, vector-borne diseases can cause the impairment of public health. The use of intensive pesticides is concerned with the protection of plant, animal and human health. Indeed, pesticide use has an unexpected effect and has caused human and environmental health problems, thus many people have been affected worldwide (Kesavachandran et al. 2009). The type and extent of health problems, arising from the use of pesticides, needs to be understood. The use of pesticides has resulted in serious health problems: acute and chronic poisoning, cancer, allergic reactions, destruction of the nervous system, learning disability, and memory loss, degradation of enzyme balances, and impairments and mutations in DNA. Globally, about 25 million agricultural workers experience symptomatic occupational pesticide poisonings each year, and pesticide poisoning cases amount to 3 million (Alavanja 2009). Consumers are exposed to pesticide residues by consuming foods, but agricultural/rural workers and their families are more affected by pesticides because of direct exposure to synthetic chemicals. Cancer, obesity, endocrine disruption and other diseases are known health problems that are the result of exposure to synthetic chemicals (Carvalho 2017, Bassil et al. 2007, Singh et al. 2016). In turn, it is claimed that organic foods are healthier than conventional ones because a lot of research shows that the secondary plant metabolites content of organic foods is higher. Plant secondary metabolites have been used for centuries in traditional medicine (Lavecchia et al. 2013). Some secondary metabolites promote public health because of their protective ability against cardiovascular disease, cancer (Schreiner et al. 2012) diabetes, osteoporosis, neurodegenerative diseases (Pandey and Rizvi 2009) and other age-related diseases, such as dementia. Flavonoids have potential in *in-vitro* antioxidant activity, and display free-radical scavenging activity (Pandey and Rizvi 2009). As a result, organic foods are popular for their health, safety, taste and environmentally friendly properties (Nandwani and Sochinwechi 2016).

2.2. Organic Agriculture and Environmental Health

In agricultural land across the globe, conventional agricultural practices pollute natural resources; deteriorate the natural balance; have adverse effects on wildlife; cause loss of biodiversity, rapid pollution of soil, erosion, desertification and drought, degradation of the quality and quantity of water resources, impairment of food quality, and chemical residue problems. All these directly and indirectly negatively impact the ecosystem.

A fundamentally important starting point for environmentally-friendly agricultural production is decreasing the use of fossil fuel energy. Conventional agriculture uses more energy overall than organic systems because of its high dependence on synthetic chemical fertilisers. Organic agriculture systems often perform better than conventional systems, contributing to reduced energy use and greenhouse gas emissions (Niggli et al. 2009).

The intensive use of synthetic chemicals in agricultural systems causes serious environmental consequences (Fig. 4). Synthetic fertilisers and pesticides in runoff from agricultural land are an important component of nonpoint source pollution. The primary impact of using synthetic chemicals is pollution of natural sources and accumulation of such substances in the ecosystem. Applying more pesticides and fertilisers to crops than necessary promotes the accumulation of heavy metals in the soil.

Many studies have been conducted to determine this interaction between the use of synthetic chemicals and heavy metal accumulation (Atafar et al. 2010, Galloway et al. 1995, Huang and Jin 2008, Parkpian et al. 2003). Some of the soil characteristics are affected by soil amendment and agricultural production systems. Bulluck Iii, Barker, and Ristaino (2002) reported that alternative fertility amendments, in comparison with synthetic soil fertility, enhanced beneficial soil micro-organisms, reduced pathogen populations, and increased soil organic matter, total carbon, and cation exchange capacity (CEC), thus improving soil quality.

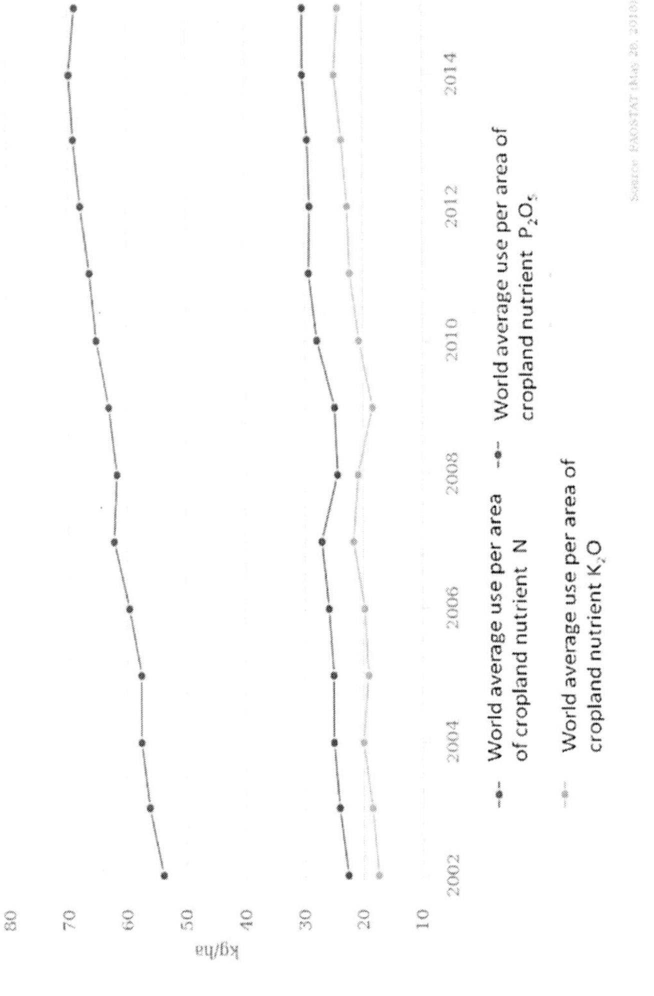

Fig. 4. World average use per area of cropland nutrient nitrogen (N), potash (K_2O) and phosphate (P_2O_5).

It is known that approximately 70% of the world's surface water resources are used by agricultural activity. Globally, water scarcity affects more than 40% of the population. Supplying clean water plays an important part in drinking, irrigation, industry, recycling in ecosystems, and in support of biodiversity. Prominent factors that cause agricultural water pollution are drainage, rainfall, microbial activity, soil and air temperature, application rate and dose, and the solubility, mobility and half-life of pesticides. The sources of nitrite in water systems, which are especially harmful to human health and cause water pollution, generally originate from high dose applications of synthetic and organic fertilisation. Fertiliser N, not recovered by the crop, can also rapidly enter the surface and groundwater pools through spreading and leaching, respectively (Vance, 2001). The U.S. Environmental Protection Agency (EPA 2002) has established a maximum contaminant level for NO_3-N in drinking water of 10 mg.L^{-1} (45 mg NO_3.L^{-1}) to protect infants and young children from methemoglobinanemia. Usually, the source of P and N to surface waters results mainly from agriculture and industrial activities. Excess P and N, sourced from agricultural and industrial activities, are closely related with hypoxia in aquatic ecosystems, and eutrophication, which is a major problem in rivers, lakes, estuaries, and coastal oceans (Carpenter et al. 1998, Galloway et al. 1995).

To reduce P and N accumulation in ground and surface waters, the following applications are proposed to emphasise the use of legumes and symbiotic N_2 fixation to improve soil N and P fertility: to develop intercrop schemes that foster efficient N and P use (Vance 2001); to avoid dumping the effluents into water reservoirs without proper pre-treatment (Agrawal, Pandey, and Sharma 2010); and to create institutional mechanisms that match scientific understanding with social realities (Gunderson and Holling 1995).

3. Organic Livestock Farming and Sustainability

Livestock products are among the main strategic products in human nutrition because of their high-value protein. In this regard, to meet the needs of the growing human population, the conventional livestock sector has grown at an unprecedented rate. It is defined as a method of production that aims at the highest productivity with more animals and more input, focuses on technologies for increased productivity, such as high-yielding breeds, modern feeding techniques and veterinary health products, and synthetic fertilisers and pesticides (van Wagenberg et al. 2017). However, this rapid development in conventional livestock production has brought with it a number of problems, such as a loss in biodiversity, animal health problems, pollution problems, and residue problems in animal products (antibiotics, hormones, growth stimulants and pesticides).

In contrast to the problems associated with conventional livestock production, organic livestock farming is an alternative production method which has set itself the goal of establishing environmentally-friendly production, sustaining animals in good health, realising high animal welfare standards, and producing high quality products (Sundrum 2001). These production systems also conserve ecosystems and their biodiversity, promote animal welfare, avoid the use of chemically synthesised substances, and offer consumers food with a high nutritional and hygienic-sanitary quality (IFOAM 2009).

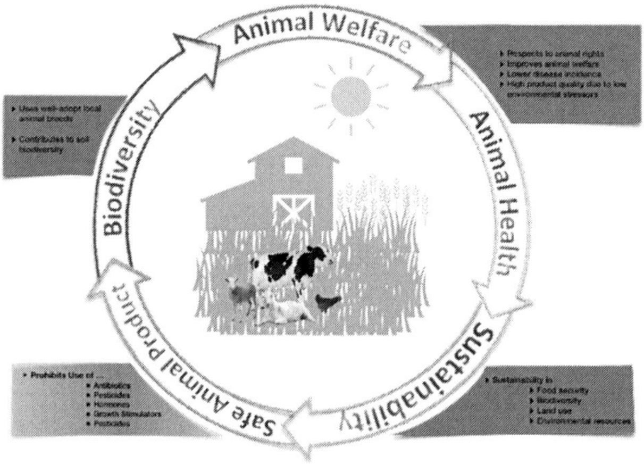

Fig. 5. Contributions of organic livestock production.

Although the objectives of organic farming are identical whether considering crop products or animal products, organic livestock farming has its own principles that are aimed at respecting the ecology by preserving biodiversity, improving animal welfare and animal health, and providing safe and healthy food to consumers.

4. Biodiversity and Organic Livestock Farming

Intensive farming systems are seen as one of the main causes of the decline in species. Studies that compared the impacts of conventional and organic farming systems verify the positive effects of organic farming on biodiversity at field and farm levels (Fuller et al. 2005, Hole et al. 2005). Recently, conventional livestock production has focused on a number of highly potent animal breeds, which are dependent on high standards of sheltering and are sensitive to adverse environmental factors. This trend caused a biodiversity loss in local animal breeds that are not being preferred in conventional animal production systems because of their low yields (Ozdemir et al. 2013). However, local animal breeds utilise lower quality feed, are more resilient to climatic stress, and are more resistant to local parasites and diseases (Van Diepen, McLean, and Frost 2007). Organic livestock production maintains breed diversity by giving a chance to the environmentally well-adapted local breeds in the production process.

Another benefit of indigenous animal breeds in organic animal production is the support it affords rural sustainability with high-quality breed-specific products. In this way, organic livestock farming also contributes to the conservation activities of endangered farm animal breeds.

Although livestock is usually positioned in the last part of the farm to be certificated as organic, it plays a vital role in organic farms in the nutrient cycling of the soil through manure and compost. In conventional animal production, the use of antibiotics, synthetic feed additives, hormone-like substrates and growth stimulators are allowed to be used in animal feeds to maximise the yield, but residues of these additives pass to the pasture soil through the animal manures and this, in turn, disrupts the natural balance of the soil. In contrast to this, the use of antibiotics and other synthetic substances to boost performance is prohibited in organic livestock farming. By providing natural manure to the pasture soil, organic animal farming contributes to enriching the soil microbial flora and also, indirectly, to vegetation diversity.

Organic agricultural methods are accepted as being eco-friendlier systems that usually operate with more crop rotation than intensive agriculture, which is dependent on the routine use of herbicides, pesticides and inorganic nutrient applications in the production of crops and animals (Bengtsson, Ahnstrom, and Weibull 2005).

The following principles are applied to organic farming, such as refusing the use of herbicides and chemically-synthesised pesticides, having fewer animals per square metre, diversified crop rotation, conservation tillage, a diversified farm structure and higher percentage of ecological areas, and promoting biodiversity in general. In brief, organic agriculture farms act as a refuge for wild plants, pollinators, arthropods, soil micro-organisms, offsetting the loss of biodiversity on conventional farms—and organic livestock farming is an important component of this system (Rundlof, Smith, and Birkhofer 2016, Tuck et al. 2014).

4.1. Respecting Animal Welfare

In conventional animal production, the pressure on animals to improve profitability is increasing day by day, and in these production techniques some measures carried out are, unfortunately, causing serious harm to animal health and welfare. Debeaking in layer chicks, feed restrictions in poultry breeding, force-feeding in ducks and geese, forced moulting of chickens, castration of cattle, pig breeding, and high stock density in all farm animal species are just some of the measures that have become routine in conventional animal production in most developing countries. In response to the maltreatment of animals under these methods, increasing awareness of animal welfare has become a priority in alternative production systems involving animals, such as organic animal farming.

Animal welfare, which means how an animal is coping with the conditions in which it lives, is a junction of the animal's biological functioning, mental state and naturalness. For animals, naturalness includes allowing animals their natural behaviour, and to meet their natural needs; natural environments; species-specific feeds; space to interact and withdraw; as much freedom of choice as possible to eat, drink, move and lie down comfortably; and harmony between animals and their surroundings (Vaarst and Alrae 2012). In this context, organic livestock farming has an express goal of improved animal well-being and health, compared with conventional livestock production; and to provide this, the process is controlled by regulations that include specific measures regarding health, comfort, nourishment, safety, humane treatment, and the ability to express natural behaviour.

Besides the negative effects of intensive agriculture on animal welfare, conventional animal production systems also threaten the health of the animals. Applications in conventional animal production, such as high stocking density, poor manure management (that causes water and air pollution), and the redundant use of antibiotics have increased the incidence of animal diseases, such as mastitis, lameness, and parasitic and infectious diseases. The use of antibiotics for the purpose of encouraging growth in animals other than for therapeutic purposes makes it difficult to treat bacterial diseases due to bacterial resistance. Moreover, feeding animals with high protein and energy content diets increases certain metabolic diseases, such as ascites and ketosis in farm animals. While maintaining farm sustainability, profitability and reliable products, optimising animal welfare and health are one of the main principles of organic livestock management. Consequently, disease prevention in organic livestock production is based on the principle that an animal that is permitted to demonstrate natural behaviour, is not subject to stress, and is given optimal feed, will have a greater ability to cope with diseases than ones raised in a conventional system (Kijlstra and Eijck 2006). The comparative studies on health problems in conventional and organic animal production highlighted that organic livestock systems have significant positive effects on reducing calf mortality, decreasing the incidence of mastitis and abortions, and reducing ectoparasite loads (Silva et al. 2014).

Although organic livestock principles favour eco-friendly measures that prioritise animal welfare and health, these principles cannot prevent the emergence of diseases in organic animal production. Diseases in organic livestock farming are often related to the outdoor access area, exposing the animals to various viral, bacterial and parasitic infections, some of which may only influence the animals' own welfare, whereas others may also endanger the health of conventional livestock (e.g. Avian Influenza), or pose a food safety (Campylobacter, Toxoplasma) problem to the consumer (Kijlstra and Eijck 2006). As a solution to the diseases in organic livestock farming, local breeds can be used as animal material, which are resistant to diseases outdoors. Furthermore, optimising the feeding programme with some organic feedstuffs, and improving rearing conditions, can also be solutions to prevent diseases.

4.2. Quality and Safety of Animal Products

In recent years, demand for organic products has been increasing steadily due to the fact that organic production is traceable, sustainable, and of low environmental impact. In particular, consumers who, to avoid the risks of dioxin in eggs, mad cow disease, avian influenza, and genetically modified organisms in animal feedstuff, prefer organic products.

Moreover, antibiotics, hormones, growth stimulants, pesticides and veterinary medicines are routinely used to maximise the yield of animals, and to ensure animal management in conventional animal production systems. The residues of these chemicals in animal products cause many diseases such as cancer in humans. In this context, consumers want to be sure that the production and processing conditions of foods obtained from animals are safe throughout the chain, which extends to transportation, packaging and retail outlets to the point of sale. Thus, the only way to meet the safe food expectations of consumers is, currently, organic animal farming. The use of chemicals such as pesticides, hormones, stimulators, fungicides, antibiotics and synthetic feed additives in organic livestock farming is largely prohibited. Consumers therefore perceive organic animal husbandry products as being much safer than conventional products because of the control mechanisms applied during the production process; and they are more popular.

Conclusion

Various agricultural techniques have been used since the dawn of civilization. In line with the increasing needs of people, agricultural techniques used to increase the productivity of agricultural products, have caused problems in environmental balance, biodiversity, food quality and safety, and also sustainability. Sufficient nourishment to sustain the growing human population, whilst minimising negative impacts on the environment, can be provided by eco-friendly, organic agriculture systems. In this chapter, the problems associated with the conventional agricultural methods have been highlighted; and how to find solutions to these problems through organic farming in plant and animal production has been discussed.

References

Acharya, CL, KM Hati, and KK Bandyopadhyay. 2005. "Mulches." In *encyclopedia of soils in the environment,* edited by D Hillel, C Rosenzweig, Ds Pawlson, KM Scow, MJ Sorger, DL Sparks and J Hatfield, 521–532. Amsterdam: Elsevier.

Agrawal, A, R S Pandey, and B Sharma. 2010. "Water pollution with special reference to pesticide contamination in India." *Journal of Water Resource and Protection* 2 (05):432.

Alavanja, M CR. 2009. "Introduction: Pesticides use and exposure, extensive worldwide." *Reviews on Environmental Health* 24 (4):303–310.

Altieri, Miguel A. 1999. "The ecological role of biodiversity in agroecosystems." In *Invertebrate Biodiversity as Bioindicators of Sustainable Landscapes,* 19–31. Elsevier.

Anttonen, M J, K I Hoppula, R Nestby, M J Verheul, and R O Karjalainen. 2006. "Influence of fertilization, mulch colour, early forcing, fruit order, planting date, shading, growing environment, and genotype on the contents of selected phenolics in strawberry (*Fragariaananassa* Duch.) fruits." *Journal of Agricultural and Food Chemistry* 54 (7):2614–2620.

Atafar, Z, A Mesdaghinia, J Nouri, M Homaee, M Yunesian, M Ahmadimoghaddam, and A H Mahvi. 2010. "Effect of fertilizer application on soil heavy metal concentration." *Environmental Monitoring and Assessment* 160 (1–4):83.

Bassil, K L, C Vakil, M Sanborn, DC Cole, J S Kaur, and KJ Kerr. 2007. "Cancer health effects of pesticides: systematic review." *Canadian Family Physician* 53 (10):1704–1711.

Bengtsson, J, J Ahnstrom, and A-C Weibull. 2005. "The effects of organic agriculture on biodiversity and abundance: a meta-analysis." *Journal of Applied Ecology* 42 (2):261–269.

Bulluck Iii, LR, KR Barker, and JB Ristaino. 2002. "Influences of organic and synthetic soil fertility amendments on nematode trophic groups and community dynamics under tomatoes." *Applied Soil Ecology* 21 (3):233–250.

Carpenter, S R, N F Caraco, D L Correll, R W Howarth, A N Sharpley, and V H Smith. 1998. "Nonpoint pollution of surface waters with phosphorus and nitrogen." *Ecological Applications* 8 (3):559–568.

Carvalho, F P. 2017. "Pesticides, environment, and food safety." *Food and Energy Security* 6 (2):48–60.

Claeys, W L, J-F Schmit, C Bragard, G Maghuin-Rogister, L Pussemier, and B Schiffers. 2011. "Exposure of several Belgian consumer groups to pesticide residues through fresh fruit and vegetable consumption." *Food Control* 22 (3–4):508–516.

Datta, S. 2012. "Organic Aquaculture—A new approach in fisheries development." In *Recent Advances in Aquaculture Technologies,* edited by S. Datta and B. K. Mahapatra, 105–120. CIFE Kolkata Centre.

De, Arnab, R Bose, A Kumar, and S Mozumdar. 2014. *Targeted delivery of pesticides using biodegradable polymeric nanoparticles*: Springer.

Drinkwater, L E, P Wagoner, and M Sarrantonio. 1998. "Legume-based cropping systems have reduced carbon and nitrogen losses." *Nature* 396 (6708):262.

EC (European Commission). 2009. "Official controls on imports of certain feed and food of non-animal origin." *Official Journal of the European Union* (L194):1–21.

EPA, U. 2002. "Edition of the drinking water standards and health advisories." *US Environmental Protection Agency Washington. E DC DC.*

FiBL, IFOAM. 2016. "The World of Organic Agriculture. Statistic and Emerging Trends 2016." *Switzerland (available at:* https://shop.fibl.org/fileadmin/documents/shop/1698-organic-world-2016. pdf).

Frieben, B, and U Kopke. 1995. "Effects of farming systems on biodiversity." Organic Farming in Land Use Systems. Proceedings of the First ENOF (The European Network for Scientific Research Co-ordination in Organic Farming) Workshop.

Fuller, RJ, LR Norton, RE Feber, PJ Johnson, DE Chamberlain, A C Joys, F Mathews, RC Stuart, MC Townsend, and WJ Manley. 2005. "Benefits of organic farming to biodiversity vary among taxa." *Biology Letters* 1 (4):431–434.

Gabriel, D, SM Sait, J A Hodgson, U Schmutz, W E Kunin, and T G Benton. 2010. "Scale matters: the impact of organic farming on biodiversity at different spatial scales." *Ecology Letters* 13 (7):858–869.

Galloway, J N, W H Schlesinger, H Levy, A Michaels, and J L Schnoor. 1995. "Nitrogen fixation: Anthropogenic enhancement-environmental response." *Global Biogeochemical Cycles* 9 (2):235–252.

Gunapala, N, and KM Scow. 1998. "Dynamics of soil microbial biomass and activity in conventional and organic farming systems." *Soil Biology and Biochemistry* 30 (6):805–816.

Gunderson, L H, and C S Holling. 1995. "Barriers and bridges to the renewal of ecosystems and institutions."

Hole, DG, AJ Perkins, JD Wilson, IH Alexander, PV Grice, and A D Evans. 2005. "Does organic farming benefit biodiversity?" *Biological Conservation*122 (1):113–130.

Huang, S-W, and J-Y Jin. 2008. "Status of heavy metals in agricultural soils as affected by different patterns of land use." *Environmental Monitoring and Assessment*139 (1–3):317.

IFOAM. 2009. "The principles of organic agriculture." http://www.ifoam.org/about ifoam/principles/index.html.

Kasperczyk, N, and K Knickel. 2006. "Environmental impacts of organic farming." *Organic Agriculture: A Global Perspective CSIRO Publishing: Collingwood: 259–294.*

Kesavachandran, C N, M Fareed, M K Pathak, V Bihari, N Mathur, and AK Srivastava. 2009. "Adverse health effects of pesticides in agrarian populations of developing countries." In *Reviews of Environmental Contamination and Toxicology Vol 200,* 33–52. Springer.

Kijlstra, A, and IAJM Eijck. 2006. "Animal health in organic livestock production systems: a review." *NJAS-Wageningen Journal of Life Sciences* 54 (1):77–94.

Kumar, K, and KM Goh. 1999. "Crop residues and management practices: effects on soil quality, soil nitrogen dynamics, crop yield, and nitrogen recovery." In *Advances in agronomy,* 197–319. Elsevier.

Kumar, S, and P Dey. 2011. "Effects of different mulches and irrigation methods on root growth, nutrient uptake, water-use efficiency and yield of strawberry." *Scientia Horticulturae* 127 (3):318–324.

Lavecchia, T, G Rea, A Antonacci, and M T Giardi. 2013. "Healthy and adverse effects of plant-derived functional metabolites: the need of revealing their content and bioactivity in a complex food matrix." *Critical Reviews in Food Science and Nutrition* 53 (2):198–213.

Le Doux, M. 2011. "Analytical methods applied to the determination of pesticide residues in foods of animal origin. A review of the past two decades." *Journal of Chromatography A* 1218 (8):1021–1036.

Luttikholt, L WM. 2007. "Principles of organic agriculture as formulated by the International Federation of Organic Agriculture Movements." *NJAS-Wageningen Journal of Life Sciences* 54 (4):347–360.

Mader, P, A Fliessbach, D Dubois, L Gunst, P Fried, and U Niggli. 2002. "Soil fertility and biodiversity in organic farming." *Science* 296 (5573):1694–1697.

McIntyre, B D. 2009. "International assessment of agricultural knowledge, science and technology for development (IAASTD): global report."

Nandwani, D., and N Sochinwechi. 2016. "Organic farming for sustainable agriculture." In, edited by D. Nandwani, 135. International Publishing Switzerland, Springer.

Niggli, U, A Fliehbach, P Hepperly, and N Scialabba. 2009. Low greenhouse gas agriculture: mitigation and adaptation potential of sustainable farming systems. Food and Agriculture Organization of the United Nations, Rome.

Pandey, K B, and S IRizvi. 2009. "Plant polyphenols as dietary antioxidants in human health and disease." *Oxidative Medicine and Cellular Longevity*2 (5):270–278.

Parkpian, P, S T Leong, P Laortanakul, and N Thunthaisong. 2003. "Regional monitoring of lead and cadmium contamination in a tropical grazing land site, Thailand." *Environmental Monitoring and Assessment* 85 (2):157–173.

Rundlof, M, H G Smith, and K Birkhofer. 2016. "Effects of organic farming on biodiversity." *eLS.*

Schreiner, M, I Mewis, S Huyskens-Keil, MAK Jansen, R Zrenner, JB Winkler, N O'Brien, and A Krumbein. 2012. "UV-B-induced secondary plant metabolites-potential benefits for plant and human health." *Critical Reviews in Plant Sciences* 31 (3):229–240.

Silva, J B, G M Fagundes, J PG Soares, A H Fonseca, and J P Muir. 2014. "A comparative study of production performance and animal health practices in organic and conventional dairy systems." *Tropical Animal Health and Production* 46 (7):1287–1295.

Singh, Z, J Kaur, R Kaur, and S S Hundal. 2016. "Toxic effects of organochlorine pesticides: a review." *American Journal of Bio Science*

4 (3–11):11–18.
Sundrum, A. 2001. "Organic livestock farming: a critical review." *Livestock Production Science* 67 (3):207–215.
Swedish Control Association of Ecological Farming. 2003. KRAV-Regler [KRAV certification].
Tuck, S L, C Winqvist, F Mota, J Ahnstrom, L A Turnbull, and J Bengtsson. 2014. "Land-use intensity and the effects of organic farming on biodiversity: a hierarchical meta-analysis." *Journal of Applied Ecology* 51 (3):746–755.
Vaarst, M, and H F Alrae. 2012. "Concepts of animal health and welfare in organic livestock systems." *Journal of Agricultural and Environmental Ethics* 25 (3):333–347.
Van Boxstael, S, I Habib, L Jacxsens, M De Vocht, L Baert, E Van de Perre, A Rajkovic, F Lopez-Galvez, I Sampers, and P Spanoghe. 2013. "Food safety issues in fresh produce: bacterial pathogens, viruses and pesticide residues indicated as major concerns by stakeholders in the fresh produce chain." *Food Control* 32 (1):190–197.
Van Diepen, P, B McLean, and D Frost. 2007. "Livestock breeds and organic farming systems."
van Wagenberg, CPA, Y de Haas, H Hogeveen, MM van Krimpen, MPM Meuwissen, CE van Middelaar, and TB Rodenburg. 2017. "Animal Board Invited review: Comparing conventional and organic livestock production systems on different aspects of sustainability." *Animal* 11 (10):1839–1851.
Vance, CP. 2001. "Symbiotic nitrogen fixation and phosphorus acquisition. Plant nutrition in a world of declining renewable resources." *Plant Physiology* 127 (2):390–397.
Yunlong, C, and B Smit. 1994. "Sustainability in agriculture: a general review." *Agriculture, Ecosystems & Environment* 49 (3):299–307.

CHAPTER 3

PLANT-MICROBE ASSOCIATIONS: A BARRIER AGAINST DISEASES AND AN INTERFACE BETWEEN UNAVAILABILITY VERSUS AVAILABILITY OF SOIL NUTRIENTS

AMJAD M. HUSAINI[1*], SAMIULLAH NAIK[1], ZAHOOR A. BABA[2]

[1]Genome Engineering and Societal Biotechnology Lab, Division of Plant Biotechnology, SKUAST-K, Shalimar, Jammu & Kashmir-190025, India
[2]Biofertilizer Research Laboratory, Faculty of Agriculture, SKUAST-K, Wadura, Sopore, Jammu & Kashmir-193201, India.
amjadhusaini@skuastkashmir.ac.in

Abstract

Plant-microbe associations are a very important component of the environment and are gaining considerable importance in the agricultural sector. These associations are of immense significance due to their roles in plant diseases, nutrient mobility, stress amelioration and bioremediation, etc. The microbes forming such associations have an immense, yet untapped, potential. The beneficial associations are exploited for achieving higher crop yields, better productivity, and improved quality of agricultural products. In recent times, there has been a renewed focus on such microbial associations. This is because of their relevance in meeting the sustainability challenges of the farming sector brought about by constrained resources and challenges posed by climate change. The beneficial products, developed from the microbes, are generally marketed as "biofertilisers": a consortium of living or latent microorganisms used to improve soil health for better crops. The relevance of these "microbial inoculants" is increasing owing to the popularisation of "organic agriculture".

Here, we present a detailed account of biofertilisers, and focus on some major issues connected with their use.

Key words: azotobacter, biofertilisers, microorganisms, mycorrhizae

Introduction

Biofertilisers, or more appropriately "Microbial inoculants", are biologically active products containing certain strains of bacteria, fungi or algae, as a single or composite culture. They are agriculturally important, living microorganisms, in terms of nutrient mobilisation, nitrogen fixation, increasing productivity of a crop, and fertility in soil. Biofertilisers were produced for the first time under the trade name "Nitragin" in 1895 by "Nobbe and Hiltner" in the USA, as a laboratory culture of *Rhizobia*, followed by the discovery of *Azotobacter*, and the blue green algae (BGA), and many other microbes. *Azospirillum* and Vesicular Arbuscular Mycorrhizae (VAM) were the latter discoveries (Hegde 2002). Most biofertilisers have the ability to convert atmospheric nitrogen into forms that are readily usable by plants. The continuous and excessive use of mineral/chemical fertilisers and agrochemicals to increase productivity causes ground water contamination and micronutrient imbalance in soil, resulting in a reduction of crop yield, and disturbed soil-plant-environment systems. Biofertilisers are an important component of integrated nutrient management for sustainable agriculture, as they are cost-effective, as well as a renewable source of nutrients to supplement and/or replace chemical fertilisers. These are preparations that contain living or latent cells of microorganisms, which are efficient in nutrient uptake for crop plants by their interactions in the rhizosphere. They accelerate microbial processes in the soil, and augment nutrient availability in a form that is easily assimilated by plants (Sheng and He 2006). The rhizospheric-efficient microorganisms promote nutrient uptake in plants through free-living bacteria, rhizobium, potassium solubilising bacteria/rhizobacteria (KSB/KSR), arbuscular mycorrhizal fungi and phosphorous solubilising bacteria. Biofertilisers differ from mineral fertilisers in that they do not supply nutrients directly to crop plants, and are only cultures of some specific efficient strains of microorganisms. Some of these rhizobacterial cultures, like *Azotobacter, Azospirillum* and *Rhizobium*, have the ability to fix atmospheric nitrogen (Khan, Zaidi, and Wani 2007). Vesicular arbuscular mycorrhizae (VAM) increase nutrient uptake of P, Zn and other micronutrients by profuse contact with roots from a larger soil volume (Basak and Biswas 2009). When biofertilisers are applied to soil, mixed with seeds, or sprayed on plant surfaces, they colonise the rhizosphere, or interior of the plant, thereby increasing the supply of nutrients to the host plant (Vessey 2003).

1. Classification of Biofertilisers

Biofertilisers fix atmospheric nitrogen in the soil or root nodules of legumes, and make it available to the plant. They solubilise the insoluble forms of phosphates (tricalcium, iron and aluminium phosphates) into available forms; secrete hormones and anti-metabolites, which promote root growth of plants; and help in the decomposition of organic matter and its subsequent mineralisation. Biofertilisers improve crop yield by 10–25%, without negatively affecting the environment (Hegde, 2002). The biofertilisers may be broadly classified into different categories: Nitrogen fixing, like *Rhizobium, Azotobacter, Azospirilum, Acetobacter*; Blue Green Algae; and *Azolla* and Phosphate solubilisers/ mobilisers, like PSM and Mycorrhizae (Fig. 1; Table 1). The major ones are discussed below:

i) *Rhizobium* spp.

These are gram-positive, symbiotic, nitrogen-fixing soil bacteria which fix atmospheric nitrogen, and assimilate it in root nodules of leguminous plants, as well as in some non-leguminous plants. Rhizobial symbiosis has been associated with regulation of root hydraulic properties (Porcel et al. 2006). A reduction in aquaporin gene expression in soybean roots, inoculated with the *Bradyrhizobium japonicum*, under both water-limited and control conditions, has been reported. It is well known that some aquaporins transport ammonium or ammonia, and hence could be involved in the transport of nitrogenous compounds between nodules and the host plant (Tyerman, Niemietz, and Bramley 2002, Uehlein et al. 2007). Mnasri et al. (2007) have reported that plants of *Phaseolus vulgaris,* when inoculated with a nitrogen-fixing, salt-tolerant bacterial strain (*Ensifermeli loti*), showed more drought tolerance than plants inoculated with a salt-sensitive bacterial strain (*Rhizobium tropici*). Swaine, Swaine, and Killham (2007) reported that a strain of *Bradyrhizobium elkanii,* isolated from a drought-prone environment, was more tolerant to *in vitro* osmotic stress than those isolated from wet environments.

Uses:

i. *Rhizobium* fixes 50–200 kg of nitrogen, per hectare, in one crop season, increasing yield by 10–25%.
ii. A mucous substance is secreted by nodules and root hairs, which improves soil fertility

iii. Nitrogenase enzyme reduces molecular nitrogen to ammonia, which is readily utilised by plants
iv. The germination of seeds gets stimulated by seed treatment and, in turn, increases crop yield potential

ii) *Azospirillum*

Azospirillum species are free-living, gram-negative, nitrogen-fixing bacteria. They possess a versatile Carbon/Nitrogen metabolism, which makes them fit for establishment in the competitive environment of rhizosphere. Ammonium, nitrite, nitrate, amino acids and molecular nitrogen serve as N-sources (Hartmann, Zimmer, and Okon 1994).

Uses:

i. *Azospirillum* spp. fix 20–40 kg of nitrogen, per hectare, resulting in 15–30% increase in crop yield.
ii. They increase water and mineral uptake, which, in turn, promotes vegetative growth and root development.
iii. They fortify soil through their metabolites and by secreting growth promoters.

They are mostly used for millets, paddy, oilseeds, vegetables, fruits, coconut, sugarcane, oil palm, cotton, tea, coffee, rubber, herbs, flower, ornaments, trees, etc.

iii) Blue Green Algae (BGA)

Blue Green Algae (Cyanobacteria) are prokaryotic, photosynthetic, free-living organisms which fix nitrogen, but some are also known to form symbiotic associations. *Azolla* represents a free-floating plant genus which belongs to the Pteridophyta and has a global distribution. *Azolla* is well known for its symbiotic relationship with the dinitrogen-fixing cyanobacterium, *Anabaena azollae*, which lives inside the leaf cavities of plants (Peters and Mayne 1974a). *Anabaena azollae* provides *azolla* with ammonium, whereas *azolla* provides it with carbon (Peters and Mayne 1974b). *Azolla* is commonly used for phytoremediation of contaminated surface water (Wagner 1997). It is used in irrigated rice fields, and is supplemented with 8–20 kg of phosphate, per hectare, for its multiplication. It improves the height, number of tillers, straw and grain yield in paddy. There is a 50% higher yield by using *Azolla-Anabaena azollae* association as a biofertiliser in irrigated rice.

BGA fix nitrogen under anaerobic conditions in specialised cells, known as heterocyst (Fleming and Haselkorn 1973). Examples of cyanobacteria are *Nostoc, Anabaena* and *Plectonema,* etc.

Uses:

i. BGA add growth promoting substances, like Vitamin B_{12}, and enhance soil aeration, as well as water-holding capacity.
ii. They provide 20–30 kg of biologically fixed nitrogen, per hectare, under normal conditions, which can be doubled if optimum quantity of phosphate and molybdenum are available in soil.
iii. They provide partial tolerance to plants against fungicides and pesticides.

iv) Phosphate Solubilisers

Phosphate solubilising bacteria (PSBs), inhabiting the rhizosphere, are promising biofertilisers, as these are known for secretion of organic acids and phosphates causing lowering of the pH in their vicinity, which in turn help in solubilisation of bound/insoluble phosphates to soluble forms (Ahemad and Khan 2011; Khan, Zaidi, and Wani 2007). Phosphate solubilisers present in soil are *Bacillus, Pseudomonas, Aspergillus, Micrococcus,* etc. The bacteria can secrete siderophores, which chelate iron present in the root zone, and thus make it non-available to harmful microbes. This protects crop plants from such harmful microbes.

Uses:

i. The insoluble phosphorus in soil is rendered soluble by the hydrolytic activities of the organic acids. These play an important role in the uptake of native/applied phosphorus
ii. Fortification of the soil occurs with bacterial metabolites
iii. They increase crop yield significantly

v) Potassium Solubilisers

Most potassium in the soil is unavailable (fixed form) to plants. This problem is compounded by an imbalance in fertiliser utilisation, as well as a depletion of soluble potassium as a result of huge increases in crop yields. Potassium deficiency has been reported in most of the crop plants (Meena, Maurya, and Verma 2014). The Potassium Solubilisers are effective in releasing K through solubilisation from inorganic and insoluble pools of total soil K (Archana et al. 2013). Inoculation of seeds and/or seedlings of plants with KSB generally shows a significant improvement in germination percentage, seedling vigour, plant yield, and K uptake under greenhouse / field conditions (Anjanadevi et al. 2016).

vi) Azotobacter

Azotobacter secretes growth regulators and antifungal compounds, which improve plant strains by the inhibition of root pathogens. It also has the ability to fix atmospheric nitrogen, which increases plant yields (Tejera et al. 2005); solubilises phosphates (Farajzadeh et al. 2012, Hayat et al. 2010); and promotes the microbial secretion of hormones, like auxins, gibberellins and cytokinins. *Azotobacter* also produces antifungal or fungistatic substances with a broad active spectrum. These substances inhibit the growth of *Fusarium, Helminthosporium, Aspergillus, Alterneria, Cephalosporium, Rhizoctonia* and *Sclerotiumrolfsii* (Farajzadeh et al. 2012, Hayat et al. 2010).

vii) Mycorrhizal fungi

The term "Mycorrhiza" was coined in 1885 by Frank. The varieties of mycorrhizae are:

a. **Ectomycorrhizae**

Ectomycorrhizae fungi are potentially excellent bio-indicators, owing to the large number of species, and vital ecological functions. Each species has a unique set of physiological characteristics (Trappe and Luoma 1992). These types of fungi are found on the roots of forest trees (e.g. pine, oak, eucalyptus, beech, etc.). They absorb phosphorus, nitrogen, potassium and calcium, besides producing growth-promoting hormones like cytokinins. Their major functions include absorption of water, and solubilisation of complex organic molecules into simplex inorganic nutrients, followed by their absorption and transfer to plant roots. Ectomycorrhizae also protect plants from attack by pathogens by secreting antimicrobial substances (Amaranthus and Perry 1987).

b. **Endomycorrhizae**

The term "endomycorrhizae" is used to describe mutualistic symbiotic associations between some fungi and plant roots, in which the fungal partner penetrates the cells of the host root and grows, mainly inside the root cortex. They do not form an external sheath or mantle, and only a small portion lives outside the root (Wu and Xia 2003). Endomycorrhiza are present in the roots of most fruit crops, as well as in crops like coffee, pepper, cardamom, and betelvine. They produce growth-promoting substances, improve phosphorus uptake, and increase resistance against pathogens (Wu, Zou, and Wang 2007).

c. **Vesicular Arbuscular Mycorrhizae (VAM)**

VAM are highly versatile, colonising 85% of plant families, and playing a major role in promoting plant growth. They enter the roots, forming arbuscules and vesicles in the cortical cells. Vesicles are thick-walled, swollen structures, whilst arbuscules are branched, haustorial branches of mycelium. VAM fungi increase drought resistance in plants by promoting a defence response through protective enzymes in host plants (Yin et al. 2010). VAM enhance antioxidant activity of enzymes, like SOD, POD, CAT, and H^+-ATPase in host plants, and reduce MDA content, as well as electrical conductivity of plasma membranes (Wu, Zou, and Wang 2007, Wu and Xia 2003).

Uses:

i. VAM enhance water uptake in plants
ii. They increase stress tolerance (especially heavy metal toxicity and high soil temperatures) and mitigate the negative effects of pathogens and pests on plants
iii. They penetrate the roots, and mobilise and supply phosphorus, as well as other micronutrients, to the plants

Piriformospora indica is a newly-described, axenically-cultivable, phyto-promotional endosymbiont, which mimics the capabilities of AMF. It has shown pronounced growth-promoting effects in strawberry plants, as well as in many other host species. *P. indica*-based *in vitro* and *ex vitro* co-culture studies have verified that it possesses pronounced plant growth-stimulating and stress amelioration capacity, most likely by inducing secondary metabolite accumulation in strawberry plants (Husaini et al. 2012, Matsubara and Husaini 2016).

1.1. Genetically Engineered Microbes (GEMs)

These are microorganisms into which a gene or genes have been introduced, using recombinant DNA technology. Genetically engineered microbes (GEMs) have shown potential for bioremediation applications in soil, groundwater, and activated sludge environments. These exhibit improved degradative capabilities over a wide range of chemical contaminants (Sayler and Ripp 2000). There are very few reports where GEMs have been applied in the elimination of recalcitrant compounds under natural conditions (De Lorenzo 2009). Efforts are being made to expand the range of compounds which could be degraded by using recombinant DNA technology (Kapley and Purohit 2009, Keasling and Bang 1998, Liu et al. 2006). The first test release of GEMs was for bioremediation. The US Environmental Protection Agency (EPA) released a variant of *Pseudomonas* in 1996, which was capable of degrading crude oil (Hill et al. 2000). A *Pseudomonas putida* strain was developed to degrade paraoxon, an organophosphorus compound. The *pnp* operon, from the *Pseudomonas sp.* strain ENV2030 encoding enzymes that transform *p*-nitrophenol into *b*-ketoadipate, was inserted into the organism with a simultaneous expression of a synthetic operon encoding a phosphodiesterase (encoded by *pde*) from *Delftiaacidovorans*; an organophosphate hydrolase (encoded by *opd*) from *Flavobacterium sp.* strain ATCC 27551; and an alkaline phosphatase (encoded by *phoA*) from *Pseudomonas aeruginosa* HN854 (de la Pena Mattozzi et al. 2006). Various GEMs, like *Pseudomonas putida, Mesorhizobium huakuii* and *Caulobacter crescentus*, accumulate Cd^{2+} by the expression of metal binding peptide (EC20), fusion proteins for phytochelatin (PC), and RsaA-6His, respectively (Sriprang et al. 2003, Patel et al. 2010, Wu et al. 2010). However, the potential risks associated with GEMs have necessitated the construction of biological containment systems which can kill bacteria in a controlled suicide process (Garbisu and Alkorta 1999).

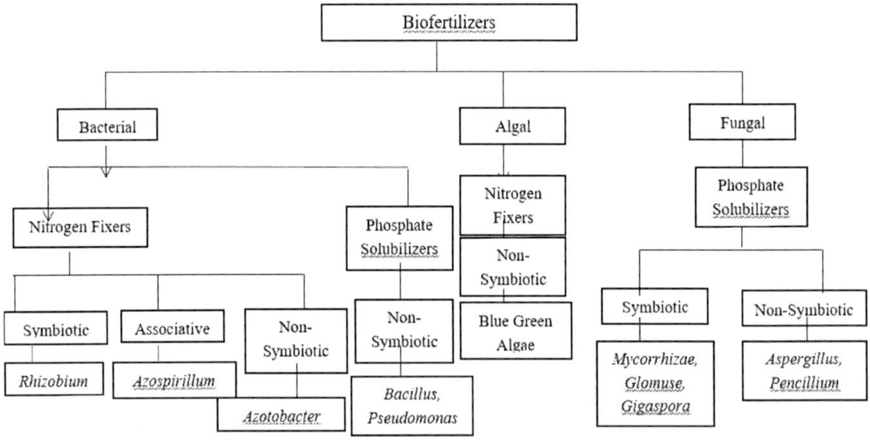

Fig. 1: Different categories of biofertilisers

Table 1: Mode of action of popular microbial fertilisers

S. No.	Microbe	Mode of action
1	*Azotobacter* sp.	Promote crop yield by fixing nitrogen, solubilising phosphates and secreting phytohormones like gibberellins, auxins and cytokinins
2	*Azospirillum* sp.	Colonising roots of host plants and fixing nitrogen
3	Vesicular Arbuscular Mycorhizza (VAM)	Water and minerals from the soil are easily available through the network of fungal filaments established inside the roots as well as surrounding rhizosphere soil, resulting in a better nutritive equilibrium. These also enhance antioxidant enzyme activity in host plants, while reducing MDA content and plasma membrane electrical conductivity, significantly.
4	Phosphate Solubilising Bacteria (PSB)	Enhance phosphorus availability by secreting organic acids like Oxalic, Citric, Succinic, Malic, Tartaric acids in the rhizosphere. These help to convert insoluble organic and inorganic bound phosphates into available forms for plant uptake and utilization. Release pathogen suppressing metabolites, phytohormones and lytic enzymes.
5	Potassium Solubilising Bacteria (KSB)	Solubilise rock potassium mineral powder, such as illite, micas and orthoclases by production and excretion of organic acids.
6	Rhizobium	The enzyme nitrogenase reduces the molecular nitrogen to ammonia which is readily utilised by the plants.
7	BGA	Cyanobacteria add growth promoting substances, including Vitamin B_{12}, and improve soil aeration and water holding capacity.

2. Production of Biofertilisers

The manure of dairy cattle is widely used as a fertiliser in organic farms (Audu, Aisien, and Eyawo 2003, Ogbeide and Aisien 2000). Animal slurry is also used as a fertiliser (Hamouda, Bahnasawy, and Ali 2016). The organic waste is generally decomposed before use as a fertiliser. For expediting the decomposition process, a bioreactor may also be used, which is an environmentally friendly technology, where anaerobic digestion of organic matter takes place. The method is affected by the concentration of a substrate, temperature, moisture content, pH, etc. (Behera et al. 2010, Jeong et al. 2010). It reduces environmental impact by capturing the methane in its sealed container instead of releasing it into the atmosphere. The burning of this methane will produce carbon-neutral carbon dioxide which, collectively, reduces the greenhouse effect. However, this process is slow and has low efficiency: to speed up the process, various chemicals and enzymes are therefore used (Lloret and Martinez-Romero 2005, Park et al. 2005, Torres and Lloréns 2008). Factors such as fermentation temperature, agitation speed, and ventilation rate are monitored as they affect the microbial, chemical and physical characteristics of a biofertiliser. These fertilisers not only promote the growth of the plants, but also provide protection against stress and diseases (Asghar et al. 2002, Paul and Lade 2014). The microorganisms are mounted onto any material to make their application in the fields easier. This material not only helps increase the shelf life of the fertiliser, but also facilitates rapid growth upon release. Examples of carrier material can be peat, charcoal, paddy, bran, etc. The material/microbe is tested for viability to make sure that it will work well with the selected organism. It is pre-sterilised and enriched with nutrients, like sucrose, maltose and molasses to ensure a longer shelf life. Such treatments provide an environment for the organisms to be maintained or grow (Aseri et al. 2008).

In order to increase their shelf life, biofertilisers can also be formulated in a liquid state, in which more nutrients can easily be added, together with cell protectants and inducers, for spore formation. The solid and liquid carrier shelf life is 6 months to 2 years, respectively. Liquid biofertilisers are more heat tolerant (as high as 55°C), but their manufacturing costs are much higher than solid fertilisers (Yardin, Kennedy, and Thies 2000).

Biofertilisers are able to mobilise important nutritional elements from non-usable to usable form. These microorganisms require soil organic matter for multiplication and activity, restore the soil's natural nutrient cycle, and promote growth of healthy plants, whilst enhancing the sustainability and the health of soil (Zhang et al. 2013). Different types of biofertilisers can be produced in the manner shown in Fig. 2:

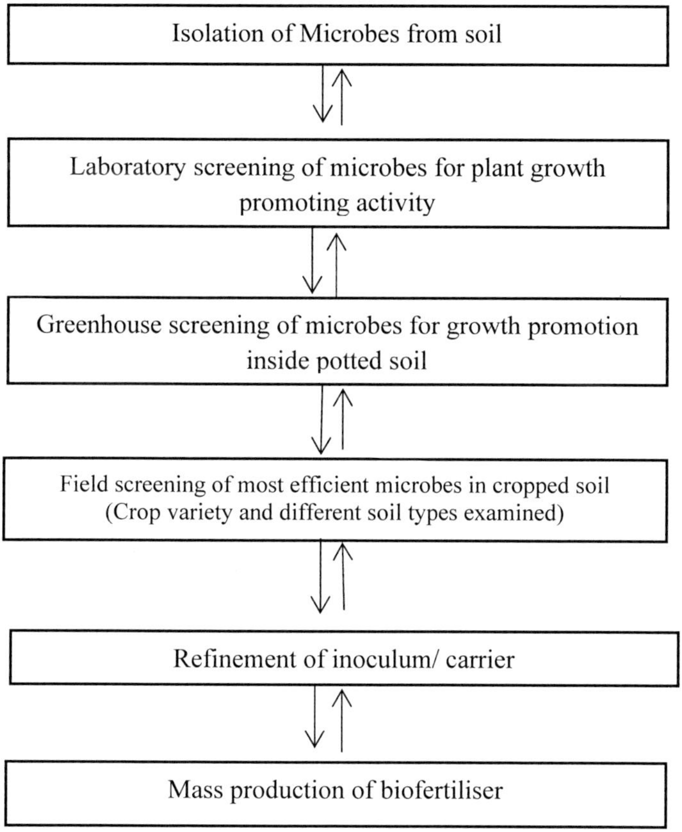

3. Multiple Roles of Plant Growth Promoting Rhizobacteria (PGPRs)

Beneficial rhizobacteria which stimulate plant growth are called plant growth promoting rhizobacteria or PGPR (Kloepper et al. 1980). PGPRs include bacterial species and strains belonging to different genera, like *Azospirillum, Acetobacter, Azotobacter, Burkholderia, Bacillus, Pseudomonas* and *Herbaspirillum* etc. (Kumar, Maurya, and Raghuwanshi 2014, Verma et al. 2014). PGPRs represent about 2–5% of the total rhizobacterial community and promote: (i) abiotic stress tolerance; (ii) nutrient fixation for easy uptake by plant; (iii) production of plant growth regulators; (iv) siderophore production; (v) production of volatile organic compounds; and (vi) production of enzymes, like glucanase, chitinase, and ACC-deaminase for plant disease prevention (García-Fraile, Menéndez, and Rivas 2015, Kloepper, Ryu, and Zhang 2004). Some bacteria support plant growth indirectly by improving antagonism against phytopathogens (Chincholkar, Chaudhari, and Rane 2007), and by induction of systemic resistance in plants (De Meyer *et al* 1999). These form an indispensable component of rhizosphere biota (Vessey 2003). They are major candidates for biofertilisers in soil ecosystems, as these are highly adaptable in a wide variety of environments; their growth rate is faster; and they have the capacity to metabolise a broad range of natural and xenobiotic compounds (Bhattacharyya and Jha 2012).

The following are the major roles of PGPRs in plant growth and development:

i) Enhancing Abiotic Stress Tolerance in Plants

The effects of climate change have emerged as a huge challenge that adversely influences crop establishment, crop quality and total crop production. To cushion crop production against these adversaries, microbial associations can come to our rescue, as reviewed by Husaini (2014) and Shabir et al. (2020). A major effect of PGPRs on plants under abiotic stress is in the improvement of leaf-water status, especially under salinity and drought stresses (Ahmad et al. 2013, Naveed et al. 2014). Sarma and Saikia (2014) have reported that *Pseudomonas aeruginosa* improved the growth of *Vignaradiata* (mung bean) under drought conditions. Marulanda et al. (2010) have reported that *Bacillus megatertum* increases the ability of maize roots to absorb water under salinity conditions. Gond et al. (2015) found similar behaviour when *Pantoea agglomerans* was inoculated into the maize roots. They found that, even under saline conditions, the ability of the maize roots to absorb water was improved. Gonzalez, Larraburu, and Llorente (2015) used *Azospirillum brasilense* during *in vitro* rooting to improve the salt tolerance of the jojoba plant. Similar findings were reported by Gabriela et al. (2015), who found that, under salt stress, inoculation with *Azospirillum sp.* extends the storage life of lettuce as well as improving its quality and yield.

ii) Improving Nutrient Availability for Plant Uptake

PGPRs have the ability to increase the availability of nutrients in the rhizosphere (Choudhary, Sharma, and Gaur 2011) by fixing nutrients, thus preventing them from leaching out. Nitrogen is the most limiting nutrient for plants, and is vital for the synthesis of amino acids and proteins. The mechanisms by which atmospheric nitrogen is added into organic forms, available to plants for assimilation, are confined to prokaryotes (Lloret and Martinez-Romero 2005, Raymond et al. 2004). A unique example of a free-living, nitrogen-fixing microbe is *Azospirillum*. It is often associated with cereals in temperate areas and improves rice yields (Tejera et al. 2005). Some PGPRs solubilise phosphate (Wani, Khan, and Zaidi 2007), resulting in the enhanced availability of phosphate ions in the soil, which are easily taken up by the plants. *Kocuria turfanensis* strain 2M4 is one such strain that was isolated from rhizospheric soil. It is a phosphate solubiliser, an IAA, and a siderophore producer (Goswami et al. 2014). Verma et al. (2010) studied the effect of PGPR strains, like *Pseudomonas putida, Pseudomonas fluorescens and Pseudomonas fluorescens* on nutrient uptake in rice. Parray et al. (2013) isolated rhizobacteria from the roots of the flowering corms of saffron. A total of 6 isolates of rhizobacteria were identified. These were: *Acinetobacter lwofii*, *Bacillus subtilis, Acinetobacter hemolyticus, Psuedomonas ssp, Klebsella ssp* and *Pantoea ssp.* The *Pantoea ssp.*, *B. subtilis* and *A. lwofii* were high IAA producers, with *B. subtilis* as the best strain for IAA production, and for enhancing the growth and development of corms. *Pseudomonas ssp.* was found to be highly efficient for phosphate solubilisation, followed by *B. subtilis*. The highest siderophore production was for *Psuedomonas ssp.* (62%), followed by *B. subtilis* (59%) (Husaini 2014).

iii) Releasing Plant Growth Regulators in soil-plant interface

According to Spaepen, Vanderleyden, and Remans (2007), and Vacheron et al. (2013), a wide range of processes in plant growth and development are controlled by exogenous IAA. High IAA levels decrease primary root length, increase root hair formation, and stimulate the formation of lateral roots, whereas low amounts of IAA can stimulate primary root elongation. Bacterial IAA increases both the root length and surface area as plants have greater access to soil nutrients. The processes of seed germination, emergence, floral induction, flowering, fruiting and stem growth involve gibberellins (GA) (Bottini, Cassán, and Piccoli 2004). Khan et al. (2014) showed that there is a significant increase in growth characteristics of tomato plants inoculated with gibberellin-producing *Sphingomonas sp.* LK11 strain. Similarly, Liu et al. (2013) reported that the oriental Thuja seedlings, inoculated with strains of cytokinin-producing *Bacillus subtilis*, had enhanced drought tolerance.

The plants synthesise aminocyclopropane-1-carboxylate (ACC) in response to exposure to stresses, like drought, cold, flooding, pathogens and heavy metals. ACC is the precursor for ethylene (Glick 2012). PGPRs, with the action to degrade ACC in the rhizosphere, shorten the deteriorating cycle, and reconstruct a healthy root system that withstands environmental stress. The enzyme ACC deaminase, involved in the primary mechanism of rhizobacteria, is utilised to degrade ethylene (Glick 2014). Ahmad et al. (2013) showed that *Rhizobium* and *Pseudomonas* ACC-deaminase-producing strains enhance the tolerance of mung bean under salt-stress.

iv) Producing Siderophores for increased uptake of Iron

The Earth's surface contains iron as a bulk mineral, yet it is unavailable to plants as it is generally present in an insoluble form (Fe^{3+}). This problem is solved by PGPRs secreting low molecular weight, iron-binding protein compounds called Siderophores. These are involved in the process of chelating ferric iron (Fe (iii)) from the environment, and providing it to plants with Fe, when it is limited. Flores-Félix et al. (2015) have shown that a siderophore-producing *Phyllobacterium* strain promotes growth and quality of strawberries.

v) Enhancing tolerance against phytopathogens

Volatile organic compounds (VOCs) are heavily involved in improving plant growth. They induce systemic resistance (ISR) towards pathogens. Several bacterial species, from diverse genera, including *Pseudomonas, Bacillus, Serratia, Stenotrophomonas* and *Arthrobacter*, produce VOCs that influence plant growth. Acetoin and 2,3-butane-diol, synthesised by *Bacillus*, are the best known compounds which significantly improve plant growth (Ryu et al. 2003). Although VOC emission is a common property of various soil microorganisms, the identity and quantity of volatile compounds emitted vary amongst them (Effmert et al. 2012, Kanchiswamy, Malnoy, and Maffei 2015).

PGPRs produce protection enzymes that promote plant growth through the control of phytopathogenic agents, primarily through antibiosis and the production of antifungal metabolites. The mechanism involves the production of hydrolytic enzymes, like chitinase and glucanase. Major fungal cell wall components are made up of chitin and beta-glucan. Hence, bacteria producing chitinases and beta-glucanases inhibit fungal growth. *Pseudomonas fluorescens* and *Sinorhizobium fredii* produce beta-glucanases and chitinase against *Fusarium udum*, causing fusarium wilt. Apart from the production of chitinase and beta-glucanases, *Pseudomonas spp*. inhibits *Rhizoctonia solani* and *Phytophthora capsici* (Arora et al. 2008).

4. Factors Affecting the Effectiveness of Bio-Fertilisers

i) Temperature

Temperature is an important factor affecting the shelf life of liquid biofertilisers. It can affect their activity before and after biofertiliser application: the optimum temperature range varies from strain to strain (Meena, Maurya, and Verma 2014). Field temperatures are usually suitable for colonisation during the cropping season, but at lower temperatures it slows down. The strains in liquid formulations grow normally at 37 ± 2°C, and can tolerate up to 45 ± 2°C for two years. However, solid-based ones have a shelf life of up to a maximum of 3 months, as a rise in temperature beyond 35°C causes a rapid decline of microbial population in the soil (Patel 2012).

ii) Effect of moisture

Moisture affects the activity, stability and storage of microbes. In carrier-based inoculums, the organisms get stressed when carriers become too dry during storage and transport. Microbial strains need a wet plant and/or soil surface for establishment as well as activity. These conditions are fulfilled by liquid formulations because humectants are present in them. The extent of the direct effect of humidity on spore-forming bacteria in liquid formulations is quite low (Kumar, Maurya, and Raghuwanshi 2014).

iii) Sunlight intensity

Ultraviolet B (280–320 nm) and Ultraviolet A (320–400 nm) are the most harmful radiations reaching the earth's surface and have a direct adverse effect on microbes. Microbes are also very sensitive to various other wavelengths. Sunscreens are added to formulations for countering the harmful effects of these radiations. These sunscreens reflect, scatter and absorb radiation selectively, converting harmful shorter wavelength radiation into relatively harmless, longer ones (Verma et al. 2014).

iv) Effect of pH

The pH has an important role in liquid inoculum preparation. The microorganisms get inactivated at extreme low and high pH. It must be stabilised within certain ranges with diluted NaOH and HCl. Therefore, a buffer is added in liquid bio-fertilisers for better shelf life (Maurya, Meena, and Meena 2014).

v) Stabilisation of liquid inoculums

The shelf life of a product between production and use in the field may vary from a few months to years. Microorganisms should remain viable during storage. There should be minimum loss of viability/activity, and there should be no deterioration in the formulation properties. Suitable additives are therefore added in liquid formulations to improve stability. These promote growth throughout the plant lifecycle by maintaining proper storage conditions prior to application. Stability is improved by addressing contamination and physical problems connected to activation time, adhesion to the target site, and the purity of the mother culture used for preparation of the product (Brar, Sarma, and Chaabouni 2012).

vi) Organic carbon content in soil

The organic fraction of soil, excluding non-decomposed animal and plant residues, is called soil organic matter (SOM). It is produced by the physical breakdown and biochemical transformation of complex organic molecules into simpler molecules, through a process called decomposition (Patel 2012). SOM contains approximately 58% C: a factor of 1.72 can therefore be used to convert OC to SOM. The main source of energy for soil microorganisms is soil organic carbon. It gets added into the soil by the decomposition of animal and plant residues, root exudates, and soil microbiota (Maurya, Meena, and Meena 2014).

5. Benefits of Biofertilisers to Crops

It is well established that microbial life occupies the rhizosphere soil (Nishanth and Biswas 2008). These microorganisms have played a vital role in nutrient recycling in agricultural ecosystems. They occur naturally in soils, but generally have a low population. So, in order to increase the crop yield, the desired microbes are isolated from rhizosphere and artificially cultured on suitable carriers, or by formation of microbial consortia through artificial culturing. Biofertilisers are preparations composed of these selected living strains of microbes, which provide nutrients to the plants through their root systems. The microbes in these fertilisers use different mechanisms to provide nutrients to the plants (Table 1). Proper packaging is ensured for a longer shelf life, and for the safety of the environment as well as the user (Muraleedharan, Seshadri, and Perumal 2010). The plant rhizospheric zone is rich in nutrients due to the accumulation of a variety of organic compounds released from roots by exudation, secretion, and deposition (Zhang et al. 2013). Because these organic compounds can be used as carbon and energy sources by microorganisms, microbial growth and activity is particularly intense in the rhizosphere. The use of bio-fertilisers is becoming popular because of their relatively lower harmful environmental impact (Balachandar 2012).

Unlike chemical fertilisers, they are a renewable source which can be actively used, with or without chemical fertilisers. They are easier to process, which generally comprises pulverisation, neutralisation, sterilisation, packaging and transport. Biofertilisers provide supplemental nutrients, along with microbes that may not be present in soil or may be present, but in small quantities. These fertilisers then speed up the process of microbial activity and provide sufficient and balanced quantities to the soil and plants. The effect of *Bacillus subtilis FZB24* on saffron corms under *ex vitro* conditions significantly increases total stigma biomass, while decreasing the time required for corms to sprout (Sharaf-Eldin et al. 2008). It has been observed that biofertilisers have a positive impact on the growth of the plant *Gladiolus* (*Gladuolus grandiflorus L.*), the presence of nitrogen in the fertiliser making the plants grow taller and heavier. The macro- and micro-nutrient absorption was also enhanced, and sprouting time was low. Another positive effect was seen in the early flowering of plants like the rose because of the early development of the auxiliary bulb to reproductive phase. Biofertilisers improved the physiological growth as well as vase life in *Galdiolus* (Mahdi et al. 2010). The weight and cormel production also improved, which was attributed to the availability of phosphorus (Ali et al. 2014).

In the strawberry, two main groups of microorganisms have been studied as biocontrol agents against strawberry fungal pathogens: (i) rhizobacteria, with *Pseudomonas* and *Streptomyces* spp. being the main members of this group; and (ii) members of the antagonistic fungal genus *Trichoderma*, the most effective fungal biocontrol agent identified to date (Carbú et al. 2016). Anandhakumar and Zeller (2008) did *in vitro* screening of about 100 rhizobacteria for controlling *Phytophthora* disease. Three bacteria, *Raoultella terrigena*, *Bacillus amyloliquefaciens* and *Pseudomonas fluorescens*, showed inhibitory effects on the mycelial growth of *Phytophthora cactorum* and *Phytophthora fragariae* var. *fragariae*. The use of the chitinolytic rhizobacterium, *Serratia plymuthica*, was evaluated as a potential strain for controlling *Verticillium* wilt and *Phytophthora* root rot of strawberry by Kurze et al. (2001). Artificial inoculation with *Serratia plymuthica* controlled the wilt and root rot of the strawberry plants, and even produced an increase in the number of buds, blossoms and fruit. Tortora, Díaz-Ricci, and Pedraza (2011) characterised the systemic resistance induced by *Azospirillum brasilense* on strawberry plants, which revealed the accumulation of salicylic acid, and induction of defence-related genes, structural cell wall modifications, increase in phenolic compounds and callose deposition. They concluded that *Azospirillum brasilense* protected the strawberry plants against anthracnose disease caused by *Colletotrichum acutatum* (Carbú et al. 2016). Li et al. (2010) have described Induced Systemic Resistance (ISR) to anthracnose and *Fusarium* wilt in mycorrhizal strawberry plants with a split root system. From the data discussed in Matsubara and Husaini (2016), it is clear that AMF colonisation does not itself stimulate a strong host antioxidative ability before pathogen inoculation; rather, the mycorrhizal strawberry plants acquire this ability after pathogen (*Fusarium*) inoculation, and maintain these higher levels for a long time.

The use of animal manure (biofertiliser) in aquaculture for the production of high-protein fish is a very effective use of bio fertilisers. Since livestock and poultry play a major role in improving the economy of a country, the waste products of these industries can be utilised as feed for the fisheries (Ajiboye, Akinremi, and Racz 2004). This not only reduces the impact of the livestock waste on the environment, but also cuts down the cost of buying feeds and fertilisers. The basic principle is that plankton feeds on the animal manure and grows (Kaur, Masud, and Khan 2015). This abundance of plankton can cause an increase in the fish yield, as they feed on this plankton (Jha, Sarkar, and Barat 2004). Improvement in fat content of fish muscles was also observed. Between the inorganically versus organically treated samples, more fat and protein were found in the organically treated fish. Several studies have shown a positive effect of biofertiliser treatments on crop plants (Table 2).

Table 2: Studies on the effect of biofertilisers on some major crops

S. No.	Biofertiliser(s) used	Crop	Major outcome	Reference
	Azospirillum, VAM, PSB	Tomato	Plant height, number of branches per plant, number of fruits per plant, fruit weight and yield were the highest.	Premsekhar and Rajashree (2009)
	Azospirillum	Chilli	The highest average yield of 136.9 q/ha which was 48.2 and 14.8 % higher than the control and 70 kg N/ha alone, respectively.	Deka, Bora, and Shadeque (1996)
	Azospirillum and organic manure	Brinjal	Fruit yield of higher girth (19.38 cm) and greater number (17.96 per plant) than control.	Naidu, Kushwah, and Dwivedi (2002)
	Azotobacter and *Azospirillum*	Cabbage	Application of *Azotobacter* + *Azospirillum* along with application of nitrogen @ 220 kg/ha increased the cabbage yield by 20% over uninoculated control.	Wange et al. (1995)
	Azotobacter	Onion	The maximum dry matter content of leaves, dry matter content of bulbs and total soluble solids were recorded higher with *Azotobacter* inoculation which was significantly superior to uninoculated control.	Rather, Ahmed, and Chattoo (2003)

	Azotobacter, *Azospirillum*, VAM and PSB	Cauliflower	Increase in plant height, leaf length, leaf width, number of leaves and increase in curd size and curd weight.	Manisha and Korla (2009)
	Azotobacter and *Azospirillum*	Tomato	Plant height, number of leaves per plant, number of fruits per plant, yield per plant, average fruit weight, chlorophyll content and protein content were recorded highest in combined inoculation than in uninoculated and singly inoculated.	Ramakrishnan and Selvakumar (2012)
	Azotobacter, *Azospirillum*, VAM and PSB	Chilli	Combined application of bio-fertilisers recorded the 36% more fruits per plant and 45% more dry matter production, as compared to control and individual inoculation of bio-fertilisers.	Sajan et al. (2002)
	Organic Farming	Tomato	Reduced growth in fruits under organic farming but titratable acidity, the soluble solids and vitamin C were respectively 29%, 57% and 55% higher at maturity. The total phenolic content was 139% higher in fruits as compared to conventional farming.	Oliveira et al. (2013)
	Azotobacter, *Azospirillum*, KSB and PSB	Tomato	Usage of PGPR increased average fruit weight, fruit length, number of total fruits, the percentage of marketable fruit and Potassium content of fruit of tomato plant, and when KSB was added to the PGPR these factors were higher as compared to the positive control and negative control treatments.	Sheikhalipour, Bolandndnazar, and Panahandeh (2016)
	Azotobacter, PSB, Vermicompost	Saffron	Recorded 57% increase in saffron yield over control with the application of 90 kg N ha-1 in conjunction with 60t FYM ha^{-1} plus *Azotobacter* inoculation.	Sofi, Kirmani, and Ansar-ul-Haq (2008)

6. Demand for Biofertilisers and Future Prospects

The injudicious use of synthetic fertilisers by farmers has led to the deterioration of soil health. This has destroyed beneficial microorganisms and friendly insects, making the crops more predisposed to diseases. Biofertilisers are a good option that can come to our rescue in the present scenario. Moreover, in order to achieve the targeted production of 321 million tonnes of food grain by 2020, the requirement of nutrients will be 28.8 million tonnes, while their availability will be only 21.6 million tonnes (Bahadur, Meena, and Kumar 2014).

In order to promote organic farming of the agriculturally important and high value crops in an eco-friendly and sustainable manner, biofertilisers are important, not only for restricting soil degradation, but also reducing environmental pollution. They are renewable, pollution-free, environmentally-friendly, and are of low cost (Bahadur, Meena, and Kumar 2014). According to the United Nations Food and Agriculture Organisation (FAO), there will be a 60% increase in demand for agricultural commodities by 2030, and more than 85% of this additional demand will be from developing countries (Mia and Shamsuddin 2010). There is thus an urgent need to integrate, and allow the use of GM crops in organic production systems (Husaini and Sohail 2018).

Biofertilisers are beneficial for increasing yields as they utilise chemical fertilisers more effectively and control soil-borne diseases. Due to these qualities, the demand for biofertilisers is increasing and is expected to be worth USD10.2 million by 2018. The Asia-Pacific region is expected to have 34% of the total demand, whereas Europe and Latin America will shift their consumption to these fertilisers due to regulations imposed on chemical fertilisers (Mazid 2008). However, one of the main limitations is the decrease in viability of biofertilisers over a period of time. Inoculants are generally available as solid carrier forms for commercial use (Maurya, Meena, and Meena 2014). A frequent observation in solid carrier-based inoculants is that the number of viable cells decreases from 10^9 to 10^7 colony-forming units (cfu) per g after 90 days of storage (Kumar, Maurya, and Raghuwanshi 2014). Hence, there is an urgent need to address the issue of the limited shelf-life of microbial fertilisers.

References

Ahemad, M, M S Khan. 2011. "Functional aspects of plant growth promoting rhizobacteria: recent advancements." *Insight Microbiology* 1(3): 39–54.

Ahmad, M, Z A Zahir, M Khalid, F Nazli, and M Arshad. 2013. "Efficacy of Rhizobium and Pseudomonas strains to improve physiology, ionic balance and quality of mung bean under salt-affected conditions on farmer's fields." *Plant Physiology and Biochemistry* 63:170–176.

Ajiboye, B, O OAkinremi, and G J Racz. 2004. "Laboratory characterization of phosphorus in fresh and oven-dried organic amendments." *Journal of Environmental Quality* 33 (3):1062–1069.

Ali, A, T Mehmood, R Hussain, A Bashir, S Najamuddin, and A Ahmad. 2014. "Investigation of biofertilizers influence on vegetative growth, flower quality, bulb yield and nutrient uptake in gladiolus (*Gladiolus grandiflorus* L.)." *International Journal of Plant, Animal and Environmental Sciences* 4 (1):94–99.

Amaranthus, M P, and D A Perry. 1987. "Effect of soil transfer on ectomycorrhiza formation and the survival and growth of conifer seedlings on old, nonreforested clear-cuts." *Canadian Journal of Forest Research* 17 (8):944–950.

Anandhakumar, J, and W Zeller. 2008. "Biological control of red stele (*Phytophthorafragariae* var. *fragariae*) and crown rot (*P. cactorum*) disease of strawberry with rhizobacteria." *Journal of Plant Diseases and Protection* 115 (2):49–56.

Anjanadevi, I P, N S John, K S John, M L Jeeva, and R S Misra. 2016. "Rock inhabiting potassium solubilizing bacteria from Kerala, India: characterization and possibility in chemical K fertilizer substitution." *Journal of Basic Microbiology* 56 (1):67–77.

Archana, D S, M S Nandish, V P Savalagi, and A R Alagawadi. 2013. "Characterization of potassium solubilizing bacteria (KSB) from rhizosphere soil." *BIOINFOLET-A Quarterly Journal of Life Sciences* 10 (1b):248–257.

Arora, N K, E Khare, J H Oh, S C Kang, and D K Maheshwari. 2008. "Diverse mechanisms adopted by fluorescent Pseudomonas PGC2 during the inhibition of Rhizoctoniasolani and Phytophthoracapsici." *World Journal of Microbiology and Biotechnology* 24 (4):581–585.

Aseri, G K, N Jain, J Panwar, A V Rao, and P R Meghwal. 2008. "Biofertilizers improve plant growth, fruit yield, nutrition, metabolism and rhizosphere enzyme activities of pomegranate (*Punicagranatum* L.) in Indian Thar Desert." *Scientia Horticulturae* 117 (2):130–135.

Asghar, H, Z Zahir, M Arshad and A Khaliq. 2002. "Relationship between in vitro production of auxins by rhizobacteria and their growth-promoting activities in *Brassica juncea* L." *Biology and Fertility of Soils*, 35(4): 231–237.

Audu, T O K, F A Aisien, and E O Eyawo. 2003. "Biogas from municipal solid waste." *The New England Journal of Medicine* 4 (1):26–30.

Bahadur, I, V S Meena, and S Kumar. 2014. "Importance and application of potassicbiofertilizer in Indian agriculture." *Research Journal of Chemical Sciences* 3 (12):80–85.

Balachandar, D. 2012. "Biofertilizers–what next." *Journal of Biofertilizers and Biopesticides* 3:e108.

Basak, BB, and D R Biswas.2009. "Influence of potassium solubilizing microorganism (*Bacillus mucilaginosus*) and waste mica on potassium uptake dynamics by sudan grass (*Sorghum vulgare* Pers.) grown under two Alfisols." *Plant and Soil* 317(1–2): 235–255.

Behera, S K, J M Park, K H Kim, and H-S Park. 2010. "Methane production from food waste leachate in laboratory-scale simulated landfill." *Waste Management* 30 (8–9):1502–1508.

Bhattacharyya, P N, and Dhruva K Jha. 2012. "Plant growth-promoting rhizobacteria (PGPR): emergence in agriculture." *World Journal of Microbiology and Biotechnology* 28 (4):1327–1350.

Bottini, R, F Cassán, and P Piccoli. 2004. "Gibberellin production by bacteria and its involvement in plant growth promotion and yield increase." *Applied Microbiology and Biotechnology* 65 (5):497–503.

Brar, S K, S J Sarma, and E Chaabouni. 2012. "Shelf-life of biofertilizers: an accord between formulations and genetics." *Journal of Biofertilizers and Biopesticides* 3 (5):1–2.

Carbú, M, V E González-Rodríguez, C Garrido, A M Husaini, and J M Cantoral. 2016. *Strawberry: Growth, Development and Diseases*. UK: CABI.

Chincholkar, S B, B L Chaudhari, and M R Rane. 2007. "Microbial siderophores in human and plant health-care." In *Microbial Siderophores*, 205–217. USA: Springer.

Choudhary, D K, K P Sharma, and R K Gaur. 2011. "Biotechnological perspectives of microbes in agro-ecosystems." *Biotechnology Letters* 33 (10):1905–1910.

de la Pena Mattozzi, M, S K Tehara, T Hong, and J D Keasling. 2006. "Mineralization of paraoxon and its use as a sole C and P source by a rationally designed catabolic pathway in Pseudomonas putida." *Applied and Environmental Microbiology* 72 (10):6699–6706.

De Lorenzo, V. 2009. "Recombinant bacteria for environmental release: what went wrong and what we have learnt from it." *Clinical Microbiology and Infection* 15 (1):63–65.

De Meyer, G, K Audenaert, and M Höfte. 1999. "*Pseudomonas aeruginosa* 7NSK2-induced systemic resistance in tobacco depends on in planta salicylic acid accumulation but is not associated with PR1a expression. "*European Journal of Plant Pathology* 105(5): 513–517.

Deka, B C, G C Bora, and A Shadeque. 1996. "Effect of azospirillum on growth and yield of chilli (*Capsicum annum* L.) cultivar "Pusa Jwala". *Haryana Journal of Horticultural Sciences* 25:44–45.

Effmert, U, J Kalderás, RWarnke, and B Piechulla. 2012. "Volatile mediated interactions between bacteria and fungi in the soil." *Journal of Chemical Ecology* 38 (6):665–703.

Farajzadeh, D, B Yakhchali, N Aliasgharzad, N Sokhandan-Bashir, and M Farajzadeh. 2012. "Plant growth promoting characterization of indigenous Azotobacteria isolated from soils in Iran." *Current Microbiology* 64 (4):397–403.

Fleming, H, and R Haselkorn. 1973. "Differentiation in *Nostocmuscorum*: nitrogenase is synthesized in heterocysts." *Proceedings of the National Academy of Sciences* 70 (10):2727–2731.

Flores-Félix, J D, L R Silva, L P Rivera, M Marcos-García, P García-Fraile, EMartínez-Molina, P F Mateos, E Velázquez, P Andrade, and R Rivas. 2015. "Plants probiotics as a tool to produce highly functional fruits: the case of Phyllobacterium and vitamin C in strawberries." *PLoS One* 10 (4):e0122281.

Gabriela, F, E M Casanovas, V Quillehauquy, A K Yommi, M G Goñi, S I Roura, and C ABarassi. 2015. "*Azospirillum inoculation* effects on growth, product quality and storage life of lettuce plants grown under salt stress." *Scientia Horticulturae* 195:154–162.

Garbisu, C, and I Alkorta. 1999. "Utilization of genetically engineered microorganisms (GEMs) for bioremediation." *Journal of Chemical Technology and Biotechnology* 74 (7):599–606.

García-Fraile, P, E Menéndez, and R Rivas. 2015. "Role of bacterial biofertilizers in agriculture and forestry." *AIMS Bioengineering* 2 (3):183–205.

Glick, B R. 2012. "Plant growth-promoting bacteria: mechanisms and applications." *Scientifica* 2012:1–15.

Glick, B R. 2014. "Bacteria with ACC deaminase can promote plant growth and help to feed the world." *Microbiological Research* 169 (1):30–39.

Gond, S K, M S Bergen, M S Torres, J F White, and RN Kharwar. 2015. "Effect of bacterial endophyte on expression of defense genes in Indian popcorn against *Fusariummoniliforme*." *Symbiosis* 66 (3):133–140.

Gonzalez, A J, E ELarraburu, and B E Llorente. 2015. "Azospirillumbrasilense increased salt tolerance of jojoba during *in vitro* rooting." *Industrial Crops and Products* 76:41–48.

Goswami, D, S Pithwa, P Dhandhukia, and J N Thakker. 2014. "Delineating Kocuriaturfanensis 2M4 as a credible PGPR: a novel IAA-producing bacteria isolated from saline desert." *Journal of Plant Interactions* 9 (1):566–576.

Hamouda, R, A Bahnasawy, and S Ali. 2016. "Some Physical and Chemical Properties of Bio-fertilizers." *Journal of Fertilizers & Pesticides* 7 (1):1–6.

Hartmann, A, W Zimmer, and Y Okon. 1994. "Physiology of Azospirillum." *Azospirillum/Plant Associations* 1:15–39.

Hayat, R, S Ali, U Amara, R Khalid, and I Ahmed. 2010. "Soil beneficial bacteria and their role in plant growth promotion: a review." *Annals of Microbiology* 60 (4):579–598.

Hegde, S V. 2002. "Liquid biofertilizers in Indian agriculture." *Bio-Fertilizer News Letter* 12:17–22.

Hill, R A, P M Chapman, G S Mann, and G S Lawrence. 2000. "Level of detail in ecological risk assessments." *Marine Pollution Bulletin* 40 (6):471–477.

Husaini, A M. 2014. "Challenges of climate change: Omics-based biology of saffron plants and organic agricultural biotechnology for sustainable saffron production." *GM Crops & Food* 5 (2):97–105.

Husaini, A M, M Z Abdin, S Khan, Y W Xu, S Aquil, and M Anis. 2012. "Modifying strawberry for better adaptability to adverse impact of climate change." *Current Science* 1:1660–1673.

Husaini, A M, and M Sohail. 2018. "Time to redefine organic agriculture: can't gm crops be certified as organics?" *Frontiers in Plant Science* 9:1–5.

Jeong, E, H-W Kim, J-Y Nam, and H-S Shin. 2010. "Enhancement of bioenergy production and effluent quality by integrating optimized acidification with submerged anaerobic membrane bioreactor." *Bioresource Technology* 101 (1):7–12.

Jha, P, K Sarkar, and S Barat. 2004. "Effect of different application rates of cowdung and poultry excreta on water quality and growth of ornamental carp, *Cyprinuscarpio*vr. *koi*, in concrete tanks." *Turkish Journal of Fisheries and Aquatic Sciences* 4:17–22.

Kanchiswamy, C N, M Malnoy, and M E Maffei. 2015. "Chemical diversity of microbial volatiles and their potential for plant growth and productivity." *Frontiers in Plant Science* 6:1–23.

Kapley, A, and H J Purohit. 2009. "Genomic tools in bioremediation." *Indian Journal of Microbiology* 49 (2):108–113.

Kaur, S, S Masud, and A Khan. 2015. "Effect of fertilization and organic manure on water quality dynamics a proximate composition of *Cyprinuscarpio*." *Journal of Fisheries & Livestock Production* 3:133–138.

Keasling, JD, and S W Bang. 1998. "Recombinant DNA techniques for bioremediation and environmentally-friendly synthesis." *Current Opinion in Biotechnology* 9 (2):135–140.

Khan, A L, M Waqas, S-M Kang, A Al-Harrasi, J Hussain, A Al-Rawahi, S Al-Khiziri, I Ullah, L Ali, and H-Y Jung. 2014. "Bacterial endophyte Sphingomonas sp. LK11 produces gibberellins and IAA and promotes tomato plant growth." *Journal of Microbiology* 52 (8):689–695.

Khan, M S, A Zaidi, and P A Wani. 2007. "Role of phosphate-solubilizing microorganisms in sustainable agriculture—a review." *Agronomy for Sustainable Development* 27 (1):29–43.

Kloepper, J W, JLeong, M Teintze, and M N Schroth. 1980. "Enhanced plant growth by siderophores produced by plant growth-promoting rhizobacteria." *Nature* 286 (5776):885–886.

Kloepper, J W, C-M Ryu, and S Zhang. 2004. "Induced systemic resistance and promotion of plant growth by Bacillus spp." *Phytopathology* 94 (11):1259–1266.

Kumar, A, B R Maurya, and R Raghuwanshi. 2014. "Isolation and characterization of PGPR and their effect on growth, yield and nutrient content in wheat (*Triticumaestivum* L.)." *Biocatalysis and Agricultural Biotechnology* 3 (4):121–128.

Kurze, S, H Bahl, R Dahl, and G Berg. 2001. "Biological control of fungal strawberry diseases by *Serratiaplymuthica* HRO-C48." *Plant Disease* 85 (5):529–534.

Li, Y, A Yanagi, Y Miyawaki, T Okada, and Y-I Matsubara. 2010. "Disease tolerance and changes in antioxidative abilities in mycorrhizal strawberry plants." *Journal of the Japanese Society for Horticultural Science* 79 (2):174–178.

Liu, F, S Xing, H Ma, Z Du, and B Ma. 2013. "Cytokinin-producing, plant growth-promoting rhizobacteria that confer resistance to drought stress in *Platycladusorientalis* container seedlings." *Applied Microbiology and Biotechnology* 97 (20):9155–9164.

Liu, Z, Q Hong, J-H Xu, W Jun, and S-P Li. 2006. "Construction of a genetically engineered microorganism for degrading organophosphate and carbamate pesticides." *International Biodeterioration and Biodegradation* 58 (2):65–69.

Lloret, L, and E Martinez-Romero. 2005. "Evolution and phylogeny of rhizobia." *Latinoam Microbiol* 47:43–60.

Mahdi, S S, G I Hassan, S A Samoon, H A Rather, S A Dar, and B Zehra. 2010. "Bio-fertilizers in organic agriculture." *Journal of Phytology* 2 (10):42–54.

Manisha, K, and B N Korla. 2009. "Effect of biofertilizers on growth and yield of cauliflower cv. PSB K-1." *Indian Journal of Horticulture* 66 (4):496–501.

Marulanda, A, R Azcón, F Chaumont, J M Ruiz-Lozano, and R Aroca. 2010. "Regulation of plasma membrane aquaporins by inoculation with a *Bacillus megaterium* strain in maize (*Zea mays* L.) plants under unstressed and salt-stressed conditions." *Planta* 232 (2):533–543.

Matsubara, Y, and A M Husaini. 2016. *Induced Systemic Resistance to Fusarium Wilt and Antioxidative Ability in Mycorrhizal Strawberry Plants. In: Strawberry-Growth, Development and Diseases*. UK: CABI.

Maurya, B R, V S Meena, and O P Meena. 2014. "Influence of Inceptisol and Alfisol's potassium solubilizing bacteria (KSB) isolates on release of K from waste mica." *Vegetos* 27 (1):181–187.

Mazid, S A. 2008. "Review on the use of biopesticides in insect pest management." *International Journal of Science and Advanced Technology* 1:169–178.

Meena, V S, B R Maurya, and J P Verma. 2014. "Does a rhizospheric microorganism enhance K+ availability in agricultural soils?" *Microbiological Research* 169 (5–6):337–347.

Mia, M A Baset, and Z H Shamsuddin. 2010. "Rhizobium as a crop enhancer and biofertilizer for increased cereal production." *African Journal of Biotechnology* 9 (37):6001–6009.

Mnasri, B, M Mrabet, G Laguerre, M E Aouani and R Mhamdi. 2007. "Salt-tolerant rhizobia isolated from a Tunisian oasis that are highly effective for symbiotic N 2-fixation with *Phaseolus vulgaris* constitute a novel biovar (bv. mediterranense) of *Sinorhizobiummeliloti*." *Archives of Microbiology*, 187(1): 79–85.

Muraleedharan, H, S Seshadri, and K Perumal. 2010. *Biofertilizer (Phosphobateria)*. Taramani, Chennai: Shri AMM MurugappaChettiar Research Centre.

Naidu, A K, S SKushwah, and Y C Dwivedi. 2002. "Influence of organic manures, chemical and bio fertilizers on growth, yield and economics of brinjal." *South Indian Horticulture* 50 (4/6):370–376.

Naveed, M, M B Hussain, Z A Zahir, B Mitter, and A Sessitsch. 2014. "Drought stress amelioration in wheat through inoculation with Burkholderiaphy to firmans strain PsJN." *Plant Growth Regulation* 73 (2):121–131.

Nishanth, D, and D R Biswas. 2008. "Kinetics of phosphorus and potassium release from rock phosphate and waste mica enriched compost and their effect on yield and nutrient uptake by wheat (*Triticumaestivum*)." *Bioresource Technology* 99 (9):3342–3353.

Ogbeide, S E, and F A Aisien. 2000. "Biogas from Cassava peelings." *Africa Journal of Environmental Studies* 1(12):42–47.

Oliveira, A B, C FH Moura, E Gomes-Filho, C A Marco, L Urban, and M R A Miranda. 2013. "The impact of organic farming on quality of tomatoes is associated to increased oxidative stress during fruit development." *PLoS One* 8 (2):e56354.

Park, C, C Lee, S Kim, Y Chen, and H A Chase. 2005. "Upgrading of anaerobic digestion by incorporating two different hydrolysis processes." *Journal of Bioscience and Bioengineering* 100 (2):164–167.

Parray, J A, A N Kamili, Z A Reshi, R Hamid, and R A Qadri. 2013. "Screening of beneficial properties of rhizobacteria isolated from Saffron (*Crocus sativus* L) rhizosphere." *African Journal of Microbiology Research* 7 (23):2905–2910.

Patel, B C. 2012. Advance method of preparation of bacterial formulation using potash mobilizing bacteria that mobilize potash and make it available to crop plant. edited by WIPO Patent Application.

Patel, J, Q Zhang, R M L McKay, R Vincent, and Z Xu. 2010. "Genetic engineering of Caulobactercrescentus for removal of cadmium from water." *Applied Biochemistry and Biotechnology* 160 (1):232–243.

Paul, D, and H Lade. 2014. "Plant-growth-promoting rhizobacteria to improve crop growth in saline soils: a review." *Agronomy for Sustainable Development* 34 (4):737–752.

Peters, G A, and B C Mayne. 1974a. "The Azolla, Anabaena azollae relationship: I. Initial characterization of the association." *Plant Physiology* 53 (6):813–819.

Peters, G A, and B C Mayne. 1974b. "The Azolla, Anabaena azollae relationship: II. Localization of nitrogenase activity as assayed by acetylene reduction." *Plant Physiology* 53 (6):820–824.

Porcel, R, R Aroca, R Azcon, and J M Ruiz-Lozano. 2006. "PIP aquaporin gene expression in arbuscular mycorrhizal *Glycine max* and *Lactucasativa* plants in relation to drought stress tolerance." *Plant Molecular Biology* 60 (3):389–404.

Premsekhar, M, and V Rajashree. 2009. "Influence of bio-fertilizers on the growth characters, yield attributes, yield and quality of tomato." *American-Eurasian Journal of Sustainable Agriculture* 3 (1):68–70.

Ramakrishnan, K, and G Selvakumar. 2012. "Effect of biofertilizers on enhancement of growth and yield on Tomato (Lycopersicumesculentum Mill.)." *International Journal of Research in Botany* 2 (4):20–23.

Rather, S A, N Ahmed, and M A Chattoo. 2003. "Response of onion to microbial inoculation and chemical nitrogen." *Haryana Journal of Horticultural Sciences* 32:270–271.

Raymond, J, J L Siefert, C R Staples, and R E Blankenship. 2004. "The natural history of nitrogen fixation." *Molecular Biology and Evolution* 21 (3):541–554.

Ryu, C-M, M A Farag, C-H Hu, M S Reddy, H-X Wei, P W Paré, and J W Kloepper. 2003. "Bacterial volatiles promote growth in Arabidopsis." *Proceedings of the National Academy of Sciences* 100 (8):4927–4932.

Sajan, K M, K K Gowda, S N Kumar, and BS Sreeramu. 2002. "Effect of bio-fertilizers on growth and yield of chilli (*Capsicum annuum* L.) cv Byadagi Dabba at different levels of nitrogen and phosphorus." *Journal of Spices and Aromatic Crops* 11 (1):58–61.

Sarma, R K, and R Saikia. 2014. "Alleviation of drought stress in mung bean by strain *Pseudomonas aeruginosa* GGRJ21." *Plant and Soil* 377 (1–2):111–126.

Sayler, G S, and S Ripp. 2000. "Field applications of genetically engineered microorganisms for bioremediation processes." *Current Opinion in Biotechnology* 11 (3):286–289.

Shabir T, Baba ZA, Dar SA, Husaini AM (2020) Improving plant growth and quality of plant-products: An interplay of plant-microbe interaction. *International Journal of Current Microbiology and Appiled Sciences* 9(11): 3759-3766.

Sharaf-Eldin, M, S Elkholy, J-A Fernández, H Junge, R Cheetham, J Guardiola, and P Weathers. 2008. "*Bacillus subtilis* FZB24® affects flower quantity and quality of saffron (*Crocus sativus*)." *Planta Medica* 74 (10):1316–1320.

Sheikhalipour, P, S A Bolandndnazar, and J Panahandeh. 2016. "Influence of KSB, PSB and NFB on fruit quality and potassium contents in tomato." *International Journal of Advanced Biological and Biomedical Research* 4 (2):170–178.

Sheng, X F, and L Y He. 2006. "Solubilization of potassium-bearing minerals by a wild-type strain of Bacillus edaphicus and its mutants and increased potassium uptake by wheat." *Canadian Journal of Microbiology* 52 (1):66–72.

Sofi, J A, N A Kirmani, and S Ansar-ul-Haq. 2008. "Effect of integrated nutrient management on saffron yield and soil fertility." *Asian Journal of Soil Science* 3 (1):117–119.

Spaepen, S, J Vanderleyden, and R Remans. 2007. "Indole-3-acetic acid in microbial and microorganism-plant signaling." *FEMS Microbiology Reviews* 31 (4):425–448.

Sriprang, R, M Hayashi, H Ono, M Takagi, K Hirata, and Y Murooka. 2003. "Enhanced accumulation of Cd2+ by a *Mesorhizobium* sp. transformed with a gene from Arabidopsis thaliana coding for phytochelatin synthase." *Applied and Environmental Microbiology* 69 (3):1791–1796.

Swaine, E K, M D Swaine, and K Killham. 2007. "Effects of drought on isolates of Bradyrhizobiumelkanii cultured from *Albiziaadianthifolia* seedlings of different provenances." *Agroforestry Systems* 69 (2):135–145.

Tejera, N, C Lluch, M V Martinez-Toledo, and J Gonzalez-Lopez. 2005. "Isolation and characterization of Azotobacter and Azospirillum strains from the sugarcane rhizosphere." *Plant and Soil* 270 (1):223–232.

Torres, M L, and M C E Lloréns. 2008. "Effect of alkaline pretreatment on anaerobic digestion of solid wastes." *Waste Management* 28 (11):2229–2234.

Tortora, María L, Juan C Díaz-Ricci, and Raúl O Pedraza. 2011. "*Azospirillumbrasilense* siderophores with antifungal activity against *Colletotrichumacutatum*." *Archives of Microbiology* 193 (4):275–286.

Trappe, James M, and Daniel L Luoma. 1992. "The ties that bind: fungi in ecosystems." *The Fungal Community: Its Organization and Role in the Ecosystem. New York: Marcel Dekker* 1:17–27.

Tyerman, S D, C M Niemietz, and H Bramley. 2002. "Plant aquaporins: multifunctional water and solute channels with expanding roles." *Plant, Cell and Environment* 25 (2):173–194.

Uehlein, Norbert, Kerstin Fileschi, Martin Eckert, Gerd Patrick Bienert, Adam Bertl, and Ralf Kaldenhoff. 2007. "Arbuscularmycorrhizal symbiosis and plant aquaporin expression." *Phytochemistry* 68 (1):122–129.

Vacheron, Jordan, GuilhemDesbrosses, Marie-Lara Bouffaud, Bruno Touraine, Yvan Moënne-Loccoz, Daniel Muller, Laurent Legendre, Florence Wisniewski-Dyé, and Claire Prigent-Combaret. 2013. "Plant growth-promoting rhizobacteria and root system functioning." *Frontiers in Plant Science* 4:1–19.

Verma, Jay Prakash, Janardan Yadav, KavindraNath Tiwari, and Durgesh Kumar Jaiswal. 2014. "Evaluation of plant growth promoting activities of microbial strains and their effect on growth and yield of chickpea

(Cicer arietinum L.) in India." *Soil Biology and Biochemistry* 70:33–37.

Verma, J P, J Yadav, K N Tiwari, S V Lavakush. 2010. "Impact of plant growth promoting rhizobacteria on crop production." *International Journal of Agricultural Research* 5(11): 954–983.

Vessey, J Kevin. 2003. "Plant growth promoting rhizobacteria as biofertilizers." *Plant and Soil* 255 (2):571–586.

Wagner, Gregory M. 1997. "Azolla: a review of its biology and utilization." *The Botanical Review* 63 (1):1–26.

Wange, S S, P L Patil, B BMeher, and M S Karkeli. 1995. "Response of cabbage to microbial inoculants and incremental levels of nitrogen." *Journal Maharashtra Agricultural Universities* 20:429–430.

Wani, Parvaze A, Mohammad Saghir Khan, and Almas Zaidi. 2007. "Synergistic effects of the inoculation with nitrogen-fixing and phosphate-solubilizing rhizobacteria on the performance of field-grown chickpea." *Journal of Plant Nutrition and Soil Science* 170 (2):283–287.

Wu, Gang, Hubiao Kang, Xiaoyang Zhang, Hongbo Shao, Liye Chu, and Chengjiang Ruan. 2010. "A critical review on the bio-removal of hazardous heavy metals from contaminated soils: issues, progress, eco-environmental concerns and opportunities." *Journal of Hazardous Materials* 174 (1–3):1–8.

Wu, Q S, and R X Xia. 2003. "Research and application on vesicular-arbuscular mycorrhiza of fruit trees." *Plant Physiology Communications* 39 (5):536–540.

Wu, Q S, Y N Zou, and G Y Wang. 2007. "Effect of inoculation with arbuscularmycorrhizal fungal on citrus under water stress conditions." *Journal of Yangtze University* 4:18–21.

Yardin, M Roseline, Ivan R Kennedy, and Janice E Thies. 2000. "Development of high quality carrier materials for field delivery of key microorganisms used as bio-fertilisers and bio-pesticides." *Radiation Physics and Chemistry* 57 (3–6):565–568.

Yin, Baozhong, Yanan Wang, Pan Liu, Jinli Hu, and Wenchao Zhen. 2010. "Effects of vesicular-arbuscular mycorrhiza on the protective system in strawberry leaves under drought stress." *Frontiers of Agriculture in China* 4 (2):165–169.

Zhang, Ai-min, Gang-yong Zhao, Tong-guo Gao, Wen Wang, Jia Li, Shuang-feng Zhang, and Bao-cheng Zhu. 2013. "Solubilization of insoluble potassium and phosphate by Paenibacilluskribensis CX-7: a soil microorganism with biological control potential." *African Journal of Microbiology Research* 7 (1):41–47.

CHAPTER 4

WHAT IS ECOLOGICAL/ORGANIC FARMING? SOIL SCIENCE AND MICROBIOLOGICAL POINT OF VIEW

IBRAHIM ORTAS

University of Çukurova, Department of Soil Science and Plant Nutrition, 01330, Adana, Turkey.
iortas@cu.edu.tr; ibrahimortas@gmail.com

Abstract

An increasing world population, which currently stands at 7.8 billion, is demanding more food production. Sufficient and healthy food production is the major challenge for the agricultural sectors. Agricultural production is presently dependent upon inputs such as inorganic fertilisers, other chemicals (pesticides, and herbicides), petrol, seed quality, heavy tillage and irrigation. Fertiliser use is required for proper and healthy crop growth. The world's rock-based fertiliser sources are diminishing, and in the near future it will be difficult to find any for manufacturing fertilisers, particularly phosphorus (P), potassium (K), and micronutrients zinc (Zn) and iron (Fe). Moreover, recent intensive use of chemical fertilisers has adversely affected human health and ecosystems. The quality of soil and water are also significantly affected by the use of chemicals in agriculture. Besides this, mineral fertilisers and disease control chemicals are expensive, which in turn leads to compromises in food safety production by developing countries that can ill afford them. In some parts of the world, such as Asia and mainly in Africa, they are facing insufficient income to afford fertilisers for adequate crop production. It has been demonstrated that plant-beneficial microorganisms, such as mycorrhizae and rhizobium species, from an endosymbiotic nitrogen-fixing association with roots, thereby significantly reducing chemical fertiliser use.

Across the globe, an elevated level of calcium carbonates, clay content, high pH, and low organic matter contents influence soil fertility and nutrient uptake. When soil is poor in the bioavailability of macro- and micro-nutrients such as N, P, K, Fe, Zn, which are diffusion-limited, a reduction occurs in the soil-to-plant yield. Also, in some areas, soil-borne diseases, such as plant-parasitic nematodes, soil-borne plant pathogens, root-rot and some weed pests are of great concern in terms of yield and food-quality reduction (nearly 25%). Farmers are seriously challenged by parasitism, insects and weed damage. In most cases, it is not easy to overcome all problems by using chemicals. Recently, there has been a great demand for ecological or organic production. Since plant roots, rhizosphere microorganisms and soil have a long-term relationship, it makes sense to find an ecological approach for better soil and crop management to boost food security and safety. Mycorrhizal symbiosis is one of the largest symbioses in the majority of plant species, and is a natural biological occurrence. As mycorrhiza is a part of the ecological system of a plant's roots it can be a very important and useful biofertiliser source for organic agriculture. It is, therefore, good judgement to use an organic fertiliser, such as mycorrhizae for organic/ecological production. Mycorrhizal inoculation or indigenous mycorrhizae spore management can thus be useful strategies for ecological farming.

It is also very important to manage the indigenous mycorrhizae when soil-borne diseases are prevalent, and the soil nutrients, especially phosphorus and zinc are limited under field conditions. For several reasons, especially for future global climate change, using plant rhizosphere mechanisms, such as management of indigenous mycorrhizae or/and selected mycorrhizal application, may contribute to agricultural sustainability. The long-term soil, crop and human benefits of using mycorrhizal inoculation in ecological farming are an alternative strategy for healthy food production.

Key words: mycorrhizae, rhizosphere, ecological farming, soil and crop management

Definition of Ecological/Organic Farming and the Effect of Rhizosphere Microorganisms in a System

An increasing world population (7.8 billion) has put intense pressure on land to produce sufficient food. At present, the agricultural system is called conventional or industrial agriculture, in which artificially produced chemicals, genetically modified seeds, and heavy tillage are used. Using synthetic chemicals (fertilisers, pesticides and herbicides) in agricultural crop production raises problems for consumers critical of food quality and food security. As a result, several alternative production systems have been suggested. Ecological, organic or bio farming are one and the same, defined as a sustainable farming system. This agricultural system avoids the use of synthetic and chemical inputs, such as fertilisers, pesticides, growth regulators, hormones, and any kind of livestock feed additives. Furthermore, ecological farming does not readily use soluble fertilisers and biocides. Organic production systems have lower concentrations of total and available soil nutrient than equivalent conventional systems, according to Gosling and Shepherd (2005). Accordingly, ecological farming systems are highly complex and integrated biological systems.

Soil is a complex system; the area where the soil and plant roots meet is a very intricate environment. This field provides sustainable food production by supplying energy continuously to the plant, like a soil generator. The energy supplied to the plants by the generator is entirely dependent on the presence and concentration of the organic carbon, which is fixed in the soil. In this context, an important part of the plant's carbon is transported to the plant's root zone (rhizosphere). In rhizosphere surroundings, there is a symbiotic association between the beneficial soil microorganisms and the plant's roots for nutrient and food exchange. Soil, microorganisms and plant root interface (rhizosphere) are directly related to the biological system. Since mycorrhizae demand carbon from plant roots, it is very important to understand the biological system's relationship to the energy supply generator (soil). However, significant differences between conventionally- and organically-managed soils are not found in the total soil organic matter, total nitrogen, or the C:N ratio. Management of the rhizosphere is also very important for an ecological crop production system. Pagano et al. (2017) indicate that soil microbial communities play a central role in ecosystem processes, and are affected by biocontrol agents and biofertilisers, such as mycorrhizae and other beneficial organisms.

What is the Role of Mycorrhizae in Ecological/Organic Farming?

Mycorrhizae are present in 92% of land plant families (263 families), occurring in 80% of species (3,617 species) (Wang and Qiu 2006). The AM fungi live in close association with approximately 85% of herbaceous plant roots, and produce arbuscules within root cells. Arbuscular Mycorrhizal fungi (AMF) belong to Glomeromycota. In this association, the fungi receive carbohydrates (sugars) from the host plant and mycorrhizae help the plant to get more nutrients and water. Symbiotic associations between AMF and plant roots improve nutrition enhance resistance to soil-borne pests and disease, tolerance of heavy metals, improved resistance to drought, and soil structure development. Shoot and root growth and nutrient acquisition are highly dependent on AM colonisation.

Ecological farming systems may be less detrimental to mycorrhizae because they exclude the use of water-soluble fertilisers and biocides (Thingstrup et al. 1998). The power of mycorrhizae might be an ability to substitute for reducing fertiliser and biocide uses in organic systems. Usually, under field conditions, the yield increase in ecological production is less than conventional agricultural practices. However, when there is a high rate of AMF colonisation, yield increase is remarkable (Ortas 2012a). It is very important to examine the benefits of the mycorrhizal association in agro-ecosystems, and how farm management practices influence the mycorrhizal association. Van der Heijden et al. (2008) indicated that under natural environmental conditions, the diversity of AM fungi is a key contributor to the diversity and productivity of plant communities. The results of Naghibi et al. (2016) showed that application of biological and organic fertilisers improved most of the studied traits of dwarf chicory yield, and *Glomus mosseae* + humic acid treatment was the best. Mycorrhizal fungi can be used in organic farming as an organic microbial fertiliser (Ortaş 2019).

Mycorrhiza Inoculated Seedlings Transplanting to Field Condition

It is widely accepted that AM technology can improve soil and horticultural crop productivity. Horticulture is an ancient tradition. The climatic origin of many important fruit and vegetable crops is Mediterranean, including pear, quince, plum, sweet and sour cherry, hazelnut, pistachio, almond, walnut, chestnut, olive, fig, pomegranate, rose, grape, lettuce, carrot, melon, watermelon, cucumber and leek (Ortas 2018). Mediterranean ecological horticultural production, specifically, is in great demand. Under ecological conditions, horticultural products are very important in the agricultural sector, and play a significant role in the economy of farmers. In general, for plant nutrition, animal manure, green manure, compost and rock-based minerals are used.

Most of the commercially grown cash crops are horticultural plants, and they are germinated under nursery conditions before being transplanted to the main field. This technique has been used for many years, and it is easy to use mycorrhizae before transplanting to field conditions. Ortas et al.

(2011a, 2011b, 2013) tested several techniques to develop suitable inoculum strategies, and also used several techniques under field conditions (Figure 1). Since AM fungi cannot be grown under laboratory conditions, production of a large quantity of the inoculum is difficult for inoculating soil under field conditions. It is thus very important to produce mycorrhizal-inoculated seedlings in vegetables before transplanting to the field.

Since horticultural plants are cultivated through seedling production, it is important to have mycorrhizal-inoculated seedling production and transplantation to the field conditions. The majority of horticultural plants, mainly vegetable crops, are potential host plants of AMF (Baum, El-Tohamy, and Gruda 2015). As most of the commercially important crops are horticultural plants, they are raised in nursery conditions before being transplanted to the main field (Ortas et al. 2011a). The inoculation of soil in the nursery would not only result in saving the cost of production of the inoculums, but would also help in the better establishment of the transplanted horticultural seedling. Recently, biotechnological techniques were applied in horticultural seedling production. Micro-propagated seedlings were produced under control conditions. Moreover, nursery production of horticultural and/or ornamental seedlings, cutting tree plants by treating the rooting, and growing media with appropriate inoculants are other important areas where mycorrhizal spores can be used with confidence (Ortaş, Rafique, and Ahmed 2017, Aka-Kacar et al. 2010, Incesu et al. 2015).

However, after transplantation to the field, there is a high mortality ratio. Mycorrhizal inoculation can be a helpful strategy (Ortas et al. 2017). Under field conditions, the effect of mycorrhizal inoculation on the mortality of the seedlings was tested (Ortas and Varma 2007). With optimum management, the transplantation of mycorrhizal-inoculated seedlings to field conditions and the management of indigenous mycorrhizae is possible for low input ecological farming.

So far, results have shown that soil-indigenous mycorrhizae successfully infect plant roots, resulting in better plant growth. The effect of mycorrhizal inoculation upon plant growth is changed by the effectiveness of inoculum and application time (Ortas et al. 2011a) (Figure 1).

Indigenous mycorrhiza-inoculated citrus seedlings successfully grown by selected exotic, mycorrhizae inoculation. For horticultural seedling quality and survival, mycorrhizae are very important.

Figure 1. Mycorrhizal inoculated horticultural seedling production and transplanting to field conditions. (Original, Ortas I).

One long-term field experiment was set up in 1996 to compare the effect of organic fertiliser on crop yield and soil quality with mineral fertiliser. The application of organic fertilisers increased several crop yields, such as maize, wheat, melon, watermelon, eggplant, bell pepper, carrot and spinach compared to mineral fertiliser (Table 1.) Over fourteen years, organic fertiliser significantly increased crop yields compared to the control and mineral fertilisers. Besides this, yields of melon, watermelon, eggplant, bell pepper, carrot, spinach, salad and radish also increased significantly with the application of organic and inorganic fertilisers, compared with the control. In most cases, plots treated with animal manure and compost produced more yield than plots using mineral fertilisers. Studies have been conducted under long term field conditions where results showed that mycorrhizal inoculation makes a substantial contribution to plant yield (Ortas 2010, 2012a). Interestingly, compost (10 tonne ha^{-1}) + mycorrhiza inoculation gives an equal yield, compared with compost (25 tonne ha^{-1}) alone application. It means mycorrhizal spore inoculation, with a small quantity of compost added, can be used for organic farming (Ortas and Lal, 2014).

Table 1. Long-term field experiment effects of organic fertilisers compared to control and inorganic fertilisers on plant species' yield kg ha^{-1} from 1996 to 2010.

		1996	1997	1998	1999	2000	2001	2002	2003	2004	2005	2006	2007	2008	2009	2010
								Yield kg ha^{-1}								
Pepper	C	10080	14000								15333					
	F	17280	15450								21730					
	C50	22080	24480								25535					
	M50	27720	25880								24540					
	C10+MZ	25392	26050								22500					
Bell Pepper	C			16543							14833					
	F			23105							18970					
	C50			25226							23868					
	M50			24565							20000					
	C10+MZ			26865							24043					
Wheat	C					3060			3254		2839	2884	2544	3095	3127	3107
	F					4780			5569		5569	5293	4843	5800	5340	5157
	C50					4740			5999		5325	4779	5219	5166	4983	5033
	M50					4151			5166		5166	5033	4823	6238	6200	5253
	C10+MZ					4462			5459		5953	5181	4992	6095	5882	5235
Maize	C				7854			7650		7458	7033	7467	7467	5573	7050	5405
	F				8544			10989		11400	12850	12900	10915	10724	10417	11879
	C50				11496			10900		13998	11267	11570	11570	8782	10883	11411
	M50				11121			10183		10217	10400	9802	9802	10504	9550	12174
	C10+MZ				10210			12467		13083	11483	11310	11310	9633	10400	12637
Cucumber	C			9277			8620									
	F			12513			15231									
	C50			18987			23456									
	M50			16832			21234									
	C10+MZ			18702			24614									
Spinach	C				58785		3299									
	F				116688		6879									
	C50				57724		3855									
	M50				63228		8014									
	C10+MZ				70476		6309									

CO: control, F: mineral fertiliser (N–P–K), C25: compost (25 tonne ha^{-1}), M25: manure (25 tonne ha^{-1}), C10 (10 tonne ha^{-1}) + MZ: compost + mycorrhizae inoculation.

Mycorrhizal Dependency

Many plants are known to be AM-dependent crops (Hamel 1996; Ortas 2018). In many agricultural (soil and plant) practices the use of fertilisers, tillage, monocultures, and the growing of non-mycorrhizal crops are detrimental to mycorrhizal fungi. Plant species show dependency on mycorrhizal fungi at different levels (Thioye et al. 2018). Several horticultural plants, such as pepper, eggplant, melon, cucumber and carrots are mycorrhizal-dependent, and can be inoculated under nursery conditions. Additionally, some tree plants, such as citrus, olive, apple, fig, tea, coffee, banana (Ortas et al. 2017, Ortas 2018) and oil-palm are mycorrhizal-dependent, and can be easily inoculated with mycorrhizal species under nursery conditions, whereas under field conditions the mycorrhizal dependency of several plants is affected by indigenous mycorrhiza which affects plant growth. Root infection has also been tested, with and without partial soil sterilisation, over a number of years (Ortas 2010, 2012a, Ortaş, Rafique, and Ahmed 2017). Ortas (2012a) suggested that for horticultural crops dependent on mycorrhizae, they can be mainly used as seedlings before being transferred to the field; however, for field crops, mycorrhizal management is also possible for mycorrhiza use in ecological farming. After three years of field experiments, it has been shown that garlic, watermelon and cucumber plants are highly mycorrhizal-dependent (Table 2) (Ortas 2012a). From the results obtained it can be suggested to farmers that arbuscular, mycorrhizal fungus-inoculated seedlings can be used under field conditions for high yield and quality (Ortas 2010, 2012a).

Table 2. The effect of mycorrhizal inoculation and P addition on garlic, watermelon and cucumber yield and inoculation effectiveness under methyl-bromide-treated (+Fumigated) and non-treated (-Fumigated) field conditions.

Treatments	Yield (kg/1000 m²)			Inoculation effectiveness (%)		
	1997#	1998	1999	1997	1998	1999
+Fumigated		GARLIC				
P0-M	63 ±14 d‼	23 ±5 d	219 ±44 d			
P0+M	217 ±141 c-d	79 ±52 b-d	294 ±41 b-c	70.8	70.8	25.6
P1-M	63 ±5d	23 ±2 d	208 ±36 d			
P1+M	155 ±23c-d	57 ±9 d-c	302 ±15 b-c	59.5	59.5	31.0
-Fumigated						
P0-M	235 ±67 b-c	186 ±110 ab	229 ±57 c-d			
P0+M	395 ±119 ab	228 ±69 a	352 ±41 ab	40.6	18.7	35.0
P1-M	239 ±24 b-c	138 ±14 a-c	241 ±71 b-c			
P1+M	406 ±6 a	235 ±4 a	393 ±18 a	41.0	41.0	38.8
+Fumigated		WATERMELON				
P0-M	174 ±25b	3206 ±2054a	1284 ±821a	64.9	7.7	39.7
P0+M	496 ±457ab	3475 ±671a	2129 ±312a			
P1-M	603± 488ab	1957 ±677ab	1202 ±865a	37.9	42.8	42.4
P1+M	971± 325a	3419 ±405 a	2086 ±599a			
-Fumigated						
P0-M	1754± 1338ab	594 ±159b	871 ±964a	8.9	1.4	30.2
P0+M	1926± 1304a	603 ±236b	1248 ±201a			
P1-M	1840± 554ab	3048 ±1699a	772 ±497a	-54.9	-44.2	31.9
P1+M	1188± 59ab	2114 ±311ab	1134 ±315a			
+Fumigated		CUCUMBER				
P0-M	1529 ±213 d	1485 ±133 c	1321 ±257 b			
P0+M	1641 ±515 c-d	1729 ±636 c	1401 ±141 b	6.9	14.1	5.7
P1-M	1676 ±289 c-d	1720 ±364 c	1355 ±166 ab			
P1+M	1835 ±439 b-d	1996 ±829 –c	1370 ±25 ab	8.7	13.8	1.1
-Fumigated						
P0-M	2062 ±200 b-c	2440 ±417 a-c	1345 ±50 b			
P0+M	2511 ±111b	2861 ±164 ab	1746 ±40 a	17.9	14.7	23.0
P1-M	2605 ±140 a	3349 ±430 a	1431 ±173 ab			
P1+M	2596 ±103 a	3396 ±107 a	1367 ±82 ab	-0.4	1.4	-4.7

Values are the averages of three samples ± standard deviation.
-M no AMF inoculum, +M AMF inoculum used, P0 = 0 kg P_2O_5/ha, P1 = 100 kg P_2O_5/ha.
‼Values in a column followed by the same letter are not significantly different (LSD, $P \leq 0.05$).
Comparison of means of LSD was calculated separately for each year.

The Role of Mycorrhizae on Nutrient Uptake under Ecological Farming Conditions

The symbiotic relationship between mycorrhizal fungi and plant roots is completely dependent on a supply of organic carbon. Since plant carbon is taken by the photosynthetic pathway, there is a significant relationship between carbon (C) fixation and nutrient uptake. Net 4–20% of photosynthate is deployed in the growth of the intra- and extra-radical mycelium to the fungus (Parniske 2008), and used in the production of both vegetative and reproductive structures to support growth and nutrient uptake (Smith and Read 2008).

Since all biological processes are controlled by carbon uptake and carbon economy in mycorrhiza development, mycorrhizal and other microorganism's colonisation needs to be considered in the context of colonisation effects on mineral nutrition. The relative cost of fungal C use is related to the benefits of nutrient uptake.

Mycorrhizae-inoculated plants can uptake nutrients directly from the soil solution or via a mycorrhiza fungal symbiont. Uptake of the nutrients can be achieved via three special pathways: 1) nutrient taken up by the mycorrhiza fungal mycelium in the soil; 2) translocation of nutrients from some distance within the hyphae to the intra-radical fungal structures, such as hyphae, arbuscules and vesicles. Hyphae, present inside and outside of the root, can transfer the nutrients to the plant cells across the complex interface between the symbionts; and 3) mycorrhizal mycelium in the soil can absorb nutrients beyond the depletion zone of root uptake pathways. This is why the effects of AM colonisation on nutrition (mainly P, Zn and Cu nutrients) uptake are great, and may have indirect effects on other aspects of plant metabolism so that the direct effects of the symbiosis on other nutrients are masked (Smith and Read 2008).

Globally, different soils have phosphorus (P), Iron (Fe) and Zinc (Zn) deficiency, and for expected ecological production organic fertiliser is needed as many soils consist of high levels of clay and lime that cause P, N, Zn, Fe and Mn deficiency (Ortas 2003). In an ecological plant production system, the provision of non-biologically fixed nutrients, such as P, Zn, Fe, K and Mg, etc., can be taken from parent-rock minerals. Mycorrhiza can help to dissolve the rock minerals and also increase the absorption of nutrients. Mycorrhiza has also been shown to improve the utilisation of both N and P in plant materials, and mineralise the organic and inorganic rock materials (Ortas 2012b). Apart from organic fertiliser, mycorrhizae are one of the most powerful organisms for nutrient uptake, mainly P, for ecological/organic farming.

Mycorrhizal fungi play an important role in plant health by improving nutrient uptake, especially inorganic P, Zn, Cu and Ca (Ryan and Graham 2002, Azcon-Aguilar and Barea 2015, Ortaş, Rafique, and Ahmed 2017). In addition, they can improve the uptake of water by their host plant and provide protection against soil-borne pathogens. Mycorrhiza can also be beneficial in increasing the resistance to foliar-feeding insects (Fontana et al. 2009) and improved drought resistance (Al-Karaki, McMichael, and Zak 2004, Augé et al. 1994).

Previously, it was shown that mycorrhizal fungi are able to absorb and transfer the major macro- and micro-nutrients (N, P, K, Ca, Mg, S, Fe, Zn, Cu, Mn, I and Se) necessary for plant growth. In order to achieve normal growth, a balanced nutrient is taken up, which is only possible if the soil has a sufficient quantity of all required nutrients, in particular, soil nutrients that are immobile in the soil, such as P, Zn and Cu (Ortas 2012b, Marschner 1996). The shoot concentrations of P, Mg, Zn and Cu are significantly higher in mycorrhizal-inoculated plants than in non-mycorrhizal plants (Liu et al. 2000). The mycorrhizal hyphae network is able to maintain P transport to the plant over greater distances, such as 15 cm away from the root surface (George et al. 1992, Li, Marschner, and George 1991, Li, George, and Marschner 1991, Ness and Vlek 2000).

Sylvia and Neal (1990) demonstrated that P had a greater effect on root colonisation in P-sufficient soils than in N-deficient soils. Also, plant P concentration was greater in ammonium than nitrate treatment. Phosphorus uptake is also controlled by nitrogen, even under mycorrhizal colonisation. Mycorrhizal hyphae may stimulate mineralisation of nitrogen from the organic compound when the crop needs it (Ortas, Harris, and Rowell 1996, Ortas and Rowell 2004).

Nitrogen fixation is also very important for ecological farming. N-fixation in legumes by *Rhizobium*, and increased uptake of phosphorus from mycorrhizal associations, can directly or indirectly reduce chemical fertiliser demand. Using N-fixing bacteria and mycorrhizae can reduce nutrients' run-off, water-table pollution caused by chemicals. Selected efficient *Rhizobium* isolates, and mycorrhizal species of various leguminous crops grown with it, can reduce fertiliser demand. According to Azcon and Barea (1997), Azcon-Aguilar and Barea (2015), mycorrhizal fungi have particular value for legumes because of their need for an adequate phosphorus supply, not only for optimum growth, but also for nodulation and nitrogen fixation. The N_2-fixing Rhizobium bacteria and mycorrhizal fungi are the most relevant representatives of beneficial plant symbionts (Barea 2015), which are directly related to nutrient uptake and nutrient cycling. The uptake of gaseous N_2 provides an extra source to plants for their N-nutrition, as well as controlling the P and other nutrients, such as Fe and Zn. Since N_2-fixing plants capture more atmospheric nitrogen, in order to balance nutrients, such as P, they require mycorrhizal infection for more P uptake. Nodulation and atmospheric nitrogen fixation capacity in legumes were also increased by AM fungi. The nodule roots tended to retain their cortical cells, with little cell material.

The Role of Mycorrhizae in Agricultural Management (Crop and Soil)

Ecological production systems have long-term benefits for soil health. Thanks to the use of organic fertilisers in ecological production, soil microbial communities perform better than the field where chemical fertiliser is used. In ecological farming systems, microbial activity is a vital factor in the proper functioning of agro-ecosystems (Gosling et al. 2010, Gosling et al. 2006, Almaca and Ortas 2010).

The use of organic fertilisers, beneficial microorganisms, and crop rotation can help nutrient cycling and soil conservation. Using beneficial microorganisms can also help in disease control (Stockdale et al. 2001). Mycorrhizal fungi can decrease the severity of diseases caused by root pathogenic fungi, bacteria and nematodes (Siddiqui and Mahmood 1995, Li et al. 2018). As mycorrhiza is a biocontrol agent it may be used in organic systems to control pathogenic fungi (Gaur et al. 2004). In organic farming, mycorrhizae are used for biological control of disease and nematode. The results of Singh et al. (2018) showed that *G. intraradices* mycorrhizae species can be used for biological management of complex root diseases under organic field conditions.

Agricultural management under low input systems includes several natural processes, such as crop rotation, and the use of organic fertilisers, etc. Crop rotation means modifying the soil micro-flora, including AMF (Atkinson et al. 2002; Marschner 2012). Through this system, soil nutrient availability is changed, and plants benefit in terms of crop health. Since crop rotation has an effect on rhizosphere dynamics, it may have an effect on root health as well. According to Atkinson et al. (2002), organic agriculture is used as a model for a low input system. Since soils in most regions are less fertile, ecological fertiliser must be carefully selected for quality production. There is a need to determine the biofertiliser requirement under several ecological management systems as many agricultural practices, such as chemical fertilisers, tillage, monoculture cropping and non-mycorrhizal crop plants, known to reduce inoculum, have an effect on AMF functions. Accordingly, anthropogenic activities (soil and crop management) strongly affect the AM fungi. According to Gosling et al. (2006), ecological management options can be employed to increase the benefits of AMF. Crop and soil management systems, such as crop rotation, and less tillage and fertiliser use, can affect mycorrhizae development and, accordingly, plants benefit from mycorrhizae. Since crop rotation helps mycorrhizal development, it may also have benefits for ecological farming. For ecological nutrient and water management, soil and crop management can help to get the maximum benefit from indigenous mycorrhizae also using selected mycorrhizal spores (Ortaş 2003; Smith and Read 2008;Ortas and Coskan 2016), or by producing mycorrhizal inoculated seedlings (Ortas et al. 2011a).

Soil and crop management also controls the nutrients supplied to crops. The management requires extensive research to find the optimal, balanced crop production. In general, soil and crop management practices have an impact on the mycorrhizal formation and effectiveness for nutrient uptake. Sometimes, especially under heavy chemical fertiliser use and tillage practices, the formation of mycorrhizal spores is negatively affected. Many studies have concluded that high levels of P fertilisation inhibit mycorrhizal efficiency and formation in several plant species, such as soybean (Ezawa, Yamamoto, and Yoshida 2000), sorghum and maize (Ortas 1996, 2003, 2015). Mozafar et al. (2000) showed that higher root colonisation was observed under reduced or no-tillage practices in Switzerland. Previously, many studies reported that the corn phenotype, with the highest shoot/root ratio, had the highest root colonisation (Ortas, Harris, and Rowell 1996, Ortas 2003, 2012a). Under ecological soil management systems, the amount of AMF inoculum and effectiveness can be enhanced (Bending et al. 2004). Since AMF are potential contributors to plant nutrition and pathogen suppression in low input agricultural systems, they are a powerful biofertiliser to use in ecological farming. Generally, in conventional soil and crop management systems, when chemicals are used, root colonisation is low. On the other hand, under the ecological system, root colonisation is higher than in mineral fertiliser use (Ortas 2012a, Ortas, Akpinar, and Lal 2013, Turgay et al. 2015).

In ecological farming, other soil management systems, such as the use of crop rotation, inter- and intra-cropping, animal manuring, green fertilising, and reduced tillage or zero tillage have a significant effect on root colonisation (Ortas and Coskan 2016). Moreover, ecological soil and crop management can enhance soil fertility and soil quality. Ecologically-managed soil and crop systems can affect the greater AMF spore numbers and root colonisation potential (Oehl et al. 2003, Ortas and Coskan 2016, Satir, Ortas, and Satir 2016, Buto et al. 2016). In drought conditions, plants also have a higher response to mycorrhizal inoculation. In semi-arid regions, the role of mycorrhiza is important in order to be able to do organic farming.

There are several mycorrhizal species, and individual species of AMF vary widely in their functional attributes to low input systems. The exact production methods of ecological farming vary considerably, but general principles include exclusion of most synthetic biocides and fertilisers, and the management of soils through the addition of biofertilisers, such as mycorrhiza inoculation. The results of Mäder et al. (2000) and Buto et al. (2016) indicated that under long-term field conditions, ecological management has been shown to stimulate mycorrhizal fungi communities. Soil and crop management have a significant effect on mycorrhizal development and species richness (Ortas 2015, Ortas and Coskan 2016). In addition, management systems have a significant effect on a number of mycorrhizal species (Oehl et al. 2010).

The Importance of Well-Developed Structure on Ecological Farming

The physical, chemical and biological fertility of the soil should be in good condition for healthy and high-quality organic crop production systems, coupled with the management of soil organic matter. Organic matter (OM) can enhance the physical, chemical and biological properties of the soil. Organic fertilisers, such as animal manure, compost, green fertiliser, biochar, vermicomposting and biofertilisers, etc., are used in soil amendment as a supplement if a substitute nutrient is not available in the soil. In most cases, it is not easy to supply all the required nutrients from the soil through organic fertilisers.

Soil structure is very important for ecological soil management. In the rhizosphere, there are various microorganisms or producers of mucilaginous substances aggregating the soil particles. Mycorrhizal fungi produce organic glomalin and glues (extracellular polysaccharides) which helps the soil to aggregate, and improves soil porosity (Ortaş, Lal, and Kapur 2017). The AMF are key microorganisms in the soil/plant system because of their positive effects on soil structure, which is affected by a proteinaceous substance called glomalin-related soil protein (Bedini et al. 2008, Ortas 2016, Ortaş, Lal, and Kapur 2017). As a result, the soil as a substrate for microbial life is, essentially, a discontinuous and heterogeneous environment, which supports life. Water, oxygen and nutrients are unevenly distributed around the soil structure. Also, mycorrhizae have a significant effect on soil-structure development (Figure 2). Just as mycorrhizae have key effects on structural development, they have positive effects on porosity, which is vital for physical quality. Under well-developed soil formation conditions, ecological growth is healthier. For ecological farming systems, the physical quality of the soil is vital for water storage in the soil profile. The long-term use of organic fertiliser sources (Celik, Ortas, and Kilic 2004, Ortas, Akpinar, and Lal 2013) has shown that organic fertiliser and mycorrhizae have positive effects on physical properties such as soil structure formation. Rillig (2004) and Ortas, Akpinar, and Lal (2013) have shown that mycorrhizal-inoculated treatments have high mean weight, diameter and water-stable indices. Since mycorrhizal fungi have a beneficial effect on soil aggregation, this may be contributing to the control of soil erosion. Extrametrical mycelia of mycorrhizae have the capacity to bind soil particles, such as clay, loam and sand. It is well known that sandy soils and many sand dune plants are mycorrhizal dependent (Sylvia 1989, Sylvia, Jarstfer, and Vosatka 1993, Sylvia et al. 2003).

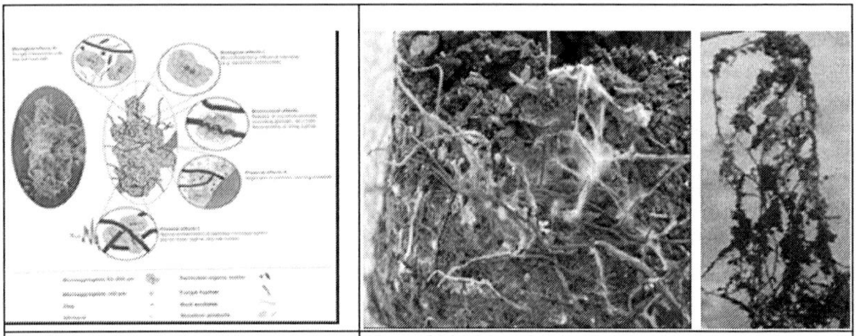

Rillig and Mummey (2006) | Ortas 2011 (unpublished picture)

Figure 2. Effect of hyphae and roots on aggregation.

The influence of the roots and rhizosphere vegetation have both direct and indirect effects on the soil structure, depending on the rhizosphere species, and how long it remains in the soil environment. The direct effect concerns the mechanical action of the root in creating spaces and pores in the aggregates, but it is commonly believed that it is the indirect effect, due to the activity of micro-organisms in the rhizosphere, that has a greater effect on the soil structure.

Mycorrhizal Inoculation Contributes to Root Development

The mycorrhizal root system increases the absorption-potential of roots up to 100 times, thereby greatly improving the ability of plants to utilise nutrients. Mycorrhizal fungi can increase root absorption capacity due to morphological and physiological changes. In particular, the root capacity of mycorrhizal-dependent, thick-root-system plants, such as the root/hyphae surface area, increases. As a result, greater longevity of absorbing roots, better utilisation of low-availability nutrients, and better storage of nutrients are achieved.

The type of root also influences microbial growth in the rhizosphere volume. Rhizosphere pH, redox potential, root exudates and microbial symbiosis are very important factors affecting nutrient uptake balance. Mycorrhizal fungi release the root exudates into the soil, which dissolve minerals and organic compounds to capture the nutrients, such as P, Zn, Cu and Fe and other bound soil nutrients. Root exudates and their processes are particularly important in plant nutrition, especially for a few micronutrients, such as Fe and Zn (Marschner 2012). The results of Koide and Kabir (2000) showed that AMF can secrete or produce extracellular phosphatases to hydrolyse phosphate from organic P compounds.

Plant roots have been shown to exude specific organic acids, such as citric, oxalic and malic, which are very important for nutrient uptake under ecological conditions. Also, these organic compounds help viable soil microorganisms, since plant roots exude the carbon compounds to the rhizosphere, in which more organisms get food. Through cell lysis, the seminal roots presumably supply more carbon (C) to the rhizosphere for microbial growth. Cheng and Gershenson (2007) reported that 5–40% of photosynthetically-fixed carbon is released by plant roots. Under these plant-exudation conditions, carbon and nitrogen produce energy for microbial life in the soil. Also, the quantities of bacteria, actinomycetes and fungi in the rhizosphere decrease, whereas populations on and in the root tissue increase with the plant's age (Sivasithamparam, Parker, and Edwards 1979). Seminal roots of wheat support a larger rhizosphere population of bacteria, actinomycetes and fungi than nodal roots. The seminal roots tend to lose their epidermal cells and some cortical cells near the root surface suffer damage. These differences were postulated to be caused by anatomical differences in the cortical root tissue.

Since organic matter is the main nutrient source for plant and soil microorganisms, accumulating carbon in the soil, as carbon sequestration and/or carbon pool, is the primary function of ecological agriculture. Pagano et al. (2017) indicate that intercropping different plant species is beneficial as this can increase carbon fixation by the transfer of carbon to the soil by plants, especially via mycorrhizas. Mycorrhiza inoculation has significant effects on carbon fixation, and keeping it in the soil (Figure 3). Since the twentieth century, conventional, even industrial agriculture has been achieved at the cost of biodiversity and increased pollution by fertilisers, which finally resulted in soil degradation and quality decreases.

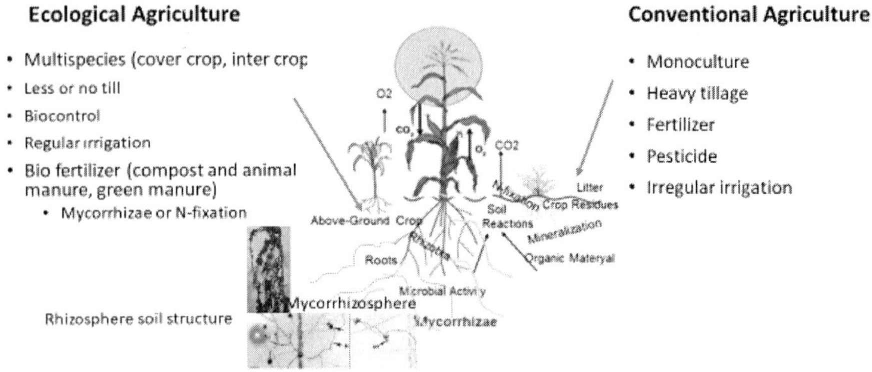

Figure 3. Some characteristics of ecological and conventional agriculture systems.

Under ecological agricultural systems, soil quality and productivity are affected by organic amendments. Also, indigenous and exotic mycorrhizal inoculation affects soil aggregation and rhizosphere organism richness (Mozafar et al. 2000).

The Importance of Rhizosphere Management on Ecological Farming

Under field conditions, rhizosphere management is also possible by using crop rotation, the addition of compost and other organic fertilisers, and using beneficial soil organisms. Bowen and Rovira (1999) suggested that a fairly managed rhizosphere would increase soil quality and plant health. Rhizosphere management can increase the population and richness of beneficial micro-organisms in the root-soil surface area. Generally, the influence of ecological farming versus conventional farming strongly influences the value of soil microbiological activity. Ecological farming can modify the soil structure, richness, evenness, and dispersion of the soil microbiota when compared with managed soils under mineral fertilisation (Fließbach et al. 2007). This is why soil quality is an important factor of productivity in ecological farms which rely on nutrient transformation processes. Under a number of farming systems, Caruso et al. (2018) showed that conventional systems, with microorganism-enriched fertilisation (CMF), resulted in the highest early and total yield, followed by organic systems with microorganism-enriched fertilisation (OMF) systems, due to their higher fruit number. They also showed that the highest total polyphenols concentration was recorded under the CMF, whereas OMF resulted in the highest flavonoids concentration and antioxidant activity. It seems that microorganism-enriched organic fertilisers increase the plant's physiological metabolisms to protect against stress factors.

There is now greater interest in the use of different microbial species and strains of rhizobacteria and mycorrhizal fungi, including *Trichoderma*, to promote plant growth, and to protect plants from stress factors, such as nutrient deficiency and diseases. Since the rhizosphere population is greater than non-rhizosphere soil (Marschner 1995), it is important to develop a strategy that reshapes the rhizosphere microbiome for bio-fertilisation; and the importance of root growth stimulation, antibiosis, induced plant systemic resistance, parasitism, and rhizoremediation is nowadays recognised (Lugtenberg 2014). Dobbelaere, Vanderleyden, and Okon (2003) determined the rhizosphere as a narrow zone of soil, specifically influenced by the root system. The rhizosphere zone, owing to the accumulation of a variety of plant exudates, such as amino acids and sugars, provides a rich source of energy and nutrients for bacteria when compared with the bulk soil (Marschner 1995, 2012, Marschner et al. 1986, Gray and Smith 2005). The rhizosphere is populated with a diverse range of microorganisms, and the bacteria-colonising habitat is generally 10–100 times higher than that in the bulk soil. Soil and crop management can change the rhizosphere dynamics, such as rhizosphere population diversity, pH and root exudates. Mainly nitrogen-formed nutrition can significantly affect rhizosphere pH, and, as a result, nutrient mineralisation and availability are high (Ortas, Harris, and Rowell 1996, Ortas 1997, Marschner 2012, Marschner et al. 1986). The effect of N forms (NH_4 and NO_3) on rhizosphere pH change was much larger when no P or soluble P was applied and reached up to 1.30 units, which extended to 3–5 mm from the root surface (Ruan, Zhang, and Wong 2000). Soil pH affects and mobilises the predominant forms of P by different mechanisms. Mycorrhiza also affects rhizospheric microbial activity, root exudates type, quality and pH. Consequently, mycorrhiza-inoculated plants have balanced nutrient uptake and are healthier. Accordingly, healthy plants have more shoot growth than root growth (Ortas 2012). Under greenhouse conditions, the contribution of AMF to P uptake extends beyond the effect of rhizosphere pH change (Ortas 1996). Since the rhizosphere pH change of NH_4^+-treated plants is greater than that of NO_3^--treated plants, plant response to P uptake is slightly different. It has been shown that some forest tree plants are able to utilise ammonium preferentially over nitrate, and ammonium net uptake rate is several times higher than nitrate uptake (Malagoli et al., 2000). Similarly, Ruan, Zhang, and Wong (2000) showed that the tea plant's leaf dry matter production was significantly greater in treatment with NH_4 than NO_3.

It is understood that the management of the rhizosphere mechanisms offers many opportunities for successful organic farming. Management of the rhizosphere involves not only introduced organisms, but also indigenous organisms living in the soil (Bowen and Rovira 1999). Rhizospheric organisms can be separated into beneficial, deleterious, infective and non-infective organisms. Beneficial, infective organisms include nitrogen-fixing symbiotic partners, actinomycetes, fungi ecto- and endo-mycorrhizae, and several other mutualistic symbioses, such as blue-green algae. Beneficial, but not infective, microorganisms include biological agents for control root disease, plant growth-promoting rhizobacteria, and yield-increasing bacteria which stimulate plant growth (Bowen and Rovira, 1999). These organisms, directly and indirectly, influence soil sustainability by improving soil biological quality. The results of Bona et al. (2018) show that inoculation with soil microorganisms can help to drastically reduce the use of chemical fertilisation and, in some cases, even inoculated microorganisms, improving the tomato fruit yield and quality. Another work of Bona et al. (2017) showed that the application of beneficial microorganisms (two Pseudomonas strains and a mixed mycorrhizal inoculum, alone or in combination) inocula, allowed the reduction of chemical inputs and positively influenced tomato quality.

The management of the rhizosphere should involve manipulation of at least two rhizosphere components such as soil, plant and microorganisms, if not all three of these factors. The loss of 10–30 % photosynthate from the root leads to a higher increase in microbial biomass, accompanied by selective stimulation of particular microorganisms. The group diversity and growth of rhizosphere microorganisms depend on plant and soil management. The exudates of each plant root and rhizosphere microorganism are different (Schippers, Bakker, and Bakker 1987). They also indicate that the negative influence of saprophytic rhizosphere organisms on crop production may be related to other metabolites of rhizosphere microorganisms, including antibiotics which are toxic to plant roots. Schippers, Bakker, and Bakker (1987) reported that with frequent (every second year) or continuous cropping of potato, there is a build-up of cyanide-producing organisms in the soil and in the rhizosphere, accounting for poor growth and lower yields; other rhizosphere organisms produce auxins, ethylene, cytokines and vitamins, and these may also affect plant root growth and root morphology. Re-plantation in old orchards generally resulted in poor growth for a range of trees, such as cherries, pears, apples, and citrus. Poor re-planted growth has been attributed to many causes, including soil toxins, nematodes, phytophthora, and other pathogenic fungi. This has been supported by the finding that soil fumigation can often rectify the problem. Utkhede and Smith (1994) have developed a new approach to overcoming the problem by drenching the soil with selected strains of *B. Subtilis* and *Enterobacteraerogenes*. The results of Utkhede, Li, and Smith (1992) showed that *G. mosseae* mycorrhizal fungus inoculation significantly increased the growth of apple seedlings in apple-replant-diseased soil, without being either pasteurised or fertilised with mono ammonium phosphate. Their results are very important for the future use of mycorrhizae in ecological farming under field conditions to reduce the disease effect on plant growth.

Interaction between Mycorrhizae and Other Soil Microorganisms on Soil Quality and Health Crop Production

AM interaction with other soil microorganisms is also beneficial for soil health and plant growth. In different studies, mycorrhizae, bacteria and trichoderma have a synergetic effect on plant growth. The findings of Amer and Abou-El-Seoud (2008) showed that *T. harzianum*-inoculated seedlings significantly increased phosphorous supply and also resulted in higher yields. Plant growth-promoting rhizobacteria (PGPR) mycorrhizae as biocontrol agents can colonise plant roots and promote plant growth. Beneficial free-living soil bacteria, usually referred to as PGPR, belonging to the genera *Pseudomonas* and *Bacillus*, have antagonistic effects on plants. PGPR are highly diverse biocontrol agents. According to Glick (Glick 1995), PGPR can promote plant growth directly and indirectly. Antagonistic rhizobacteria (PGPR) produce several substances, such as siderophores and antibiotics, against soil-borne pathogens, and indirectly promote plants growth. PGPR can either provide or synthesise phytohormones or indirectly facilitate the uptake of certain nutrients from the environment. Also, PGPR can prevent the deleterious effects of phytopathogenic organisms on plant diseases. The ability of bacterial siderophores and antibiotics to suppress soil-borne phytopathogens could have significant agronomic importance. Maksimov, Abizgil'Dina, and Pusenkova (2011) indicated that studies using PGPR rhizobacteria are able to produce siderophores and metabolites, contributing to antibiosis. Recently, microbial-based fertilisers, N-fixing bacteria, mycorrhizae and P solubilising bacteria, and PGPR have also been used in plant growth. This is very important for future commercial PGPR application for nitrogen and phosphorus nutrition in ecological farming.

Since plant growth-promoting bacteria (PGPB) have the potential to improve soil health and plant growth, they are also regarded as enriching the soil. Under controlled greenhouse conditions, PGPB provide mineral nutrients and organic substances to plants. Also, microbes eliminate the pathogenic organisms found in the soil. Since the same organisms exist in field soil, their effect and role are not great. Majeed, Muhammad, and Ahmad (2018) indicated that PGPB increased potato yield. Johansson, Paul, and Finlay (2004) reported that bacterial communities promote mycorrhizal spore germination, which will increase the rate and extent of root colonisation. Also, such interactions among bacteria and AM fungi might affect plant and soil microbial activity by stimulating the production of root exudates, such as phytoalexins and phenolic compounds, for healthy plant growth (Norman and Hooker 2000). Generally, after mycorrhizal inoculation, as mycorhizosphere is expended, the microbial community is increased.

Conclusion

Mycorrhizal fungi are the most important group of soil organisms which affect a majority of plant roots; play a critical role in nutrient and water uptake; reduce stress factors; and protect the plant against soil-borne diseases and pathogens. In many field experiments, it has been demonstrated that mycorrhizal inoculation has reduced fertiliser use. Some other plant-beneficial microorganisms also contribute significantly to plant growth, and reduce chemical application levels. Mycorrhizae increase the ability of plants to survive shock under non-sterile soil conditions. Mycorrhizae can also increase micro-propagated seedling survival after transplanting to field conditions. A number of research results suggested that the ecological farming system leads to an increase in the inoculum level of AMF, with greater crop colonisation and, as a result, plants can benefit from mycorrhizae to enhance nutrient uptake. Also, indigenous mycorrhizae can successfully affect many horticultural plant roots, resulting in better plant growth and nutrient (P and Zn) uptake. Also, beneficial microorganisms, such as *Pseudomonas* strains and mycorrhizae species, increase the yield of many horticultural plants. Several experiments concluded that, under field conditions, mycorrhizal spores effectively infected many horticulture plants and increased a plant's health and yield. The high mycotrophic field and horticultural crops are the most dependent on mycorrhizal colonisation for their growth. For ecological farming, soil and plant management systems are recommended, but for horticultural plants, mycorrhizal inoculation is the more beneficial practice for health production. The effect of the rhizosphere mechanism on plant growth is greater than our expectations. In this context, rhizosphere management such as root exudate and pH change are very important agricultural strategy for ecological/organic farming.

Acknowledgements This research was funded by the Çukurova University Research Fund.

References

Aka-Kacar, Y, C Akpinar, A Agar, Y Yalcin-Mendi, S Serce, and I Ortas. 2010. "The effect of mycorrhiza in nutrient uptake and biomass of cherry rootstocks during acclimatization." *Romanian Biotechnological Letters* 15 (3):5246–5252.

Al-Karaki, G, BZAKJ McMichael, and J Zak. 2004. "Field response of wheat to arbuscular mycorrhizal fungi and drought stress." *Mycorrhiza* 14 (4):263–269.

Almaca, A, and I Ortas. 2010. "Growth response of maize plants (Zea mays L.) to wheat and lentil pre-cropping and to indigenous mycorrhizae in field soil." *Spanish Journal of Agricultural Research* 8:S131–S136.

Amer, MA, and II Abou-El-Seoud. 2008. "Mycorrhizal fungi and *Trichodermaharzianum* as biocontrol agents for suppression of *Rhizoctoniasolani* damping-off disease of tomato." *Communications in Agricultural and Applied Biological Sciences* 73 (2):217–32.

Atkinson, D, JA Baddeley, N Goicoechea, J Green, M Sanchez-Diaz, and CA Watson. 2002. "Arbuscular mycorrhizal fungi in low input agriculture." In *Mycorrhizal technology in agriculture*, 211–222. Springer.

Augé, RM, X Duan, RC Ebel, and AJW Stodola. 1994. "Nonhydraulic signalling of soil drying in mycorrhizal maize." *Planta* 193 (1):74–82.

Azcon-Aguilar, C, and JM Barea. 2015. "Nutrient cycling in the mycorrhizosphere." *Journal of Soil Science and Plant Nutrition* 15 (2):372–396.

Azcon, R, and JM Barea. 1997. "Mycorrhizal dependency of a representative plant species in Mediterranean shrublands (Lavandulaspica L.) as a key factor to its use for revegetation strategies in desertification-threatened areas." *Applied Soil Ecology* 7 (1):83–92.

Barea, JM 2015. "Future challenges and perspectives for applying microbial biotechnology in sustainable agriculture based on a better understanding of plant-microbiome interactions." *Journal of Soil Science and Plant Nutrition* 15 (2):261–282.

Baum, C, W El-Tohamy, and N Gruda. 2015. "Increasing the productivity and product quality of vegetable crops using arbuscular mycorrhizal fungi: A review." *Scientia Horticulturae* 187:131–141. doi: 10.1016/j.scienta.2015.03.002.

Bedini, S, C Cristani, L Avio, C Sbrana, A Turrini, and M Giovannetti. 2008. "Influence of organic farming on arbuscular mycorrhizal fungal populations in a Mediterranean agro-ecosystem."

Bending, GD, MK Turner, F Rayns, M-C Marx, and M Wood. 2004. "Microbial and biochemical soil quality indicators and their potential for differentiating areas under contrasting agricultural management regimes." *Soil Biology and Biochemistry* 36 (11):1785–1792.

Bona, E, V Todeschini, S Cantamessa, P Cesaro, A Copetta, G Lingua, E Gamalero, G Berta, and N Massa. 2018. "Combined bacterial and mycorrhizalinocula improve tomato quality at reduced fertilization." *Scientia Horticulturae* 234:160–165. doi: 10.1016/j.scienta.2018.02.026.

Bona, E, S Cantamessa, N Massa, P Manassero, F Marsano, A Copetta, G Lingua, G D'Agostino, E Gamalero, and G Berta. 2017. "Arbuscular mycorrhizal fungi and plant growth-promoting pseudomonads improve yield, quality and nutritional value of tomato: a field study." *Mycorrhiza* 27 (1):1–11.

Bowen, GD, and AD Rovira. 1999. "The rhizosphere and its management to impove plant growth." *Advances in Agronomy, Vol 56* 66:1–102.

Buto, T, K Suzuki, T Kaidzu, T Narisawa, OC Turgay, I Ortas, N Harada, and M Nonaka. 2016. "Arbuscular mycorrhizal fungal community of wheat under long-term mineral and organic amendments in semi-arid Mediterranean Turkey." *Arid Land Research and Management* 30 (4):479–489. doi: 10.1080/15324982.2016.1177749.

Caruso, G, VVStoleru, NC Munteanu, VM Sellıtto, GC Telıban, M Burducea, T Ioan, G Morano, and M Butnarıu. 2018. "Quality Performances of Sweet Pepper under Farming Management." *Notulae Botanicae Hort iAgrobotanici Cluj-Napoca* 47 (1).

Celik, I, I Ortas, and S Kilic. 2004. "Effects of compost, mycorrhiza, manure and fertilizer on some physical properties of a Chromoxerert soil." *Soil & Tillage Research* 78 (1):59–67. doi: 10.1016/j.still.2004.02.012.

Cheng, W, and A Gershenson. 2007. "Carbon fluxes in the rhizosphere." In *The Rhizosphere*, 31–56. Elsevier.

Dobbelaere, S, J Vanderleyden, and Y Okon. 2003. "Plant growth-promoting effects of diazotrophs in the rhizosphere." *Critical Reviews In Plant Sciences* 22 (2):107–149.

Ezawa, T, K Yamamoto, and S Yoshida. 2000. "Species composition and spore density of indigenous vesicular-arbuscular mycorrhizal fungi under different conditions of P-fertility as revealed by soybean trap culture." *Soil Science and Plant Nutrition* 46 (2):291–297.

Fließbach, A, H Rudolf Oberholzer, L Gunst, and P Mäder. 2007. "Soil organic matter and biological soil quality indicators after 21 years of organic and conventional farming." *Agriculture, Ecosystems and Environment* 118 (1–4):273–284.

Fontana, A, M Reichelt, S Hempel, J Gershenzon, and SB Unsicker. 2009. "The effects of arbuscular mycorrhizal fungi on direct and indirect defense metabolites of *Plantagolanceolata* L." *Journal of Chemical Ecology* 35 (7):833–843.

Gaur, R, N. Shani, K, BN Johri, P Rossi, and M Aragno. 2004. "Diacetylphloroglucinol-producing pseudomonads do not influence AM fungi in wheat rhizosphere." *Current Science*: 453–457.

George, E, KU Haussler, D Vetterlein, E Gorgus, and H Marschner. 1992. "Water and nutrient translocation by hyphae of *Glomus mosseae*." *Canadian Journal of Botany-Revue Canadienne De Botanique* 70 (11):2130–2137.

Glick, BR 1995. "The enhancement of plant growth by free-living bacteria." *Canadian journal of microbiology* 41 (2):109–117.

Gosling, P, A Hodge, G Goodlass, and GD Bending. 2006. "Arbuscular mycorrhizal fungi and organic farming." *Agriculture Ecosystems and Environment* 113 (1–4):17–35.

Gosling, P, and M Shepherd. 2005. "Long-term changes in soil fertility in organic arable farming systems in England, with particular reference to phosphorus and potassium." *Agriculture, Ecosystems and Environment* 105 (1–2):425–432.

Gosling, P, A Ozaki, J Jones, M Turner, F Rayns, and G D Bending. 2010. "Organic management of tilled agricultural soils results in a rapid increase in colonisation potential and spore populations of arbuscular mycorrhizal fungi." *Agriculture Ecosystems and Environment* 139 (1–2):273–279. doi: 10.1016/j.agee.2010.08.013.

Gray, EJ, and DL Smith. 2005. "Intracellular and extracellular PGPR: commonalities and distinctions in the plant–bacterium signaling processes." *Soil Biology and Biochemistry* 37 (3):395–412.

Hamel, C. 1996. "Prospects and problems pertaining to the management of arbuscular mycorrhizae in agriculture." *Agriculture, Ecosystems and Environment*60 (2–3):197–210.

Incesu, M, T Yesiloglu, B Cimen, B Yilmaz, C Akpinar, and I Ortas. 2015. "Effects on growth of persimmon (*Diospyrosvirginiana*) rootstock of arbuscular mycorrhizal fungi species." *Turkish Journal of Agriculture and Forestry* 39 (1):117–122. doi: 10.3906/tar-1405-134.

Johansson, JF, LR Paul, and RD. Finlay. 2004. "Microbial interactions in the mycorrhizosphere and their significance for sustainable agriculture." *FEMS Microbiology Ecology*48 (1):1–13.

Koide, RT, and Z Kabir. 2000. "Extraradical hyphae of the mycorrhizal fungus *Glomus intraradices* can hydrolyse organic phosphate." *New Phytologist* 148 (3):511–517. doi: 10.1046/j.1469–8137.2000.00776.x.

Li, F, MJ Christensen, P Gao, Y Li, and T Duan. 2018. "An arbuscular mycorrhizal fungus and *Epichloëfestucae* var. lolii reduce Bipolarissorokiniana disease incidence and improve perennial ryegrass growth." *Mycorrhiza* 28 (2):159–169.

Li, X L, E George, and HMarschner. 1991. "Extension of the phosphorus depletion zone in va-mycorrhizal white clover in a calcareous soil." *Plant and Soil* 136 (1):41–48.

Li, XL, H Marschner, and E George. 1991. "Acquisition of phosphorus and copper by va-mycorrhizal hyphae and root-to-shoot transport in white clover." *Plant and Soil* 136 (1):49–57.

Liu, A, C Hamel, RI Hamilton, BL Ma, and DL Smith. 2000. "Acquisition of Cu, Zn, Mn and Fe by mycorrhizal maize (*Zea mays* L.) grown in soil at different P and micronutrient levels." *Mycorrhiza* 9 (6):331–336.

Lugtenberg, B. 2014. *Principles ofplant-microbe interactions: microbes for sustainable agriculture*: Springer.

Mäder, P, S Edenhofer, T Boller, A Wiemken, and U Niggli. 2000. "Arbuscular mycorrhizae in a long-term field trial comparing low-input (organic, biological) and high-input (conventional) farming systems in a crop rotation." *Biology and fertility of Soils* 31 (2):150–156.

Majeed, A, Z Muhammad, and H Ahmad. 2018. "Plant growth promoting bacteria: role in soil improvement, abiotic and biotic stress management of crops." *Plant Cell Reports*:1–11.

Maksimov, IV, RR Abizgil'Dina, and LI Pusenkova. 2011. "Plant growth promoting rhizobacteria as alternative to chemical crop protectors from pathogens." *Applied Biochemistry and Microbiology* 47 (4):333–345.

Marschner, H. 1995. *Mineral Nutrition of High Plants*. Second Edition ed. London: Academic Press

Marschner, H. 1996. "Mineral nutrient acquisition in nonmycorrhizal and mycorrhizal plants." *Phyton-Annales Rei Botanicae* 36 (3):61–68.

Marschner, H, V Romheld, W J Horst, and P Martin. 1986. "Root-induced changes in the rhizosphere–importance for the mineral-nutrition of plants." *Zeitschrift Fur Pflanzenernahrung Und Bodenkunde* 149 (4):441–456.

Marschner, P. 2012. *Marschner's Mineral Nutrition of Higher Plants*. Third Edition ed. London, UK: Academic Press.

Mozafar, A, T Anken, R Ruh, and E Frossard. 2000. "Tillage intensity, mycorrhizal and nonmycorrhizal fungi, and nutrient concentrations in maize, wheat, and canola." *Agronomy Journal* 92 (6):1117–1124.

Naghibi, R, P Rezvani Moghaddam, A Balandari, and R Ghorbani. 2016. "The Effects of Organic Fertilizers and Mycorrhizae Inoculation (*Glomus mosseae* and *G. intraradices*) on Quantitative and Qualitative Yield of Dwarf Chicory (*Cichoriumpumilum* Jacq.) in Different Cuttings." *Majallah-i 'Ulum-iBāghbānī* 29 (2):302–313.

Ness, R, L Lange, and PLG Vlek. 2000. "Mechanism of calcium and phosphate release from hydroxy-apatite by mycorrhizal hyphae."

Norman, JR, and JE Hooker. 2000. "Sporulation of *Phytophthorafragariae* shows greater stimulation by exudates of non-mycorrhizal than by mycorrhizal strawberry roots." *Mycological Research* 104 (9):1069–1073.

Oehl, F, E Laczko, A Bogenrieder, K Stahr, R Bosch, M van der Heijden, and E Sieverding. 2010. "Soil type and land use intensity determine the composition of arbuscular mycorrhizal fungal communities." *Soil Biology & Biochemistry* 42 (5):724–738. doi: 10.1016/j.soilbio.2010.01.006.

Oehl, F, E Sieverding, K Ineichen, P Mäder, T Boller, and A Wiemken. 2003. "Impact of land use intensity on the species diversity of arbuscular mycorrhizal fungi in agroecosystems of Central Europe." *Applied and Environmental Microbiology* 69 (5):2816–2824.

Ortas, I 2016. "Role of Mycorrhizae and Biochar on Plant Growth and Soil Quality." In *Biochar, A Regıonal Supply Chain Approach In View Of Climate Change Mitigatıon.*, edited by VJ Bruckman, EA Varol, BB Uzun and JF Liu, 398. Cambridge. UK.: Cambridge University Press.

Ortas, I, and A Varma. 2007. "Field Trials of Bioinoculants." In *Modern Tools and Techniques*, edited by R Oelmüller and A Varma, 397–413. Springer-Verlag.

Ortas, I. 1996. "The influence of use of different rates of mycorrhizal inoculum on root infection, plant growth, and phosphorus uptake." *Communications in Soil Science and Plant Analysis* 27 (18–20):2935–2946. doi: 10.1080/00103629609369753.

Ortas, I. 1997. "Determination of the extent of rhizosphere soil." *Communications in Soil Science and Plant Analysis* 28 (19–20):1767–1776. doi: 10.1080/00103629709369914.

Ortas, I. 2003. "Effect of selected mycorrhizal inoculation on phosphorus sustainability in sterile and non-sterile soils in the Harran Plain in South Anatolia." *Journal of Plant Nutrition* 26 (1):1–17. doi: 10.1081/pln-120016494.

Ortas, I. 2010. "Effect of mycorrhiza application on plant growth and nutrient uptake in cucumber production under field conditions." *Spanish Journal of Agricultural Research* 8: S116-S122.

Ortas, I. 2012a. "The effect of mycorrhizal fungal inoculation on plant yield, nutrient uptake and inoculation effectiveness under long-term field conditions." *Field Crops Research* 125:35–48. doi: 10.1016/j.fcr.2011.08.005.

Ortas, I. 2015. "Comparative analyses of Turkey agricultural soils: Potential communities of indigenous and exotic mycorrhiza species' effect on maize (*Zea mays* L.) growth and nutrient uptakes." *European Journal of Soil Biology* 69:79–87. doi: 10.1016/j.ejsobi.2015.05.006.

Ortas, I. 2018. "Role of mycorrhizae on mineral nutrition of fruit trees." In *Viii International Symposium on Mineral Nutrition of Fruit Crops*, edited by T. Mimmo, Y. Pii and F. Scandellari, 271–283.

Ortas, I, and A Coskan. 2016. "Precipitation as the most affecting factor on soil-plant environment conditions affects the mycorrhizal spore numbers in three different ecological zones in Turkey." *Acta Agriculturae Scandinavica Section B-Soil and Plant Science* 66 (4):369–378. doi: 10.1080/09064710.2015.1132005.

Ortas, I, PJ Harris, and DL Rowell. 1996. "Enhanced uptake of phosphorus by mycorrhizal sorghum plants as influenced by forms of nitrogen." *Plant and Soil* 184 (2):255–264. doi: 10.1007/bf00010454.

Ortas, I, M Rafique, C Akpinar, and YA Kacar. 2017. "Growth media and mycorrhizal species effect on acclimatization and nutrient uptake of banana plantlets." *Scientia Horticulturae* 217:55–60. doi: 10.1016/j.scienta.2017.01.025.

Ortas, I, and DL Rowell. 2004. "Effect of ammonium and nitrate on indigenous mycorrhizal infection, rhizosphere pH change, and phosphorus uptake by sorghum." *Communications in Soil Science and Plant Analysis* 35 (13–14):1923–1944. doi: 10.1081/lcss-200026820.

Ortas, I, N Sari, C Akpinar, and H Yetisir. 2011a. "Screening mycorrhiza species for plant growth, P and Zn uptake in pepper seedling grown under greenhouse conditions." *Scientia Horticulturae* 128 (2):92–98. doi: 10.1016/j.scienta.2010.12.014.

Ortas, I, N Sari, C Akpinar, and H Yetisir. 2011b. "Screening Mycorrhizae Species for Increased Growth and P and Zn Uptake in Eggplant

(*Solanummelongena* L.) Grown under Greenhouse Conditions." *European Journal of Horticultural Science* 76 (3):116–123.

Ortas, I, N Sari, C Akpinar, and H Yetisir. 2013. "Selection of ArbuscularMycorrhizal Fungi Species for Tomato Seedling Growth, Mycorrhizal Dependency and Nutrient Uptake." *European Journal of Horticultural Science* 78 (5):209–218.

Ortas, I. 2012b. "Mycorrhiza in citrus: growth and nutrition." In *Advances in Citrus Nutrition*, 333–351. Springer.

Ortas, I, C Akpinar, and R Lal. 2013. "Long-Term Impacts of Organic and Inorganic Fertilizers on Carbon Sequestration in Aggregates of an Entisol in Mediterranean Turkey." *Soil Science* 178 (1):12–23. doi: 10.1097/SS.0b013e3182838017.

Ortaş, I, R Lal, and S Kapur. 2017. "Carbon Sequestration and Mycorrhizae in Turkish Soils." In *Carbon Management, Technologies, and Trends in Mediterranean Ecosystems*, 139–149. Springer.

Ortaş, I. 2019. "Role of Microorganisms (Mycorrhizae) in Organic Farming." In *Organic Farming*, 181–211. Elsevier.

Ortaş, I, M. Rafique, and I AM Ahmed. 2017. "Application of Arbuscular Mycorrhizal Fungi into Agriculture." In *Arbuscular Mycorrhizas and Stress Tolerance of Plants*, 305–327. Springer.

Pagano, MC, EJA Correa, N F Duarte, B Yelikbayev, A O'Donovan, and VK Gupta. 2017. "Advances in Eco-Efficient Agriculture: The Plant-Soil Mycobiome." *Agriculture-Basel* 7 (2):12. doi: 10.3390/agriculture7020014.

Parniske, M. 2008. "Arbuscular mycorrhiza: the mother of plant root endosymbioses." *Nature Reviews Microbiology* 6 (10):763.

Rillig, M C. 2004. "Arbuscular mycorrhizae, glomalin, and soil aggregation." *Canadian Journal of Soil Science* 84 (4):355–363.

Ruan, J, F Zhang, and MH Wong. 2000. "Effect of nitrogen form and phosphorus source on the growth, nutrient uptake and rhizosphere soil property of *Camellia sinensis* L." *Plant and Soil* 223 (1–2):65–73.

Ryan, MH., and JH Graham. 2002. "Is there a role for arbuscular mycorrhizal fungi in production agriculture?" In *Diversity and Integration in Mycorrhizas*, 263–271. Springer.

Satir, NY, I Ortas, and O Satir. 2016. "The ınfluence of mycorrhizal species on sour orange (*Citrus aurantium* L.) growth under saline soil conditıons." *Pakistan Journal of Agricultural Sciences* 53 (2):399–406. doi: 10.21162/pakjas/16.2425.

Schippers, B, AW Bakker, and P Bakker. 1987. "Interactions of deleterious and beneficial rhizosphere microorganisms and the effect of cropping practices." *Annual Review of Phytopathology* 25:339–358. doi:

10.1146/annurev.py.25.090187.002011.
Siddiqui, ZA, and I Mahmood. 1995. "Role of plant symbionts in nematode management: a review." *Bioresource Technology* 54 (3):217–226.
Singh, R, S Tiwari, RP Patel, SK Soni, and A Kalra. 2018. "Bioinoculants and AM fungus colonized nursery improved management of complex root disease of *Coleus forskohlii* Briq. under field conditions." *Biological Control* 122:11–17.
Sivasithamparam, K, CA Parker, and CS Edwards. 1979. "Rhizosphere micro-organisms of seminal and nodal roots of wheat grown in pots." *Soil Biology and Biochemistry* 11 (2):155–160.
Smith, SE, and DJ Read. 2008. *Mycorrhizal Symbiosis*. Third ed. San Diego, CA: Academic Press.
Stockdale, EA, NH Lampkin, M Hovi, R Keatinge, EKM Lennartsson, DW Macdonald, S Padel, FH Tattersall, MS Wolfe, and CA Watson. 2001. "Agronomic and environmental implications of organic farming systems."
Sylvia, DM. 1989. "Nursery inoculation of sea oats with vesicular-arbuscular mycorrhizal fungi and outplanting performance on Florida beaches." *Journal of Coastal Research* 5 (4):747–754.
Sylvia, DM, AK Alagely, M E Kane, and NL Philman. 2003. "Compatible host/mycorrhizal fungus combinations for micropropagated sea oats–I. Field sampling and greenhouse evaluations." *Mycorrhiza* 13 (4):177–183. doi: 10.1007/s00572-003-0232-y.
Sylvia, DM, AG Jarstfer, and M Vosatka. 1993. "Comparisons of Vesicular-Arbuscular Mycorrhizal Species and Inocula Formulations in a Commercial Nursery and on Diverse Florida Beaches." *Biology and Fertility of Soils* 16 (2):139–144. doi: 10.1007/bf00369416.
Sylvia, DM, and LH Neal. 1990. "Nitrogen affects the phosphorus response of VA mycorrhiza." *New Phytologist* 115 (2):303 310.
Thingstrup, I, G Rubaek, E Sibbesen, and I Jakobsen. 1998. "Flax (*Linumusitatissimum*L.) depends on arbuscular mycorrhizal fungi for growth and P uptake at intermediate but not high soil P levels in the field." *Plant and Soil* 203 (1):37–46.
Thioye, B, A Kane, SM de Faria, D Fall, D Sanogo, C Ndiaye, KB Sanon, A Soule, R Duponnois, and S NSylla. 2018. "Growth Response of Different Species and Provenances of Jujube Seedlings to Inoculation with Arbuscular Mycorrhizal Fungi." In *Root Biology*, 501–519. Springer.
Turgay, OC, D Buchan, B Moeskops, B de Gusseme, I Ortas, and S de Neve. 2015. "Changes in Soil Ergosterol Content, Glomalin-Related Soil Protein, and Phospholipid Fatty Acid Profile as Affected by Long-Term

Organic and Chemical Fertilization Practices in Mediterranean Turkey." *Arid Land Research and Management* 29 (2):180–198. doi: 10.1080/15324982.2014.944246.

Utkhede, RS, TSC Li, and EM Smith. 1992. "The effect of Glomus-Mosseae and enterobacter-aerogenes on apple seedlings grown in apple replant disease soil." *Journal of Phytopathology-Phytopathologische Zeitschrift* 135 (4):281–288. doi: 10.1111/j.1439–0434.1992.tb04313.x.

Utkhede, RS, and EM Smith. 1994. "Development of biological control of apple replant disease." In *Third International Symposium on Replant Problems*, edited by RS Utkhede, 129–134.

van der Heijden, M, V Rinaudo, E Verbruggen, C Scherrer, P Bàrberi, and M Giovannetti. 2008. "The significance of mycorrhizal fungi for crop productivity and ecosystem sustainability in organic farming systems."

Wang, B, and YL Qiu. 2006. "Phylogenetic distribution and evolution of mycorrhizas in land plants." *Mycorrhiza* 16 (5):299–363.

CHAPTER 5

MUTATIONS LEADING TO A BETTER TOMORROW IN THE PRODUCTION OF IMPROVED CROP VARIETIES

RIDA REHMAN,[1,2] AMEEMA TARIQ,[3] NAMRA HAQ,[1] ALVINA GUL,[3] VOLKAN ALTAY,[4] MUNIR OZTURK[5*]

[1] School of Mechanical and Manufacturing Engineering (SMME), National University of Sciences and Technology (NUST), Pakistan
[2] Dept. of Neurology, Ulm University, Germany
[3] Atta Ur Rahman School of Applied Biosciences (ASAB), National University of Sciences and Technology (NUST), Pakistan
[4] Biology Department, Faculty of Science and Arts, Hatay Mustafa Kemal University, Hatay, Turkey
[5] Botany Department and Centre for Environmental Studies, Ege University, Izmir, Turkey
munirozturk@gmail.com

Abstract

To meet the increased demands of sustenance, worldwide crop improvement serves as a *sine qua non* to confront the challenges of the agricultural world. Crop improvement through induced mutation adds credence. Currently, crop improvement via induced mutation has been reviewed. The present review suggests that for crop improvement, induced mutation is a formidable approach. Studies and surveys suggest that crop improvement has contributed enormously towards the economic step up of both developing and developed parts of the world. To serve the purpose of crop improvement, various techniques are employed, intended to improve crops in a holistic context.

Key words: crop improvement, environment, induced mutation

1. Introduction

Plant breeding necessitates variation of useful traits on a genetic level for crop improvement. Mutagenic agents, such as chemicals and radiation, can be employed for inducing mutations, causing variations on a genetic level from which desired mutations can be selected (Novak and Brunner 1992). Over the past seventy years, more than 2,250 varieties have been released worldwide. The basic purpose of breeding, based on mutation, is to improve the already-adjusted plant varieties by causing manipulation in either a single or a number of traits, which can thus serve our purpose. Better quality starch, protein and modified oil; production of drought and salinity-tolerant varieties; and the generation of disease-resistant varieties, to name a few, serve the purpose of sustainable agriculture. Initial studies were aimed at finding the best possible combination of mutagens, and their respective effective dose to obtain the desired results. Both the chemical and physical mutagens were tested on a number of crops, including wheat, barley, rice, tobacco and many other fruit crops and vegetables (Cavanagh et al. 2008).

2. Mutagenic Treatment—the History

X-ray induced mutations in *Drosophila melanogaster* (fruit fly) (Muller 1927) and in barley (Stadler 1928), provided a grounding for crop improvement. It was then employed as a useful tool for gene location on chromosomes, gene structure identification, regulation and expression, and also for thoroughly understanding the genome. Later on, radiation-induced mutation was employed for altering traits that can generate varieties with desirable traits (Ahloowalia, Maluszynski, and Nichterlein 2004).

Among the 2,252 mutant varieties which have been released so far, nearly 60% have been released since 1985. Table 1 shows the percentage of mutant varieties released in different countries.

Amongst the varieties released, most were direct mutants in the sense that the selected mutant was multiplied, and ultimately released as a new variety (Ahloowalia, Maluszynski, and Nichterlein 2004).

Table 1 Percentage of varieties released in different countries in the past 70 years.

No	Country	Percentage of varieties released
1	China	26.8%
2	India	11.5%
3	USSR and Russia	9.3%
4	Netherlands	7.8%
5	USA	5.7%
6	Japan	5.3%

2.1. Single Mutagen

Radio phosphorus was reported to be an effective mutagen that can work single-handedly. Radio phosphorus-treated maize resulted in multiple nucleolar buds, indicating that either the radio phosphorus had deleted, or, at least, inactivated the nucleolar organiser. Irradiated population of wheat was frequently searched for haploids. Later on, it was revealed that a particular dose, at a particular time, administered at the inflorescence stage, before anthesis, can result in useful haploid production. Additionally, P32 and S35 were reported to be potent mutagens for haploid production (Natarajan and Swaminathan 1958).

In rice, dwarf populations were developed using chemical and physical mutagens (Singh, Siddiq, and Swaminathan 1979).

2.2. The Combined Effect of Mutagens

Studies have revealed that both physical and chemical mutagens differently cause lesions in DNA. This led to an investigation of the combined effects of mutagens. When wheat was treated with a combination of UV and X-ray radiations, it revealed dose-dependent effects. When seeds were pre-treated with UV, it reduced the frequency of mutation at low X-ray doses but increased at higher doses (Swaminathan and Natarajan 1959). A similar effect was observed using two chemical mutagens: Ethyl methane sulfonate (EMS) and hydroxylamine, (HA) in wheat. Furthermore, studies revealed that when wheat was treated with HA preceding EMS treatment, the mutation rate was noticeably reduced in wheat plant, suggesting that HA may be involved in the mutation repair process (Mahajani and Chopra 1973).

Recently, evolutionary dynamics has been understood more precisely, using a computer simulation model which integrates the effects of drift, selection and mutation. It was reported that the empirical patterns of mutation accumulations signify adaptive evolution (Cuevas et al. 2015).

Molecular markers can be defined as genetic loci which can be effortlessly traced and quantified in some individuals, within a sexually reproducing population. These are possibly linked with a specified trait and gene of interest. These variations are triggered either by INDELS (insertion/deletions), mutations, duplications, translocations or inversions (Hayward et al. 2015).

3. Sustainable Agriculture

Biotechnology and plant breeding massively contribute towards environmentally-friendly agriculture. A phenomenal gene pool has been established for crop improvement. Advancements have been made to conserve the gene pool, both efficiently and effectively.

According to the USDA, sustainable agriculture is a system with integrated features for both plant and animal production practices, which have a site-specific application for basic necessities, such as food and fibre needs of living organisms. The following serve as essentials towards sustainable agriculture:

a. Fulfilment of nutritional requirements
b. Improvement in quality of life
c. Ensuring productive natural resources
d. Decreasing the agricultural adversities by reducing natural and socio-economic factors

Sustainable agriculture is an ultimate requirement for current and future development of economic, social and environmental status (Haron 2013).

4. Advanced Techniques Using Ion Beams

Plant mutagenesis via ionising radiation was employed for inducing mutation in maize and barley by Stadler in 1928, using X-rays. He revealed that when ionic beams fall on the target, either densely or locally, it induces single- or double-stranded DNA breakage with damaged end groups. The repair process of this broken DNA was low. It can thus be inferred that ion beams produce alterations in DNA as inversion, deletion and translocation, rather than as point mutations (Tanaka 2013).

5. Irradiation Method

In TIARA, ion beams with characteristic features for biological endpoints have been investigated using *Arabidopsis*, barley, rice, etc. (Tanaka 2011). The AVF-cyclotron generates all the ions in TIARA. The sample is placed under the beam window and irradiated in the atmosphere. For example, Kapton films are used for sandwiching 100–3000 *Arabidopsis* or tobacco seeds to make a monolayer seed for homogenous irradiation. The side of the embryo faces the ion beams in the case of relatively large seeds, such as seeds of rice, barley, or soybean (Tanaka 2013).

UVB mutation rates of different moss lines that have defects in DNA repair (ppmre11, ppku70, pplig4, pprad50), and in certain clones resistant to 2–Flouroadenine that were mutated in APT (adenosine phosphor transferase gene), were analysed. Analysis was carried out by inducing mutations by sequencing, and DNA break repair was followed. It was found that UVB causes gigantic, sequence-specific, error-prone bypass repair, which accounts for an increased mutation rate due to comparatively slow, nonetheless error-free, deletion of phospho products by nucleotide excision repair (NER) (Holá, Vagnerova, and Angelis 2015).

6. Mutation via Thermal Neutrons

Bacterial leaf blight (BLB) is one of the deadliest agricultural diseases of Asian countries. It was noted that thermal neutrons and ion-beam have generated a higher frequency of resistant mutants compared with gamma rays and other chemical mutagens. High linear energy transfer (LET) radiations have greater effectiveness towards mutant-resistant inductions against diseases. The mutants obtained as a result of this treatment were resistant to many types of BLB races from Japan and the Philippines. The mutant is known as M95 mutant because of a high resistance level or complete resistance towards BLB. The M95 gene was worked out to be located at the edge of chromosome 1. As far as BLB resistance is concerned, this gene has never been found in the gene sources. However, the M95 mutant containing a durable gene, which causes resistance, could not be released directly because this was a dwarf variety. Crossing tests between M95 and the original variety revealed that resistance and dwarf genes were closely linked. Later studies also revealed that the M95 mutant could be easily used for crossing purposes (Nakai 2013).

7. Plant Growth Promoter (PGP) From Natural Polymers

Studies have revealed that radiation caused degradation in the development of plant growth promoters from naturally occurring polymers. These polymers increase the yield. Radiation-induced degradation results in the production of Oligo-chitosan. Crab/shrimp shell and oligo-carrageenan obtained from seaweed, which acts as oligo-chitosan, displayed a significant increase in crop yields. Thus, it can help towards crop improvement in terms of better yield, and reduced use of chemicals, such as pesticides and fertilizers (Machi 2013).

Promising improvement exhibited by some prominent varieties has been reviewed in the following discussion.

8. Tomato

The tomato is regarded as an important crop across the globe. Attempts have been made to improve both its quality and shelf life. The important variations are discussed below.

8.1. Increasing the ascorbic acid content

Ascorbic acid (Vit C) is both an antioxidant and a cofactor for collagen synthesis. Three quantitative trait loci (QTL) have been identified, which are obtained by crossing varieties of cultivated tomatoes i.e. *Solanumlycopersicum* accessions and wild tomato species and subspecies (Stevens et al. 2007). The detection of QTL requires the following four elements: a population of plants that is genetically variable for the targeted phenotype; genotyping of the population by using marker systems; methodologies of reproducible, quantitative phenotyping; and, for the detection and location of QTL, the application of appropriate experimental and statistical methods (Cavanagh et al. 2008). The drawback of this method is that owing to the number of breeding generations required, the task is cumbersome. Nevertheless, the difference in phenotype between the initial and selected population (or between the low and high selections) may be more than available in alternative mapping populations, indicating that a potential increase in the QTL number can be identified (Cavanagh et al. 2008).

Ascorbic acid undergoes recycling and goes through a specific pathway. An unstable monodehydroascorbate is a reduced product of ascorbate and its reduction is via ascorbic acid. Its dissociation products include ascorbate and dehydroascorbate. Dehydroascorbate is unstable *per se*, and undergoes degradation, ultimately depleting the ascorbate collection. This occurs if the oxidized radicals are not recovered by monohydroascorbate reductase (MDHAR) and dehydroascorbate reductase (DHAR) (Noctor and Foyer 1998, Smirnoff and Wheeler 2000). DHAR activity controls the amount of ascorbate in tissues. Increased expression of the enzyme was reported to increase the ascorbic acid level significantly in tobacco (Green and Fry 2005). The ascorbate level in cells is controlled by synthesis, recycling, degradation and transport of this molecule in cells (Horemans, Foyer, and Asard 2000). Nevertheless, the information regarding the process is insufficient.

It has been reported that wild species are prolific sources of ascorbic acid as compared with commercial varieties which are reported to contain nearly five times less. This ingrained, characteristic feature of wild species can be exploited to enrich the commercial varieties. These attributes are invariably controlled by genes and also the respective ambiance of it. Through molecular markers, the loci for quantitative traits can be identified. Additionally, genetic maps can aid the process. Thirty QTL regions have been reported. Positional cloning or candidate gene approach can be employed for QTL region identification (Stevens et al. 2007).

Three populations of tomato were derived from crossing three different species and sub species. The first population was the product of the introgression lines (IL-pen), obtained through crossing each *Solanumpennellii*, with a unique genome fragment belonging to the wild. The said population was employed for locating potential genes and QTLs for many metabolites (Schauer et al. 2006). The next population was a further backcross population (BC-hab) obtained through crossing a cultivated tomato and *Solanumhabrochiates* accession. The third population was the product of crossing a cherry tomato and a large-fruited line—the recombinant inbred line population (RIL-cherry) (Saliba-Colombani et al. 2001). An increased level of ascorbic acid and better fruit quality was observed in all these attempts.

8.2. *Hp*-2 Mutants

Tomato mutants which carry high pigment (*hp*) mutations are typified by their intense response to light. In *Hp*-2^{dg} fruit, increased amounts of isoprenoids and phenolic compounds were observed. The fruit was characterised by its overproduction of various metabolites. These overproduced metabolites also included the photo protective or antioxidant ones, possibly contributing towards curbing light stress (Bino et al. 2005).

8.3. Tomato PPARγ-mediated Gene Expression

The PPARγ-mediated gene expression was conducted *in vitro* to control the expression of genes in tomatoes. The receptor is a peroxisome proliferator-activated receptor and its initiation is associated with positive effects. A study using thirty compounds was conducted to validate the hypothesis. Among these thirty compounds, seven were carotenoids and three polyphenols including PPARγ-mediated luciferase expression. Elevated flavonoids content was found in two varieties that were sturdy, and the opposite was observed in two weak varieties. It can thus be concluded that tomato extracts possess the *in vitro* ability of inducing PPARγ- mediated gene expression (Gijsbers et al. 2013).

A non-transgenic tomato variety, Tomelo, resistant to powdery mildew fungal pathogen, was generated using CRISPR/Cas9 technology. It was found, using whole genome sequencing, that Tomelo didn't carry any foreign DNA sequence, but has only a deletion that was indistinguishable from naturally occurring mutations (Nekrasov et al. 2017). In tomato florigen paralog and flowering repressor, SELF-PRUNING 5G (SP5G) were responsible for failure to day length-dependent flowering. CRISPR/Cas9-engineered mutations in SP5G resulted in accelerated flowering and enhanced the well-defined development pattern of tomatoes in the field, ultimately producing an early yield owing to speedy flower production (Soyk et al. 2017). Parthenocarpy is a significant attribute of horticulture based crops that contributes towards agriculture designed for numerous industrial purposes, in addition to a direct eating attribute. Based on the CRISPR/Cas9 system, a breeding strategy was designed to yield tomatoes of this kind, thus finally aiding mutant-based breeding intended for crop improvement (Ueta et al. 2017).

Moreover, Indoxacarb is a significant active ingredient broadly employed for controlling *Tutaabsoluta*, a pest typical of the tomato, thus prompting procedures intended to manage insecticide resistance. The indoxacarb resistance mutations-detection via continuous advancement in diagnostics have allowed immediate detection to resistance, thus devising a suitable alternative, and consequently suspending the resistance stretch (Roditakis et al. 2017).

The greatest obstacle to *Solanumlycopersicum* (tomato) sexual reproduction was the moving sensitivity of the microsporogenesis method at fairly extreme temperatures. A parthenocarpic mutant was designed to yield in heat stress. It was found that a mutation in SlAGAMOUS-LIKE 6 (SlAGL6) was accountable for the parthenocarpic phenotype. For this next-generation sequencing, marker-assisted mapping and CRISPR/Cas9 gene knockout were employed. The resultant mutant, in addition to being endowed with heat-sustaining attributes, exceptionally retained its already useful attributes, such as normal weight and shape. Moreover, pollen viability remained unaffected and sexual reproduction capacity was retained, thus making Slagl6 an attractive gene for facultative parthenocarpy (Klap et al. 2017).

Examining the tissue and stage-specific promoters of specific genes can play a significant role in crop improvement. The tomato and *Arabidopsis* established transgenic lines with 5'-deletion were made to characterise the areas of RIP1 (Ripening induces protein 1) promotor that was necessary for its specificity and maximum activity in fruits. That a full-length promoter was essential for functioning was revealed by categorisation of transgenic plants (Agarwal et al. 2017).

Tomato plants are resistant to PMMoV (Pepper mild mottle virus) as they possess tm-1 protein that inhibits the viral RNA replication. A study was conducted in which numerous mutant PMMoVs were constructed to know the viral replication status in each one of them. It was observed that terminal mutations in replication proteins forestalled the PMMoV mutants' development meant for countering tm-1-mediated resistance (Mizumoto et al. 2014).

9. Rice

Recently, considerable insight has been gained regarding rice domestication. The cultivated rice genomes have been shaped by extensive introgression between artificial and natural selections. Evolutionary forces arose through the interplay of these cultivated species, *Oryzasativa*, with divergent genomic backgrounds from *O. rufipogon* and *O. nivara*, two wild species (Sang and Ge 2013). In rice, direct mutants selected from mutated populations were developed from the majority of mutant varieties. The other 667 varieties of crop were derived through crosses with induced mutants. Induction of mutations via radiation was the most frequently used method for the development of direct varieties of mutants (89%). Chemicals used as mutagens were relatively infrequent. Table 2 elaborates further information regarding mutagens (Ahloowalia, Maluszynski, and Nichterlein 2004).

Table 2 Percentages of mutagens employed for inducing mutations.

No	Mutagens	Employing Percentage
1	Gamma rays	64%
2	X-rays	22%
3	Chemicals	Infrequently used

Nearly 50% of the global population uses rice as a staple crop. In the Asia-Pacific region, these mutation techniques have played a significant role in increasing production of rice. Table 3 shows the number of species which were targeted to induce mutations in rice.

Table 3 Number of rice species subjected to mutagenic treatment by different sources.

No	Source	No of Species
1	Gamma rays	225
2	X-rays	16
3	Fast neutrons	7
4	Other radiation sources	12

9.1. Radiation treatment

Radiation is the most efficient method of inducing mutation intended for crop improvement. The literature reveals that more than 3,000 better plant varieties were developed using this method. A new banana variety was developed in the Philippines under the FNCA project. This banana variety is better resistant to Bunch top viral disease. This helps in the reduction of pesticide usage which, in turn, protects the environment from pollution. The next focus of the FNCA is better rice varieties (Machi 2013). According to FAO/IAEA data there were at least 3,212 mutant species in 2011. Studies also showed that Asia is the centre of breeding mutations. As far as mutagens are concerned, gamma rays, being the most convenient ones, contribute to 60% of all the mutagens used (Nakai 2013).

Furthermore, ionising radiations were found to induce genetic variability in rice. A study was conducted to assess the effects of gamma rays involved in mutation, resulting in genetic variability in Basmati rice. For increasing a particular breed, treatments with mutagens having a low effect on physiology and a high effect on the genetic factors are required. Mutagenic treatment has the ultimate aim of inducing mutation to improve specific traits. Practically, when mutagenic treatment is applied, 50% reduction in fertility of M1 plants serves as a criterion for treatment with promising results. A study conducted on Basmati rice species using gamma radiation resulted in LD50. This resulted in maximum chlorophyll mutations observed in M1 and M2 panicles and seedlings, respectively. This showed that mutation via radiation results in better crops for breeding purposes, with low physiological effects and high vital mutation rate (Cheema 2003).

9.2. Mutation Breeding Leading to Varietal Improvement under Minimal Water Supplies

Gamma rays at 300 Gy were used to induce mutation in the MR219 variety to generate genotypes with superior qualities with minimal water requirements. At M4 generation, after several selection and fixation procedures, 12 lines with required traits were found. Nevertheless, only two lines having potential traits were selected and designated as MR219-4 and MR219-9. Under a controlled environment, the yield of MR219-4 and MR219-9 was observed to be 4.8 t/ha and 5.5 t/ha respectively, nevertheless, the original variety MR219, produced 5.8 t/ha. For MR219-4 and MR219-9 under flooded conditions and saturated soil conditions, the yield was observed to be 5.9 t/ha and 6.8 t/ha as well as 7.2 t/ha and 6.1 t/ha, respectively. Submergence screening was used in these two lines and a new adaptive trait was noticed where both species were found to be tolerant to submergence. This showed that these two strains possess dual adaptations, i.e. flooded and water stress regime. This also showed that this rice species possesses tolerance to submergence. Increasing the capacity of an antioxidative system, the tolerance to submergence stress was improved in rice. Further studies showed that mutant MR219-4 increased efficiency in growth; as well, better adaptation and higher yield performance were reported (Ibrahim, Harun, and Ahmad 2013).

9.3. Methylated Arsenic Rice Species

Inorganic arsenic and dimethylarsinic acid (DMA) are predominantly responsible for rice speciation. In Asia, rice showed stronger relationships with a slope of 0.78 between inorganic arsenic and overall arsenic concentration. An erratic relationship with DMA was observed in United States rice, but European rice species showed a variable relationship. In grain arsenic speciation, a significant genetic variation is observed. Environmental factors possibly contribute towards these variations. Other evidence reveals that methylated arsenic rice species are soil-derived, while rice plants lack the ability of arsenic methylation. DMA also acts as a weak carcinogen, but as a result of rice ingestion, the value of its consumption is much lower as compared to the concerned value (Zhao, Zhu, and Meharg 2013). Phytotoxicity can occur owing to arsenic (As) contamination in a paddy environment and elevated as accumulation in rice (*Oryzasativa*). It was found that the OsCLT1 contribute towards glutathione homeostasis, eventually curbing the toxicity levels in rice (Yang et al. 2016).

9.4. Salinity Stress in Rice

Salinity stress is a major threat to plants. The potential of plants to withstand salt was studied by investigating various features, coupled with salinity tolerance in the salt-tolerant rice mutant (Till-II-877). Discrepant responses of plants subjected to salt stress, were due to variations in different metabolic pathways governing the metabolism of amino acid, carbohydrate, lipid and energy metabolism of photosynthetic organisms (Yu et al. 2016).

Moreover, it was reported that K1 Lrp protein might have developed its role as a virulence factor in rice because of a frameshift mutation (Kondo et al. 2017). Rice is a vital crop to humans. Agronomical trait improvement lies at the core of most crop improvement attempts. Recently, the functions of more than a thousand genes were scrutinized. 1,022,836 known genes were assessed using present FST lines, representing FST genes as potential candidates for functional genomic studies (Chandran et al. 2016).

10. The Melon

The melon (*Cucumismelo*) is a salubrious edible fruit. The Spanish initiative with regards to whole genome sequencing is remarkable. The various available modern tools have made it possible to identify important genes, including those involved in introducing plant resistivity and enhancing the fruit quality (Gonzalez-Ibeas et al. 2007).

Ripening and consequent tissue senescence is a unique developmental process found in fleshy fruits. Fruit plants undergo noticeable alterations at various physiological, biochemical and structural levels that ultimately serve the purpose of seed dispersal. Additionally, over-ripening and senescence, marked by softening of tissues and consequent degradation, makes it an opportunity for opportunistic pathogens. Though the terms ripening and senescence are used interchangeably, ripening precedes senescence in fleshy fruits (Gapper, McQuinn, and Giovannoni 2013).

10.1. Mutations via Antimitotic Agents

The transference of gene of interest into the plants from wild relatives is limited by the interspecific barrier of hybridisation. Antimitotic agents in breeding programmes are meant for crop improvement. Seeds treated with oryzalin and colchicine showed that such plants, when planted directly in field spots, have significantly longer leaves with increased areas, longer internodes, early flowering and a higher percentage of fruit setting; nevertheless, treated seeds took an extended time to germinate. Problems, such as lower survival rate, are also observed. This can be attributed to colchicine treatment which either results in some type of physiological disturbance or chromosomal anomaly. Nevertheless, colchicine treatment when optimized showed better performance and improved traits compared with untreated seeds. Such improved lines can be backcrossed with traditional varieties. The product of backcross with a useful trait can then make way in conventional breeding for crop improvement (Udensi et al. 2011).

10.2. EcoTILLING leading to transcription factor improvement

TILLING is a widely used reverse genetic approach, and it identifies nuclease cleavage of mismatches of DNA. It is a process for detection of induced polymorphism. EcoTILLING is a relatively similar approach that is involved in the identification of natural polymorphisms in target genes of any crop or fruit crop. EcoTILLING has made it possible to identify allelic variants within the natural population. EcoTILLING also played an important role in identifying new allelic variants of elf4E. Functional analyses after EcoTILLING suggested that new elf4E allele is responsible for resistance to MNSV (Nieto et al. 2007).

10.3. EMS mediated mutation

TILLING uses EMS mutagenesis along with detection of specific genes for single nucleotide mutations (Minoia et al. 2010). A new melon population was derived by introducing mutations in it with the help of ethyl methanosulfonate (EMS). Diversity at different stages of fruit development was observed. Nevertheless, three genes responsible for disease resistance and overall fruit quality were screened (González et al. 2011).

11. Wheat

Wheat is an important ingredient of diet worldwide. Its ability to withstand diverse environments and its boundless uses has surprisingly given the diversity bottlenecks expected from the recent domestications and selection of polyploid plants (Dubcovsky and Dvorak 2007). History shows that "Creso", a durum wheat cultivar, was developed in Italy, in 1974, by crossing "Cappelli" mutant "Cp B144" and, in a short time period, it became popular among the Italian masses (Ahloowalia, Maluszynski, and Nichterlein 2004). A study revealed that this cultivar contributed immensely both in terms of land and currency. Furthermore, five other cultivars were developed using Creso and were cultivated extensively (Scarascia-Mugnozza et al. 1993).

11.1. Radiometric approaches

Studies showed that genes closer to the centromere were more prone to mutagenic treatment than others.

11.1.1. Physical mutagens

Studies showed that among physical mutagens, thermal neutrons had the maximum effect, followed by 35S, 32P and X-rays (Bhatia and Swaminathan 1963). The sensitivity to mutagenic treatment of *Triticummonococcum* was found to be greater when compared with polyploids. *T. dicoccum* (tetraploid) had greater resistivity towards fast X-rays and thermal neutrons, compared with *T. aestivum* (hexaploid) (Bhaskaran and Swaminathan 1960).

11.1.2. Chemical mutagens

Chemical mutagens are very different to physical agents. Among all the chemical mutagens, alkylating agents, especially EMS, were found to be the most potent. A linear behaviour of mutagenic response was observed with the dose. It was also observed that the tolerance level of polyploidy is greater than diploids.

11.1.2a Resistance in Wheat

Locus *Lr34/Yr18* is known to contribute towards resistance against stripe and leaf rust in wheat. Additionally, the locus contributes to conferring resistance to powdery mildew and a leaf tip necrosis phenotype. Several mutant lines were generated which carry *Lr34/Yr18* with single chromosome substitution using gamma irradiation in order to bestow the transgenic plant with resistance against upcoming pathogenic attack (Spielmeyer et al. 2008).

11.1.2b Low-energy ion beam effect

Studies revealed that the low energy nitrogen ion beam can potentially induce chromosomal aberrations in wheat. Different varieties responded differently to ion implantation, and a marked effect on both germination and survival rate of the exposed seed was observed with higher doses of ion implantation. Cytological analysis showed that ion beams effectively caused chromosomal aberrations. A direct relationship was observed between the frequencies of meiotic and mitotic cells with chromosome aberrations and increasing doses. Morphological changes and mutants were also observed in exposed seeds derived plants (Wu and Yu 2001).

11.1.2c Synthesis of Heat Shock Protein

Heat shock proteins give plants thermal tolerance. A correlation has been found between the synthesis of specific low molecular weight heat shock proteins and the extent of their thermal tolerance, as these allow subsequent exposure to increased temperatures (Krishnan, Nguyen, and Burke 1989).

11.1.2d Rapid Genome Evolution Induced via Allopolyploidy

A study was conducted into the elimination rate and time of eight sequences of DNA in F1hybrids and newly-formed allopolyploids of *Aegilops* and *Triticum*. A total of 35 interspecific and intergeneric F1hybrids and 22 derived allopolyploids, along with their parental parts, were used. It was observed that in all diploid species of Triticeae, the sequence under consideration occurs only in one homologous pair, i.e. sequences specific to chromosomes CSS, or in that very genome as quite a few pairs, i.e. sequences specific to genomes GSSs in the polyploidy wheat. In newly-synthesized allopolyploids, frequent removal of CSSs and

GSSs is generally prominent. In the F1 plants, CSSs removal was started and accomplished in 2nd and 3rd generation allopolyploids; nevertheless, the removal was a reproducible and non-random event whose direction was detected by the genetically combined hybrid or the allopolyploid. This removal was noticed in a substantial portion of the genome and apparently non-coding regions. This aided in augmenting the differentiation of polyploidy level homologous chromosomes, thereby uncovering the meiotic behaviour as in diploids of newly-formed allopolyploids. Studies further revealed that the genome adjustment at this rapid pace may have contributed to newly-formed allopolyploids as new wheat species, as successful establishments (Ozkan, Levy, and Feldman 2001). Nevertheless, in a recent study on 2,735 mutant lines of tetraploid and hexaploid wheat, more than 10 million mutations were sequenced and catalogued, typically in regions coding proteins. Noticeable mutations in both tetraploid and hexaploid were recorded (Krasileva et al. 2017).

Furthermore, it was reported that the course of accumulation of novel dominant alleles at all VRN1 loci was expressively augmented in cultivated wheat species, particularly in common hexaploid wheat *T. aestivum*, owing to artificial selection of spring forms revised for diverse climatic circumstances, and comprising numerous combinations of VRN1 homoeoalleles (Liang et al. 2017). On the other hand, an efficient genome editing method for bread wheat using CRISPR/Cas9 ribonucleoproteins (RNPs) was described. Deep sequencing disclosed that the off-target mutation's chance in wheat cells is considerably reduced in RNP mediated genome editing as compared to editing with CRISPR/Cas9 DNA. Consistent with this finding, in plants endowed with mutations no off-track mutations were noticed. Since no far-off, distant DNA was employed in CRISPR/Cas9 RNP mediated genome editing, the mutants obtained were Transgene-free. This technique can thus be employed for creating genome-edited crop plants (Liang et al. 2017). Additionally, it was found that the SS mutants demonstrated a significant genetic tool for research on the wheat spikelet's development, and for the identification of genes that regulate meristem events (Dobrovolskaya et al. 2017).

11.1.2e Potage

POPSEQ Ordered *Triticumaestivum* Gene Expression (POTAGE) is an online tool that speeds up the due course of action necessary to isolate and recognise potential genes intended for quantitative trait loci (QTL) in hexaploidwheat. The potential genes isolated and recognized include TaMKK3, utilising POTAGE. The said genes, being isolated and recognised were reported as underlying genes responsible for inducing seed dormancy in wheat, and reduction in pre-harvest sprouting was reported via mutation in its barley ortholog (Suchecki, Watson-Haigh, and Baumann 2017).

12. Modern Tools for Crop Improvement

Recently, some advanced tools have also been employed for crop improvement.

a) Next Generation Sequencing (NGS): NGS has been recognised for studies associated with resistance, pesticide use, and associated problems. Furthermore, NGS-based studies for analysing increased numbers of mutation have better prospects compared with molecular techniques (Délye et al. 2015).

b) Targeting Induced Local Lesions in Genomes (TILLING): TILLING, based on reverse genetics, is the method of choice to identify unidentified single, nucleotide polymorphisms (SNP) in genes of interest using enzymatic digestion. This method not only allows the high-throughput detection of allelic series of mutants, but also helps to understand the revised functions of a particular gene (Wang and Shi 2015).

c) Oligodeoxynucleotide inhibition (ODN): ODN is a high-throughput and selective screening method. ODN has been devised for screening transformed cells and other purposes (Xie et al. 2014).

13. The Economic Value of New Mutant Variety

The economic value of a newly-developed variety can be evaluated using parameters such as crop area occupied by the newly-developed and already-present variety; improvement in both quality and quantity; reduction in the use of pesticides and fungicides; reduced water requirements; positive responses towards photoperiod; nutritional value; high lysine index; improved vitamins content; increased shelf life of oil and reduced toxin levels; increased yield of essential oils; novel and designer crop development; convenient threshing and harvesting; and stepping up exports and stepping down imports. But besides these pronounced improvements, induced mutations also need improvement in their ingrained attributes (Cavanagh et al. 2008).

14. Conclusion

Mutation-induced cultivars have contributed massively towards the economic uplift of not only developing, but also developed, countries (Ahloowalia, Maluszynski, and Nichterlein 2004). Currently, the world is facing an increased demand for food and, rice being a staple of many, requires a variety with yield. Technology-aided identification of such genes is indispensable (Ikeda et al. 2013). Micronutrient deficiency is a global issue and, therefore, development of nutritionally rich staple food is absolutely essential (Bhullar and Gruissem 2013).

Up to 70% increase in food production will be required in the next four decades owing to population explosion. Moreover, population explosion also results in capricious climatic changes, for example, arable land depletion, water resource exhaustion, etc. The only way left is the development of improved crops capable of coping with the impending challenges of a changing world (Mba 2013).

References

Agarwal, P, R Kumar, A Pareek, and A K Sharma. 2017. "Fruit preferential activity of the tomato RIP1 gene promoter in transgenic tomato and Arabidopsis." *Molecular Genetics and Genomics* 292 (1):145–156.

Ahloowalia, BS, M Maluszynski, and K Nichterlein. 2004. "Global impact of mutation-derived varieties." *Euphytica* 135 (2):187–204.

Bhaskaran, S, and MS Swaminathan. 1960. "Polyploidy and radiosensitivity in wheat and barley." *Genetica* 31 (1):449–480.

Bhatia, C, and MS Swaminathan. 1963. "Frequency and spectrum of mutations induced by radiations in some varieties of bread wheat." *Euphytica* 12 (1):97–112.

Bhullar, NK, and W Gruissem. 2013. "Nutritional enhancement of rice for human health: the contribution of biotechnology." *Biotechnology Advances* 31 (1):50–57.

Bino, R J, CH Ric De Vos, M Lieberman, RD Hall, A Bovy, HH Jonker, Y Tikunov, A Lommen, S Moco, and I Levin. 2005. "The light-hyperresponsive high pigment-2dg mutation of tomato: alterations in the fruit metabolome." *New Phytologist* 166 (2):427–438.

Cavanagh, C, M Morell, I Mackay, and W Powell. 2008. "From mutations to MAGIC: resources for gene discovery, validation and delivery in crop plants." *Current Opinion in Plant Biology* 11 (2):215–221.

Chandran, AKN, G-S Lee, Y-H Yoo, U-H Yoon, B-O Ahn, D-W Yun, J-H Kim, H-K Choi, G An, and T-H Kim. 2016. "Functional classification of rice flanking sequence tagged genes using MapMan terms and global understanding on metabolic and regulatory pathways affected by dxr mutant having defects in light response." *Rice* 9 (1):17.

Cheema, AA. 2003. "Radiosensitivity studies in Basmati rice." *Pakistan Journal Botany* 35 (2):197–207.

Cuevas, JM, A Willemsen, J Hillung, MP Zwart, and SF Elena. 2015. "Temporal dynamics of intrahost molecular evolution for a plant RNA virus." *Molecular Biology and Evolution* 32 (5):1132–1147.

Délye, C, R Causse, V Gautier, C Poncet, and S Michel. 2015. "Using next-generation sequencing to detect mutations endowing resistance to pesticides: application to acetolactate-synthase (ALS)-based resistance in barnyard grass, a polyploid grass weed." *Pest Management Science* 71 (5):675–685.

Dobrovolskaya, OB, Y Amagai, KI Popova, AE Dresvyannikova, P Martinek, AA Krasnikov, and N Watanabe. 2017. "Genes WHEAT FRIZZY PANICLE and SHAM RAMIFICATION 2 independently regulate differentiation of floral meristems in wheat." *BMC Plant Biology* 17 (2):252.

Dubcovsky, J, and J Dvorak. 2007. "Genome plasticity a key factor in the success of polyploid wheat under domestication." *Science* 316 (5833): 1862–1866.

Gapper, NE, RP McQuinn, and JJ Giovannoni. 2013. "Molecular and genetic regulation of fruit ripening." *Plant Molecular Biology* 82 (6):575–591.

Gijsbers, L, HDLM van Eekelen, LHJ de Haan, JM Swier, NL Heijink, SK Kloet, H-Y Man, AG Bovy, J Keijer, and JMMJG Aarts. 2013. "Induction of peroxisome proliferator-activated receptor γ (PPARγ)-mediated gene expression by tomato (*Solanum lycopersicum* L.) extracts." *Journal of Agricultural and Food Chemistry* 61 (14):3419–3427.

Gonzalez-Ibeas, D, J Blanca, C Roig, M González-To, B Picó, V Truniger, P Gómez, W Deleu, A Caño-Delgado, and P Arús. 2007. "MELOGEN: an EST database for melon functional genomics." *BMC Genomics* 8 (1):306.

González, M, M Xu, C Esteras, C Roig, AJ Monforte, C Troadec, M Pujol, F Nuez, A Bendahmane, and J Garcia-Mas. 2011. "Towards a TILLING platform for functional genomics in Piel de Sapo melons." *BMC Research Notes* 4 (1):289.

Green, MA, and SC Fry. 2005. "Vitamin C degradation in plant cells via enzymatic hydrolysis of 4-O-oxalyl-L-threonate." *Nature* 433 (7021): 83–87.

Haron, S. 2013. "Strategies and technologies in enhancing sustainable agriculture." FUNCA Mutation Breeding Project Open Seminar, Kajang, Malaysia.

Hayward, A C, R Tollenaere, J Dalton-Morgan, and J Batley. 2015. "Molecular marker applications in plants." In *Plant Genotyping*, 13–27. Springer.

Holá, M, R Vagnerova, and K Angelis. 2015. "Mutagenesis during plant responses to UVB radiation." *Plant Physiology and Biochemistry* 93:29–33. doi: doi.org/10.1016/j.plaphy.2014.12.013.

Horemans, N, CH Foyer, and H Asard. 2000. "Transport and action of ascorbate at the plant plasma membrane." *Trends in Plant Science* 5 (6):263–267.

Ibrahim, RB, AR Harun, and ZB Ahmad. 2013. "Success story of FNCA mutation breeding project in Malaysia." FNCA Mutation Breeding Project Open Seminar, Malaysia.

Ikeda, M, K Miura, K Aya, H Kitano, and M Matsuoka. 2013. "Genes offering the potential for designing yield-related traits in rice." *Current Opinion in Plant Biology* 16 (2):213–220.

Klap, C, E Yeshayahou, AM Bolger, T Arazi, SK Gupta, S Shabtai, B Usadel, Y Salts, and R Barg. 2017. "Tomato facultative parthenocarpy results from SlAGAMOUS-LIKE 6 loss of function." *Plant Biotechnology Journal*15 (5):634–647.

Kondo, M, H Hirai, T Furukawa, Y Yoshida, A Suzuki, T Kawaguchi, and F-SChe. 2017. "Frameshift Mutation Confers Function as Virulence Factor to Leucine-Rich Repeat Protein from Acidovorax avenae." *Frontiers in Plant Science*7:1988.

Krasileva, K V, H A Vasquez-Gross, T Howell, P Bailey, F Paraiso, L Clissold, J Simmonds, RH Ramirez-Gonzalez, X Wang, and P Borrill. 2017. "Uncovering hidden variation in polyploid wheat." *Proceedings of the National Academy of Sciences* 114 (6):E913–E921.

Krishnan, M, H T Nguyen, and JJ Burke. 1989. "Heat shock protein synthesis and thermal tolerance in wheat." *Plant Physiology* 90 (1):140–145.

Liang, Z, K Chen, T Li, Y Zhang, Y Wang, Q Zhao, J Liu, H Zhang, C Liu, and Y Ran. 2017. "Efficient DNA-free genome editing of bread wheat using CRISPR/Cas9 ribonucleoprotein complexes." *Nature Communications* 8:14261.

Machi, S. 2013. "Nuclear techniques for sustainable agriculture." FNCA Mutation Breeding Project Open Seminar, Kajang, Malaysia.

Mahajani, S, and VL Chopra. 1973. "Interaction of formaldehyde and X-rays on mutation production in *Drosophila melanogaster*." *Current Science* 42 (13):473–474.

Mba, C. 2013. "Induced mutations unleash the potentials of plant genetic resources for food and agriculture." *Agronomy* 3 (1):200–231.

Minoia, S, A Petrozza, O D'Onofrio, F Piron, G Mosca, G Sozio, and F Carriero. 2010. "A new mutant genetic resource for tomato crop improvement by TILLING technology." *BMC Research Notes*, 3(1): 69.

Mizumoto, H, Y Morikawa, K Ishibashi, K Kimura, K Matsumoto, M Tokunaga, A Kiba, M Ishikawa, T Okuno, and Y Hikichi. 2014. "Functional characterization of the mutations in Pepper mild mottle virus overcoming tomato tm-1-mediated resistance." *Molecular Plant Pathology*15 (5):479–487.

Muller, H J. 1927. "Artificial transmutation of the gene." *Science* 66 (1699):84–87.

Nakai, H. 2013. "World success and prospect of mutation breeding." FNCA Mutation Breeding Project Open Seminar, Malaysia.

Natarajan, AT, and MS Swaminathan. 1958. "Haploidy induced by radiations in wheat." *Experientia* 14 (9):336–337.

Nekrasov, V, C Wang, J Win, C Lanz, D Weigel, and S Kamoun. 2017. "Rapid generation of a transgene-free powdery mildew resistant tomato by genome deletion." *Scientific Reports* 7 (1):482.

Nieto, C, F Piron, M Dalmais, CF Marco, E Moriones, ML Gómez-Guillamón, V Truniger, P Gómez, J Garcia-Mas, and MA Aranda. 2007. "EcoTILLING for the identification of allelic variants of melon eIF4E, a factor that controls virus susceptibility." *BMC Plant Biology* 7 (1):34.

Noctor, G, and CH Foyer. 1998. "Ascorbate and glutathione: keeping active oxygen under control." *Annual Review of Plant Biology* 49 (1):249–279.

Novak, FJ, and H Brunner. 1992. "Plant breeding: Induced mutation technology for crop improvement." *IAEA Bulletin* 4:25–33.

Ozkan, H, AA Levy, and M Feldman. 2001. "Allopolyploidy-induced rapid genome evolution in the wheat (*Aegilops–Triticum*) group." *The Plant Cell* 13 (8):1735–1747.

Roditakis, E, KMavridis, M Riga, E Vasakis, E Morou, JL Rison, and J Vontas. 2017. "Identification and detection of indoxacarb resistance mutations in the para sodium channel of the tomato leafminer, *Tuta absoluta*." *Pest Management Science* 73 (8):1679–1688.

Saliba-Colombani, V, M Causse, D Langlois, J Philouze, and M Buret. 2001. "Genetic analysis of organoleptic quality in fresh market tomato. 1. Mapping QTLs for physical and chemical traits." *Theoretical and Applied Genetics* 102 (2–3):259–272.

Sang, T, and S Ge. 2013. "Understanding rice domestication and implications for cultivar improvement." *Current Opinion in Plant Biology* 16 (2):139–146.

Scarascia-Mugnozza, GT, FD Amato, S Avanzi, D Bagnara, ML Belli, Bozzini A, and Martini G. 1993. "Mutation breeding for durum wheat improvment in Italy." *Mutation Breeding Review*:1–28.

Schauer, N, Y Semel, U Roessner, A Gur, I Balbo, F Carrari, T Pleban, A Perez-Melis, C Bruedigam, and J Kopka. 2006. "Comprehensive metabolic profiling and phenotyping of interspecific introgression lines for tomato improvement." *Nature Biotechnology* 24 (4):447–454.

Singh, VP, EA Siddiq, and MS Swaminathan. 1979. "Mode of inheritance of dwarf stature and allelic relationships among various spontaneous and induced dwarfs of cultivated rice *Oryza sativa* L." *Theoretical and Applied Genetics* 55 (3–4):169–176.

Smirnoff, N, and GL Wheeler. 2000. "Ascorbic acid in plants: biosynthesis and function." *Critical Reviews in Biochemistry and Molecular Biology* 35 (4):291–314.

Soyk, S, NA Müller, SJ Park, I Schmalenbach, K Jiang, R Hayama, L Zhang, J van Eck, J M Jiménez-Gómez, and ZB Lippman. 2017. "Variation in the flowering gene SELF PRUNING 5G promotes day-neutrality and early yield in tomato." *Nature Genetics* 49 (1):162.

Spielmeyer, W, RP Singh, H McFadden, CR Wellings, J Huerta-Espino, X Kong, R Appels, and ES Lagudah. 2008. "Fine scale genetic and physical mapping using interstitial deletion mutants of Lr34/Yr18: a disease resistance locus effective against multiple pathogens in wheat." *Theoretical and Applied Genetics* 116 (4):481–490.

Stadler, LJ. 1928. "Mutations in barley induced by X-rays and radium." *Science* 68 (1756):186–187.

Stevens, R, M Buret, P Duffé, C Garchery, P Baldet, C Rothan, and M Causse. 2007. "Candidate genes and quantitative trait loci affecting fruit ascorbic acid content in three tomato populations." *Plant Physiology* 143 (4):1943–1953.

Suchecki, R, NS Watson-Haigh, and U Baumann. 2017. "POTAGE: A visualisation tool for speeding up gene discovery in wheat." *Scientific Reports* 7 (1):14315.

Swaminathan, MS, and AT Natarajan. 1959. "Effect of ultraviolet pretreatment on yield of mutations by X-rays in wheat." *Science* 130 (3386):1407–1409.

Tanaka, A. 2011. "Ion-Beam Breeding of Plants charged Particle and Photon Interactions with Matter." In *Applications to Biotechnology*, 943–957. New York: CRC Press.

Tanaka, A. 2013. "Advance techniques for mutation breeding by using ion beams." FNCA Mutation Breeding Project Open Seminar, Malaysia.

Udensi, O, VO Ntui, PA Otu, and PN Chukwurah. 2011. "Sensitivity of some quantitative and yield characters of Egusi melon (*Colocynthis citrullus* L.) to treatment with microtubule inhibitors." *African Journal of Plant Science* 5 (13):759–766.

Ueta, R, C Abe, T Watanabe, SS Sugano, R Ishihara, H Ezura, Y Osakabe, and K Osakabe. 2017. "Rapid breeding of parthenocarpic tomato plants using CRISPR/Cas9." *Scientific Reports* 7 (1):507.

Wang, N, and L Shi. 2015. "Screening of mutations by TILLING in plants." In *Plant Genotyping*, 193–203. Springer.

Wu, L, and Z Yu. 2001. "Radiobiological effects of a low-energy ion beam on wheat." *Radiation and Environmental Biophysics* 40 (1):53–57.

Xie, Z, JF Sundström, Y Jin, C Liu, C Jansson, and C Sun. 2014. "A selection strategy in plant transformation based on antisense oligodeoxynucleotide inhibition." *The Plant Journal* 77 (6):954–961.

Yang, J, M-X Gao, H Hu, X-M Ding, H-W Lin, L Wang, J-M Xu, C-Z Mao, F-J Zhao, and Z-C Wu. 2016. "OsCLT1, a CRT-like transporter 1, is required for glutathione homeostasis and arsenic tolerance in rice." *New Phytologist* 211 (2):658–670.

Yu, J, Z Meng, W Liang, S Behera, JKudla, MR Tucker, Z Luo, M Chen, D Xu, and G Zhao. 2016. "A rice $Ca2^{+}$ binding protein is required for tapetum function and pollen formation." *Plant Physiology* 172 (3):1772–1786.

Zhao, F-J, Y-G Zhu, and A A Meharg. 2013. "Methylated arsenic species in rice: geographical variation, origin, and uptake mechanisms." *Environmental Science and Technology* 47 (9):3957–3966.

CHAPTER 6

ROLE OF ARBUSCULAR MYCORRHIZAL FUNGI AND BIOCHAR IN REMEDIATION OF HEAVY METALS

FAZLI WAHID,[1] AMJAD ALI,[2] MUNIR OZTURK,[3] MUHAMMAD SHARIF,[4] M. AZIM KHAN,[5] BENGU TURKYILMAZ UNAL[6]

[1]Department of Agriculture, The University of Swabi-23561, KPK Pakistan
[2]College of Natural Resources & Environment, Northwest Agriculture & Forestry University, Yangling,
Shaanxi, P.R. China
3 Centre for Environmental Studies & Botany Department, Ege University, Izmir, Turkey
[4]Department of Soil and Environmental Sciences, The University of Agriculture, Peshawar, Pakistan.
[5]Department of Weed Science, The University of Agriculture, Peshawar, KPK 25130, Pakistan
[6] Nigde Omer Halisdemir University, Faculty of Science and Arts, Dept. of Biotechnology, Nigde-Turkey
fazliwahid@aup.edu.pk

Abstract

Both natural and anthropogenic activities have increased heavy metal (HM) contamination in the soil. HM has negative effects on the environment, agricultural productivity and human health through the soil-crop-food chain. The remediation of HM is needed to ensure sustainable crop growth, and maintain the natural ecosystem on unpolluted soils. Many methods for the remediation of polluted soils have been developed, but they are expensive and laborious. Moreover, secondary pollution also affects the soil microbial community. The exploitation of biological and organic amendments, like arbuscular mycorrhiza fungi (AMF) and biochar (BC), are cost-effective and environmentally-friendly approaches to plant growth and phytoremediation in polluted soils. The individual role of AMF and/or BC for the remediation of contaminated soils has been reported in the literature, but studies regarding their combined effects are still limited, and there remains a large knowledge gap regarding their connection. This chapter highlights the potential benefits, tolerance and implementation of AMF and BC interactive use in the remediation of polluted soils and plant nutrition in sustainable agriculture.

Keywords: biochar, heavy metals, mycorrhiza, nutrition, organic agriculture, phytoremediation.

Introduction

Heavy metals (HMs) are continuously released into natural ecosystems due to industrial installations, agricultural practices, weathering of rocks, mining activities and other technical innovations (Sheikh et al. 1976, Ozturk 1989, Aksoy and Ozturk 1996, 1997, Ozturk, Mermut, and Celik, 2011a, Paz-Ferreiro et al. 2013, Ali et al. 2017). High concentrations of HMs (zinc, lead, cadmium, copper, chromium, mercury and nickel) cause toxicity in soil, water and plants. They pose a serious threat, not only to human health (Mimi and Mcdonald 2013, Yang et al. 2012, Ashraf et al. 2015), but also to the taxonomic and functional diversity of soil (Vacca et al. 2012, Aziz et al. 2016). The toxic forms of HMs react with the biomolecules within the human body to form hardly dissociated stable bio-toxic compounds (Duruibe, Ogwuegbu, and Egwurugwu, 2007). Compared with the by-products of petroleum and pesticides, the HMs are more persistent in the environment, and can become mobile by leaching to the aquifer and thus bioavailable to living organisms (Santona, Castaldi, and Melis, 2006, Ashraf, Ozturk, and Ahmad, 2010a). The toxic metals in polluted soil can accumulate in plants in different quantities, causing severe health hazards to humans and animals through the soil-crop-food chain (Ozturk et al. 2011b, Morgan 2013, Hakeem et al. 2015).

The soils polluted by HMs can be remediated by adopting physical methods (soil washing, thermal desorption, electrokinetic remediation, incineration and stabilisation); chemical methods (mobilisation, oxidation and chelation); and/or biological methods (AMF, bacteria, algae) (Ashraf, Ozturk, and Ahmad, 2010b, Ozturk, Sakcali, and Celik, 2013a, Ozturk et al. 2015, Ozturk, Altay, and Karahan, 2017, Sabir et al. 2015, Ali et al. 2017).

However, some of these methods are expensive and time-consuming. Therefore, implementing cost-effective and environmentally-friendly methods and techniques is essential for the effective cleanup, removal, immobilisation and stabilisation of toxic metals from polluted soils (Henry 2000, Ozturk et al. 2013b, Hakeem et al. 2011, 2017). Currently, scientists are trying to remediate polluted soils by using cost-effective methods to enhance efficiency of microbes, increase soil fertility and plant growth. Therefore, the use of soil microbes, like arbuscular mycorrhizal fungi (AMF) (Qiao et al. 2015, Warnock et al. 2007, Erdinçet al. 2017, Bernaola et al. 2018) and soil amendments, like biochar (BC) is receiving more attention (Gadkar et al. 2001, Kim et al. 2015, Houben, Evrard, and Sonnet 2013, Hmid et al. 2015). The AMF form an association with the roots of terrestrial plants, and this phenomenon is widely recognised in plant growth promotion studies on contaminated soils. It has been found that AMF play an important role in the tolerance and accumulation of HMs in polluted soils (Karimi et al. 2011). BC is a carbon-rich material produced by pyrolysis of organic material (Lehmann et al. 2011). It is useful for the immobilisation of HMs due to its higher surface area, pH and CEC (Abdelhafez, Li, and Abbas 2014, Ahmad et al. 2014). It is also one of the emerging technologies for increasing soil fertility and reducing the bioavailability, accumulation and toxicity of pollutants in soil and plants (Fellet, Marmiroli, and Marchiol 2014). The combined use of BC with AMF improves the soil quality and mycorrhizal abundance in the plant rhizosphere (Warnock et al. 2007, Qiao et al. 2015). It has been proposed that the inoculation of AMF with BC alters the bioaccumulation of toxic metals in the soil (Hu et al. 2014). Several researchers have demonstrated the individual role of AMF and BC in polluted soils, but there is a lack of knowledge regarding the combined effect of AMF and BC on remediation of HMs from polluted soils. In this chapter, we will highlight the potential benefits, tolerance and implementation of AMF and BC interactive use for remediation of polluted soils and plant nutrition in sustainable agriculture.

Sources of HMs in the Environment

Anthropogenic activities such as industries, mines and residential areas discharge their waste and contaminate the environment, which ultimately degrades surrounding ecosystems. Moreover, different anthropogenic activities are also responsible for the disposal of organic/inorganic pollutants, which in turn have hazardous effects on living systems. Different natural and anthropogenic sources of HMs in the environment are presented in Fig. 1. The global discharge of HMs from geogenic and anthropogenic sources has increased the concentration of HMs in the soil system. Moreover, the industrial revolution has increased their concentrations in soil and water ecosystems (Gray, Mclaren, and Roberts 2003).

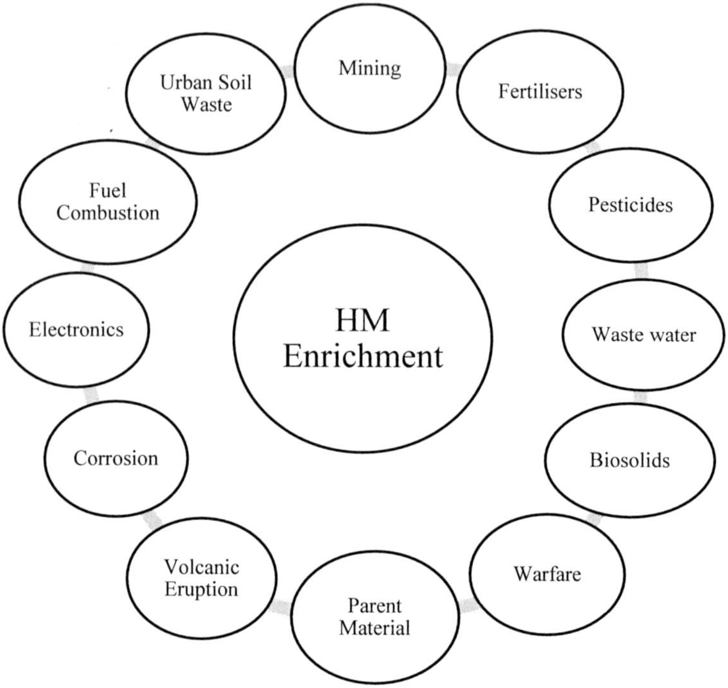

Figure 1. Natural and anthropogenic sources of HMs in the environment (Original).

Toxicity of Heavy Metals

The elements having higher density (5 g cm^{-3}) and atomic number (>20) are considered to be HMs. The most common of these found in soil are: Zn, Pb, Cd, Cu, Cr, Hg, Ni and Co (Jaishankar et al. 2014). Some (Cu, Fe, Mn, V, Zn and Mo) are vital for organism functions and are accepted as micronutrients. The hazardous effects of HMs on human health and plant growth have been studied at length (Abdelhafez, Li, and Abbas 2014, Karami et al. 2011, Aleksandra and Urszula 2008, Ainza, Trevors, and Saier 2010). The study of the toxic effects of HMs on flora and fauna, water and air quality, and the health of humans and animals have been key topics for environmental researchers. HMs accumulation is toxic to soil, plant and human health through the soil-crop-food chain (Cheng et al. 2016, Hua et al. 2012). Some of these are carcinogenic and mutagenic in nature, and consequently cause diseases in humans and animals at high levels of exposure (Jaishankar et al. 2014). A list of some HMs is given in Table 1.

Table 1. List of selected heavy metals with higher densities and molecular weights

Potentially Toxic Elements	Density (g cm^{-3})	Atomic number	Molecular weight (g mol^{-1})
Zinc (Zn)	7.14	30	65.39
Lead (Pb)	11.34	82	207.2
Cadmium (Cd)	8.65	48	112.41
Copper (Cu)	8.96	29	63.55
Arsenic (As)	5.72	33	74.92
Mercury (Hg)	13.53	80	200.59
Cobalt (Co)	8.90	27	58.93
Nickle (Ni)	8.90	28	58.69
Chromium (Cr)	7.19	24	51.99
Aluminium (Al)	2.70	13	26.98
Manganese (Mn)	7.21	25	54.93

Effects of Heavy Metals on Plant Growth

The optimum level of HMs in soil affects plants, both directly and indirectly. They are responsible for direct oxidative stress, and disturb the metabolic activities, cytoplasmic enzymes and functions of cell structure (Jadia and Fulekar 2009). At the same time, HMs retard plant growth and decrease biosynthesis of chlorophyll, reduce respiration, limit antioxidant activities (POD, PPO, PAL, CAT and SOD), and even cause lipid peroxidation (MDA) in plants inhabiting polluted soils (Xian, Wang, and Chen 2015, Wang et al. 2014). HMs lower the soil's enzymatic activities; reduce the formation of soil organic matter, decomposition (mineralisation), humus production and the transformation of essential nutrients (Ma et al. 2015, Cao et al. 2016). They also degrade air and water quality and soil functional diversity, as well as ecosystem services in close proximity to the source point (Schneider, Bundschuh, and Araujo do Nascimento 2016). Hg, Se, Cd, As and Pb make no biochemical contribution to plant physiology, but Cu, Mn, Fe, Mo, Ni and Zn are involved in normal growth, metabolism and many biological functions of plants as essential elements. However, at higher concentrations these too can cause toxicity and interfere with important plant functions (Rascio and Navariizzo 2011). The release of metal-enriched solid and liquid wastes into the soil, water and air deteriorate the physical, chemical and biological properties of ecosystems and ultimately affect the vegetation and other agricultural activities (Ginocchio et al. 2004, Navarro et al. 2008).

Several researchers have worked on the mechanisms of HMs uptake by plants. Among these mechanisms, some are very efficient and highly specific to acquiring the low levels of essential nutrients from the environment. The plant roots, some chelating agents, redox reaction, and some induced pH changes enhance the solubility and uptake of low level and insoluble precipitated micronutrients. Furthermore, plants also use the same mechanisms for storage, translocation and uptake of toxic metals. Sinha, Herat, and Tandon (2007) have demonstrated that some plants (accumulators) adopt various mechanisms on contaminated sites for their survival by accumulating these in their tissues; other plants (excluders) restrict these in their biomass to avoid their toxicity. Similarly, HMs influence the activities of soil microorganisms and replace essential nutrients at exchange sites, thereby affecting plant growth and productivity. They also reduce soil-beneficial microbes, decrease soil organic matter decomposition and hamper the enzymatic activities, leading to plant death (Violante et al. 2010).

Zinc (Zn) is an important micronutrient, which is mostly involved in the metabolic processes of plants. The high concentration of Zn in soil is responsible for stunted growth, plant senescence, leaf chlorosis (reduced chlorophyll a and b), and production of ROS and limited shoot and root growth (Fontes and Cox 1998). Lead (Pb) exerts its adverse effects on plant growth, morphology, as well as photosynthetic processes. Moreover, the hazardous effects of Pb cause necrosis, chlorosis, stunted root-shoot growth and biomass reduction (Malar et al. 2014). Similarly, a high concentration of Cd in the soil causes growth inhibition, the browning of root tips, chlorosis and, finally, plant death. Generally, Cd affects the uptake and transport of nutrients and water by plants (Das, Samantaray, and Rout 1997), interferes with the absorption and transport of nitrate from roots to shoots, and prevents nitrate reductase activity (Hernandez, Carpenalnuiz, and Gárate 1996). Moreover, Cd also reduces NH_3 assimilation and nodule formation (Balestrasse et al. 2003). Cu has various adverse effects on growth, metabolism, assimilation and nutrient transport in plants, leading to leaf chlorosis, variations in the photosynthesis, disturbance in respiratory processes, inactivation of enzymatic activities, disruption of protein structures, and growth retardation (Karimi et al. 2011, Mackie et al. 2015). Cr is also hazardous, and its toxicity reduces the plant's chlorophyll content and growth, leads to oxidative damage and, finally, affects enzymatic activities (Singh et al. 2013). The increased concentration of Nickel in soil is also harmful to plant growth, although a low concentration of it is considered better for plant growth (Singh et al. 2012).

Remediation Techniques for HMs Eradication in Soil

The remediation techniques include physical, chemical and biological methods applied to remediating the HMs in soil, and reducing the risks to plant and human health. These techniques are presented in Fig. 2. Within the physical and biological techniques, biochar and AMF, for the remediation of HMs, are considered very effective.

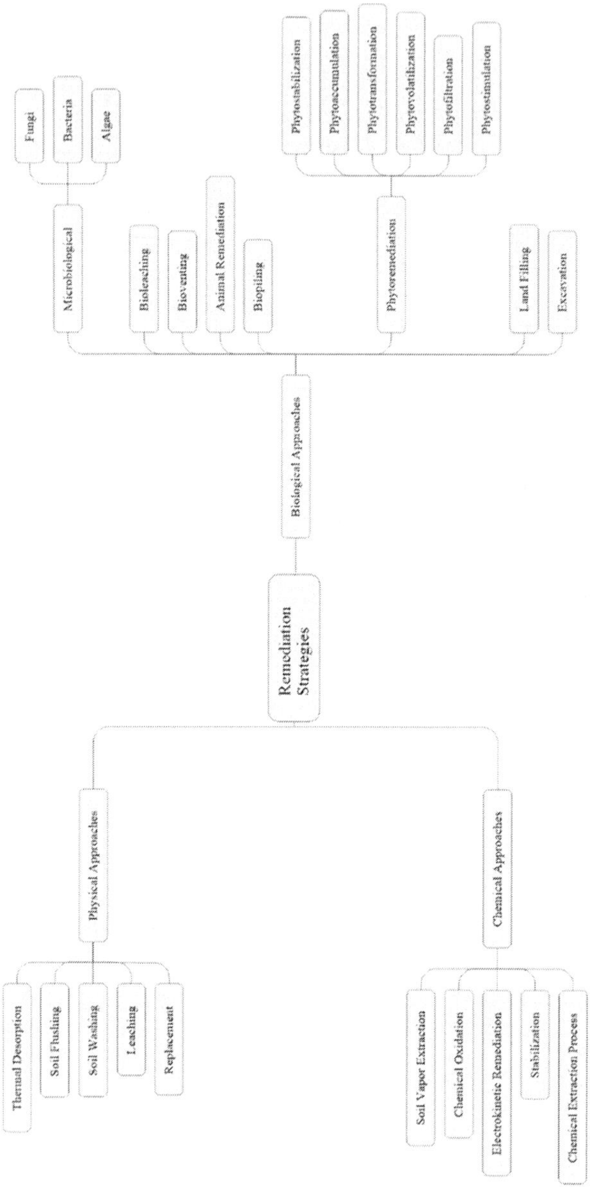

Figure 2. Summary of the remediation techniques for HMs eradication in soil.

Significance of AMF in HMs Polluted Soil

Arbuscular mycorrhizal fungi (AMF) belong to the phylum *Glomeromycota*, which form a widespread symbiotic association with approximately 80 per cent of the terrestrial plants in the ecosystem (Schussler et al. 2001). AMF form direct symbiotic associations and physical links between roots of different higher plants and soil. They can be used in several ways, including as biofertilisers, bioprotectants and biodegraders. Furthermore, AMF improve the health and growth of plants in contaminated soils by increasing mineral nutrition and providing tolerance against HMs stress (Hildebrandt, Regvar, and Bothe 2007, Turnau and Haselwandter 2002). The AMF transfer nutrients (N, P) to their host by extending their hyphae outside the root zone and exchanging photosynthetically-fixed C to the host plant (Wahid et al. 2016). The AMF symbiosis with other plants has been reported in almost all habitats and under different climatic conditions, including fertile soil, soils affected by mining and other disturbed soils (Enkhtuya, Rydlova, and Vosatka 2000).

AMF colonisation in plants grown on mine-polluted soils depict its symbiosis and HMs tolerance. AMF absorb HMs from the contaminated soil through their hyphae, and channel them to the host plant to facilitate the phytoremediation. They behave differently, sometimes facilitating higher metal uptake in shoot and root (phytoextraction), and sometimes immobilising and stabilising the metals in the soil (phytostabilization). AMF also hinder soil erosion by forming stable aggregates to reduce the transfer of HMs to the plant shoot, which also increases the re-vegetation of sites (waste, coal mine and degraded) containing high levels of HMs (Lakshman and Channabasava 2013).

Mechanism of AMF in Metal Contaminated Soils

The presence of AMF has been reported in human-disturbed ecosystems, e.g. soils contaminated with HMs. There are certain "classic" ecological principles demonstrating the phenomenon of enhancing resistance or tolerance of a community (Boivin et al. 2002). Resistance is the ability of microbes to survive the adverse effects of pollutants, whilst tolerance is the capability of microorganisms to adapt themselves to the prevailing conditions of persistent pollutants. Published scientific reports have confirmed the resistance of AMF to HMs (Hildebrandt, Regvar, and Bothe 2007). The positive effect of mycorrhizal symbiosis on the remediation of soils contaminated with HMs is important for future evaluation of lands, because mycorrhizal infection is equally important in extracting metals as hyper-accumulator plants, i.e. non-mycorrhizal. The mycorrhizal-colonised plants also improve the phytostabilization of HMs (Zn, Cu, Cd) by restricting them in the hyphae and roots, and hampering their translocation to other aerial parts of the plant. Consequently, the HMs' bioavailability is reduced and, finally its hazardous effects are minimised (Leyval 2002). In soils, the AMF facilitates the dissolution of metals by proton- or ligand-promoted mechanisms, and the released organic acids cause solubility of metal chelating anions (Finlay 2008).

The presence of AMF in metal-contaminated sites is essential for re-vegetation; however, inoculation in these areas is a useful approach in enhancing mycorrhizal herbaceous species. They play a crucial role in modulating plant HM acquisition in a wide range of soil and metal concentrations, and so are considered to be a key element in the improvement of micronutrient concentrations in crops and the phytoremediation of polluted soils (Ferrol, Elisabeth, and Vargas 2016). The presence of high concentrations of HMs in the soil affects AMF spores and colonisation. Metal-tolerant AMF, isolated from contaminated areas, cope better with metal toxicity than those isolated from uncontaminated soils (Cornejo et al. 2013). Several metal-tolerant AMF have been isolated and identified in contaminated soils, which recommends their adaptation and propagation to the indigenous environment. The metallophyte plants, like *Viola calaminaria*, grow on highly polluted soils and can support the colonisation and association of AM fungi (Tonin et al. 2001). AM fungi, isolated from the roots of this plant, have been applied to the maize plant in polluted soil; the results have shown that ability of metal uptake in shoot and root by isolated fungi was more than the common fungi and in the control (Kaldorf et al. 1999). Furthermore, Val, Barea, and Azcónaguilar (1999) have conducted long-term experiments on soil amended with sewage sludge containing different HMs, and observed that the total number of AMF spores decreased with high concentration of metals, but that their total number never diminished in the soil, which recommended the adaptation of AM fungi to metal-polluted soils.

Several methods are used to evaluate the metal tolerance of both endo- and ectomycorrhizal fungi. These include the number of AMF spores, root colonisation, and abundance of ectomycorrhizal (ECM)-fruiting bodies. However, these methods are not helpful in providing information regarding the threshold levels, limitations, growth, genetics, resistance, tolerance and survival of AMF. The real interaction between metal and AMF in the rhizosphere is obvious from the fact that AMF coexists with other microbes and plant roots that can easily accumulate and tolerate metals in contaminated soils (Karimi et al. 2011). Biró et al. (2005) conducted a twelve-year experiment under controlled conditions to evaluate the stress-buffering effect and colonisation behaviour of AMF in metal-polluted soil. They found that most of the metals decreased AMF spore formation, whilst a few metals enhanced spore formation. AMF interact with metals in the plant rhizosphere for uptake, tolerance and resistance, but the proper mechanism of potential interaction on the cellular and molecular levels is still unknown. It has been reported that metal and plant encoded transporters play a key role in tolerance and uptake of HMs (Hildebrandt, Regvar, and Bothe 2007) from extra- and intra-cellular media (Gaither and Eide 2001).

AMF Assisted Phytoremediation of Heavy Metals

The use of vegetation for *in situ* treatment of contaminated soil is called phytoremediation. Among different microbes in the soil ecosystem, the symbiotic association of AMF has been mainly studied in the phytoremediation of HMs (Warnock et al. 2010, Meier et al. 2015, Qiao et al. 2015). Many researchers have reported that AM fungi can grow and adapt themselves to soil largely polluted with HMs (Shetty et al. 1994), and under such conditions mycorrhiza evolve different ways and mechanisms to cope with the prevailing metals (Joner and Leyval 1997). Among different remediation methods, phytoremediation is a promising technique because, by this method, the contaminants are physically removed or kept in a suitable proportion that does not affect plant growth and development (Paz-Ferreiro et al. 2013). The main advantage of AMF-assisted phytoremediation of HMs is that AMF get the C directly from their host for growth in contaminated soils. The amount of C taken by AMF is also finally available to bacteria present in association with mycorrhiza, and probably due to this, they enhance the bioremediation of metals in polluted soils (Finlay 2008).

The phytoremediation technology is cost-effective and can be successfully launched on large areas polluted by HMs. Inoculation of AM fungi not only enhances the growth but also improves uptake of some metals, like Zn, Pb, Cd and Cu by plant shoots, which signifies the potential role of AMF in phytoextraction of HMs. Root biomass and uptake of most of the elements in sorghum has shown significant increase in mycorrhiza and heavy metal-resistant bacteria treatment groups in comparison with the control (Dhawi, Datta, and Ramakrishna 2016). AMF inoculation of Pb and Cd-contaminated soils has resulted in an increase in growth and production of marigold (Tabrizi et al. 2015) and golden berry (Hristozkova et al. 2017) as compared with control plants. Similarly, inoculation of AMF has also reduced the Ni uptake from root to shoot in fescue plant (Shakeel and Yaseen 2015). AMF has significantly decreased Cd concentrations, both in roots and shoots in rice. This suggests that AMF could convert Cd into inactive forms (Li et al. 2016). It has been hypothesised that Cd is chelated with some specific polar (such as hydroxyl or carboxyl) material (Fu and Wang 2011). AMF decreases the concentration of Uranium (U) and Thorium (Th) in the shoots of plants, indicating the contribution of AMF in phytostabilization of both U and Th in the soil (Roos and Jakobsen 2008). Decreases in Zn uptake, and their concentrations in plant parts were identified in tomato (Watts-Williams et al. 2014) and red clover (Li and Christie 2001). The carotenoids/chlorophyll ratio, a broadly-used marker of plant physiological status, generally increases under various stress conditions due to chlorophyll photodegradation and/or stimulated synthesis of carotenoids with a photoprotective function (Martínez-Peñalver, Reigosa, and Sanchez-Moreiras 2011). Doubková and Sudova (2016) have proved that AMF-inoculated *Agrostis capillaris* plants show higher chlorophyll contents and a lower carotenoids/chlorophyll ratio. *Calendula officinalis* plants inoculated with AMF are reported to have higher total phenol, flavanoid and carotenoid ratios, and increased antioxidant capacity compared with the non-mycorrhizal plants at higher Cd and Pb levels (Hristozkova et al. 2016). The accumulation of HMs in the soil also leads to the disappearance of plant cover, and increases the production of reactive oxygen species (ROS). ROS accumulation enhances expression of genes encoding antioxidant enzymes and redox homeostasis proteins. Fuentes et al. (2016) have demonstrated that inoculation of *Solanum lycopersicum* contributes to the reduction of plant stress by improving defence mechanisms and homeostasis.

There are certain factors which contribute to the uptake of HMs by plants: competition between AMF strains; fungal genotypes; the extent and uptake of metal through symbiosis; root length density; soil properties; metal properties and concentration in the soil; soil conditions (contaminated and non-contaminated soil); interactions of P with existing metals; experimental and plant growth conditions; and plant species and size (Giasson, Karam, and Jaouich 2008). Metal uptake in contaminated soil appears to be plant or AMF-specific as AMF is effective on metal and root association. The inoculation of plants with AMF increases root-length density and probably increases hyphae, thus assisting plants to explore more widely in the soil for metal uptake (Rusan, Pan, and Kennedy 1995, Lasat 2002). Some plants inoculated with AMF under HM stress may result in the expression of specific genes, which are responsible for the production of resistance proteins (metallothioneins, etc.) to stress (Ferrol, Elisabeth, and Vargas 2016). Furthermore, the symbiotic association of AMF with plants suggests that AMF can be used for remediation of metals on contaminated soils.

The Significance of Biochar in Phytoremediation

Biochar (BC) is a porous and carbon-rich biological charcoal produced through the pyrolysis process of organic materials (crop straw, animal litter, wood chip and sludge) within a temperature range of between 300–1000 °C (Verheijen et al. 2010, Ali et al. 2017). BC is used as a soil amendment (Ahmad et al. 2014, Hu et al. 2014). It resists decomposition and is predicted to remain in the soil for a million years. The persistent nature and high stability of BC in soil determines its suitability for use as a C sink to reduce the amount of CO_2 (Lai et al. 2013, Malghani, Gleixner, and Trumbore 2013) and greenhouse gas emissions in the atmosphere (Stewart et al. 2013). The addition of BC to the soil is not only suitable for C sequestration, but can also be used as a soil amendment and conditioner to improve soil fertility (Sohi et al. 2010), increase CEC and pH of acidic soils, as well as to immobilise HMs (Zwieten et al. 2010). Moreover, due to its high cation-adsorption capacity, BC is also used for nutrients and water retention, as well as reduction of N losses by leaching in soils (Knowles et al. 2011, Novak et al. 2010). It restores the soil contaminated with inorganic and organic pollutants, like HMs and pesticides (Uchimiya et al. 2011, Vithanage et al. 2015, Cao et al. 2009).

The addition of BC to the soil alters its quality (Paz-Ferreiro et al. 2013) and physical and chemical properties; has the potential to enhance agricultural production (Abdelhafez, Li, and Abbas 2014, Ahmad et al. 2014); promotes soil biological activities; and improves the soil's microbial biomass (Lehmann et al. 2011, Paz-Ferreiro et al. 2013). Moreover, it acts as an energy source; provides a niche for soil micro-flora and fauna (Lehmann et al. 2011, Wang et al. 2014, Mackie et al. 2015); reduces phytoavailability and mobility of HMs (Beesley et al. 2011, Houben, Evrard, and Sonnet 2013, Ahmad et al. 2014); and provides resistance to many crops against diseases and soil-borne pathogens (Uzoma et al. 2011). Akhter et al. (2015) have reported that application of BC can change mycorrhization, root exudation and growth of *Fusarium oxysporum* ssp. *lycopersici (Fol)* in tomato.

The Phytostabilization of HMs with Biochar Amendments

This is a cost-effective and environmentally-friendly phytoremediation approach. The use of BC for phytostabilization of HMs is an important and modern phytotechnology to remediate and restore polluted soils (Fellet, Marmiroli, and Marchiol 2014, Ahmad et al. 2014). The addition of BC to contaminated soils is important for HMs stability, speciation, degradation and removal by plants of microorganisms. It reduces the phytoavailability of contaminants in polluted soils (Houben, Evrard, and Sonnet 2013, Karami et al. 2011). It can remediate the habitat by reducing the pollutants in soil, as well as in the surrounding environment (Ahmad et al. 2014). As such, its application is recommended for promoting revegetation on HM-polluted soils due to long-term phytostability (Beesley et al. 2011, Fellet, Marmiroli, and Marchiol 2014). An application rate of 10% BC significantly reduces HMs (Zn, Pb, and Cd) and increases the biomass of rapeseed in a phytoremediation test (Houben, Evrard, and Sonnet 2013). Lu et al. (2017) have shown that $CaCl_2$-and DTPA- extractable Cd, Cu, Pb and Zn concentrations are significantly lower in the bamboo and rice straw biochar-treated soils than those in the non-amended soil. Lu et al. (2014) have also found that accumulation of Cd, Cu, Pb and Zn in *Sedum plumbizincicola* is significantly lower in the presence of both bamboo and rice straw biochar than in the control. An increase in the enzymatic activities and biochemical properties of soil, as well as the goal of phytoremediation by *Amaranthus tricolor* has been attained after BC application to Cd contaminated soil (Lu et al. 2015). Similarly, used in combination with *Streptomyces pactum*, BC increases the metal uptake and enzymatic activities (soil and plant) in *Sorghum* grown in mines and industrial polluted soils (Ali et al. 2017). Biochar application in phytoremediation of HMs is an important approach as it reduces the soil pollutants (organic and inorganic) and increases the growth of accumulator plants (Cao et al. 2009, Vithanage et al. 2015, Paz-Ferreiro et al. 2013). HMs availability is not only affected by BC type and feedstock, but also by the plant and soil. Fellet, Marmiroli, and Marchiol (2014) have evaluated the effect of different types of BC for phytostabilization of mining tailings containing HMs (Cd and Pb). They report that BC resulted in significant changes in soil properties, as well as bioavailability of HMs by reducing their uptake in shoot. Chicken manure and green waste BC have been tested for immobilisation and phytotoxicity reduction of HM (Cd, Pb and Cu) polluted soil. BC has proved effective in immobilisation and lowering the phytotoxicity of HMs (Park et al. 2010). The surface area of BC is large and porous, which adsorbs HMs due to the presence of different functional groups i.e. carboxyl, hydroxyl, phenolic and aromatic hydroxyl (Houben, Evrard, and Sonnet 2013, Kim et al. 2015, Uchimiya et al. 2012, Al-Wabel et al. 2014). Sugar cane bagasse and orange peel BC have been tested for stabilisation of HMs (As and Pb) in contaminated soil. The addition of BC significantly increases soil aggregation, water-holding capacity and organic matter content in soil. Sugarcane bagasse BC decreases the solubility of Pb and transforms the labile Pb into less available forms (Abdelhafez, Li, and Abbas 2014).

The Mechanisms Involved in Heavy Metal Remediation by Biochar

Before using BC as a soil amendment for the remediation of polluted soil, the type of HMs in soil, temperature, pyrolysis condition, type of feedstock, residence time and moisture of feedstock all need to be considered (Paz-Ferreiro et al. 2013). It has been evaluated that electrostatic interactions, surface sorption, increasing pH, and formation of phosphate as hydroxide or carbonate precipitates are involved in HMs remediation through BC amendments (Jiang et al. 2012, Lu et al. 2012, Houben, Evrard, and Sonnet 2013). For pesticide degradation, hydrolysis is considered an important mechanism (Sarmah and Sabadie 2002). In the case of acidic-contaminated soils amended with BC, the exchangeable cations are released with HMs during sorption process and thus enhance the HMs stabilisation. BC increases the CEC and reduces the HMs availability and translocation (Al-Wabel et al. 2014). According to Lu et al. (2012), the exchange of HMs with certain cations (Ca^{2+}, Mg^{2+}, Na^{+} and K^{+}) in the presence of sludge-derived BC is the main mechanism for adsorption of HMs in contaminated soils. According to Cao et al. (2009), the phosphate and carbonate minerals in BC precipitate with HMs and contribute to reducing bioavailability and increasing stability. The alkaline nature of BC is responsible for decreasing the concentration of HMs in BC amended soils. The variation in the pH of BC depends on pyrolysis temperature and the presence of ash content (Al-Wabel et al. 2014, Houben, Evrard, and Sonnet 2013). The addition of BC decreases zeta potential and increases cation exchange capacity due to which more negative charges develop on the soil surface, and consequently the electrostatic attraction between positively charged HMs also increases. Jiang et al. (2012) has studied the effects of rice straw BC on the adsorption of surface charge and Pb by using two Ultisol and one Oxisol soil in a 30-day incubation experiment. BC increased CEC and adsorption of Pb, whereas both electrostatic and non-electrostatic mechanisms were involved. The role of the non-electrostatic mechanism was more prominent due to the development of surface complexes on BC.

Soil Microbial Interaction with Biochar

Biochar incorporation in the soil provides a habitat and refuge for different soil microorganisms, such as bacteria, fungi and protozoa (Warnock et al. 2007, Gul et al. 2015). It improves the content of organic carbon, which in turn stimulates microbial activities and alters microbial abundance in the soil environment (Lehmann et al. 2011, Gul et al. 2015).

BC contains macro-, micro- and mesopores. The macropores provide shelter against predators and niche to bacteria, whilst both micro- and mesopores store H_2O and dissolve substances essential for microbial metabolism. The porous and storing (organic carbon) nature of BC facilitates microbiota to feed on organic substances and remain protected from nematodes, protozoa and collembolan attacks (Gul et al. 2015). This leads to high microbial biomass and activities in soil amended with BC (Kolb, Fermanich, and Dornbush 2009, Purakayastha, Kumari, and Pathak 2014). There are also reports that no increase in microbial biomass occurs with the addition of BC prepared from hardwood (Schomberg et al. 2012).

BC increases the diversity of micro- and macro-biota in HM contaminated soils (Beesley et al. 2011). In such polluted soils, BC reduces phytoavailability and increases microbial community structure (Lehmann et al. 2011). A significant decrease in Cd and Pb uptake in rice has been reported after BC incorporation into the polluted soil in a two-year experiment in southern China. Meanwhile, the population of fungi and actinomycetes also significantly increase as compared to the control (Cui et al. 2013).

AMF and Biochar Interaction in HM Contaminated Soil

An application of AMF and BC has been widely studied in soils contaminated by heavy metals. Both have proved to be very effective in the phytoremediation. The combined use of AMF and BC is very limited and there is a lack of knowledge on the connection. The limited published data has reported variable findings on the combined use of AMF and BC in contaminated soils. BC reduces HM uptake in corn, whilst AMF shows no such effect on the bioavailability of HM (Qiao et al. 2015, Warnock et al. 2007, Warnock et al. 2010). The efficiency of AMF to solubilise and translocate HMs can be improved by soil organic amendments, like compost and BC (Akhter et al. 2015, Warnock et al. 2007, Wahid et al. 2016) as shown in Table 2. Incorporation of BC into soil has a substantial effect by serving as a refuge for AMF hyphae, and providing protection against fungal grazers (Warnock et al. 2007). BC not only increases endomycorrhizal colonisation, but also plays a potential role in enhancing ectomycorrhizal colonisation. Some researchers have reported negative effects of BC on root colonisation due to high ash content in BC, which releases high amounts of nutrients to plants (Mukherjee and Zimmerman 2013). As AMF colonisation is more effective under nutrient (P) deficient conditions, addition of BC leads to nutrient saturation and decreases root colonisation (Gryndler et al. 2006). The alteration in mycorrhizal colonisation after the addition of BC might be due to high water and nutrients, especially P availability to plant (Gryndler et al. 2006), or may be due to the direct effects of high content of salts or HMs on mycorrhiza. Another important mechanism which favours AMF mycorrhization, spore formation, and growth of hyphae is the signalling process of BC on mycorrhiza (Paszkowski 2006). Several researchers have reported the effects of BC on AMF. BC addition to the soil reportedly increases the access of AMF to P, and enhances its colonisation in bean plants, as reported from a greenhouse experiment (Vanek and Lehmann 2015). Similarly, an increase in root colonisation of tomato plants after BC amendment has been recorded (Akhter et al. 2015). Hammer et al. (2014) applied monoxenic culture of *Rhizophagus irregularis* to carrot roots and observed the growth of AMF on the inner and outer surfaces of BC. They recorded growth and access of AMF hyphae to microsites within BC and P uptake.

Table 2. Interactive role of AMF and biochar

Biochar Used	AMF Inoculated	Parameters Studied	Expt type	Experiment Duration	Plants Used	Remarks	References
Peanut shell pellets, *Mangifera indica*, lodgepole pine	Native *Plantago lanceolata* host species	AMF hyphal length, Soil P	Pot and field	Pots 30 days and field 7 months	*Plantago lanceolata*	Positive effects on nutrients uptake ant root colonisation	(Warnock et al. 2010)
Maple and hickory sawmill waste	*Glomus clarum*	P uptake, AM colonisation, specific root length (SRL), AM, neutral lipids and microbial biomass P.	Greenhouse study	28–42 days	*Phaseolus vulgaris*	Positive as well as no effects	(Vanek and Lehmann 2015)
Rice straw	*Glomus caledonium*	Growth and Cd accumulation, mycorrhizal colonisation, plant biomass	Greenhouse pot trial	4 months	Kangkong (*Ipomoea aquatic Forsk.*), Alfred Stonecrop (*Sedum alfredii* Hance)	Positive and no effects p	(Hu et al. 2014)

Corn stalks	*Glomus intraradices*	Heavy metals (Cd, Cr, Ni, Cu, Pb, Zn) bioavailability	Growth chamber experiment	6 weeks	Sweet corn, Golden Cross Bantam (Hybrid)	Positive for biochar application but very little or no effect for the combined application of AMF and biochar	(Qiao et al. 2015)
Hardwood dust	*G. monosporum, G. clarum, G. brasilianum, G. intraradices, Gigaspora margarita*	AM fungi colonisation and Fusarium crown and root rot	Greenhouse and field study	Two years	Asparagus	Biochar was found significant in controlling allelopathic effects	(Elmer and Pignatello 2011)
Eucalyptus globolus trees	*G. mosseae*	Ca, B, Cu, Mn, Na or Zn Nutrient uptake, Root colonisation, plant growth and fruit yield	Field study	Whole tomato season	*Solanum lycopersicum*	Combined application was not significant	(Nzanza, Marais, and Soundy 2012)

Chicken manure and oat hull	*Glomus genus, being Glomus aff. intraradices*	Root colonization, spore number, plant biomass weight, respiration, and dehydrogenase activity, AM fungi, and glomalin production	Pots greenhouse	6 months	*Oenotherapicensis*	Biochar decreased Cu uptake and increased microbes habitat	(Meier et al. 2015)
Jarrah sawdust, *Acacia saligna*	Native AMF	Fe, N, P concentration, Root colonisation and root length	Pot glasshouse	6 weeks	*Triticum aestivum* L.	Biochars Containing Magnetic Iron Nanoparticles increased N and P uptake, plant growth and colonisation	(Joseph et al. 2015)

Ash poor wood biochar and ash-rich chicken manure biochar	*Rhizophagus irregularis*	Nutrients uptake	Laboratory study	Three experiments with variable time	*Daucus carota L.*	Both biochar and AMF enhanced the adsorption of nutrients for plant uptake	(Hammer et al. 2014)
Wood feedstock	*G. intraradices* and *G. mosseae*	Plant growth, bio-physiological nutritional responses and root mycorrhizatio n	Pots in greenhouse conditions	3 months	*Pelargonium plants*	Biochar is useful but its proportion in Peat mixture does not exceed 30%	(Conversa et al. 2015)
Acacia saligna, Simcoa jarrah and Wundowi jarrah	*G. intraradices*	Ca, Mg, Zn, Mn, Cu, Fe, B and Mo, N, P, K, S and Zn uptake mycorrhizal colonisation and growth	Glass house and field experiment	8 weeks, 3 months	*Triticum aestivum L. Sorghum bicolour*	Both experiments confirmed positive effects of biochar and AMF on nutrients uptake by wheat and sorghum	(Blackwell et al. 2015)

Cow dung	AMF	Maize growth and phosphorus uptake	Glasshouse study	6 months	*Zea mays* L.	Combination of AMF and biochar proved better effects on maize	(Mau and Utami 2014)
Apple wood (*Malus domestica*) sawdust	*G. intraradices*	Plant fungal symbiosis and nutrients uptake	Greenhouse study	4 weeks	*Sorghum bicolour*	Soil N control the ability of biochar to influence AM fungi symbiosis	(Le Croy et al. 2013)
Spent brewer's grains	Native mycorrhizal species	Plant growth, root symbionts, arbuscular mycorrhiza abundance and root nodulation	Greenhouse study	12 weeks	*Medicago sativa*	Biochar improved soil aggregation and root nodulation in the presence of AMF	(George et al. 2012)
Rice-hull charcoal	Native *Glomus* sp.	AM colonisation, growth and soil physico-chemical status	Pot experiments	6 weeks	*Tagetes patula L.*	Roots colonisation and plant growth was increased	(Ezawa, Yamamoto, and Yoshida 2002)

Bark charcoal	Indigenous arbuscular mycorrhizal species	crop yield and soil chemical properties	Open garden experiments	3 months	*Zea mays*, *Vigna unguiculata* and peanut	Improved nutrients contents, soil properties, colonisation rate and growth,	(Yamato et al. 2006)

AMF and BC in Plant Nutrition

The mycorrhizal fungi have an important place in terms of soil fertility and plant nutrition. The fungi in AMF strengthen the morphological structure of the plant with physiological and microbial changes in the rhizosphere by increasing nutrient uptake, and suppressing fungal root diseases and nematodes by changing chemical compounds in plant tissues. The most important benefit of AMF for host plants is usually an increase in P uptake (Thompson 1987, Smith and Read 2010, Graham and Abbott 2000). AMF, in some cases, provide the use of P efficiency and increases the total uptake (Koide 2000). This can be correlated with increased growth and yield (Koide 2000, Vosátka 1995, Ibijbijen et al. 1996). Glaser et al. (2015) have found that uptake of P, K, Mg and Zn in corn plants is affected by biochar. Lehmann et al. (2003) and Uzoma et al. (2011) have reported significantly higher P uptake in cowpea and corn, respectively. It is thought that AMF may be important for host plants in the uptake of other nutrients, as well as phosphorus. The uptake of Zn, Cu, Fe, N, K, Ca and Mg have also been reported to have increased (Smith and Read 1997, Clark and Zeto 2000). Higher P, K, and Zn; more or less equal N, Mg, Ca, Mn, Co, Cr and Pb; and Na, Cu, Ni and Cd uptake by maize plants were determined when biochar was added, compared to corresponding fertilisers (Glaser et al. 2015). AMF increase nitrogen uptake from organic sources (Hodge, Campbell, and Fitter 2001), but further studies need to be carried out to understand the mechanism (Read and Perez-Moreno 2003).

Colonisation with AMF may result in a decrease in the uptake of other nutrients whilst increasing the uptake of a nutrient by the host plant (Kothari, Marschner, and George 1990). There is a reduction in host plant manganese (Mn) uptake after AMF colonisation, whilst other nutrient uptake is reported to have increased (Kothari, Marschner, and Romheld 1991). It has also been expressed that nutrient uptake in plants increases after biochar application (Lehmann et al. 2003). BC can improve plant productivity directly, as a consequence of its nutrient content, as well as indirectly, via: (i) improved conservation of nutrients (Lehmann et al. 2003, Wardle, Zackrisson, and Nilsson 1998); (ii) increased soil cation exchange capacity (Liang et al. 2006); (iii) improvements in soil pH (Rondon et al. 2007); (iv) improved physical properties of soil (Chan et al. 2008), including an increase in soil water content (Laird et al. 2010); and (v) alteration of soil microbial populations (Pietikäinen Kiikkila, and Fritze 2000).

Use of AMF and BC in Organic Agriculture

Organic agriculture is controlled and certified at every stage from production to consumption, and limited quantities of chemical input are used (Zengin 2007). It is known that many plants found in nature are produced under healthy conditions without any chemical input. The aim is to protect our soil, water resources and air from pollution, and protect plant, animal and human health. The most important items that are restricted as input in organic vegetable production are: “Soil Remedies and Fertilisers” and “Plant Protection Products”. There is a problem, however, with achieving the permitted inputs, and with their costs. In addition, the products used are not very effective (Erzurumlu and Kara 2014). One of the most important production inputs is fertilisers. Use of mycorrhizal fungi can replace fertiliser applications in agriculture and will increase soil fertility as well as product quality, reduce input costs and agricultural pollution. The application of mycorrhiza to plants, especially those with low phosphorus stress, is important for natural remediation of soil conditions (Erzurumlu and Kara 2014). Affirmative properties of biochar, together with its nutrient value, supplied either directly by providing nutrients to plants, or indirectly by increasing soil quality, can improve the efficiency of fertiliser use (Yin and Xu 2012). A number of organic compounds belonging to various chemical classes, including benzoic acids, diols, hydroxy and acetoxy acids, triols, and n-alkanoic acids and phenols have been identified in biochar (Graber et al. 2010). AMF and BC can be used both for the remediation of HMs, and for solving the problem of plant nutrition in organic agriculture.

References

Abdelhafez, AA, J Li, and MH Abbas. 2014. “Feasibility of biochar manufactured from organic wastes on the stabilization of heavy metals in a metal smelter contaminated soil.” *Chemosphere* 117:66–71.

Ahmad, M, AU Rajapaksha, JE Lim, M Zhang, N Bolan, D Mohan, M Vithanage, SS Lee, and YS Ok. 2014. “Biochar as a sorbent for contaminant management in soil and water: a review.” *Chemosphere* 99 (3):19-23.

Ainza, C, J Trevors, and M Saier. 2010. “Environmental Mercury Rising.” *Water, Air, and Soil Pollution* 205 (1):47–48.

Akhter, A, K Hage-Ahmed, G Soja, and S Steinkellner. 2015. “Compost and biochar alter mycorrhization, tomato root exudation, and development of *Fusarium oxysporum* f. sp. *lycopersici*.” *Frontiers in Plant Science* 6:529.

Aksoy, A, M Ozturk. 1996. "*Phoenix dactylifera* as a biomonitor of heavy metal pollution in Turkey." *Journal for Trace and Microprobe Techniques* 14 (3): 605–614.

Aksoy, A, M Ozturk. 1997."*Nerium oleander* as a biomonitor of lead and other heavy metal pollution in mediterranean environments." *Science of the Total Environment* 205:145–150.

Al-Wabel, MI, ARA Usman, AH El-Naggar, AA Aly, HM Ibrahim, S Elmaghraby, and A Al-Omran. 2014. "Conocarpus biochar as a soil amendment for reducing heavy metal availability and uptake by maize plants." *Saudi Journal of Biological Sciences* 95 (4):503–511.

Aleksandra, DC, and B Urszula. 2008. "The impact of nickel on human health." *Journal of Elementology* 13 (4):685–696.

Ali, A, D Guo, A Mahar, P Wang, F Ma, F Shen, R Li, and Z Zhang. 2017. "Phytoextraction of toxic trace elements by *Sorghum bicolor* inoculated with *Streptomyces pactum* (Act12) in contaminated soils." *Ecotoxicology Environmental Saftey* 139:202–209.

Ashraf, M, M Ozturk, MSA Ahmad. (Eds.) 2010a. Plant Adaptation and Phytoremediation. Science+Business Media, NY, 481 pp.

Ashraf, M, M Ozturk, MSA Ahmad 2010b. Toxins and their phytoremediation. In: Plant adaptation and phytoremediation (Eds. Ashraf et al.), pp:1–34. Science+Business Media, NY.

Ashraf, MY, M Roohi, Z Iqbal, M Ashraf, M Ozturk, S Gücel. 2015. Cadmium (Cd)and Lead (Pb) induced inhibition in growth and alteration in some biochemical attributes and mineral accumulation in mung bean [*Vigna radiate* (L.) Wilczek]. *Communications in Soil Science and Plant Analysis* 1532-2416.

Aziz, MA, HR Ahmad, DL Corwin, M Sabir, M Ozturk and MKR Hakeem. 2016. Influence of farmyard manure on retention and availability of nickel, zinc and lead in metal-contaminated calcareous loam soils. *Journal of Environmental Engineering and Landscape Management* 25 (3): 289–296.

Balestrasse, KB, MP Benavides, SM Gallego, and ML Tomaro. 2003. "Effect of cadmium stress on nitrogen metabolism in nodules and roots of soybean plants." *Functional Plant Biology* 30 (1):57–64.

Beesley, L, E Moreno-Jiménez, JL Gomez-Eyles, E Harris, B Robinson, and T Sizmur. 2011. "A review of biochars' potential role in the remediation, revegetation and restoration of contaminated soils." *Environmental Pollution* 159 (12):3269-3282.

Bernaola, L, M Cosme, RW Schneider, and M Stout. 2018. "Below ground inoculation with arbuscular mycorrhizal fungi increases local and systemic susceptibility of rice plants to different pest organisms." *Frontiers in Plant Science* 9, 747, 1–16.

Biró, B, K Posta, A Füzy, I Kádar, and T Németh. 2005. "Mycorrhizal functioning as part of the survival mechanisms of barley (*Hordeum vulgare* L) at long-term heavy metal stress." *Acta Biologica Szegediensis* 49 (1):65–67.

Blackwell, P, S Joseph, P Munroe, HM Anawar, P Storer, RJ Gilkes, and ZM Solaiman. 2015. "Influences of biochar and biochar-mineral complex on mycorrhizal colonisation and nutrition of wheat and sorghum." *Pedosphere* 25 (5):686–695.

Boivin, M-Y, AM Breure, L Posthuma, and M Rutgers. 2002. "Determination of field effects of contaminants—significance of pollution-induced community tolerance." *Human and Ecological Risk Assessment an International Journal* 8 (5):1035–1055.

Cao, S, W Wang, F Wang, J Zhang, Z Wang, S Yang, and Q Xue. 2016. "Drought-tolerant Streptomyces pactum Act 12 assist phytoremediation of cadmium-contaminated soil by *Amaranthus hypochondriacus*: great potential application in arid/semi-arid areas." *Environmental Science and Pollution Research* 23 (15):14898-14907.

Cao, X, L Ma, B Gao, and W Harris. 2009. "Dairy-manure derived biochar effectively sorbs lead and atrazine." *Environmental Science & Technology* 43 (9):3285.

Chan, KY, L Van Zwieten, I Meszaros, A Downie, and S Joseph. 2008. "Agronomic values of green waste biochar as a soil amendment." *Soil Research* 45 (8):629–634.

Cheng, J, H Qiu, Z Chang, Z Jiang, and W Yin. 2016. "The effect of cadmium on the growth and antioxidant response for freshwater algae *Chlorella vulgaris*." *SpringerPlus* 5 (1):1290.

Clark, RB, and SK Zeto. 2000. "Mineral acquisition by arbuscular mycorrhizal plants." *Journal of plant Nutrition* 23 (7):867–902.

Conversa, G, A Bonasia, C Lazzizera, and A Elia. 2015. "Influence of biochar, mycorrhizal inoculation, and fertilizer rate on growth and flowering of Pelargonium (*Pelargonium zonale* L.) plants." *Frontiers in Plant Science* 6:429.

Cornejo, P, J Pérez-Tienda, S Meier, A Valderas, F Borie, C Azcón-Aguilar, and N Ferrol. 2013. "Copper compartmentalization in spores as a survival strategy of arbuscular mycorrhizal fungi in Cu-polluted environments." *Soil Biology and Biochemistry* 57:925-928.

Cui, L, J Yan, Y Yang, L Li, G Quan, C Ding, T Chen, Q Fu, and A Chang. 2013. "Influence of biochar on microbial activities of heavy metals contaminated paddy fields." *Bioresources* 8 (4):5536–5548.

Das, P, S Samantaray, and GR Rout. 1997. "Studies on cadmium toxicity in plants: A review." *Environmental Pollution* 98 (1):29–36.

Dhawi, F, R Datta, and W Ramakrishna. 2016. "Mycorrhiza and heavy metal resistant bacteria enhance growth, nutrient uptake and alter metabolic profile of sorghum grown in marginal soil." *Chemosphere* 157:33–41.

Doubková, P, and R Sudová. 2016. "Limited impact of arbuscular mycorrhizal fungi on clones of Agrostis capillaris with different heavy metal tolerance." *Applied soil ecology* 99:78–88.

Duruibe, JO, MC Ogwuegbu, and JN Egwurugwu. 2007. "Heavy Metal Pollution and Human Biotoxic Effects." *International Journal of Physical Sciences* 2 (5):112–118.

Elmer, WH, and J Pignatello. 2011. "Effect of biochar amendments on mycorrhizal associations and Fusarium crown and root rot of Asparagus in replant soils." *Plant Disease* 95 (95):960–966.

Enkhtuya, B, J Rydlová, and M Vosátka. 2000. "Effectiveness of indigenous and non-indigenous isolates of arbuscular mycorrhizal fungi in soils from degraded ecosystems and man-made habitats." *Applied Soil Ecology* 14 (3):201–211.

Erdinç, Ç, ED Durak, A Ekincialp, S Şensoy, and S Demir. 2017. Variations in response of determinate common bean (*Phaseolus vulgaris* L.) genotypes to arbuscular mycorrhizal fungi (AMF) inoculation. *Turkish Journal of Agriculture and Forestry*, 41: 1–9

Erzurumlu Sandal, G, and E Erman Kara. 2014. "Mikoriza Konusunda Türkiye'de Yapılan Çalışmalar." *Türk Bilimsel Derlemeler Dergisi* 7 (2):55–65.

Ezawa, T, K Yamamoto, and S Yoshida. 2002. "Enhancement of the effectiveness of indigenous arbuscular mycorrhizal fungi by inorganic soil amendments." *Soil Science and Plant Nutrition* 48 (6):897–900.

Fellet, G, M Marmiroli, and L Marchiol. 2014. "Elements uptake by metal accumulator species grown on mine tailings amended with three types of biochar." *Science of the Total Environment* s 468–469:598–608.

Ferrol, N, T Elisabeth and V Paola. 2016. "The heavy metal paradox in arbuscular mycorrhizas: from mechanisms to biotechnological applications." *Journal of Experimental Botany*: 67(22):6253-6265.

Finlay, R. 2008. "Ecological aspects of mycorrhizal symbiosis—With special emphasis on the functional diversity of interactions involving the extraradical mycelium." *Journal of Experimental Botany* 59 (5):1115–1126.

Fontes, RLF, and FR Cox. 1998. "Zinc toxicity in soybean grown at high iron concentration in nutrient solution." *Journal of Plant Nutrition* 21 (8):1723–1730.

Fu, F, and Q Wang. 2011. "Removal of heavy metal ions from wastewaters: a review." *Journal of environmental management* 92 (3):407–418.

Fuentes, A, L Almonacid, JA Ocampo, and C Arriagada. 2016. "Synergistic interactions between a saprophytic fungal consortium and *Rhizophagus irregularis* alleviate oxidative stress in plants grown in heavy metal contaminated soil." *Plant and soil* 407 (1–2):355–366.

Gadkar, V, R David-Schwartz, T Kunik, and Y Kapulnik. 2001. "Arbuscular mycorrhizal fungal colonization. factors involved in host recognition." *Plant Physiology* 127(4), 1493–1499.

Gaither, LA, and DJ Eide. 2001. "Eukaryotic zinc transporters and their regulation." *BioMetals* 14 (3):251–270.

George, C, M Wagner, M Kücke, and MC, Rillig. 2012. "Divergent consequences of hydrochar in the plant–soil system: Arbuscular mycorrhiza, nodulation, plant growth and soil aggregation effects." *Applied Soil Ecology* 59:68–72.

Giasson, P, A Karam, and A Jaouich. 2008. Arbuscular mycorrhizae and alleviation of soil stresses on plant growth. Mycorrhizae: Sustainable Agriculture and Forestry 99–134.

Ginocchio, R, G Carvallo, I Toro, E Bustamante, Y Silva, and N Sepúlveda. 2004. "Micro-spatial variation of soil metal pollution and plant recruitment near a copper smelter in central Chile." *Environmental Pollution* 127 (3): 343–352.

Glaser, B, K Wiedner, S Seelig, H-P Schmidt, and H Gerber. 2015. "Biochar organic fertilizers from natural resources as substitute for mineral fertilizers." *Agronomy for Sustainable Development* 35 (2):667–678.

Graber, ER, YM Harel, M Kolton, E Cytryn, A Silber, DR David, L Tsechansky, M Borenshtein, and Y Elad. 2010. "Biochar impact on development and productivity of pepper and tomato grown in fertigated soilless media." *Plant and soil* 337 (1–2):481–496.

Graham, JH, and LK Abbott. 2000. "Wheat responses to aggressive and non-aggressive arbuscular mycorrhizal fungi." *Plant and Soil* 220 (1–2):207–218.

Gray, CW, RG Mclaren, and AH Roberts. 2003. "Atmospheric accessions of heavy metals to some New Zealand pastoral soils." *Science of the Total Environment* 305 (1–3):105–115.

Gryndler, M, J Larsen, H Hršelová, V Řezáčová, H Gryndlerová, and J Kubát. 2006. "Organic and mineral fertilization, respectively, increase and decrease the development of external mycelium of arbuscular mycorrhizal fungi in a long-term field experiment." *Mycorrhiza* 16 (3):159–166.

Gul, S, JK Whalen, BW Thomas, V Sachdeva, and H Deng. 2015. "Physico-chemical properties and microbial responses in biochar-amended soils: Mechanisms and future directions." *Agriculture, Ecosystems and Environment* 206:46–59.

Hakeem KR, A Ahmad, M Iqbal, S Gucel, M Ozturk. 2011. "Nitrogen efficient rice genotype can reduce nitrate pollution." *Environmental Science and Pollution Research*18:1184–1193.

Hakeem, KR, M Sabir, M Ozturk, A Mermut. 2015. Soil remediation and plants: prospects and challenges. Academic Press, Elsevier, London, USA, 724 pp.

Hakeem, KR, M Ozturk, M Sabir, MS Akhtar. 2017."Nitrate and nitrogen oxides: Sources, health effects and their remediation". *Reviews of Environmental Contamination and Toxicology* 242: 183–217.

Hammer, EC, Z Balogh-Brunstad, I Jakobsen, PA Olsson, SLS Stipp, and MC Rillig. 2014. "A mycorrhizal fungus grows on biochar and captures phosphorus from its surfaces." *Soil Biology and Biochemistry* 77 (7):252–260.

Henry, JR. 2000. "An overview of the phytoremediation of lead and mercury." National Network of Environmental Management Studies (NNEMS). Environmental Protection Agency, Washington, DC, 1-31.

Hernandez, LE, R Carpena-Ruiz, and A Gárate. 1996. "Alterations in the mineral nutrition of pea seedlings exposed to cadmium 1." *Journal of Plant Nutrition* 19 (12):1581–1598.

Hildebrandt, U, M Regvar, and H Bothe. 2007. "Arbuscular mycorrhiza and heavy metal tolerance." *Phytochemistry* 68 (1):139–46.

Hmid, A, Z Al Chami, W Sillen, A De Vocht, and J Vangronsveld. 2015. "Olive mill waste biochar: a promising soil amendment for metal immobilization in contaminated soils." *Environmental Science and Pollution Research* 22 (2):1444–1456.

Hodge, A, CD Campbell, and AH Fitter. 2001. "An arbuscular mycorrhizal fungus accelerates decomposition and acquires nitrogen directly from organic material." *Nature* 413 (6853):297.

Houben, D, L Evrard, and P Sonnet. 2013. "Mobility, bioavailability and pH-dependent leaching of cadmium, zinc and lead in a contaminated soil amended with biochar." *Chemosphere* 92 (11):1450–1457.

Hristozkova, M, M Geneva, I Stancheva, M Boychinova, and E Djonova. 2016. "Contribution of arbuscular mycorrhizal fungi in attenuation of heavy metal impact on Calendula officinalis development." *Applied Soil Ecology* 101:57–63.

Hristozkova, M, M Geneva, I Stancheva, I Iliev, and C Azcón-Aguilar. 2017. "Symbiotic association between golden berry (*Physalis peruviana*) and arbuscular mycorrhizal fungi in heavy metal-contaminated soil." *Journal of Plant Protection Research* 57 (2):173–184.

Hu, J, F Wu, S Wu, CL Lam, X Lin, and HW Ming. 2014. "Biochar and *Glomus caledonium* influence Cd accumulation of upland Kangkong (*Ipomoea aquatica* Forsk.) intercropped with *Alfred Stonecrop* (*Sedum alfredii* Hance)." *Scientific Reports* 4:4671.

Hua, M, S Zhang, B Pan, W Zhang, L Lv, and Q Zhang. 2012. "Heavy metal removal from water/wastewater by nanosized metal oxides: A review." *Journal of Hazardous Materials* 211:317–331.

Ibijbijen, J, S Urquiaga, M Ismaili, BJR Alves, and RM Boddey. 1996. "Effect of arbuscular mycorrhizal fungi on growth, mineral nutrition and nitrogen fixation of three varieties of common beans (*Phaseolus vulgaris*)." *New phytologist* 134 (2):353–360.

Jadia, CD, and MH Fulekar. 2009. "Phytoremediation of heavy metals: Recent techniques." *African Journal of Biotechnology* 8 (6):921–928.

Jaishankar, M, T Tseten, N Anbalagan, BB Mathew, and KN Beeregowda. 2014. "Toxicity, mechanism and health effects of some heavy metals." *Interdisciplinary Toxicology* 7 (2):60-72.

Jiang, TY, J Jiang, RK Xu, and Z Li. 2012. "Adsorption of Pb(II) on variable charge soils amended with rice-straw derived biochar." *Chemosphere* 89 (3):249–256.

Joner, EJ, and C Leyval. 1997. "Uptake of 109 Cd by roots and hyphae of a *Glomus mosseae*/*Trifolium subterraneum* mycorrhiza from soil amended with high and low concentrations of cadmium." *New Phytologist* 135 (2):353–360.

Joseph, S, HM Anawar, P Storer, P Blackwell, C Chia, Y Lin, P Munroe, S Donne, J Horvat, J Wang, and ZM Solaiman. 2015. "Effects of enriched biochars containing magnetic iron nanoparticles on mycorrhizal colonisation, plant growth, nutrient uptake and soil quality improvement." *Pedosphere* 25 (5):749–760.

Kaldorf, M, AJ Kuhn, WH Schröder, U Hildebrandt, and H Bothe. 1999. "Selective element deposits in maize colonized by a heavy metal tolerance conferring arbuscular mycorrhizal fungus." *Journal of Plant Physiology* 154 (5–6):718–728.

Karami, N, R Clemente, E Moreno-Jimnez, NW Lepp, and L Beesley. 2011. "Efficiency of green waste compost and biochar soil amendments for reducing lead and copper mobility and uptake to ryegrass." *Journal of Hazardous Materials* 191 (1–3):41-48.

Karimi, A, H Khodaverdiloo, M Sepehri, and MR Sadaghiani. 2011. "Arbuscular mycorrhizal fungi and heavy metal contaminated soils." *African Journal of Microbiology Research* 5 (13):1571–1576.
Kim, H-S, K-R Kim, H-J Kim, J-H Yoon, JE Yang, YS Ok, G Owens, and K-H Kim. 2015. "Effect of biochar on heavy metal immobilization and uptake by lettuce (*Lactuca sativa* L.) in agricultural soil." *Environmental Earth Sciences* 74 (2): 1249–1259.
Knowles, OA, BH Robinson, A Contangelo, and L Clucas. 2011. "Biochar for the mitigation of nitrate leaching from soil amended with biosolids." *Science of the Total Environment* 409 (17):3206-3210.
Koide, RT. 2000. "Functional complementarity in the arbuscular mycorrhizal symbiosis." *The New Phytologist* 147 (2):233–235.
Kolb, SE, KJ Fermanich, and ME Dornbush. 2009. "Effect of charcoal quantity on microbial biomass and activity in temperate soils." *Soil Science Society of America Journal* 73 (73):1173–1181.
Kothari, SK, H Marschner, and E George. 1990. "Effect of VA mycorrhizal fungi and rhizosphere microorganisms on root and shoot morphology, growth and water relations in maize." *New Phytologist* 116 (2):303–311.
Kothari, SK, H Marschner, and V Römheld. 1991. "Contribution of the VA mycorrhizal hyphae in acquisition of phosphorus and zinc by maize grown in a calcareous soil." *Plant and soil* 131 (2):177–185.
Lai, WY, CM Lai, GR Ke, RS Chung, CT Chen, CH Cheng, CW Pai, SY Chen, and CC Chen. 2013. "The effects of woodchip biochar application on crop yield, carbon sequestration and greenhouse gas emissions from soils planted with rice or leaf beet." *Journal of the Taiwan Institute of Chemical Engineers* 44 (6):1039–1044.
Laird, DA, P Fleming, DD Davis, R Horton, B Wang, and DL Karlen. 2010. "Impact of biochar amendments on the quality of a typical Midwestern agricultural soil." *Geoderma* 158 (3–4):443–449.
Lakshman, H, and A Channabasava. 2013. "Mycorrhizoremediation of mine spoil by using foxtail millet inoculated with *Rhizophagus fasciculatus*: an ex-situ solid waste management." *International Journal of Current Science* 8:85–92.
Lasat, MM. 2002. "Phytoextraction of Toxic Metals: A Review of Biological Mechanisms." *Journal of Environmental Quality* 31 (1):109.
LeCroy, C, CA Masiello, JA Rudgers, WC Hockaday, and JJ Silberg. 2013. "Nitrogen, biochar, and mycorrhizae: Alteration of the symbiosis and oxidation of the char surface." *Soil Biology and Biochemistry* 58:248–254.
Lehmann, J, J P da Silva, C Steiner, T Nehls, W Zech, and B Glaser. 2003. "Nutrient availability and leaching in an archaeological Anthrosol and a

Ferralsol of the Central Amazon basin: fertilizer, manure and charcoal amendments." *Plant and Soil* 249 (2):343–357.

Lehmann, J, MC Rillig, J Thies, CA Masiello, WC Hockaday, and D Crowley. 2011. "Biochar effects on soil biota—A review." *Soil Biology and Biochemistry* 43 (9):1812–1836.

Leyval, C. 2002. "Potential of arbuscular mycorrhizal fungi for bioremediation." *Mycorrhizal technology in agriculture*. Birkhäuser, Basel, 175–186.

Li, H, N Luo, LJ Zhang, HM Zhao, YW Li, QY Cai, MH Wong, and CH Mo. 2016. "Do arbuscular mycorrhizal fungi affect cadmium uptake kinetics, subcellular distribution and chemical forms in rice?" *Science of the Total Environment* 571:1183–1190.

Li, X, and P Christie. 2001. "Changes in soil solution Zn and pH and uptake of Zn by arbuscular mycorrhizal red clover in Zn-contaminated soil." *Chemosphere* 42 (2):201–207.

Liang, B, J Lehmann, D Solomon, J Kinyangi, J Grossman, B O'Neill, JO Skjemstad, J Thies, FJ Luizao, and J Petersen. 2006. "Black carbon increases cation exchange capacity in soils." *Soil Science Society of America Journal* 70 (5):1719–1730.

Lu, H., Z Li, S Fu, A Méndez, G Gascó, and J Pazferreiro. 2015. "Combining phytoextraction and biochar addition improves soil biochemical properties in a soil contaminated with Cd." *Chemosphere* 119:209-216.

Lu, H, W Zhang, Y Yang, X Huang, S Wang, and R Qiu. 2012. "Relative distribution of Pb 2+ sorption mechanisms by sludge-derived biochar." *Water Research* 46 (3):854–862.

Lu, K, X Yang, G Gielen, N Bolan, YS Ok, NK Niazi, S Xu, G Yuan, X Chen, and X Zhang. 2017. "Effect of bamboo and rice straw biochars on the mobility and redistribution of heavy metals (Cd, Cu, Pb and Zn) in contaminated soil." *Journal of Environmental Management* 186:285–292.

Lu, K, X Yang, J Shen, B Robinson, H Huang, D Liu, N Bolan, J Pei, and H Wang. 2014. "Effect of bamboo and rice straw biochars on the bioavailability of Cd, Cu, Pb and Zn to *Sedum plumbizincicola*." *Agriculture, Ecosystems and Environment* 191:124–132.

Ma, SC, HB Zhang, ST Ma, R Wang, GX Wang, Y Shao, and CX Li. 2015. "Effects of mine wastewater irrigation on activities of soil enzymes and physiological properties, heavy metal uptake and grain yield in winter wheat." *Ecotoxicol Environmental Safety* 113:483–90.

Mackie, KA, S Marhan, F Ditterich, HP Schmidt, and E Kandeler. 2015. "The effects of biochar and compost amendments on copper immobilization and soil microorganisms in a temperate vineyard." *Agriculture, Ecosystems and Environment* 201:58–69.

Malar, S, SS Vikram, PJC Favas, and V Perumal. 2014. "Lead heavy metal toxicity induced changes on growth and antioxidative enzymes level in water hyacinths [*Eichhorniacrassipes* (Mart.)]." *Botanical Studies* 55 (1):1–11.

Malghani, S, G Gleixner, and SE Trumbore. 2013. "Chars produced by slow pyrolysis and hydrothermal carbonization vary in carbon sequestration potential and greenhouse gases emissions." *Soil Biology and Biochemistry* 62 (5):137–146.

Martínez-Peñalver, A, MJ Reigosa, and AM Sánchez-Moreiras. 2011. "Imaging chlorophyll a fluorescence reveals specific spatial distributions under different stress conditions." *Flora-Morphology, Distribution, Functional Ecology of Plants* 206 (9):836–844.

Mau, AEE, and SR Utami. 2014. "Effects of biochar amendment and arbuscular mycorrhizal fungi inoculation on availability of soil phosphorus and growth of maize." *Journal of Degraded and Mining Lands Management* 1 (2): 69-74.

Meier, S, G Curaqueo, N Khan, N Bolan, J Rilling, C Vidal, N Fernández, J Acuña, ME González, P Cornejo, and F Borie. 2015. "Effects of biochar on copper immobilization and soil microbial communities in a metal-contaminated soil." *Journal of Soils and Sediments*: 1–14.

Mimi, R, and LM Mcdonald. 2013. "Metal uptake inplants and health risk assessments in metal-contaminated smelter soils." *Land Degradation and Development* 26 (8):785–792.

Morgan, RK. 2013. "*Soil, Heavy Metals, and Human Health.*" CRC Press, 74–97.

Mukherjee, A, and AR Zimmerman. 2013. "Organic carbon and nutrient release from a range of laboratory-produced biochars and biochar-soil mixtures." *Geoderma* s 193–194 (2):122–130.

Navarro, MC, C Pérez-Sirvent, MJ Martínez-Sánchez, J Vidal, PJ Tovar, and J Bech. 2008. "Abandoned mine sites as a source of contamination by heavy metals: A case study in a semi-arid zone." *Journal of Geochemical Exploration* 96 (2–3):183–193.

Novak, JM, WJ Busscher, DW Watts, DA Laird, MA Ahmedna, and MAS Niandou. 2010. "Short-term CO_2 mineralization after additions of biochar and switchgrass to a Typic Kandiudult." *Geoderma* 154 (3–4):281–288.

Nzanza, B, D Marais, and P Soundy. 2012. "Yield and nutrient content of tomato (*Solanum lycopersicum* L.) as influenced by *Trichoderma harzianum* and *Glomus mosseae* inoculation." *Scientia Horticulturae* 144 (3):55–59.

Ozturk, M., (Ed.) 1989. Plants and pollutants in Developed and developing countries. Ege Univ. Press, Izmir, Turkey, 759 pp.

Ozturk, M, Mermut, A, Celik, A. 2011a. Land Degradation, Urbanisation, Land Use and Environment, NAM S. & T. (Delhi-India), 445 pp.

Ozturk, M, S Gucel, S Sakcali, A Guvensen. 2011b. An overview of the possiblities for wastewater utilization for agriculture in Turkey. *Israel Journal of Plant Sciences* 59: 223–234.

Ozturk, M, S Sakcali, A Celik. 2013a. A Biomonitor of heavy metal on ruderal habitats in Turkey—*Diplotaxistenuifolia* (L.) DC. Sains Malaysiana 42(10): 1371–1376.

Ozturk, M, S Gucel, S Sakcali, S Baslar. 2013b. Nitrate and edible plants in the mediterranean region of Turkey: An Overview. In: Nitrate in leafy vegetables-toxicity and safety measures (eds. Umar, S. et al.), pp: 17–51. I.K. Intern. Publ. House Pvt. Ltd. New Delhi-Bangalore (India)

Ozturk, M, M Ashraf, A Aksoy, MSA Ahmad (Eds.) 2015a. Phytoremediation for green energy. Springer Science+Business Media, NY, 191 pp.

Ozturk, M, M Ashraf, A Aksoy, MSA Ahmad (Eds.) 2015b. Plants, pollutants and remediation. Springer Science+Business Media, NY, 407 pp.

Ozturk, M, V Altay, F Karahan. 2017. "Studies on trace elements in *Glycyrrhiza*taxa distributed in Hatay-Turkey." *International Journal of Plant and Environment*, 3 (2): 01–07.

Park, JH., D Lamb, P Paneerselvam, G Choppala, N Bolan, and JW Chung. 2010. "Role of organic amendments on enhanced bioremediation of heavy metal(loid) contaminated soils." *Journal of Hazardous Materials* 185 (2–3):549–574.

Paszkowski, U. 2006. "A journey through signaling in arbuscular mycorrhizal symbioses 2006." *New phytologist* 172 (1):35–46.

Paz-Ferreiro, J, H Lu, S Fu, A Méndez, and G Gascó. 2013. "Use of phytoremediation and biochar to remediate heavy metal polluted soils: a review." *Solid Earth Discussions* 5 (2):2155–2179.

Pietikäinen, J, O Kiikkilä, and H Fritze. 2000. "Charcoal as a habitat for microbes and its effect on the microbial community of the underlying humus." *Oikos* 89 (2):231–242.

Purakayastha, TJ, S Kumari, and H Pathak. 2014. "Characterisation, stability, and microbial effects of four biochars produced from crop residues." *Geoderma* 239–240:293–303.

Qiao, Y, D Crowley, K Wang, H Zhang, and H Li. 2015. "Effects of biochar and Arbuscular mycorrhizae on bioavailability of potentially toxic elements in an aged contaminated soil." *Environmental Pollution* 206:636–643.

Rascio, N, and F Navariizzo. 2011. "Heavy metal hyperaccumulating plants: how and why do they do it? And what makes them so interesting?" *Plant Science an International Journal of Experimental Plant Biology* 180 (2):169–181.

Read, DJ, and J Perez-Moreno. 2003. "Mycorrhizas and nutrient cycling in ecosystems–a journey towards relevance?" *New Phytologist* 157 (3):475–492.

Rondon, MA, J Lehmann, J Ramírez, and M Hurtado. 2007. "Biological nitrogen fixation by common beans (*Phaseolus vulgaris* L.) increases with bio-char additions." *Biology and Fertility of Soils* 43 (6):699–708.

Roos, P, and I Jakobsen. 2008. "Arbuscular mycorrhiza reduces phytoextraction of uranium, thorium and other elements from phosphate rock." *Journal of Environmental Radioactivity* 99 (5):811–819.

Rusan, MJM, W Pan, and AC Kennedy. 1995. "Wheat responses to vesicular-arbuscular mycorrhizal fungal inoculation of soils from eroded toposequence." *Soil Science Society of America Journal* 59 (4):1086–1090.

Sabir, M, EA Waraich, KR Hakeem, M Ozturk, HR Ahmad, M Shahid. 2015. Phytoremediation: mechanisms and adaptations. Chapter 4 In: Soil remediation and plants: prospects and challenges (Eds.: Hakeem et al.), pp: 85–105. Academic Press, Elsevier, New York-USA.

Santona, L, P Castaldi, and P Melis. 2006. "Evaluation of the interaction mechanisms between red muds and heavy metals." *Journal of Hazardous Materials* 136 (2):324.

Sarmah, AK, and J Sabadie. 2002. "Hydrolysis of sulfonylurea herbicides in soils and aqueous solutions: A review." *Journal of Agricultural & Food Chemistry* 50 (22):6253–6265.

Schneider, J, J Bundschuh, and CW Araújo do Nascimento. 2016. "Arbuscular mycorrhizal fungi-assisted phytoremediation of a lead-contaminated site." *Science of the Total Environment* 572:86–97.

Schomberg, HH, J W Gaskin, K Harris, KC Das, JM Novak, WJ Busscher, DW Watts, RH Woodroof, IM Lima, M Ahmedna, D Rehrah, and B Xing. 2012. "Influence of biochar on nitrogen fractions in a coastal plain soil." *Journal of Environmental Quality* 41 (4):1087–1095.

Schussler, A, H Gehrig, D Schwarzott, and C Walker. 2001. "Analysis of partial Glomales SSU rRNA gene sequences: implications for primer design and phylogeny." *Mycological Research* 105 (1):5–15.

Shakeel, M, and T Yaseen. 2015. "An Insight into phytoremediation of heavy metals from soil assisted by ancient fungi from Glomeromycota-Arbuscular Mycorrhizal Fungi." *Science, Technology and Development* 34 (4):215–220.

Sheikh, KH, M Ozturk, O Secmen, Y Vardar.1976. "Field studies of the effects of cement dust on the growth and yield of olive trees in Turkey." *Environmental Conservation* 3:117–121.

Shetty, KG, BAD Hetrick, DAH Figge, and AP Schwab. 1994. "Effects of mycorrhizae and other soil microbes on revegetation of heavy metal contaminated mine spoil." *Environmental Pollution* 86 (2):181–188.

Singh, HP, P Mahajan, S Kaur, DR Batish, and RK Kohli. 2013. "Chromium toxicity and tolerance in plants." *Environmental Chemistry Letters* 11 (3):229–254.

Singh, S, M Zacharias, S Kalpana, and S Mishra. 2012. "Heavy metals accumulation and distribution pattern in different vegetable crops." *Journal of Environmental Chemistry and Ecotoxicology* 4(10):170–177.

Sinha, RK., S Herat, and PK Tandon. 2007. Phytoremediation: Role of plants in contaminated site management: pp 315-330 In: Singh S.N., Tripathi R.D. (eds) Environmental bioremediation technologies. Springer, Berlin, Heidelberg.

Smith, SE, and D Read. 2010. Mycorrhizal symbiosis: *Academic Press, London.*

Smith, SE, and DJ Read. 1997. "Mycorrhizal symbiosis." *Academic Press, London.*

Sohi, SP, E Krull, E Lopezcapel, and R Bol. 2010. "A review of biochar and its use and function in soil." *Advances in Agronomy* 105 (1):47–82.

Stewart, CE, J Zheng, J Botte, and MF Cotrufo. 2013. "Co-generated fast pyrolysis biochar mitigates green-house gas emissions and increases carbon sequestration in temperate soils." *GCB Bioenergy* 5 (2):153–164.

Tabrizi, L, S Mohammadi, M Delshad, and B Motesharezadeh. 2015. "Effect of arbuscular mycorrhizal fungi on yield and phytoremediation performance of pot marigold (*Calendula officinalis* L.) under heavy metals stress." *International journal of phytoremediation* 17 (12):1244–1252.

Thompson, JP. 1987. "Decline of vesicular-arbuscular mycorrhizae in long fallow disorder of field crops and its expression in phosphorus deficiency of sunflower." *Australian Journal of Agricultural Research* 38 (5):847–867.

Tonin, C, P Vandenkoornhuyse, EJ Joner, J Straczek, and C Leyval. 2001. "Assessment of arbuscular mycorrhizal fungi diversity in the rhizosphere of *Viola calaminaria* and effect of these fungi on heavy metal uptake by clover." *Mycorrhiza* 10 (4):161–168.

Turnau, K, and K Haselwandter. 2002. Arbuscular mycorrhizal fungi, an essential component of soil microflora in ecosystem restoration. In Mycorrhizal technology in agriculture. Birkhäuser, Basel, 137-149.

Uchimiya, M, KB Cantrell, PG Hunt, JM Novak, and S Chang. 2012. "Retention of heavy metals in a Typic Kandiudult amended with different manure-based biochars." *Journal of Environmental Quality* 41 (4):1138–1149.

Uchimiya, M, LH Wartelle, KT Klasson, CA Fortier, and IM Lima. 2011. "Influence of pyrolysis temperature on biochar property and function as a heavy metal sorbent in soil." *Journal of Agricultural and Food Chemistry* 59 (6):2501–2510.

Uzoma, KC, M Inoue, H Andry, H Fujimaki, A Zahoor, and E Nishihara. 2011. "Effect of cow manure biochar on maize productivity under sandy soil condition." *Soil Use & Management* 27 (2):205–212.

Vacca, A, MR Bianco, M Murolo, and P Violante. 2012. "Heavy metals in contaminated soils of the Rio Sitzerri Floodplain (Sardinia, Italy): Characterization and Impact on Pedodiversity." *Land Degradation and Development* 23 (4):350–364.

Val, CD, JM Barea, and C Azcónaguilar. 1999. "Diversity of arbuscular mycorrhizal fungus populations in heavy-metal-contaminated soils." *Applied and Environmental Microbiology* 65 (2):718.

Vanek, SJ., and J Lehmann. 2015. "Phosphorus availability to beans via interactions between mycorrhizas and biochar." *Plant and Soil* 395 (1):105–123.

Verheijen, F, S Jeffery, AC Bastos, M van der Velde, and I Diafas. 2010. Biochar application to soils: A critical scientific review of effects on soil properties, processes and functions. EUR 24099 EN. Luxembourg (Luxembourg): European Commission. 162.

Violante, A, V Cozzolino, L Perelomov, AG Caporale, and M Pigna. 2010. "Mobility and bioavailability of heavy metals and metalloids in soil environments." *Journal of Soil Science and Plant Nutrition* 10 (3):266–290.

Vithanage, M, AU Rajapaksha, M Zhang, S Thielebruhn, SS Lee, and YS Ok. 2015. "Acid-activated biochar increased sulfamethazine retention in soils." *Environmental Science and Pollution Research* 22 (3):2175–2186.

Vosátka, M. 1995. "Influence of inoculation with arbuscular mycorrhizal fungi on the growth and mycorrhizal infection of transplanted onion." *Agriculture, Ecosystems and Environment* 53 (2):151–159.

Wahid, F, M Sharif, S Steinkillner, MA Khan, and KB Marwat. 2016. "Inoculation of Arbuscular mycorrhizal fungi in presence of rock phosphate improve phosphorus uptake and growth of maize." *Pakistan Journal of Botany* 48 (2):739–747.

Wang, Y, F Pan, G Wang, G Zhang, Y Wang, X Chen, and Z Mao. 2014. "Effects of biochar on photosynthesis and antioxidative system of Malus *hupehensis* Rehd. seedlings under replant conditions." *Scientia Horticulturae* 175 (1):9–15.

Wardle, DA, O Zackrisson, and M-C Nilsson. 1998. "The charcoal effect in Boreal forests: mechanisms and ecological consequences." *Oecologia* 115 (3):419–426.

Warnock, DD, DL Mummey, B Mcbride, J Major, J Lehmann, and MC Rillig. 2010. "Influences of non-herbaceous biochar on arbuscular mycorrhizal fungal abundances in roots and soils: results from growth-chamber and field experiments." *Applied Soil Ecology* 46 (3):450–456.

Warnock, DD., J Lehmann, TW Kuyper, and MC Rillig. 2007. "Mycorrhizal responses to biochar in soil—concepts and mechanisms." *Plant and Soil* 300 (1):9–20.

Watts-Williams, SJ, TW Turney, AF Patti, and TR Cavagnaro. 2014. "Uptake of zinc and phosphorus by plants is affected by zinc fertiliser material and arbuscular mycorrhizas." *Plant and Soil* 376 (1/2):165–175.

Xian, Y, M Wang, and W Chen. 2015. "Quantitative assessment on soil enzyme activities of heavy metal contaminated soils with various soil properties." *Chemosphere* 139:604–608.

Yin, CK, Z Xu. 2012. "Biochar: nutrient properties and their enhancement." In *Biochar for environmental management*, 99–116. Routledge.

Yamato, M, Y Okimori, IF Wibowo, S Anshori, and M Ogawa. 2006. "Effects of the application of charred bark of *Acacia mangium* on the yield of maize, cowpea and peanut, and soil chemical properties in South Sumatra, Indonesia." *Soil Science and Plant Nutrition* 52 (4):489–495.

Yang, D, DH Zeng, J Zhang, LJ Li, and R Mao. 2012. "Chemical and microbial properties in contaminated soils around a magnesite mine in northeast China." *Land Degradation and Development* 23 (3):256–262.

Zengin, M. 2007. Organic Agriculture. Hasad Publishing, Istanbul, Turkey.

Zwieten, LV, S Kimber, S Morris, KY Chan, A Downie, J Rust, S Joseph, and A Cowie. 2010. "Effects of biochar from slow pyrolysis of papermill waste on agronomic performance and soil fertility." *Plant and Soil* 327 (1):235–246.

CHAPTER 7

GREEN FERTILIZER AND ALGAL BIOFERTILIZER

CEMIL ISLEK

Department of Biotechnology, Faculty of Art and Sciences, Nigde Omer Halisdemir University, Nigde, Turkey.
cislek@ohu.edu.tr

Abstract

As one of the agricultural inputs, fertilizer is essential for producing foodstuffs needed for the nutritional health of an increasing global population. It is of great importance in terms of organic substance, soil productivity and sustainable agriculture. Moreover, there is a worldwide need for organic fertilizer to conduct organic agriculture. Organic fertilizer, which is a nutritional source for plants, comprises vegetal, animal and human remnants or waste. In recent years, green fertilization has been regarded as an alternative to mineral fertilizer for increasing productivity. Green fertilization, which is carried out by burying green plant materials in the ground to enrich the soil, is important in organic agriculture. Different plants are used for green fertilization, however leguminous plants are generally preferred to non-leguminous plants and are accepted as the best for green fertilization. Green fertilization may increase the organic substance content of the soil. Biofertilizers contain fungal or algal strains and enhance the productivity of the soil. The use of algal biofertilizer provides an eco-friendly and non-polluting approach in improving the productivity of crops by both nitrogen fixation and photosynthesis.

Keywords: organic agriculture, soil, organic fertilizer, algae, biofertilizer.

Introduction

The transformation of agriculture has accelerated with the development of technology and industry. As a specific result of the rapid increase in population, the transformation termed the "green revolution" began in the 1960s–70s (Davies 2003). This transformation aimed solely at an increase in productivity and led to an increase in the use of synthetic chemical pesticides and mineral fertilizers. The negative impacts of these were first seen in developed countries, where they were used intensively from the time of their introduction, and, following on from this, the pursuit of alternatives to conventional agricultural methods commenced at the beginning of the twenty-first century. The first study on this issue, from the perspective of organic agriculture, was published in England in the 1910s (Aksoy and Altındişli 1999).

The area of agricultural land worldwide has nearly reached the point of depletion. Because of rising populations, increased food production became a necessity (Gomiero 2016). Consequently, obtaining more products from a unit area was vital and, in turn, the use of chemical compounds increased. As a result of the intensive use of chemical fertilizer productivity and increased amounts produced, sustainable soil productivity and natural balances were negatively affected. The increase in the use of chemical fertilizers and environmental pollution has had an impact on all living things and has begun to pose a serious threat to them. As a result, many countries, particularly those with strong economies, have started to prefer agricultural products that are non-toxic, produced by methods that do not harm the environment (Kavuk 2013).

The primary purpose of organic agriculture is the addition of natural organic-based substances that maintain vital activities in the natural cycle without any chemical additives, to protect soil vitality and productivity (Yağmur 2001, Engin 2009, Morshedi et al. 2017).

Green Manure

Organic agriculture is a production form involving human- and eco-friendly production systems to re-establish the natural balance in the ecological system due to erroneous practices. Besides prohibiting the use of synthetic chemical drugs and fertilizers, it is a form of production aimed at raising the quality of the product, recommending the use of organic and green manuring, alternation, conservation of soil, and increasing resistance of crop, parasitoid, and predator within a closed system (İlter 1996). Soil-fertility management is one of the most important issues in organic agriculture (Biswas and Gresshoff 2014). This agricultural system has an obligation to fertilize with natural and organic substances. "Green manuring" is one of the ways to improve the physical, chemical and biological properties of soil by enriching the content of organic substances, to prevent erosion and protect soil moisture (Soyergin 2006, Duyar 2014).

Organic substances formed by the disintegration of crop residues and animal remains contain some nutrients required for crops. Fertilization of soil with chemical fertilizers is often inadequate. The soil must be enriched with organic substances (Karakurt 2009), which is only possible using organic fertilizers—barn stubble, compost, and green manure. Green manure is more effective in increasing organic substances in the soil and is more easily adaptable to different agricultural systems (Karakurt 2009). The various leguminous or non-leguminous green manure crops are shredded appropriately at the end of the flowering period and mixed into the soil. This process is called "green manuring" (Leonard 1986, Yılmaz 2010).

The Inca grew corn by burying fish heads with a few corn grains. Around the third century BC, the Chinese used nails and horns; the Greeks and Romans used plaster, lime, and marl as fertilizer, and ploughed beans into the ground as green manure. In the late seventeenth century in Germany and in the early eighteenth century in England, buckwheat, oats, and rye started to be used for green manuring. In America, around the middle of the eighteenth century, people began to grow triplets and legumes as green manure crops (Açıkgöz 2001).

Green manure enriches the soil in terms of beneficial nutrients, such as nitrogen, for new crops. It enhances the yield of soil by adding organic substances and nitrogen to the soil; protects the soil from erosion; improves the physical conditions of the soil, allowing the crops to benefit more from the nutrients; facilitates the accumulation of nutrients by preventing washing and enhances the biological activity of the soil. Such crops may provide a 100% increase in the crop yield for the next season (Soyergin 2006).

In green manuring it is suitable to choose crops that accumulate nitrogen in their roots, such as vetch, pods, soybean filaments, and stone flake (Kacar and Katkat 1999). In many trials, it has been observed that these crops increase the yield of the product after themselves by 20–100% (Tüzel et al. 2013).

Legumes provide the necessary nitrogen for a crop by binding the nitrogen in the air with the nodules in the roots (Santi, Bogusz, and Franche 2013). Besides legumes, other crops are used as green fertilizer. However, they cannot supply nitrogen to the soil (Crews and Peoples 2004). Approximately 90% of the green manure crops are broken up properly in the period of flowering, before being ploughed under and incorporated into the soil. *Rhizobium* and *Bradyrhizobium* bacteria release free nitrogen in the air into the crop through symbiosis with legumes (Biswas and Gresshoff 2014).

Green manuring is the process of burying green crops underground when they have completed a certain period of development of the grown crops to provide the organic substances that the soil needs (Rosenfeld and Rayns 2011). The best green manure crops are leguminous crops (Talgre 2013, Rosenfeld and Rayns 2011). Rose barley, rye, oat, wheat, turf, mustard, colza and corn can also be used in green manure from non-leguminous crops. Some reasons for preferring these crops are that the seeds of green fertilizer crops can be easily found; they are easily cultivated in the soil and easily adapt to natural life (Okur 2008). Green fertilization is an application generally used in areas with insufficient water content, such as sand soils; in unproductive areas; in conditions where artificial fertilizers are not effective; in heavy soil; in sloping terrain and places with extreme winds (Yılmaz 2010).

Green fertilizer plants are grown as main plants, sub-plants, and stubble plants. The main plant is only grown as a green manure plant in the vegetation period in the green manure field. But in this system, the green manure must be buried in the soil by the end of spring at the latest, otherwise, not enough water will be found to break it down, and problems will arise in autumn plantations (Okur 2008, Mooleki et al. 2016).

Growing a green manure plant as a sub-plant is particularly good in areas with abundant rainfall and well-characterised soil (Hirel et al. 2011). In the application of this system, as well as the characteristics of water and soil, it is also necessary to consider the development period of the main plant. The main plant must leave the field after completing its development as soon as possible.

The stubble plant is a system in which green manure plants are planted on the stubble after the harvest of the main plant. According to the next planting, the soil is mixed in autumn or early spring (Yılmaz 2010, Mooleki et al. 2016).

The Importance of Green Manuring for Soil Fertility

The most important use of green manure is to enrich the soil with organic substances. Especially in places where barn stubble is not available, the level of organic substances in the soil is increased by green manure. Depending on the nitrogen content of the crop applied as green manure, nitrogen is distributed in the soil by green manuring. If leguminous crops are used as green manure, relatively more nitrogen is provided to the soil. Ali (1999), in his work on the use of green manure in rice production in the tropics, stated that green manure crops are more profitable than mineral and organic nitrogen when they are regarded as short-lived food and feed crops. Considering the increase in yield of rice, green manuring does not contribute in the short term, but increases its usefulness in the long term (Yılmaz 2010).

The primary and most important benefit of green manuring, which helps to increase the yield of the soil and to obtain more and better quality products, is to enrich the organic substances in the soil (Adekiya et al. 2017). Especially in places where barnyard manure is rare, green manuring considerably increases the content of organic substances in the soil. Organic substances are one of the most important factors affecting the physical, chemical, and biological properties of soil and improves soil productivity directly, by positively affecting physical properties, such as soil structure restoration, aggregate durability, water infiltration and aeration (Gardner, Laryea, and Unger 1999).

The effect of green manuring lasts not only for that year, but for several years. However, this effect varies according to climate and soil conditions. It takes longer in humid and cool climates than in hot and arid climates. In sandy soils, it is effective for a shorter time than in clay soils (Florentín et al. 2010). Leguminous green manure crops have fixed the free nitrogen in the air into their structures and add nitrogen to the soil (Chen et al. 2014).

For effective fertilization, the time at which the green manure crop is mixed with the soil is crucial. Many factors influence the timing of the crop's mixing with the soil: the characteristics of the green manure crop, the weather conditions, the characteristics of the soil, and the characteristics of the cultivated crop to be added to the field after the green manure crop. Although these factors affect the mixing time of the green manure crop to the soil, we can generally say that the most appropriate time is the flowering period, in which the nitrogen and organic matter are more balanced in the crop. However, the most important point is that the mineral fertilization of the green manure crop should be completed when the cultivated crop affected by the mixing of the green manure crop is cropped on the soil (Cavigelli and Thien 2003, Florentín et al. 2010).

Benefits of Green Manuring

Green manuring prevents soil erosion and crop nutrients (K, Ca, Mg, NO3) from soil washing (Soyergin 2006); improves the physical properties of soil (Watson et al. 2002); and increases the amount of organic substances

(Sullivan 2003, Duyar 2007) and the amount of nutrients (Sullivan 2003, Dinnes et al. 2002). It also helps to control weeds (Sullivan 2003), positively affecting the quantity and function of microorganisms in the soil by providing nutrients to soil microorganisms (Açıkgöz 2001, Urzua, Urzua, and Pizarro 2001). With green manuring, nematodes and fungal pathogens in the roots are suppressed, and the amount of toxic chemicals used in soil fumigation is reduced (Duyar 2014).

Rotation, like all other practices in organic agriculture, aims to protect soil health. Soil organic substances and biodiversity are indispensable for a successful organic farming system (Mohler and Johnson 2009). For this reason, crop rotation practices and green manuring are among the most effective methods used in organic farming (Kaya et al. 2010). In green manuring, it is recommended that the different complementary crops are cultivated in succession to ensure the highest quality production, taking into consideration the ecological conditions of the region (Duman et al. 2013).

Approximately 90% of the green manure crops are ploughed under and incorporated into the soil while green, or shortly after flowering. *Rhizobium* and *Bradyrhizobium* bacteria fix atmospheric nitrogen in a form that crops can use at the end of the symbiosis with legumes. By grafting the legumes, significant quantities of nitrogen can be harvested (Yılmaz 2010).

Green manuring is regarded as an important management practice in the organic cultivation system to maintain and increase soil microbial activity and an abundance of microbes in different microbial populations (Edesi et al. 2013).

Green manure considerably increases the level of organic substances in places where there is less barnyard manure (Anaç and Okur 1996). Green manuring reduces the need for barnyard manure and adds up to 3–5% of organic substances to the soil (Sorensen and Thorup-Kristensen 2006, Göktekin 2015).

Use of Microalgae in Agriculture

Organic agriculture practices have gradually been gaining more importance. Products that are made using organic fertilizers and without pesticide applications are in demand. These preferences reveal that a more detailed examination of the opportunities to use various organic materials in agriculture is required (İnal, Sözüdoğru, and Erden 1996). One of these is the use of certain algae types as fertilizers (Yazıcı and Kaynak 2001, Namlı et al. 2016).

Algalization technology affects the nitrogen content of the crop, the number of seeds and the number of organs (Roger 1985). The main element in the biomass yield of crops is the nitrogen requirement. This requirement of the crop is obtained from soil, rain, atmosphere and biological nitrogen fixation (Chatterjee et al. 2017). Industrial nitrogen fixation is approximately 85 million tonnes per year, which requires a lot of energy input. 65% of the nitrogen used in agriculture is provided in biological processes (Albrecht 1998). Biological nitrogen fixation has been reported to provide nitrogen (75kg N / ha) to the paddy product in each cultural period (Irisarri, Gonnet, and Monza 2001).

Microalgae are used in agriculture as soil conditioners and biofertilizers, that is, cyanobacteria that detect nitrogen used in the production of tropical and flat land rice (Kannaiyan et al. 1997, De Pauw, Morales, and Persoone 1984, Querijero-Palacpac, Martinez, and Boussiba 1990, Sharma and Mittra 1988). Cyanobacteria fix CO_2 and N_2 using light, and thus contribute to crop growth (Thomas, Zaritsky, and Boussiba 1991). This effect includes increasing the amount of organic substances through photosynthesis, stabilization of soil masses, and crop growth regulators; and it is reported that the capacity of the *Anabaena* species to fix nitrogen offers the potential for use in the production of green fertilizer (Metting 1990, Roger and Watanabe 1986, Boussiba 1991, Ghosh and Saha 1993, Dönmez et al. 2001). Thomas, Zaritsky and Boussiba (1991) have stated that different cyanobacteria species release crop hormones such as gibberellin and auxin, and that culture filtrates enhance seed germination and the development of rice paddies. Kumar, Ghosh, and Bhat (1999) proposed that the long-term use of inorganic fertilizers has caused some disorders in soil health, and Pillai, Kundu, and Subbaiah (1990) point out that organic sources of nutrients play an important role in the restoration of soil health (Okmen 2005).

Use of Marine Algae in Agriculture

Marine algae, which are a natural resource, have wide-ranging uses in agriculture. Certain marine algae, which have been washed ashore for millennia, have been used as fertilizers for generations. The oldest use of marine algae for fertilizing was in the Far East. In the twelfth century, these considerations appeared in countries with lengthy shorelines, such as France, Ireland and England. France began to benefit from marine algae in the seventeenth century. In England, marine algae had begun being collected by 1720, and it is recorded that by the end of the eighteenth century annual algae production had reached 20,000 tonnes of dry algae. This value is accepted as approximately 400,000 tonnes of wet algae (Abetz 1980, Yazıcı and Kaynak 2001, Duan 2013).

Among the brown algae used in European countries were *Fucus*, *Ascophyllum* and *Laminaria*. In America, brown algae with a large thallus, such as Macrocystis and Nereocystis were used (Güner and Aysel 1996, Kavuk 2013).

Even though marine algae have been used as fertilizers for soil in the past, it is only in the last 40–50 years that productivity and product quality have increased, with the application of marine algae extract to leaves by spraying, in comparison to fertilizing the soil directly (Yazıcı and Kaynak 2001).

In recent times, marine algae have been applied in many countries as liquid extracts or by directly mixing them into the soil. The marine algae which are applied directly to the soil are applied to regulate the soil structure and protect soil productivity for a longer period (Özenç and Şen 2017).

The red algae, which are known as "maerly" on the French coast, are used in acidic soils and peats for regulating soil pH instead of lime as they include 80% calcium carbonate (Şimşek 1995).

Generally, liquid extracts of brown algae are marketed for use in agriculture and for houseplants. Most of these extracts are produced from the algae named *Ascophyllum nodosum*. Besides this, in France, *Fucus serratus* and *Laminaria* are used in the production of the extract named "SM3". Kelpak 66, which is marketed today, is produced from *Ecklonia maxima* in South Africa. Moreover, "Algiert" is produced by a Norwegian company under the name of "Seagro" in New Zealand and "Seasol" in Australia. These products are produced from hot water extracts of both dry and wet marine algae (Blunden, Whapham, and Jenkins 1992).

Marine algae extracts are used frequently in many countries for greenhouse cultivation, pomiculture and ornamental cultivation (Güner and Aysel 1996).

Marine algae absorb vitamins, minerals and trace elements in oceans and seas. For this reason, marine algae have been widely employed in agriculture from the past to the present, as well as in other areas (Dring 1991).

Marine algae have important characteristics in organic agriculture for plant growth and development. By providing a strong root development, they enable plants to receive more nutrients and water. By accelerating chlorophyll production, they increase green components and, accordingly, enable substances, such as carbohydrate, protein, etc., to be produced. At the same time, they enable plants to be more resistant against illness and pests, and endure environmental stress. Marine algae are the micro- and macro-sources of nutrition for plants. By enabling the take-up of micro- and macro-nutritional elements, they increase productivity, regulate quality and increase market and export value (Blunden, Whapham, and Jenkins 1992, Kavuk 2013).

As fertilizer material, not only brown algae but also green and red algae are used. Brazilian fishermen have collected *Hypnea* which is common on coasts, and used it as a fertilizer for coconut and palm trees to produce stronger roots. Again, in Brazil, the green algae *Ulva* and *Enteromorpha* have been collected and used for the same purpose (Güner and Aysel 1996).

Farmers who use marine algae regularly have achieved high productivity and quality in clover, soy, cauliflower, cucumber, tomato, potato, and strawberry production. It was recorded that marine algae extracts increased fruit set in citrus fruits, apples, peaches, cherries, grapes and tomatoes (Kumbul 2000).

Applications of marine algae in the form of leaf spray also prevent the damage that can occur in oranges, limes, apples, cucumbers, and tomatoes during harvesting (Blunden 1991).

It was found that marine algae extracts increase the height and dry weight of the plants as a result of both leaf and soil applications. It was recorded that marine algae extracts alter microorganism numbers in the soil under normal circumstances (Allwright 1992).

Himanthalia elongate, which is a brown alga, is used in artichoke cultivation by Breton farmers. Again, brown algae extracts are used for germination on cultivated land in the form of spraying, to bury the seeds in the soil for a long period and allow them to retain water (Allwright 1992).

The impact of the liquid algae extract, Kelpak, on the growth and development of lettuce was examined and it was recorded that it increases the size of the product and the amount of Ca, K and Mg in leaves (Crouch, Beckett, and Van Staden 1990, Yazıcı and Kaynak 2001).

When marine algae stay in the soil for a long period, they produce nitrogen (N) and calcium (Ca) by easily disintegrating in natural circumstances. Furthermore, their positive impact on the productivity of the soil arises from compounds such as macro- and micro-elements, plant growth regulators and betaines (Hong et al. 1995).

As a result of the use of marine algae extracts in agriculture, various effects have been noted, such as an increase in germination; provision of better root development; increased storage life; larger and darker leaf and flower formation; increased resistance against stress conditions, such as illness and pests, frost and drought; greater intake of nutritional elements by the soil; and enabling plants to stay young for longer periods of time (Hong et al. 1995).

Among algae, the cultivation of other algae types is more difficult and uncontrolled in comparison to microalgae which can be cultivated more conveniently and in a more controlled manner. For this reason, the use of microalgae as biofertilizers has proliferated, and is continually increasing. It has been shown in studies that microalgae can be used as a fertilizer, in addition to their use as soil regulators and plant growth regulators (PGR) (Namlı et al. 2016).

Microalgae have become one of the most discussed groups of organisms due to their high protein, vitamin, mineral, fatty acids and pigment content. Among the microalgae, the types most beneficial for plants are those which can be fixed to nitrogen (Roger and Reynaud 1982, Namlı et al. 2016).

References

Abetz, P. 1980. "Seaweed extracts: have they a place in Australian agriculture or horticulture?" *Journal of the Australian Institute of Agricultural Science* 46 (1):23–29.

Açıkgöz, E. 2001. "Yem bitkileri. *Uludağ Üniversitesi Ziraat Fak. Tarla Bit. Böl. Uludağ Üniversitesi Basımevi, Bursa.*

Adekiya, A O, T M Agbede, C M Aboyeji, O Dunsin, and J O Ugbe. 2017. "Green manures and NPK fertilizer effects on soil properties, growth, yield, mineral and vitamin C composition of okra (*Abelmoschus esculentus* (L.) Moench)." *Journal of the Saudi Society of Agricultural Sciences.*

Aksoy, U, and A Altındişli. 1999. "Dünya'da ve Türkiye'de ekolojik tarım ürünleri üretimi, ihracatı ve geliştirme olanakları." *İstanbul Ticaret Odası Yayınları* (1999–70).

Albrecht, S L 1998. *Eukaryotic algae and cyanobacteria, Principles and Applications of Soil Microbiology.* America: Prentice-Hall Incorporated.

Ali, M. 1999. "Evaluation of green manure technology in tropical lowland rice systems." *Field Crops Research* 61 (1):61–78.

Allwright, K J. 1992. "Effect of Seaweed Extracts on Growth of wheat, and Soil-BorneDiseases." Abstract of the 14th International Seaweed Symposium, Brest and St Malo, France.

Anaç, D, and B. Okur. 1996. "Increasing Soil Efficacy by Natural Ways." *Ecological Agriculture, Org. Association.*

Biswas, B, and P M Gresshoff. 2014. "The role of symbiotic nitrogen fixation in sustainable production of biofuels." *International Journal of Molecular Sciences* 15 (5):7380–7397.

Blunden, G. 1991. "Agricultural uses of seaweeds and seaweed extracts." *Seaweed Resources in Europe*: 65–81.

Blunden, G, C Whapham, and T Jenkins. 1992. "Seaweed extracts in agriculture and horticulture: their origins, uses and modes of action."

School of Pharmacy and Biomedical Science and School of Biological Sciences, University of Portsmouth, King Henry John Street, Portsmouth, Hampshire P01 UK:202.

Boussiba, S. 1991. "Nitrogen fixing cyanobacteria potential uses." *Plant and Soil* 137 (1):177–180.

Cavigelli, Michel A, and Steve J Thien. 2003. "Phosphorus bioavailability following incorporation of green manure crops." *Soil Science Society of America Journal* 67 (4):1186–1194.

Chatterjee, A, S Singh, C Agrawal, S Yadav, R Rai, and L C Rai. 2017. "Role of Algae as a Biofertilizer." In *Algal Green Chemistry*, 189–200. Elsevier.

Chen B, E Liu, Q Tian, C Yan, and Y Zhang. 2014. "Soil nitrogen dynamics and crop residues. A review." *Agronomy for Sustainable Development* 34 (2):429–442.

Crews, T E, and M B Peoples. 2004. "Legume versus fertilizer sources of nitrogen: ecological tradeoffs and human needs." *Agriculture, Ecosystems and Environment* 102 (3):279–297.

Crouch, I J, R P Beckett, and J Van Staden. 1990. "Effect of seaweed concentrate on the growth and mineral nutrition of nutrient-stressed lettuce." *Journal of Applied Phycology* 2 (3):269–272.

Davies, W P. 2003. "An historical perspective from the green revolution to the gene revolution." *Nutrition Reviews* 61 (s6).

De Pauw, N, J Morales, and G Persoone. 1984. "Mass culture of microalgae in aquaculture systems: progress and constraints." *Hydrobiologia* 116 (1):121–134.

Dinnes, D L, D L Karlen, D B Jaynes, T C Kaspar, J L Hatfield, T S Colvin, and C A Cambardella. 2002. "Nitrogen management strategies to reduce nitrate leaching in tile-drained Midwestern soils." *Agronomy Journal* 94 (1):153–171.

Dönmez, G, S Taban, F Varol, and S Dönmez. 2001. "Effect of Cyanobacteria isolated from Rice fields on growth of rice plants." 1th Symposium of Alg Technology, Turkey.

Dring, M J. 1991. *The biology of marine plants*: Cambridge University Press.

Duan, E. 2013. "The production of the fermented liquid organic fertilizer of some macro algae (*Ulva* sp. *Cystosiera* sp. ve *Corallina* sp.) and determine the effect of yield bean (*Phaseolus vulgaris*)." Master Thesis, Graduate School of Natural and Applied Sciences, Department of Biology, Giresun University.

Duman, İ, S Kaya, E Düzyaman, U Aksoy, LAlbitar, C A Nazik, E Bilen, M Unal, and N Özsoy. 2013. "The effect of green fertilizer by organic

production (*Vicia sativa*) on efficiency and soil properties of some vegetables."
Duyar, H. 2007. "The effects of green manure fertilization on organic vegetable production in greenhouse." Ph.D. Thesis, Horticulture Science, Ege University.
Duyar, H. 2014. "Effect of green manure and poultry manure on yield and fruit quality in organic greenhouse tomato production." *Journal of Agricultural Sciences* 20 (1): 10–18.
Edesi, L, M Jarvan, E Lauringson, E Akk, and K Tamm. 2013. "The effect of solid cattle manure on soil microbial activity and on plate count microorganisms in organic and conventional farming systems." *International Journal of Current Microbiology and Applied Sciences* 2:476–488.
Engin, Y Ö. 2009. "Use of seaweeds as organic fertilizers." Ph.D. Thesis, Department of Aquaculture, Ege University.
Florentín, M A, M Peñalva, A Calegari, R Derpsch, and M J McDonald. 2010. "Green manure/cover crops and crop rotation in conservation agriculture on small farms." *Integrated Crop Management* 12:97.
Gardner, C M K, K B Laryea, and P W Unger. 1999. *Soil physical constraints to plant growth and crop production*: Land and Water Development Division, Food and Agriculture Organization.
Ghosh, T K, and K C Saha. 1993. "Effects of inoculation with N 2–fixing cyanobacteria on the nitrogenase activity in soil and rhizosphere of wetland rice (*Oryza sativa* L.)." *Biology and Fertility of Soils* 16 (1):16–20.
Gomiero, T. 2016. "Soil degradation, land scarcity and food security: Reviewing a complex challenge." *Sustainability* 8 (3):281.
Göktekin, Z. 2015. "Effects of farm manure, green manure, microbial fertılizer and plant actıvator on yield and quality of grown tomato." M.Sc. Thesis, Department of Horticultural Sciences, Süleyman Demirel University.
Güner, H, and V Aysel. 1996. *Cryptogamic plant systematics* Turkey: Ege University Faculty of Science Book Series.
Hirel, B, T Tétu, P J Lea, and F Dubois. 2011. "Improving nitrogen use efficiency in crops for sustainable agriculture." *Sustainability* 3 (9):1452–1485.
Hong, Y P, C C Chen, H L Cheng, and C H Lin. 1995. "Analysis of auxin and cytokinin activity of commercial aqueous seaweed extract." *Gartenbauwissenschaft (Germany)*.
İlter, E. 1996. "Ecological agriculture and principles." *Ecological Agriculture Organization*:1–6.

İnal, A, S Sözüdoğru, and D Erden. 1996. "The content of chicken manure and fertilizer value." *Journal of Agricultural Sciences* 2 (3):45–50.

Irisarri, P, S Gonnet, and J Monza. 2001. "Cyanobacteria in Uruguayan rice fields: diversity, nitrogen fixing ability and tolerance to herbicides and combined nitrogen." *Journal of Biotechnology* 91 (2–3):95–103.

Kacar, B, and V Katkat. 1999. *Fertilizers and fertilization techniques* Turkey: Uludağ Üniversitesi Vakfı.

Kannaiyan, S, S J Aruna, S M P Kumari, and D Hall. 1997. "Immobilized cyanobacteria as a biofertilizer for rice crops; Intl. Conference on Applied Algology, Knysna, South Africa, April 1996." *Journal of Applied Phycology* 9 (2):167–174.

Karakurt, E. 2009. "Green fertilizers and fertilization in terms of soil fertility." *Journal of Field Crops Central Research Institute* 18 (1–2).

Kavuk, E. 2013. "Manufacturing of suspended organic fertilizer from some marine macroalgae (*Ulva* sp., *Cystoseira* sp., *Corallina* sp.) and determination of its effect by biotreatment." Ph.D., Department of Biology, Giresun University

Kaya, S, İ Duman, E Düzyaman, C A Nazik, E Bilen, M Ünal, U Aksoy, and L Al-Bitar. 2010. "Determination of the effect of yield and quality on some summer vegetables of brokkolinin cultivated as fallow and preliminary plant in organic farming." Turkey IV Organic Agriculture Symposium, Turkey, June 28–July 1.

Kumar, V, B C Ghosh, and R Bhat. 1999. "Recycling of crop wastes and green manure and their impact on yield and nutrient uptake of wetland rice." *The Journal of Agricultural Science* 132 (2):149–154.

Kumbul, B. 2000. "Usage areas of marine algae in garden plants." Thesis, Department of Horticulture, Akdeniz University.

Leonard, D. 1986. *Soils, Crops, and Fertilizer Use: A Field Manual for Development Workers*: United States Peace Corporation.

Metting, B. 1990. "Microalgae applications in agriculture." *Developments in Industrial Microbiology* 31:265–270.

Mohler, Charles L, and Sue Ellen Johnson. 2009. *Crop rotation on organic farms: a planning manual*: Natural Resource, Agriculture, and Engineering Service (NRAES) Cooperative Extension Ithaca.

Mooleki, S P, Y Gan, R L Lemke, R P Zentner, and C Hamel. 2016. "Effect of green manure crops, termination method, stubble crops, and fallow on soil water, available N, and exchangeable P." *Canadian Journal of Plant Science* 96 (5):867–886.

Morshedi, L, F Lashgarara, S J F Hosseini, and M Omidi Najafabadi. 2017. "The role of organic farming for improving food security from the perspective of fars farmers." *Sustainability* 9 (11):2086.

Namlı, A, O C Turgay, T Tekinay, N T Gök, and B Ertit. 2016. *Development of biological fertilizer from microalgae*. Turkey: Ankara Unıversıty.

Okmen, K G. 2005. "Influence of environmental factors on nitrogenase activity of cyanobacteria isolated from rice fields." Ph.D. Thesis, Department of Biology, Ankara University.

Okur, B 2008. "To increase agricultural production and soil efficacy; green fertilization." *Ege University Agricultural Application and Research Center Bulletin*.

Özenç, D, and O Şen. 2017. "The effect on tomato growth and some quality properties of seaweed fertilizer applied at different development stages." *Academic Journal of Agriculture* 6:235–242.

Pillai, K G, D K Kundu, and S V Subbaiah. 1990. "An integrated approach to nutrient management in rice." *Indian Farming* 40 (9):15–18.

Querijero-Palacpac, N M, M R Martinez, and S Boussiba. 1990. "Mass cultivation of the nitrogen-fixing cyanobacterium *Gloeotrichia natans*, indigenous to rice-fields." *Journal of Applied Phycology* 2 (4):319–325.

Roger, P A. 1985. "Blue-green algae in rice fields." *A technical document issued by the International atomic energy agency*:9–128.

Roger, P-A, and P-A Reynaud. 1982. "Free living blue green algae in tropical soils." In *Microbiology of tropical soils and plant productivity*, 147–168. Springer.

Roger, P-A, and I Watanabe. 1986. "Technologies for utilizing biological nitrogen fixation in wetland rice: potentialities, current usage, and limiting factors." In *Nitrogen Economy of Flooded Rice Soils*, 39–77. Springer.

Rosenfeld, A, and F Rayns. 2011. "Sort Out Your Soil: A practical guide to green manures."

Santi, C, D Bogusz, and C Franche. 2013. "Biological nitrogen fixation in non-legume plants." *Annals of Botany* 111 (5):743–767.

Sharma, A R, and B N Mittra. 1988. "Effect of green manuring and mineral fertilizer on growth and yield of crops in rice-based cropping on acid lateritic soil." *The Journal of Agricultural Science* 110 (3):605–608.

Sorensen, J N, and K Thorup-Kristensen. 2006. "An organic and environmentally friendly growing system for greenhouse tomatoes." *Biological Agriculture and Horticulture* 24 (3):237–256.

Soyergin, S. 2006. "Conservation of Soil Fertility in Organic Agriculture, Fertilizers and Organic Soil Treatment." In *Achieving Sustainable Competitive Advantage Organic Agriculture Sector, Sectoral Strategies and Applications*, edited by İ H Eraslan and F Şelli. Turkey: International Competition Research Institution Association.

Sullivan, P. 2003. "Overview of cover crops and green manures: Fundamentals of sustainable agriculture." https://attra.ncat.org/publication.html.

Şimşek, Z. 1995. "The effects of bracelet, iron compounds and seaweed extract applications on yield and quality of Klemantine Mandarin." Master Thesis, Department of Horticulture, Akdeniz University.

Talgre, L. 2013. "Biomass production of different green manure crops and their effect on the succeeding crops yield." PhD Thesis, Eesti Maaülikool.

Thomas, S P, A Zaritsky, and S Boussiba. 1991. "Ammonium excretion by a mutant of the Nitrogen-fixing cyanobacterium *Anabaena siamensis*." *Bioresource Technology* 38 (2–3):161–166.

Tüzel, Y, H Duyar, G B Öztekin, Ö Kılıç, D Anac, N Madanlar, and Z Yoldaş. 2013. "Effects of winter green manuring on organic cucumber production in unheated greenhouse conditions." *Turkish Journal of Agriculture and Forestry* 37 (3):315–325.

Urzua, H, J M Urzua, and R Pizarro. 2001. "Pre-selection of *Rhizobium leguminosarum* bv. viceae stains in forage vetch for use as green manure." *Ciencia e Investigación Agraria* 28 (1):3–6.

Watson, C A, H Bengtsson, M Ebbesvik, A-K Løes, A Myrbeck, E Salomon, J Schroder, and E A Stockdale. 2002. "A review of farm-scale nutrient budgets for organic farms as a tool for management of soil fertility." *Soil Use and Management* 18 (s1):264–273.

Yağmur, B. 2001. "Applications in Organic Agriculture and Organic Agriculture." In *Agricultural Environment and Water Pollution*. Turkey: Turkey Ministry of Food, Agriculture and Livestock.

Yazıcı, K, and L Kaynak. 2001. "Usage of sea moss in organic farming." Turkey 2nd Symposium on Ecological Agriculture, Turkey November 14–16.

Yılmaz, D. 2010. "The importance of green manure on mineral nitrogen input nitrogen economy of culture plants and environment." M.Sc. Thesis, Institute of Natural and Applied Sciences, Çukurova University.

CHAPTER 8

VERMICOMPOSTING: A BIOTECHNOLOGICAL PROCESS THAT SPEEDS UP THE ORGANIC WASTE RECYCLING PROCESS

ESRA KOC

Ankara University, Faculty of Science, Department of Biology, Ankara, TURKEY
ekoc@science.ankara.edu.tr, esrakoc.es@gmail.com

Abstract

Industrial agriculture involving the intensive use of agrochemicals has a negative effect on the environment, and can cause serious health problems for humans and other species. Producers who advocate sustainable and organic agricultural models have therefore looked to the development and use of alternative organic-based products that can be used instead of chemical fertilizers and pesticides. Organic agriculture has become one of the fastest-growing sectors in global agriculture, creating greater profits and healthier products. It plays an important role in the nutrition of the rapidly growing world population—the rapid population growth requires the efficient use of all organic resources including large quantities of organic waste. Organic fertilizers originate from plant and animal waste and are recycled in the soil. Vermicompost is a biotechnological method where worms are employed to cooperate with micro-organisms in order to convert organic waste into a valuable product. Vermicomposting enhances degradation of organic waste and enables the reduction of environmental pollution. It protects environmental and human health, and at the same time increases the sustainability of soil fertility. Vermicompost is the method that provides the highest economic benefit among the methods that support sustainability in agriculture today.

Keywords: Earthworms, organic waste, recycling, vermicompost

A "Green Revolution" movement emerged and spread around the world like an "epidemic" after World War II, encouraging the use of chemical fertilizers and pesticides (DDT in particular) in agricultural production. This revolution was aimed at providing increased yields in the short-term (Schuman and Simpson 1997; Şimşek Erşahin 2007). In 1962, Rachel Carson reported that excessive and uncontrolled pesticide use negatively affected the environment and had the capacity to cause irreparable damage. Scientific studies conducted in the following years showed that industrial agriculture involving the intensive use of pesticides had negative effects on the environment, and might cause serious health problems for humans and other species (Akhtar and Malik 2000).

In many Asian countries, providing people with sufficient food has always been the greatest challenge owing to the high population numbers in these regions. As a result of a limited and gradually shrinking area of agricultural land per capita, it is necessary to take measures to increase agricultural production so that the demands of a growing population can be satisfied. There is also a need for rich and healthy soil, which allows plants to have necessary nutrients to maintain the food production on these shrinking agricultural lands (Ozturk et al. 2008). Intensive and careless use of chemical fertilizers to increase agricultural yield reduces the quality of soil and causes a gradual deterioration in its health, as well as increasing the resilience of pathogens and negatively affecting environmental safety and human health. Mineral fertilizers can only provide the soil with plant nutrients; however, they are not effective for other physical, chemical, and biological aspects of soil health (Chattopadhyay 2012). Detecting chemical fertilizer residues in underground and surface waters, detecting pesticide residues in human and animal food products, and determining the carcinogenic effects of such residues has therefore led to the process of reviewing industrial and traditional agricultural methods (Baier-Anderson and Anderson 2000; Şimşek Erşahin 2007).

In this sense, organic materials play a much more positive role. Studies report that both various organic materials used by farmers such as cow manure, fowl manure, and farm manure and inorganic fertilizers improve plant production-yield and help maintain soil fertility (Chukwu et al. 2012; Sharma and Chetani 2017). Organic farming has therefore become one of the fastest-growing sectors in global agriculture. The main purpose of organic farming is to maintain the balance between interconnected systems such as soil organisms, plants, animals, and humans. Organic fertilizers are derived from biological or living materials. These may take various forms, including fertilizers obtained from chicken, goat, and other farm animals, green fertilizers obtained from plants such as various types of legumes, and composts consisting of organic waste materials such as hay, cornstalk, or degraded waste (Sharma and Chetani 2017). In this context, the concept of an integrated plant nutrition system involving both inorganic and organic sources of nutrition has been suggested and has gained in popularity (Chattopadhyay 2005 and 2012).

Soils from many tropical and subtropical countries are weak in terms of organic substances (especially organic carbon content) owing to high temperatures and high microbial activity; the integration of mineral fertilizers and organic fertilizers has thus become vital for most Asian countries. It is therefore necessary to add organic materials to the soil on a regular basis, and in large quantities, in order to protect soil health by maintaining organic substance content at an optimum level (Manna et al. 2005). Although the use of organic substances is very significant for tropical soils, the use of organic substances obtained from traditional sources on a large scale has been a great challenge. Millions of tonnes of solid waste produced in modern societies enter storage areas every day, and the management and monitoring of this waste by local governments to ensure environmental safety causes massive economic and environmental problems (such as greenhouse gas emission, poisonous gases, leakage to underground water and soil). A serious contemporary concern is the emission of greenhouse gases such as methane (CH_4) and nitrogen oxide (N_2O) released during the disposal of solid waste in storage areas or using conventional composting systems. "Biodegradable" organic waste constitutes a very large portion of solid waste (e.g. potato skins, cafeteria waste, sericultural waste, sugar cane, agricultural waste, onion, garlic waste, paper factory sludge, etc.). The rapid growth in human population requires the efficient use of all organic resources including large quantities of organic waste. Organic waste can be treated to become a biofertilizer that is 4–5 times stronger than traditional composts and chemical fertilizers in terms of plant improvement, yield and safe food production (Edwards 1998; Pandit et al. 2012). Successfully recycling this waste will not only significantly increase plant nutrients, but also contribute to the improvement of soil health by adding organic substances to the soil.

Hence, producers who advocate sustainable and organic agricultural models realize the necessity of developing organically-based alternative products to replace chemical fertilizers and pesticides. One of the primary new approaches in this field is the use of aerobic (thermophilic) compost products to increase the organic substance content of the soil. The fact that these compost products provide plants with required nutrients and inhibit plant pathogens has led to their extensive use in organic farming (Hoitink et al. 1997; Hoitink and Boehm 1999; Lim et al. 2015; Eker 2016). The other effective factor in the rapid spread of compost applications in the 1980s was that compost is an economic, sustainable, and environmentally-friendly alternative to treating urban waste, which became a significant environmental problem in parallel with increased urbanization. Observations and studies on degradation behaviours and the activities of various organic substances in the mid-twentieth century gave rise to the idea of using living organisms to compost a wide variety of organic waste (Senapati 1993; Chattopadhyay 2012). Later, studies on composting have shown that the use of the vermicomposting (mesophilic composting) method in the recovery of urban and industrial organic waste was superior to the use of aerobic composting in terms of both process and product (Dominguez et al. 1997; Boran 2015).

In the process where approaches which emphasize the concept of sustainability in agricultural production worldwide and promoting organic production methods have become popular, the realization that earthworms have the capacity to turn organic waste into high-quality and valuable products in a short span of time has led to increased interest in these organisms (Darwin 1837). Actually, the positive effects of worms on the soil have long been known. Charles Darwin described worms as “the unheralded soldiers of mankind”, since they can digest a wide variety of organic substances, while Aristotle described them as “the gut of the earth” (Martin 1976; Darwin and Seward 1903; Pathma and Sakthivel 2012). Earthworms regulate the physical properties of the soil and contribute to organic substance degradation and humus formation. In short, they have positive effects on soil fertility. Earthworms are found at depths of 30–45 cm, which are particularly rich in nutrients, and help maintain soil temperature and moisture at optimum levels. Their activities increase in spring, and they move to deeper areas during summer when the soil is dry. Among the organisms which are important for soil fertility, earthworms perform their activities during the night, and help ventilate the soil by constantly digging new channels, allowing rain water to accumulate. Worm faeces contain materials with antibiotic properties, amino acids, and vitamins generated in the digestive system, which improve the biological properties of fertilizers. Also, they contain growth regulators, such as humic acid and fulvic acid, which are important for plant nutrition.

Worm manure is a very useful material for many plants, as well as a very valuable soil regulator (Yıldız et al. 2005; Demir et al. 2010). All these properties of earthworms led to a new sector of agricultural production—called vermiculture—in Europe, India, and the United States over the last

40 years. Vermiculture is a process whereby all sorts of biodegradable waste such as farm waste, kitchen waste, market waste, and the biowaste from agriculture-based industries are transformed into a nutritious and rich vermicompost while passing through the worm's gut (Adhikary 2012).

Although thermophilic composts are more commonly chosen as an organic fertilizer today, studies conducted in recent years have led to a gravitation towards vermicompost. The reason for this may be increased environmental pollution, because this pollution and the waste that accumulates negatively affect human and environmental health. In addition, vermicompost production takes much less time compared to thermophilic compost. Vermicompost products have superior physical, chemical, and biological properties and a higher economic value compared to thermophilic compost products (vermicompost end-products are odourless and non-toxic). Vermicompost end-products do not contain pathogens that threaten human health; the applicator may touch the vermicompost with bare hands, even if the main material is sewage waste (Şimşek Erşahin 2007; Namlı et al. 2014).

Vermicompost is a type of biotechnological compost obtained from certain worm species to speed up the waste recycling process and achieve a better final product (similar to humus) (Boran 2015). For example, one million worms, which double in number every two months, can be 64 million worms at the end of the year (Eker 2016). Considering that an adult worm (*Eisenia fetida* in particular) consumes an amount of organic waste equal to its own weight, 64 million worms (weighing 64 tonnes) would consume 64 tonnes of waste and produce 30–32 tonnes of vermicompost with a conversion rate of 40–50%. Composting is an agricultural system aimed at eliminating environmental pollution. It protects environmental and human health, as well as improving the sustainability of soil fertility. Vermicompost takes place by feeding household, industrial, animal, and similar waste which would cause environmental pollution to certain species of worms, and environmental pollution is thus eliminated by contributing to the environment (Pandit et al. 2012).

The most commonly preferred species in commercial vermicomposting establishments include *Eisenia spp* and the Californian Redworm *Lumbricus rubellus* (Dickerson 2004). The term "vermicompost" is used for the product obtained from the organic waste composting process performed with worms such as *Lumbricus rubellus* and *Eisenia foetida*, while the vermicompost product is usually referred to as "vermicast", or "cast" for short. Fertilizer worm species widely used in the vermiculture industry include *Eisenia fetida* (tiger worm), *Eisenia andrei* (red tiger worm), *Eudrilus eugeniae* (African night crawler), *Dendrobaena veneta*, *Lumbricus rubellus* (red worm), *Fletcherodrilus spp., Herteroporodrilus spp., Perionyx excavatus* (Indian blue worm), and *Pheretoma excavatus*. While *E. fetida*, *E. andrei* and *D. veneta* adapt well to regions in the mild temperate zone, *L. rubellus* and *P. excavatus* are more commonly seen in warm tropical regions. These five species give the best results in vermicomposting activities performed to degrade organic waste (Edwards and Bohlen 1996; Şimşek Erşahin 2007).

During their passage through the earthworm's gut, they remove fragments and produce bacteria-rich excrement, thereby homogenizing the organic material. The term "biofertilizer", or more appropriately "microbial inoculants", may generally be defined as preparation containing live or latent cells of efficient strains of nitrogen fixing micro-organisms (Tiunov and Scheu 2000; Aira et al. 2003; Pandit et al. 2012). Vermicomposting is a simple biotechnological composting process which uses some worms and micro-organisms to improve the waste conversion process and achieve a better product. It is a mesophilic process, which uses worms and micro-organisms which are active at 10°C to 32°C (Adhikary 2012). While micro-organisms are responsible for the biochemical degradation of organic substances, earthworms fragment the substrate, thereby increasing the surface area for microbial degradation, which constitutes the active phase of vermicomposting (Munnoli et al. 2010; Pathma and Sakthivel 2012). In this process, worms act as mechanical blenders, and by comminuting the organic substance, they modify its physical and chemical status, gradually reducing its C:N ratio. Waste feed materials ingested are finely ground with the aid of stones in their muscular gizzard into small particles and passed on to the gut for enzymatic actions. Worms secrete enzymes protease, lipase, amylase, cellulase, and chitinase in their gizzard and gut which allows for rapid biochemical conversion of cellulosic and protein-containing materials in the waste organics. Worms thus ingest food materials, eliminate harmful micro-organisms, and deposit them mixed with minerals (such as phosphor and nitrogen) and beneficial microbes as vermicasts in the soil. This end product, vermicompost, is a valuable and marketable plant-growth medium similar to natural soil humic materials. Worms host millions of decomposer micro-organisms (bacteria, actinomycetes, fungi) in their guts (Haritha Devi et al. 2009; Pathma and Sakthivel 2012). Earthworms ingest these micro-organisms with food material and the ingested material is reported to increase while passing through the gut. Minerals such as nitrogen and phosphor are used by micro-organisms for multiplication and action. If the ratio between elements such as oxygen (O), and hydrogen (H), carbon (C) and nitrogen (N) is not well maintained, the microbial activity and the composting process will therefore stop.

Vermicompost is a nutritious organic fertilizer that is rich in NPK (nitrogen, potassium, and phosphor), micronutrients, useful soil microbes such as nitrogen-fixing bacteria, and mycorrhizal fungi. It also contains useful amounts of magnesium (Mg), zinc (Zn), calcium (Ca), and manganese (Mn) (Yasir et al. 2009). Additionally, vermicompost contains enzymes such as amylase, lipase, cellulase, and chitinase, which continue to decompose organic substances in the soil even after being applied to the soil (Munnoli et al. 2010). A sufficient annual vermicompost application causes a significant increase in the activities of soil enzymes such as "phosphomonoesterase", and "phosphodiesterase", and the soil has significantly higher electrical conductivity and comes closer to a neutral pH (Sinha et al. 2012; Amouei et al. 2017). Vermicompost contains antibiotics and actinomycetes which help improve the "biological resistance" against plant mites and soil-borne diseases. Hahn (2011) reported that when vermicompost was dissolved in water and turned into the solution called "vermicompost tea", it could be used as a spray to repel pests and suppress plant diseases. Vermicompost contains plant growth hormones such as auxin, gibberellin, and cytokinin, which are generated by micro-organisms during the vermicomposting process (Padmavathiamma and Kumari 2008; Singh et al. 2008; Lim et al. 2015). Grappelli et al. (1985) compared the effects of the aqueous vermicompost extract derived from vermicompost and hormones such as auxin, gibberellin, and cytokinin directly added to the soil on the growth of ornamental plants, and found that they had similar effects. Humic acid contained in vermicompost improves root growth, lateral root formation, and food intake by increasing the permeability of root cell membranes. Previous studies have shown that vermicomposting led to a higher decrease in pathogens after storage, while thermophilic compost samples had higher pathogen levels after three months. Several studies demonstrated that there were biologically accumulated and decomposed various organic and inorganic chemicals in the environment including "heavy metal", "organochlorine pesticide", and "polycyclic aromatic hydrocarbon" (PAH) residues (Kothari et al. 2012). The addition of vermicompost to the soil improves its water-retention capacity, minimizes evaporation losses, and serves as a good absorber for the moisture in the atmosphere due to colloidal materials (worm mucus) in its composition (Sinha 2011 and 2012; Singh 2012). Vermicompost has very high porosity, ventilation, drainage, and water retention capacity. Vermicompost does not contain weed seed, is completely natural, and does not cause toxic effects to the plant. For this reason, it can be safely applied to all plants. The bacteria in it compete against harmful bacteria in the soil, which improves the resistance of the plant. Its organic nature regulates soil pH and soil structure, and it is useful for the soil for prolonged periods of time.

Studies have shown that excretion products (vermicast) obtained by treating waste using worms are effective for plant growth, and compared to inorganic fertilizers, vermicompost applications were more effective in seed germination, seedling growth, inflorescence, and fructification (in terms of size and taste) in primary products such as wheat, rice, corn, sugar cane, tomato, potato, okra, spinach, strawberry, grape, and apple (Ansari 2008; Sinha et al. 2009; Dominguez et al. 2014; Hanc and Chadimova 2014; Zuo Y. 2018). It was also found that the storage value of fruits increased and the soil quality improved. Vermicompost applications on pot plants and field crops were found to show a superior performance compared to fertilizers and traditional composts in terms of plant height, flower and fruit count, leaf colour and texture, and fruit structure. In all growth studies, plants were observed to give the best growth responses when vermicompost made up a small ratio of the total plant growth medium volume (10–20%). Compared to traditional fertilizers and chemical fertilizers, grains and vegetables cultivated on vermicompost were found to demonstrate extraordinary improvement (Sinha 2011 and 2012; Sing 2012).

Organic waste must be compatible with the reasons for its agricultural use, and negative effects on plant growth must be avoided. Primary sewage sludge, waste treatment sludge derived from various facilities including dairy farms, sugar industry waste, textile waste, swine manure, water hyacinth, product residues, farm animal manure, paper waste, kitchen waste, and other similar waste were tested for their suitability for vermicomposting. Commonly used biodegradable organic waste was used as compost material in vermicomposting. This included manure from, e.g. sheep, horses, goats, and poultry; agricultural waste such as plant leaves, stems, crusts, the leafy waste materials from fruit orchards, processed food waste, sugar cane waste, and rye bread waste; forest waste such as wood shavings, bark, sawdust, and wood pulp; urban leaf waste such as leaves found in streets or settlements, waste from mango, guava, and some weed, paper and cotton waste, kitchen waste; industrial waste such as waste obtained from sericulture, vegetable oil industry waste, potato and corn chips industry waste, mining waste (sulfur waste), aromatic oil extraction waste, food processing waste, distillery waste, and biogas waste. Many researchers examined degradation by worms and compost formation, and reported this organic waste to be excellent feeding material for worms in the vermicomposting process (Kale et al. 1998: Lakshmi et al. 2000; Kothari et al. 2012; Pandit et al. 2012). Urban and rural waste cause continual pollution, which poses a threat to environmental and social health. The source of poisonous materials on the soil surface is essentially solid waste containing pesticides used for health and agriculture and heavy metals released by industrial facilities. The accumulation of toxic materials in worm tissue has great ecological importance, since these animals constitute a significant component of the food chain for various birds and mammals. This is very important for minimizing soil pollution (Kothari et al. 2012). From a technological point of view, worms ingest heavy metals in fly ash from coal plants (which causes serious disposal problems owing to the hazardous heavy metals in its composition), and transform these into vermicompost (Saxena et al. 1998).

It is necessary to spread organic fertilizer use instead of chemical inputs in order to ensure a durable, environmentally-friendly, and sustainable agricultural system that protects natural resources. Vermicomposting is a biotechnological process where worms are used. Worms are natural bioreactors which play a significant role in degrading organic substances and ensuring soil fertility. It is possible to prepare vermicompost without large-scale investments. Today, this is the method that has the highest economic benefit while also supporting sustainable agriculture. Owing to rapid global urbanization and industrialization, the nature and volume of waste have changed significantly. Millions of tonnes of food waste, which have become a great environmental problem with rapid industrial development and population growth, cause serious costs and environmental damage every year. This technology reduces the large-scale environmental effects of solid waste management. It is also an innovative technology which has the potential to minimize the organic load in storage areas through recycling. It is the most economic and sustainable option for organic waste management in nature. Vermicompost turns many types of organic waste into nutritious soil for plants. For this reason, vermicompost is considered, first of all, as medication for soil health, and secondly, as a food provider for crops. Prospective studies on vermicompost use in agriculture must be guided to certain fields of research (Sinha et al. 2012) which include reducing greenhouse gas emission and global warming by atmospheric withdrawing "carbon" into soil through vermicompost during the organic waste vermicomposting process; reducing irrigation water use by improving the water retention capacity of the soil; improving food quality and the storage values of fruits and vegetables and protein values of grains beans through vermicomposting; and using vermicompost tea to repel pests and suppress diseases instead of chemical pesticides used against insects, pests, and micro-organisms, which harm products and cause diseases.

References

Adhikary, S. 2012. "Vermicompost, the story of organic gold: A review." *Agricultural Sciences* 3 (7):905–917.

Aira, M, F Monroy, and J Domínguez. 2003. "Effects of two species of earthworms (*Allolobophora* spp.) on soil systems: a microfaunal and biochemical analysis." *Pedobiologia* 47:877–881.

Akhtar, M, and A Malik. 2000. "Roles of organic soil amendments and soil organisms in the biological control of plant-parasitic nematodes: a review." *Bioresource Technology* 74 (1):35–47.

Amouei, A I, Z Yousefi, and T Khosravi. 2017. "Comparison of vermicompost characteristics produced from sewage sludge of wood and paper industry and household solid wastes." *Journal of Environmental Health Science and Engineering* 15 (5):2–6.

Ansari, A A. 2008. "Effect of Vermicompost on the Productivity of potato (*Solanum tuberosum*) spinach (*Spinacia oleracea*) and turnip (*Brassica campestris*)." *World Journal of Agricultural Sciences* 4 (3):333–336.

Baier-Anderson, C, and R S Anderson. 2000. "The effects of chlorothalonil on oyster hemocyte activation: Phagocytosis, reduced pyridine nucleotides, and reactive oxygen species production." *Environmental Research* 83 (1):72–78.

Boran, D. 2015. "Determination of quality parameters of vermicompost under different thermal techniques." Master Thesis, Department of Soil Science and Plant Nutrition, Ankara University Graduate School of Natural and Applied Sciences.

Chattopadhyay, G N. 2005. *Vermicomposting as a biotechnological tool for recycling organic wastes*. Edited by T K Ghosh, T Chakrabarti and G Tripathi. Vol. 1, *Biotechnology in environmental management*. New Delhi: A.P.H. Publishing Corporation.

Chattopadhyay, G N. 2012. "Use of vermicomposting biotechnology for recycling organic wastes in agriculture." *International Journal of Recycling of Organic Waste in Agriculture* 1 (8):2–6.

Chukwu, L I, A O Ano, and D O Asawalam. 2012. "Effects of Poultry Manure and NPK fertilizer on soil properties and Nutrient uptake of maize (*Zea mays* L.) plants growth in an ultisol." Proceedings of the 36th Annual Conference of the Soil Science Society of Nigeria (SSSN) on 7th–11th March. University of Nigeria Nsukka.

Darwin, C. 1837. "On the formation of vegetable mould." *Transactions of the Geological Society of London* 2 (5):505–509.

Darwin, F, and A C Seward. 1903. "More letters of Charles Darwin." In *A record of his work in series of hitherto unpublished letters*, edited by M John, 508. London.

Demir, H, E Polat, and İ Sönmez. 2010. "A new organic fertilizer for our country: Vermicompost." *Tarım Aktüel* (14): 54–60.

Dickerson, G W. 2004. Vermicomposting. Cooperative extension service. College of agriculture and home Economics. New Mexico State University. http://www.cahe.nmsu.edu/Pubs/_h/h_164.pdf.

Dominguez, J, C A Edwards, and S Subler. 1997. "A comparison of vermicomposting and composting." *Biocycle* 38 (4):57–59.

Domínguez, J, H Martínez-Cordeiro, M Álvarez-Casas, and M Lores. 2014. "Vermicomposting grape marc yields high quality organic biofertiliser and bioactive polyphenols." *Waste Management and Research* 32 (12):1235–1240.

Edwards, C A. 1998. *The Use of Earthworms in the Breakdown and Management of Organic Wastes*. Boca Raton: CRC Press.

Edwards, C A, and P J Bohlen. 1996. *Biology and Ecology of Earthworms*. 3rd ed. New York, USA: Chapman and Hall.

Eker, M. 2016. "The research of effect to growth of different outdoor ornimental plants of vermicost and the other some organic fertilizer." Master Thesis, Main Science Division of Soil Science and Plant Nutrition, Namık Kemal University Graduate School of Natural and Applied Sciences.

Grappelli, A, V Tomati, E Galli, and B Vergari. 1985. "Earthworm casting in plant propagation." *Horticultural Science* 20:874–876.

Hahn, G. 2011. "Chitin rich vermicompost repels pests & suppress plant diseases." *California Vermiculture Limited Liability Company*.

Hanc, A, and Z Chadimova. 2014. "Nutrient recovery from apple pomace waste by vermicomposting technology." *Bioresource Technology* 168:240–244.

Haritha Devi, S, K Vijayalakshmi, K Pavana Jyotsna, S K Shaheen, K Jyothi, and M Surekha Rani. 2009. "Comparative assessment in enzyme activities and microbial populations during normal and vermicomposting." *Journal of Environmental Biology* 30:1013–1017.

Hoitink, H A J, and M J Boehm. 1999. "Biocontrol within the context of soil microbial communities: A substrate-dependent phenomenon." *Annual Review of Phytopathology* 37 (1):427–446.

Hoitink, H J, W Zhang, D Y Han, and W A Dick. 1997. "Making compost to suppress plant disease." *BioCycle (USA)* 38 (4):40.

Kale, R D, and N S Sunitha. 1995. "Efficiency of earthworms (*E. eugeniae*) in converting the solid waste from aromatic oil extraction into vermicompost." *Journal of Internal and Emergency Medicine* 22 (1):267–269.

Kothari, R, S Verma, and V V Tyagi. 2012. "Organic fertilizers types, production and environmental fact "In *Chapter 5: Role of various vermicomposting parameters in gren sustainable aprooach*, edited by RP Sing, 85–95. New York: Nova Science Publishers.

Lakshmi, B L, and G S Vizaylakshmi. 2000. "Vermicomposting of sugar factory filter pressmud using African earthworm species (*Eudrillus eugeniae*)." *Journal of Pollution Research* 19 (3):481–483.

Lim, S, T Y Wu, P N Lim, and Y Y Shak. 2015. "The use of vermicompost in organic farming: overview, effects on soil and economics." *Journal of the Science of Food and Agriculture* 95 (6):1143–1156.

Manna, M C, A Swarup, R H Wanjari, H N Ravankar, B Mishra, M N Saha, Y V Singh, D K Sahi, and P A Sarap. 2005. "Long-term effect of fertilizer and manure application on soil organic carbon storage, soil quality and yield sustainability under sub-humid and semi-arid tropical India." *Field Crops Research* 93:264–280.

Martin, J P. 1976. *Darwin on earthworms: the formation of vegetable moulds*. Ontario: Bookworm Publishing.

Munnoli, P M, J A T da Silva, and B Saroj. 2010. "Dynamics of the soil-earthworm-plant relationship: a review." *Dynamic Soil, Dynamic Plant* 4 (1):1–21.

Namlı, A, O Akça, C Perçimli, S Beşe, Ş Gür, H Arıkan, İ Eser, E İzci, E Gümüşay, and G Tunca. 2014. "Vermicomposting of domestic and municipal sewage sludge with earthworm (*Eisenia fetida*)." *Toprak Bilimi ve Bitki Besleme Dergisi* 2 (2):46–56.

Ozturk, M., A Guvensen, S Sakcali, S Celik. 2008. Agriculture and Ecological Interactions. In: Ecological / Organic Agriculture and Environment (Edit. I. AK), pp:14–26, Ozsan Press, EKODER-Bursa, Turkey.

Padmavathiamma, P K, L Y Li, and U R Kumari. 2008. "An experimental study of vermi-biowaste composting for agricultural soil improvement." *Bioresource Technology* 99 (6):1672–1681.

Pandit, N P, N Ahmad, and S K Maheshwari. 2012. "Vermicomposting biotechnology: An eco-loving approach for recycling of solid organic wastes into valuable biofertilizers." *Journal of Biofertilizers and Biopesticides* 3 (1):3–8.

Pathma, J, and N Sakthivel. 2012. "Microbial diversity of vermicompost bacteria that exhibit useful agricultural traits and waste management potential." *Springer Plus* 1 (1):26.

Saxena, M, A Chauhan, and P Asokan. 1998. "Flyash Vermicompost from non-ecofriendly organic wastes." *Pollution Research Paper* 17:5–11.

Schuman, S H, and W Simpson. 1997. "A clinical historical overview of Pesticide health issues." *Occupational Medicine (State of the Art Rev.)* 12 (2):203–207.

Senapati, B A. 1993. "Earthworm gut contents and its significance." In *Earthworm resources and vermiculture*, edited by AK Ghosh, 97–99. Kolkatta: Zoological Society of India.

Sharma, A, and R Ronak Chetani. 2017. "A Review on the effect of organic and chemical fertilizers on plants." *International Journal for Research in Applied Science and Engineering Technology* 5 (2):677–680.

Şimsek Erşahin, Y. 2007. "Acquiring Vermicompost Products and Their Application Alternatives through Agricultural Production." *Gaziosmanpaşa University, Ziraat Fakültesi Dergisi (Journal of Agricultural Faculty)* 24 (2):99–107.

Sing, R P. 2012. "Organic fertilizers types, production and environmental fact." In. U.S.A.: Nova Science Publishers. https://www.researchgate.net/publication/274896698, (accessed April 2018).

Singh, R, R R Sharma, S Kumar, R K Gupta, and R T Patil. 2008. "Vermicompost substitution influences growth, physiological disorders, fruit yield and quality of strawberry (*Fragaria x ananassa* Duch.)." 99:8507–8511.

Sinha, RK. 2011. *Vermiculture Revolution: The technological revival of Charles Darwin's unheralded soldiers of mankind.* U.S.A.: Nova Science Publication.

Sinha, R K, P K Singh, D Valani, and S Agarwal. 2009. Earthworms Vermicompost: an economically viable and environmentally sustainable alternative to destructive chemical fertilizers: some experimental studies on potted and farmed cereal and vegetable crops. In *Advances in Environmental Research - 9 (Chapter 2)*, edited by Justin A Daniels. Newyork, USA: Nova Science Publishers; ISBN: 978-1-61728-999-6.

Sinha, R K, D Valani, K Chauhan, B K Soni, and S R Agarwal. 2012. "Organic fertilizers types, production and environmental fact." In *Chapter 9: Earthworms vermicompost: a nutritive biofertilizer and powerful biopesticide for promoting organic framing while protecting farm soils and mitigating global warming*, edited by RP Sing, 163–207. Newyork: Nova Science Publishers.

Tiunov, A V, and S Scheu. 2000. "Microfungal communities in soil, litter and casts of *Lumbricus terrestris* L.(Lumbricidae): a laboratory experiment." *Applied Soil Ecology* 14 (1):17–26.

Yasir, M, Z Aslam, S W Kim, S W Lee, C O Jeon, and Y R Chung. 2009. "Bacterial community composition and chitinase gene diversity of vermicompost with antifungal activity." *Bioresource Technology* 100 (19):4396–4403.

Yıldız, M, M O Gürkan, C Turgut, Ü Kaya, and G Ünal. 2005. "Environmental problems caused by pesticides used in agricultural war." VI. Technical Agriculture Congress, Ankara, Turkey.

Zuo, L, J Zhang, R Zhao, H Dai, and Z Zhang. 2018. "Application of vermicompost improves strawberry growth and quality through increased photosynthesis rate, free radical scavenging and soil enzymatic activity." *Scientia Horticulturae* 233:132–140.

CHAPTER 9

PLANT METABOLOMICS: ROLE IN CROP IMPROVEMENT

SAMEEN RUQIA,[1] SYEDA NAMRA HAQ,[1] MEHMOONA ILYAS,[2] AROOBA ILYAS,[3] ALVINA GUL,[1] MUNIR OZTURK,[4] VOLKAN ALTAY[5]

[1]Atta-ur-Rahman School of Applied Biosciences, National University of Sciences and Technology, Islamabad, Pakistan
[2]Department of Botany, Government Gordon College, Rawalpindi, Pakistan
[3]Department of Chemistry, University of Wah, Wah, Pakistan
[4]Botany Department & Centre for Environmental Studies, Ege University, Izmir, Turkey
[5]Biology Department, Faculty of Science and Arts, Hatay Mustafa Kemal University, Hatay, Turkey
munirozturk@gmail.com

Abstract

Metabolites reflect the integration of gene expression, protein interaction and other different regulatory processes and are, therefore, closer to the phenotype than mRNA transcripts or proteins alone. Spatial and temporal characterisation of metabolome is accomplished through metabolomics. Transversal metabolomics has been successfully applied to find particular patterns associated with biotic and abiotic stress tolerance. Integrated analysis of transcriptome and metabolome unearths new gene functions, circadian clock functions and natural product biosynthesis pathways. In quality control, yield optimisation and investigating environmental and ecological interactions, metabolic profiling of phytomedicinal species is a powerful tool. It can act as a vital tool for crop improvements in breeding programmes, because the phenotype leans on intricate metabolic interactions of an organism. This review discusses the crucial role of metabolomics in plant growth, and the development and response of plants to different biotic and abiotic environmental constraints.

Key words: circadian rhythms, crop improvement, metabolite profiling, crop breeding, environmental stresses, *Arabidopsis thaliana*.

Introduction

Plant biotechnology is a dynamic, expanding science, incorporating high-throughput sciences. Earlier approaches played a very vital role towards achieving the long-standing goal of food security; nevertheless, this ambitious goal has not yet been achieved, and an emerging science, metabolomics, under the umbrella of plant biotechnology, has emerged to resolve the issue of global food insecurity (Hakeem, Ahmad, and Ozturk 2013, Hakeem et al. 2012, Hakeem, Tombuloğlu, and Tombuloğlu 2016, Kumar et al. 2017).

An advanced array of technologies has led to the immersion of metabolomics in the field of omics, and its recognition in the plant sciences was initiated in the mid-1990s. Metabolomics encompasses spatial and temporal characterisation of metabolic levels in the metabolome. A metabolome represents a critical aspect of a plant's physiology, growth characteristics and, ultimately, its economic value. According to an estimate based on genome size, different species of plants contain 5,000 to 25,000 metabolites. In order to identify and quantify a specific class or classes of chemically-related metabolites through effective metabolite analysis techniques, metabolite profiling will prove to be of great help (Hakeem, Tombuloğlu, and Tombuloğlu 2016). Plant metabolomics has been extensively used for crop improvement in plant breeding experiments and for evaluation of genetically modified crops. Metabolomics also assists in genome annotation. Metabolomics is also utilised in the discovery of new gene functions, circadian clock functions and natural product biosynthesis. Plant metabolomics also play an important role under biotic and abiotic environmental constraints. In the case of quality control, yield optimisation and investigations to be followed on the environmental and ecological interactions, metabolite profiling of phyto-medicinal species can serve as a powerful tool. For an analysis of species lacking a sequenced genome, metabolomics can be evaluated successfully (Ashraf, Ahmad, et al. 2012, Ashraf, Öztürk, et al. 2012, Hakeem, Tombuloğlu, and Tombuloğlu 2016, Hegeman 2010).

Metabolomics can contribute significantly towards crop improvements in breeding programmes for the simple reason that, as mentioned above, the phenotype leans on intricate metabolic interactions of an organism (Hakeem, Tombuloğlu, and Tombuloğlu 2016, Matsuda et al. 2012). In the recent era of functional genomics, metabolomics is regarded as the most novel and useful omic science and technique. It has been used as a powerful tool for precise phenotyping, particularly for industrial application (Fukusaki and Kobayashi 2005). Metabolomics was first defined in 1998, as a study of low molecular mass compounds, synthesised and modified by living cells or organisms (Resham et al. 2014). Later on, it was further stated that as metabolites are essential elements of plant biomass and its unique architectural organisation, metabolomics, together with plant morphophysiological analyses, can potentially serve in achieving goals pertaining to crop plant improvement (Turner et al. 2016).

Metabolite profiling can also allow us to dissect the complex polygenic inheritance of agronomic traits (Baker et al. 2006, Catchpole et al. 2005, Hakeem, Tombuloğlu, and Tombuloğlu 2016). Nuclear magnetic resonance is a promising tool for identification of primary and secondary metabolites involved in herbivore resistance. Significant differences in metabolomes of thrips-resistant and susceptible hybrids of *Senecio* have been identified through nuclear magnetic resonance. Copious amounts of secondary defensive metabolites, like pyrrolizidine alkaloids (PA), jacobine, jaconine, and kaempferol glucoside can be detected in resistant hybrids. Accumulation of defensive primary and secondary metabolites is higher in younger leaves than older leaves (Leiss, Choi, et al. 2009, Leiss, Maltese, et al. 2009).

An ultra-performance liquid chromatography/time-of-flight mass spectrometry (UPLC-TOFMS)-based metabolomics approach identified the close association of secondary metabolites with growth stages in *Angelica sinensis* (Qian et al. 2013). Metabolomic profiling of thrips-resistant chrysanthemums identified higher quantities of phenylpropanoids, chlorogenic acid and feruloyl quinic acid. Both phenylpropanoids are well known for their inhibitory effect on herbivores as well as pathogens (Leiss, Choi, et al. 2009, Leiss, Maltese, et al. 2009).

Gas-chromatography mass-spectroscopic metabolite profiling helped in identification of the association of metabolites with flavour and pigmentation traits in potato tubers (Beckmann et al. 2007). The importance of plant metabolomics in different fields of crop plant is highlighted below. Detailed profiling of all the nutrients in plants will help in directly elucidating plant metabolomic processes (Sawada et al. 2008).

Mass spectrometry-based metabolomics workflows can be employed for plant immunity assessment purposes, since plant metabolism serves as an essential built-in tool of the plant system, conferring both physiological and biochemical disease resistance to a plant; nevertheless, the vast number of metabolites working towards building up a plant's defence mechanism entails the application of various chemical detection platforms, and putting into practice the non-targeted approach, to get a holistic view that could serve our goal of understanding plant immunity against disease in the context of characterising the metabolome; and broadening our understanding of the underlying mechanisms controlling plant-pathogen interactions (Heuberger et al. 2014).

Though mass spectrometry is a powerful, indomitable and awe-inspiring tool, there is a probability that problems such as ion suppression could result in onerous quantification or identification of metabolites present in meagre concentrations; or there are cases where ionisation of the specific metabolites is not possible. To resolve these problems, the best possible way is to pair mass spectrometry with other separation techniques, particularly with either liquid chromatography or gas chromatography. This pairing up of mass spectrometry with either of these techniques will provide us with more precise and accurate results (Lopes et al. 2017).

Additionally, metabolomic approaches, together with genomic approaches, will help us to understand the underpinning role played by particular genetic determinants towards a particular metabolic pathway. How they affect the different regulatory pathways will ultimately provide us with understanding of overall plant performance, and how we can manipulate our understanding strategically towards crop genetic improvement (Wen et al. 2015). Nevertheless, the information and knowledge garnered through metabolomics and other post-genomic technologies such as transcriptomics and proteomics, and with particular reference to plants, their metabolome comprises an enormous and varied quantity of chemicals. It is estimated that the number of metabolites present in the plant kingdom exceeds 200,000. For managing such a wealth of information and knowledge we need to exploit web resources for better identification and quantification of metabolites. For more effective management we can opt for suitable software for better comprehension of a plant metabolome (Perez de Souza et al. 2017).

Plant metabolomics, though a relatively new research area, has contributed immensely towards a better understanding of various different plant systems; this emerging field has a lot more scope for providing us with further understanding of plant systems that can ultimately serve the purpose of improved quality crops and increased yields (Hong et al. 2016).

Nevertheless, plant metabolome exhibits a phenomenal assortment and is often considered essentially to be a strong link between plant genome and phenome. Based on the results gathered through genome-wide association studies conducted on cereal plants, focusing on rice plants in particular, it was revealed that the integrative approach employed in the said study, working on interactive rice functional genomics and metabolomics, throws light on the sum total of genetic and biochemical regulation of both metabolic and agronomic traits, thus serving the goal of crop improvement, both in terms of quality and quantity, and ultimately achieving the age-old goal of food security that lies at the heart of plant biotechnology (Chen et al. 2016).

Integrating the Omic Approach in *Arabidopsis*

Functional genomics, involving transcriptomes and metabolomes, leads to the demarcation of genes and the discovery of molecular networks involved in biological processes of *Arabidopsis* (Fukushima, Kusano, Redestig, et al. 2009). According to Guilt by Association theory, the changes in gene expression and metabolite accumulation are coordinated via metabolic profile (Saito, Hirai, and Yonekura-Sakakibara 2008). Potential new genes, narrowed down through correlation studies, can be affirmed via reverse genetics. The integrated omics approach was first applied in *Arabidopsis*, in order to investigate its stress response to sulphur starvation (Albinsky, Sawada, et al. 2010, Hirai et al. 2005, Hirai and Saito 2008, Hirai et al. 2007, Hirai et al. 2004, Sawada et al. 2009).

Integrated omic techniques not only increase understanding of transcriptome and metabolome responses, but also lead to the discovery of new genes, enzymes and transcriptional factors. Genes involved in flavonoid biosynthesis have been identified by studying gain of function mutant of the MYB transcriptional factors (Tohge et al. 2005). Co-expression analysis of genes and metabolites under different developmental and stress conditions can be performed via the integrated metabolomic approach in a single set of experiments. It is a powerful tool in identification of new genes involved in different biological processes (Okazaki et al. 2009, Yonekura-Sakakibara et al. 2012, Yonekura-Sakakibara et al. 2008, Yonekura-Sakakibara et al. 2007).

The intricate network involved in gene expression and/or in metabolite accumulation, can be helpful in efficiently understanding the cellular processes, gene mutations and the response of cells to different biotic and abiotic stresses. Examples can be highlighted, as the impact of the global environment on *Arabidopsis* metabolome has been investigated with respect to its amino acid biosynthesis mutants (Diaz et al. 2011, Fukushima et al. 2011, Kusano et al. 2007, Kusano et al. 2010, Watanabe, Kusano, et al. 2008, Watanabe, Mochida, et al. 2008); its flavonoid biosynthesis (Kusano et al. 2007); its biological clock (Fukushima, Kusano, Nakamichi, et al. 2009); and its responses to abiotic stress, such as drought (Maruyama et al. 2009, Urano et al. 2009), nutrients (Kusano, Fukushima, et al. 2011, Kusano, Tabuchi, et al. 2011), and UV-B light (Kusano, Tohge, et al. 2011).

Metabolomic studies on *Arabidopsis thaliana* have helped in the exploration of sulphur as an important macro-element in plant and animal nutrition. It has been observed that plants can assimilate inorganic sulphate into two sulphur-containing amino acids, cysteine and methionine (Nikiforova et al. 2005). The effects of cold acclimation on metabolome were measured using the non-targeted metabolic fingerprinting approach on *Arabidopsis thaliana.* It was observed that the reprogramming of metabolism occurs globally during cold acclimation in *Arabidopsis*. The leaves which are shifted to low temperatures present a metabolic profile that is continuously changing, but the leaves which are developed at low temperatures constitute a stable complement of components (Gray and Heath 2005).

Metabolite annotation via integrated omics

Integrated analysis of *Arabidopsis* transcriptome with metabolome permits the identification of related genes and metabolites. All the peaks detected in tandem mass-spectrophotometry have enabled the metabolite annotation in *Arabidopsis* (Matsuda et al. 2009). For the characterisation of detectable metabolites, untargeted metabolomics based on liquid chromatography-tandem mass spectrometry (LC-MS/MS) is an exigent task. The prerequisite in the field of fuel, pharmaceuticals, safety assessment, and breeding is an extensive characterisation of metabolite accumulation patterns (Patti, Yanes, and Siuzdak 2012, Sawada and Hirai 2013). By following the LC-MS/MS strategy, thousands of metabolites can be dissected in an organism. In whole-genome sequencing the gene annotations are assigned to representative biosynthesis enzyme classes (hydroxylation, methylation, glycosylation, acetylation, etc.). These modification reactions (Sawada and Hirai 2013) generate multiple derivatives having diverse chemical properties and specific bioactivities (Sankawa 1999). Plants are the largest producers of phytochemicals on the planet. The plant's interactions with other organisms as well as microorganisms are controlled by these metabolites. In the body of a predator, the ingested phytochemicals are metabolised and undergo further changes (Barrett and Heil 2012, Sawada and Hirai 2013).

Thus, for an elucidation of complex metabolic systems among organisms, metabolite profiling is an imperative attribute. More comprehensive analysis of these biological samples has been carried out following recent advances in both hardware and software. It is possible to integrate metabolomics with other omics approaches (Klie et al. 2014). However, there is broad scope for interpreting the untargeted data in the connections between metabolites and gene functions (Matsuda et al. 2010). The reason is that the effects of multiple levels dynamically change the metabolite levels in an organism (e.g. genome, transcript, protein, and metabolite) (Sawada and Hirai 2013). Integrated LC-MS/MS-based plant metabolomics allow detection of several hundred metabolites in a sample and integrated omics analyses, e.g. large-scale reverse genetics, linkage mapping, and association mapping provide a powerful tool for candidate structure selection or rejection (Sawada and Hirai 2013).

Metabolic approaches to improve rice quality

The rice genome has been sequenced and comprises 32,000 genes (Sato et al. 2010). Approximately more than half of the sequenced genes comprise unidentified biological functions (Itoh et al. 2007). Novel genes in rice have been identified through gain and loss of function approaches. Quantitative trait loci (QTL) analysis of mapping populations and genetic stock has been used to investigate different relationships between metabolite composition, phenotypes and genotypes. The gain of function approach is based on the FOX hunting system (full length cDNA over expresser gene hunting system). The FOX hunting system allows functional analysis of genes in due of ectopic expression of any plant FL-cDNA library, even in heterologous plant systems. Rice FOX *Arabidopsis* lines (Transgenic *Arabidopsis* lines over expressing rice FL-cDNAs) have been used for metabolic fingerprinting (Suzuki et al. 2010) and metabolic profiling (Albinsky, Kusano, et al. 2010) in order to identify functional genes in rice. Rice FOX *Arabidopsis* lines are screened through Fourier Transform-near-infrared Spectroscopy and Gas Chromatography-time-of-flight MS (GC-TOF-MS). Fourier Transform-near-infrared Spectroscopy (FT-NIR) is a non-destructive analytical approach for analysis of FOX seeds (Suzuki et al. 2010).

FT-NIR analysis of 3000 FOXES genes identified seven lines with altered metabolite fingerprints (Kusano and Saito 2012). Metabolite profiling through GC-TOF-MS provides comprehensive details of primary metabolites and intermediates of secondary metabolites (Kusano, Fukushima, et al. 2011). Metabolite fingerprinting of 350 FOX seeds through GC-TOF-MS identified twenty-six candidate lines for gene characterisation. A close association between lateral organ boundary domain asymmetric leaves—like Os-LBD37/ASL39 and nitrogen metabolism in rice—has been identified via metabolite profiling of FOX seeds (Kusano and Saito 2012).

Tos-17 retrotransposon and Ds-transposon-inserted mutant lines have been used for the characterisation of novel genes in rice (Hirochika et al. 2004, Kolesnik et al. 2004); the glutamine synthetase gene involved in ammonium assimilation is characterised through Tos-17-knockout lines. Gene encoding for glutamine synthetase *(OsGS1:1)* is imperative for normal growth and grain filling (Tabuchi et al. 2005).

Metabolic profiling of mutants showed a significant increment of sugar and sugar phosphates, and a declined pattern for amino acid and leaf tricarboxylic acid intermediates as compared to wild-type rice. The metabolite profile of rice is seen to differ spatially for root tissues and leaf tissues in the presence of ammonium (Kusano and Saito 2012). The Rice diversity research set was genotyped with restriction fragment length polymorphism (RFLP) markers in order to investigate the relationship of metabolites with agronomically important attributes (Hall 2006). Untargeted high coverage metabolomic characterisation of rice kernels identified a fine-gained correlation between genetic diversity and metabolite abundance (Kusano and Saito 2012).

Metabolite fingerprinting of backcross inbred lines developed by crossing of *O. sativa* L. ssp. *japonica* 'Sasanishiki' and *O. sativa* L. ssp. *indica* 'Habataki' has helped in the identification of 759 metabolite signals. QTL mapping results in the identification of 802 mQTL from 759 metabolites traits and chromosome 3, which is characterised as hotspot loci owing to the coordinated control of amino acids, triacylglycerols and some other metabolites. Secondary metabolites, such as lysophosphatidylcholines, oryzanols, and flavone glycosides are also shown to have high broad sense heritability (Kusano and Saito 2012).

A metabolomic study conducted on two rice cultivars, namely japonica and indica rice, not only revealed a correlation between metabolic phenotype and the geographic origin of the rice seeds, but it also revealed that both these subspecies varied as per their relative metabolite concentrations, and also in their related metabolic networks. This provides us with a clue regarding the relative adaptation of the two subspecies depending on their circumstances, thus enabling us to conduct further improvement programmes on rice quality using metabolic engineering (Hu et al. 2014).

Metabolomic approaches to improve wheat quality

The comparative metabolomic approach, applied to transgenic and parental wheat lines, showed that B73-6-1, which has the highest transgene expression, and its parental lines L88-6, show some separation. Comparison between the spectra of these two lines showed that there are increased levels of maltose or sucrose in the transgenic line and that there is therefore a difference between amino acid sequences, which may be the cause of separation between the transgenic and parental lines. Through metabolome analysis it is also observed that both the transgenic and parental lines tested have a larger amount of proline and gamma aminobutyric acid. These tests reveal that the environment affects the metabolomes, and the differences between transgenic and parental lines are within the range, as compared to the differences observed between these lines when grown in different places (Baker et al. 2006).

Semolina has been analysed metabolomically to identify its compositions from raw semolina to refined pasta. 76 metabolites are detected in it. It has been observed that there are significant differences between raw and refined forms. These differences are in phytosterols, policonasols, unsaturated fatty acids, amino acids, carotenoids and many minerals. It can be concluded that a number of nutrients are metabolised in the processing (Beleggia et al. 2011).

Metabolomic studies have been carried out on varying sulphur nutrition on metabolite patterns of different organs of ears of winter wheat, *Triticum aestivum*, at final maturity. It was seen that sulphur deficiency decreases the bulk of metabolites in straws of wheat, which results in an increased metabolite concentration in husk, raschis and grains. It was also observed that N-acetyl glucosamine, lysine, ferulic acid and beta aminoisobutyric acid are most responsible for organ specific differences in the metabolite profile of *Triticum aestivum.* Sulphur deficient conditions result in the transfer of these chemicals from source tissues to ears and grains of plants (Zörb et al. 2013).

Metabolomic studies revealed that metabolites in wheat grain prominently affect the nutritional value of bread wheat. Metabolite concentrations vary greatly in a developing wheat grain. Thus metabolomic profiling can be employed in breeding programmes that work towards improving the overall quality of the wheat grain (Zhen et al. 2016).

Metabolomic studies on resistance to *Fusarium* head blight of wheat has revealed that resistance to this disease is associated with salicylic acid and jasmonic acid signalling, which is involved in the activation of deposition of cell wall appositions, consisting of phenolic glucosides, flavonoids, lignans and hydroxycinnamic acid amides that are usually synthesised through the shunt phenylpropanoid pathway (Gunnaiah et al. 2012).

Metabolomic analysis of *Fusarium* head-blight resistant Fhb locus reveals certain resistant-related metabolites. The key components of resistant-related metabolites are those which belong to phenylpropanoids and flavonoid biosynthetic pathways. A several-fold increase in hydroxycinnamic acid amides (HCAA's), such as N-Caffeoylputrescines and 4-coumaroylagmatine is also observed in resistant plants upon inoculation of a pathogen. These HCAA's are named as anti-funal acid amides; some amino acids, which have a significant role in plant defence against fungal attacks, are also observed to increase several-fold (Karre 2014).

Metabolic profiling applied on kernels of tetraploid wheat to analyse the effects of domestication and breeding on wheat revealed that there is high heritability for all of the metabolites with a mean of 92%, and a range between 29% for alpha tocopherol, to 100% for alanine, proline, and quinic acid (Beleggia et al. 2013).

Metabolomics have also been employed in understanding plant genomics in a more precise manner. Metabolic profiling has been used in understanding the role of genes towards controlling biochemical pathways governing trait variation. In the study being conducted on wheat aneuploid lines, and homeologous group 3 ditelosomic lines, some specific compounds revealed significant variations that can be attributed to genes responsible for encoding enzymes governing biochemical pathways. By employing wheat genome survey sequencing, it found the genetic components responsible for metabolite variation; this metabolite analysis was untargeted metabolite analysis. The results showed significant differences in metabolite profiles between Chinese spring and ditelosomic lines. Differences in metabolite concentration were observed between Chinese spring and at least one ditelosomic line. A change in the concentration of metabolites in both provides evidence of either direct or indirect control of genes in biosynthetic pathways, thus influencing the regulation of a metabolome. Metabolomics can, therefore, provide us with an opportunity to study metabolomic-genomics networks in a more practical context in contrast to other theoretical methods, and it provides us with an appropriate experimental genetic system (Francki et al. 2016).

Metabolomic approaches to improve *Solanum lycopersicum* quality

The quality and productivity of a tomato fruit largely depends on the quantity and composition of metabolites (Iijima and Aoki 2009). *Solanum lycopersicum*, also known as tomato, was analysed by metabolic profiling to determine the functions and level of flavonoids that are involved in different functions. The metabolic profiling revealed that flavonoids play a key role in many biological processes, such as pigmentation of flowers, fruits and vegetables, interaction between plant and pathogens, fertility and protection against ultra violet radiations. Three approaches to metabolic profiling were applied to expose the functions of flavonoids: increasing endogenous tomato flavonoids using regulatory genes; blocking steps in flavonoids pathways by RNA interference; and producing novel tomato flavonoids (Bovy, Schijlen, and Hall 2007).

Metabolite composition of tomato fruit carp, measured by gas chromatography, showed that a greater number of quantitative trait loci are present in high fruit load cultivations as compared to low fruit load cultivations. It was also observed that metabolite levels are not correlating with levels of genes associated with metabolism (Do et al. 2010). Compatible interaction between tomato and tomato mosaic virus, when studied metabolomically, showed that amino acids, organic acids, sugars, Phenylpropanoids, flavonoids and some other compounds are involved in *Solanum tuberosum* reaction to attack by the tomato mosaic virus (ToMV). It was observed that when the virus is inoculated on leaves, flavonoids are increased at the place of inoculation, whereas Phenylpropanoids are seen to increase in inoculated leaves (López-Gresa et al. 2012).

Metabolite profiling of mature green tomato fruit carp shows increased levels of arabinose, fructose-6-phosphate, valine and shikimic acid, which proves that these compounds are associated with heat-shock-induced chilling tolerance as their levels are also seen to be affected in chilled temperatures (Luengwilai, Saltveit, and Beckles 2012). Metabolite profiling of tomato leaves revealed that photoautotrophic leaves of tomato at the rooting stage are characterised by an accumulation of urea and erythritol, whereas photomixotrophic leaves are characterised by an accumulation of metabolites which are involved in primary metabolism, including catecholamines, proline, hydroxyproline, asparagines, gamma aminobutyric acid (GABA), soluble sugars and myo-inositol (Badr, Angers, and Desjardins 2011).

Cadmium treatment was applied to tomato plants for 90 days. After cadmium treatment, the plants were analysed metabolomically. It was observed that cadmium-treated plants have a lower concentration of proline and total ascorbate content in their leaves. Mature leaves are shown to have decreased levels of carotenoid and chlorophyll, which is related to the reduced expression of genes associated with the expression of essential isoprenoid and carotenoid accumulation. It is observed that plants can acclimatise a 20 micro-molar concentration of cadmium, but if the concentration is increased to 100 micro-molar then it leads to plant growth limitation and fruit abortion (Hediji et al. 2010).

The metabolomic profiling approach was also employed for studying diel compositional variations occurring in those tomato fruits affected by source-sink relationships. For putting the analysis in a practical context, biochemical phenotyping of vital compounds, in addition to the use of proton nuclear magnetic resonance and mass-spectrometry metabolic profiling were employed to estimate the constituents of nearly seventy metabolites occurring in leaves and sixty metabolites in the fruits, along with ecophysiological estimations being conducted. Metabolite data was analysed using various algorithmic approaches. The results obtained were significant: those trees grown in shady conditions compared to the control plants revealed several adaptations by plants; for example, leaf area was adjusted, sink carbon requirements were minimised and compositional adjustments were noticed. In leaves, more pronounced diel fluctuations in metabolite concentrations were observed than in fruits. Amongst the metabolites which exhibited noticeable diel fluctuations, organic acids were prominent. The metabolmic profiling clearly showed that relationships between compositional variations in fruits and leaves are in accordance with instant sucrose provision. The metabolmic profiling successfully asserted the existence of diel patterns in tomato leaf composition. This study will definitely play an important role towards further defining of ideotypes, and also towards amelioration of traditional practices (Bénard et al. 2015).

Metabolomic approaches to improve *Solanum tuberosum* quality

Solanum tuberosum, commonly known as potato, when tested metabolomically showed that there exists a large range of metabolites, which includes asparagines, fructose and glucose, that are really important for the production of tubers (Dobson et al. 2010). The Gas-chromatography mass-spectrometry-based technique was used to perform metabolic profiling of suberin polymer, a chemical which prevents plants from pathogen and desiccation. The technique was performed on wound healing in the potato tuber. It was observed that the metabolite profile of the potato tuber is separated, with clear temporal differences of different suberization stages in nonpolar and polar profiles. The nonpolar profiles showed suberin-associated aliphatic contribution to cluster formation, whereas a relatively broad range of metabolites influence the cluster formation among polar profiles. This wide range of metabolites consists of organic acid, sugar, amino acids and phenylpropanoids (Yang and Bernards 2007).

Comparative subcellular metabolite profiling of potato tubers was performed to find out about the regulation of compartmentation of metabolites in them. The study revealed that the hexoses in tubers accumulate in vacuole independent of their site of production; cytosolic invertase expression leads to a large increase in cytosolic glucose concentration and a decrease in cytosolic sucrose concentration; and three out of a total of four compounds found in cytosolic over expressers accumulate in the same compartment (Farre, Fernie, and Willmitzer 2008).

The differences among the metabolite profile of potato tubers are considered in terms of taxonomy and breeding history of the cultivars. These differences may be due to possible influences from factors, like developmental stages of potato tubers (Dobson et al. 2008). Metabolite profiling of six different cultivars of potato revealed there does not exist any correlation between metabolite content of freshly harvested tubers, starch content or cooking type. Storage of potato is seen to decrease fructose and sucrose and increase proline concentration in tubers (Uri et al. 2014).

A comparative analysis of healthy potato and potato suffering from late blight was conducted; metabolic examination revealed that the compounds which differ in susceptible potatoes and resistant potato varieties are responsible for conferring resistance against disease. These compounds belong to phenylproponoids and flavonoids. Hence, it was concluded that these compounds are key to achieving resistance against late blight (Doddaraju et al. 2013).

Objective evaluation of genetically modified crops by metabolomics approach

Metabolomics gives a snapshot of the ongoing biochemical status of cells and can be used to investigate the metabolic composition of genetically modified food. 86% of chemical diversity was detected in genetically modified tomatoes over expressing miraculin (Sun et al. 2007). Drought-sensitive cultivar, Josefina, showed a significant decrease in hydroxycinnamic acids and favonoid glycosides in response to water deficiency, whereas the tolerant Zarina did not show any such response. Grafting of a sensitive cultivar on a tolerant cultivar is an imperative tool to improve fruit quality in a water deficient environment (Wahid et al. 2007).

The nutritional equivalence of genetically modified organisms can be evaluated on the basis of plant metabolomics. Genetically modified organisms showed remarkable similarities to conventional cultivars owing to the occurrence of reproducible metabolic signatures, and a significant variation of more than 92% of compounds in green and red stages of tomatoes. The post-genomic functional approach, involving transcriptome and metabolome, can reduce the time required for selection in plant breeding and unearths novel genes in crops, vegetables and medicinal plants (Bunsupa et al. 2012, Yonekura-Sakakibara and Saito 2009).

Compositional analysis of genetically modified and non-genetically-modified potatoes was conducted using mass spectrophotometer fingerprinting and detailed quantitative profiling by Catchpole et al. (2005). Strong expression of the phytase gene had no significant effect on the overall gene expression of wheat cultivars (Gregersen, Brinch-Pedersen, and Holm 2005). Minor differences in the flavonoid profile of genetically modified crops and their conventional cultivars were detected by Ioset et al. (2007). Gene expression of genetically engineered glyphosate-tolerant varieties of soybean was significantly different from their closest conventional varieties (Cheng et al. 2008).

Targeted and non-targeted analytical approaches appear feasible for compositional analysis of genetically modified and non-genetically modified crops. Genetically modified tobacco plants showed altered patterns of shikimic acid, phenylalanine and malonic acid secondary metabolites as compared to control plants (Broeckling, Li, and Xie 2012).

The Discovery of Metabolites Induced by Fungal Infections

Plants susceptible to various different types of biotic stresses exhibit several biochemical defence responses, such as enzyme synthesis, cell wall deposition of lignin, suberin and accumulation of specific metabolites (Daayf et al. 2000). The metabolite-involved defence mechanism is mainly associated with plants' secondary metabolites when we look at diverse plant defence mechanisms. Phenylpropanoids are considered as biomarkers of fungal infected plants. Fungal infected *Brassica rapa* plants have shown higher levels of phenylpropanoids, like sinapoyl- and feruloyl-malate (Abdel-Farid et al. 2009). Phenylpropanoid concentration is increased upon the infection of roots with *P. sylvaticum and P. syringae* (Tan et al. 2004). Hydroxycinnamic acid esters play an imperative role in resistance in cucumber and chick pea against powdery mildew fungi (Daayf et al. 2000) and *Sclerotium rolfsii* (Sarma and Singh 2003). The best-known pathogen-induced metabolites of the Brassicaceae family are glucosinolates, cysteine-rich polypeptides and phenolics (Abdel-Farid et al. 2009, Kliebenstein 2004, Ludwig-Müller et al. 1997, Tan et al. 2004, Thevissen et al. 2003).

Significant differences in the glucosinolate of tolerant and susceptible cultivar of cabbage have been detected. Tolerant varieties significantly enhanced the expression of aromatic compounds whereas sensitive varieties increased the aliphatic compounds and indole glucosinolates (Brader, Tas, and Palva 2001). In the leaves of Chinese cabbage, increased levels of anthocyanins are observed in response to the fungal pathogen, *A. brassicae* (Abdel-Farid et al. 2009, Rostás, Bennett, and Hilker 2002). Higher levels of kaempferol and quercetin analogues have been found in methyl jasmonate-treated *Arabidopsis* (Hendrawati et al. 2006). In *Brassica rapa* (Herfstraap and Oleifera) an accumulation of aliphatic and indole glucosinolates is found in the samples infected with *F. oxysporum* and in *A. brassicae* infected *Brassica napus* (Doughty et al. 1991). Chinese cabbage, after treatment with jasmonic acid, and infected with *A. brassicae* or *P. brassicae*, shows a significant increase in aliphatic and indole glucosinolates (Ludwig-Müller et al. 1997). An enhanced expression of aliphatic and indole glucosinolates is also reported in *Brassica oleracea* infected with *P. parasitica* (Abdel-Farid et al. 2009, Ménard et al. 1999).

The Role of Plant Metabolomics in Quantitative Analysis

Large populations of tomatoes were screened on the basis of carotenoids content through matrix-assisted laser desorption ionisation time-of-flight mass spectrometry (Fraser et al. 2007). QTL mapping of recombinant inbred population and introgression lines derived from Columbia x C24 identified six biomass QTLs and 157 metabolite QTLs. Two of the biomass QTLs were co-segregating with metabolite QTLs (Meyer et al. 2007).

QTL analysis of Ler and Cvi, two ecotypes of *Arabidopsis*, identified six loci for total aliphatic glucosinolate accumulation; six loci controlling total indolic glucosinolate concentration; and three loci regulating benzylic glucosinolate levels (Kliebenstein, Gershenzon, and Mitchell-Olds 2001). Candidate gene approach through SNPs identified an association between allelic variants of water soluble carbohydrates, nitrogen and dry matter digestibility within an alkaline invertase candidate gene LpcAI in *Lolium perenne* (Skot et al. 2007).

QTL mapping of tomato introgression lines (ILs) identified 889 quantitative fruit metabolic loci, and 326 loci that modify yield-associated traits (Schauer et al. 2006). Spatial and temporal differences in multiple enzyme activities, biomass production and metabolite contents of *Arabidopsis* accessions have been identified (Cross et al. 2006). A highly significant correlation was observed in the intermediates of the hexose phosphate pool of the Col × C24 recombinant inbred line (RIL) and near isogenic line populations (Keurentjes et al. 2008).

F_1 population of potato was genotyped through AFLP, RFLP and phenotyped across six environments. QTL analysis of this population identified QTL for glucose, fructose and sucrose content across all the chromosomes of potato (Menéndez et al. 2002). QTL mapping of recombinant inbred lines of *Arabidopsis* derived from Bay × Sha identified 438 QTLs, affecting 243 metabolites. Eleven genomic regions were recognised as QTL hotspots. (Eckardt 2008).

Schauer et al. (2008) identified dominant, additive and recessive inheritance of QTLs of fruit metabolites of introgression lines of heterozygous tomatoes (*Solanum lycopersicum*). Harrigan et al. (2007) identified significant environmental differences in sugars, organic acids, and amino acids in three different wheat inbred lines grown at three different sites. Four robust metabolite quantitative trait loci (mQTL), associated with rate-limiting steps in flavonoid biosynthesis, were mapped in two full-sib poplar families, *Populus deltoides* cv. S9-2 × *P. nigra* cv. Ghoy and *P. deltoides* cv. S9-2 × *P. trichocarpa* cv. V24 (Morreel et al. 2006).

The Role of Metabolomics in Biotic Stress

Plants possess secondary metabolites in order to overcome the ravages of biotic stress (González-Lamothe et al. 2009). The compounds constituting defence roots are from alkaloids, isoprenoids and shikimates. Alkaloids are synthesised via the citric acid cycle or shikimate pathway. Isoprenoids are synthesised through the acetate- mevalonate or methylerythritol phosphate pathways (Großkinsky, van der Graaff, and Roitsch 2012).

Metabolite profiling has become a standard research tool for higher level screening of crop germplasm collections in crop breeding programmes. A plant's metabolome plays an imperative role in physiological processes involved in defence mechanisms of various biotic constraints. *Fusarium* head blight is an urgent, devastating anomaly in wheat and barley. More than one hundred quantitative trait loci controlling resistance to fusarium head blight have been identified on seven chromosomes of wheat and barley. QTL involved in detoxification of the mycotoxins deoxynivalenol (DOH) into less toxic glucoside has been mapped on the short arm of chromosome 3B (Lemmens et al. 2005).

Bollina et al. (2010) identified 496 metabolites in barley. Most of the identified metabolites were derived from the phenylpropanoid, flavonoid, fatty acids and terpenoid metabolite pathways. Precursors of kaemferol also play an imperative defensive role in resistant cultivars as compared to susceptible ones. *Fusarium* head blight resistance in wheat is due to regulation of the phenylpropanoid pathway. Resistant wheat cultivars over-accumulate the jasmonic acid amide, conjugates of phenol-poliamines, such as coumaroyl putrescine, agmatine and feruloylputrescine/agmatine (Gunnaiah et al. 2012).

Benzoxazinones play an important role in resistance in maize against *Setosphaeria turcica* (Ahmad et al. 2011). Rice plants infected with *Magnaporthe grisea* showed enhanced expression of alonine, proline, histidine, cysteine, tryptophan, amino acids, sucrose, malate, fructose and glucose (Jones et al. 2011). Benzoxazinoids are well known anti-herbivore secondary metabolites of cereals (Niemeyer 2009).

Benzoxazinoids also have an imperative regulatory role in innate immunity. BX1 mutants were less resistant to the fungal pathogen *Setosphaeria turcica* and cereal aphid *Rhopalosiphum padi* (Ahmad et al. 2011). Benzoxazinoids also protect a plant body from nematode attack (Zasada et al. 2005). Flavonoids, such as the C-glycosyl flavones, maysin, apimaysin and Isoorientin inhibit the growth of corn earthworm larvae (Lee et al. 1998). A 55% to 65% phenotypic variation in corn earthworms was attributed to maysin and api maysin, respectively. These two QTLs did not interact with each other, indicating the existence of an independent biosynthetic pathway. Hordeninie and gramine are well recognised feeding deterrents of the grasshopper (Hinks and Olfert 1992). Glucosinolates alter insect herbivory (Li et al. 2000) nematode survival (Donkin, Eiteman, and Williams 1995) and fungal resistance (Manici, Lazzeri, and Palmieri 1997).

Role of metabolomics in abiotic stresses

Plants are prone to various different types of abiotic stresses that negatively influence not only the growth of the plant, but also its net productivity. In response to abiotic stresses, metabolic activities of plants get principally affected. Metabolic activities of plants get affected for various reasons, such as that metabolic enzymes become inhibited; substrate deficit; some compounds are overly required by the plant system; or a combination of these, to name a few. Ensuring a face-lift to the metabolic network is thus indispensable, not only for carrying out the normal metabolic activities of plants but also to ensure the plants' acclimatisation in damaging circumstances. Metabolomics can thus be used as a powerful tool in understanding these metabolic networks inside the plant cell responsible for plant metabolic processes, which can potentially lead us towards solutions to the problems encountered. The term metabolomics refers to both the comprehensive and the quantitative study of the total number of small molecules occurring in a biological system, exclusively in the present context (Fiehn 2001).

Oxidative stress, induced by a plethora of biotic and abiotic constraints, results in the over-production of antioxidant compounds such as ascorbic acid, glutathione and polyphenols. All these metabolites protect the plant body against reactive oxygen species (Munns and Tester 2008, Urano et al. 2009). Drought stress enhances the expression of amino acids, carbohydrates and carbohydrate alcohol. All these metabolites, along with proline, play an osmoprotective role, and scavenge the reactive oxygen species (Breckle 2002).

Augmentation of carbohydrates in drought-stricken plants increases root-to-shoot ratio (Nishizawa, Yabuta, and Shigeoka 2008). Carbohydrates play an important role in stress tolerance because of their direct involvement in photosynthesis. Metabolites stabilise the cell membrane by prevention of protein degradation. Metabolites are key molecules for turgor maintenance (Sharp et al. 2004). Polyols act as scavengers of reactive oxygen species. Polyamines play important regulatory, protective and reactive oxygen, species-scavenging role upon exposure of drought stress. Exogenous application of polyamines alleviates stress pressure by reducing hydrogen peroxide concentration. Sugars and polyamines play important roles in acclimatisation of cold stress. Roles of different metabolites under biotic and abiotic stresses are highlighted in Table 1. Forty flavonoids were identified in *Citrus* under a soil flooding environment. Phenylpropanoids, precursors of lignin, increase stress tolerance (D'Auria and Gershenzon 2005).

Methylcylate and methylbenzoate play defensive roles under stress conditions (Roetschi et al. 2001, Kang and Udvardi 2012). Salicylic acid plays an important role in host pathogen defence (Hara, Harazaki, and Tabata 2013). Salicylic acid also plays an important role under abiotic stress tolerance. Drought stress not only increases carbohydrates and amino acid, but also changes the expression of phenylpropanoids, leading to differential flavonoid profiles (Rasmussen, Parsons, and Jones 2012). Tolerant model crop organisms contain high concentration of caffeoylquinic acid and phenylalanine, whereas sensitive plants have high a concentration of sinapic acid and quercetin flavonoids (Lugan et al. 2009).

Further studies on this subject will help towards a breakthrough in the fields of metabolic engineering and synthetic biology, enabling us to attain the goal of food security (Obata and Fernie 2012).

The study of metabolites reflects the integrated expression of genes, protein interaction and various other regulatory processes occurring in plant cells, thus providing a more precise phenotype than mRNA transcripts or protein studies conducted individually. Metabolomics is by far the most practical approach as compared with other omics studies, as it potentially requires very minute, or sometimes even no, modification. It has been employed with potential success to analyse and understand the molecular phenotypes of plants during stress conditions, thus enabling us to understand the patterns related to stress tolerance. Additionally, the occurrence, elimination or relative accumulation of different metabolites, together with gene expression studies, provide us with some valuable markers for identifying tolerant crops that can be further used in breeding programmes (Arbona et al. 2013).

Table 1. Role of metabolites under biotic and abiotic stress.

Metabolites	Types of stresses	Functions
Proline	Salt stress Soil flooding Drought Extreme Temperature	I. Maintenance of turgor through osmotic adjustment II. Scavenging of reactive oxygen species III. Stabilisation of membranes, protein and DNA (Arbona et al. 2013).
Polyamines These are positively-charged nitrogenous, aliphatic molecular compounds.	Abiotic stresses	I. Scavenging of reactive oxygen species II. Regulation of amino acid metabolism III. Role in Signalling
Phenolic compounds These are most diverse secondary metabolites which include	Abiotic stress Biotic Stress	I. Provides protection against pathogens. II. Defence mechanism against abiotic stress.

Phenylpropanoids and derivative polyphenols.		
Glucosinolates These are nitrogen and sulphur containing compounds derived from amino acids.	Biotic and abiotic stress	Glucosinolates protects plant body against different types of biotic and biotic stresses.
Carotenoids and other terpenoid derivatives	Heat stress	Stabilisation of lipid phase of thyllakoid membranes.

The role of plant metabolites in natural product biosynthesis

The alkaloid vinblastine was first characterised from the leaves of *Catharanthus roseus* in the 1950s by Robert Noble and Charles Beer (Moudi et al. 2013). At present, investigations into different parts of *C. roseus* have revealed the presence of more than 130 monoterpenoid indole alkaloids (MIA) (Van der Heijden et al. 2004). The vincristine and vinblastine are used clinically as anticancer agents as dimeric indole (bisindole) alkaloids (Abdel-Farid et al. 2009). Amino acid, organic acid, sugar and phenylpropanoid levels are changed in both Octadecanoid-derivative Responsive *Catharanthus* AP2-domain and geraniol 10-hydroxylase transgenic, as compared to the controls (Pan et al. 2012).

A classic example of the ability of plant taxa to produce chemotherapeutics is *Taxus brevifolia* (Pacific yew), used to produce Taxol (paclitaxel). The latter is a complex and highly functionalised terpenoid. It has been approved for the treatment of refractory ovarian and metastatic breast cancer. The production of taxol is based on a semisynthetic route. In this route there is a chemical conversion of a paclitaxel intermediate isolated from the needles or cell cultures of various *Taxus* species (Croteau et al. 2006). *Artemisia annua* (Asteraceae family)—commonly known as sweet wormwood—is evaluated for the production of Artemisinin, an antimalarial drug, which is highly effective against multidrug resistant strains of *Plasmodium falciparum* (Chemler and Koffas 2008).

As a subfamily of isoprenoids, the carotenoids are the most widely distributed yellow, orange, and red natural pigments synthesised in bacteria, algae, and fungi. The lycopene, b-carotene, and astaxanthin are the commercially available carotenoids. These are used as food colourants, animal fodder supplements, and for nutritional and cosmetic purposes. This group of compounds has recently started attracting the attention of researchers for their significant antioxidant activities. They also play important roles in inhibiting the onset of chronic diseases (Chemler and Koffas 2008, Rao and Rao 2007).

Another diverse group of plant secondary metabolites derived from the condensation of several malonyl-CoA molecules are phenylpropanoids. These are starter units derived from phenylalanine or tyrosine. Together with flavonoids and stilbenes, they possess extraordinary antioxidant and estrogenic, antiviral, antibacterial, and anticancer activities (Chemler and Koffas 2008). The stilbenes as another class of phenylpropanoids are also attracting great attention. A major one is resveratrol, touted as an antioxidant. It is reported to have the ability to increase the lifespan of yeast, fruit fly, round worm (Chemler and Koffas 2008), and fish by activating siruin deacetylases (Chemler and Koffas 2008).

The Role of Plant Metabolomics in the Circadian Clock Function

Diverse categories of phenotypes within species are based on genetic variations associated with quantitative traits of development, metabolism and pathogenic resistance (Fukushima et al. 2009). Because of their quantitative and polygenic inheritance, it is very complicated to understand the molecular and genetic basis of quantitative traits (Fukushima et al. 2009). Identification of molecular systems through the system biological approach has been the thrust of biologists.

The biological clock of an organism coordinates its metabolism and development to optimally coincide with the local day/night cycle (Fukushima et al. 2009). A natural variation in plant circadian clocks helps plant survival under different environmental conditions via its effect on transcriptomes (Fukushima et al. 2009). Analysis of natural variations within the circadian clock requires the development of new experimental tools, whereas individual phenotypes can be measured across time to estimate clock output (Fukushima et al. 2009).

Expression quantitative trait loci analysis has been conducted on large populations via microarrays in order to identify genomic regions associated with circadian clocks (eQTL) (Fukushima et al. 2009). In order to assign the genes to time bins based upon coincidence of peak expression, pre-existing data can be used within a circadian period. The microarray data obtained from a sample harvested at a single time point can then be used in the analysis of these bins to estimate the individual's internal circadian time (CT), and identify altered circadian clock function (Fukushima et al. 2009).

Twenty-four circadian time phase groups were identified in recombinant inbred populations of *Arabidopsis thialiana* using existing single time point microarray. These circadian quantitative trait loci were linked to pre-existing metabolomics quantitative trait loci (Fukushima et al. 2009), and analysis identified that circadian clock output was altered by natural variation in *Arabidopsis thaliana* secondary metabolism. Specifically, genetic manipulation of a secondary metabolic enzyme led to altered free-running rhythms (Fukushima et al. 2009).

Conclusion

Metabolomics, at the current time, lies at the heart of plant sciences and applied biotechnology. It has recently started to serve as a nucleus, thus serving in the unparalleled goals of plant biotechnology, i.e. crop improvement, mass production of crops and, to some extent, satiating its associated aesthetic requirements (Saito, Hirai, and Yonekura-Sakakibara 2008; Kumar et al. 2017). Metabolomic approaches are fascinating, which enables the parallel assessment of various different levels of range of metabolites. Recently, metabolomic approaches have been used for the evaluation of natural variance apparent in metabolite composition (Fukushima et al. 2009). Metabolomics can be regarded as an advanced level of post-genomic analysis as it aims to demonstrate the changes in metabolite fluxes, which are controlled by very minute changes within the gene expression, measured using transcriptomics and proteomics, that reveals post translational control over enzyme activity (Allwood, Ellis, and Goodacre 2008).

Metabolomics is providing assistance in locating the targets for plant breeding by linking gene expression and allelic variation to variation in metabolite complement, that is, functional genomics. Additionally, metabolomic studies are used for assessing the potential impacts of climate change and reduced input agricultural systems on crop composition (Davies et al. 2010). There has been constant effort under the umbrella of plant biotechnology to deal with the issue of food insecurity; nevertheless, the target still needs to be achieved. We need to exploit the exotic germplasm, a prolific source of valuable genes, that can serve both our goals of crop improvement and food security, on a mass scale (Wang et al. 2017).

The challenges are immense regarding the integration of metabolomics with phenotypic and genomic databases, which will provide us with an overall picture of plant metabolomics, thus aiding in furthering our application of this knowledge in genome-assisted breeding programmes (Fernie, Tadmor, and Zamir 2006).

Looking ahead, the expansion of metabolomics will lead to achieving the goals of improvement in comprehensive coverage of plant metabolome; the facilitation of the comparison of results between laboratories and experiments; and enhancement of integration of the metabolomic data with other functional genomics techniques. Having achieved these goals and targets, metabolomic science will lead to achieving the goals of secure and sustainable agriculture (Bino et al. 2004), which are the cruxes of plant biotechnology and its development.

References

Abdel-Farid, I B, M Jahangir, CAMJJ van den Hondel, H K Kim, Y H Choi, and R Verpoorte. 2009. "Fungal infection-induced metabolites in *Brassica rapa*." *Plant Science* 176 (5):608–615.

Ahmad, S, N Veyrat, R Gordon-Weeks, Y Zhang, J Martin, L Smart, G Glauser, M Erb, V Flors, and M Frey. 2011. "Benzoxazinoid metabolites regulate innate immunity against aphids and fungi in maize." *Plant Physiology*: pp. 111.180224.

Albinsky, D, M Kusano, M Higuchi, N Hayashi, M Kobayashi, A Fukushima, M Mori, T Ichikawa, K Matsui, and H Kuroda. 2010. "Metabolomic screening applied to rice FOX *Arabidopsis* lines leads to the identification of a gene-changing nitrogen metabolism." *Molecular Plant* 3 (1):125–142.

Albinsky, D, Y Sawada, A Kuwahara, M Nagano, A Hirai, K Saito, and M Y Hirai. 2010. "Widely targeted metabolomics and coexpression analysis as tools to identify genes involved in the side-chain elongation steps of aliphatic glucosinolate biosynthesis." *Amino Acids* 39 (4):1067–1075.

Allwood, J W, D I Ellis, and R Goodacre. 2008. "Metabolomic technologies and their application to the study of plants and plant–host interactions." *Physiologia Plantarum* 132 (2):117–135.

Arbona, V, M Manzi, C Ollas, and A Gómez-Cadenas. 2013. "Metabolomics as a tool to investigate abiotic stress tolerance in plants." *International Journal of Molecular Sciences* 14 (3):4885–4911.

Ashraf, M, M S A Ahmad, M Öztürk, and A Aksoy. 2012. "Crop improvement through different means: challenges and prospects." In *Crop Production for Agricultural Improvement*, 1–15. Springer.

Ashraf, M, M Öztürk, M S A Ahmad, and A Aksoy. 2012. *Crop Production for Agricultural İmprovement*: Springer Science & Business Media.

Badr, A, P Angers, and Y Desjardins. 2011. "Metabolic profiling of photoautotrophic and photomixotrophic potato plantlets (*Solanum tuberosum*) provides new insights into acclimatization." *Plant Cell, Tissue and Organ Culture (PCTOC)* 107 (1):13–24.

Baker, J M, N D Hawkins, J L Ward, A Lovegrove, J A Napier, P R Shewry, and M H Beale. 2006. "A metabolomic study of substantial equivalence of field-grown genetically modified wheat." *Plant Biotechnology Journal* 4 (4):381–392.

Barrett, L G, and M Heil. 2012. "Unifying concepts and mechanisms in the specificity of plant–enemy interactions." *Trends in Plant Science* 17 (5):282–292.

Beckmann, M, D P Enot, D P Overy, and J Draper. 2007. "Representation, comparison, and interpretation of metabolome fingerprint data for total composition analysis and quality trait investigation in potato cultivars." *Journal of Agricultural and Food Chemistry* 55 (9):3444–3451.

Beleggia, R, C Platani, G Laidò, D Rau, F Nigro, A M Mastrangelo, P De Vita, and R Papa. 2013. "Evolutionary metabolomics of durum wheat domestication." *Molecular Biology and Evolution* 33 (7):1740–1753.

Beleggia, R, C Platani, R Papa, A Di Chio, E Barros, C Mashaba, J Wirth, A Fammartino, C Sautter, and S Conner. 2011. "Metabolomics and food processing: from semolina to pasta." *Journal of Agricultural and Food Chemistry* 59 (17):9366–9377.

Bénard, C, S Bernillon, B Biais, S Osorio, M Maucourt, P Ballias, C Deborde, S Colombié, C Cabasson, and D Jacob. 2015. "Metabolomic profiling in tomato reveals diel compositional changes in fruit affected by source–sink relationships." *Journal of Experimental Botany* 66 (11):3391–3404.

Bino, R J, R D Hall, O Fiehn, J Kopka, K Saito, J Draper, B J Nikolau, P Mendes, U Roessner-Tunali, and M H Beale. 2004. "Potential of metabolomics as a functional genomics tool." *Trends in Plant Science* 9 (9):418–425.

Bollina, V, G K Kumaraswamy, A C Kushalappa, T M Choo, Y Dion, S Rioux, D Faubert, and H Hamzehzarghani. 2010. "Mass spectrometry-based metabolomics application to identify quantitative resistance-related metabolites in barley against *Fusarium* head blight." *Molecular Plant Pathology* 11 (6):769–782.

Bovy, A, E Schijlen, and R D Hall. 2007. "Metabolic engineering of flavonoids in tomato (*Solanum lycopersicum*): the potential for metabolomics." *Metabolomics* 3 (3):399.

Brader, G, E Tas, and E T Palva. 2001. "Jasmonate-dependent induction of indole glucosinolates in *Arabidopsis* by culture filtrates of the nonspecific *Pathogenerwinia carotovora*." *Plant Physiology* 126 (2):849–860.

Breckle, S-W. 2002. "Salinity, halophytes and salt affected natural ecosystems." In *Salinity: Environment-Plants-Molecules*, 53–77. Springer.

Broeckling, C D, K-G Li, and D-Y Xie. 2012. "Comparative metabolomics of transgenic tobacco plants (*Nicotiana tabacum* var. Xanthi) reveals differential effects of engineered complete and incomplete flavonoid pathways on the metabolome." In *Transgenic Plants-Advances and Limitations*. InTech.

Bunsupa, S, K Katayama, E Ikeura, A Oikawa, K Toyooka, K Saito, and M Yamazaki. 2012. "Lysine decarboxylase catalyzes the first step of quinolizidine alkaloid biosynthesis and coevolved with alkaloid production in Leguminosae." *The Plant Cell*: tpc. 112.095885.

Catchpole, G S, M Beckmann, D P Enot, M Mondhe, B Zywicki, J Taylor, N Hardy, A Smith, R D King, and D B Kell. 2005. "Hierarchical metabolomics demonstrates substantial compositional similarity between genetically modified and conventional potato crops." *Proceedings of the National Academy of Sciences* 102 (40):14458–14462.

Chemler, J A, and M A G Koffas. 2008. "Metabolic engineering for plant natural product biosynthesis in microbes." *Current Opinion in Biotechnology* 19 (6):597–605.

Chen, W, W Wang, M Peng, L Gong, Y Gao, J Wan, S Wang, L Shi, B Zhou, and Z Li. 2016. "Comparative and parallel genome-wide association studies for metabolic and agronomic traits in cereals." *Nature Communications* 7:12767.

Cheng, K C, J Beaulieu, E Iquira, F J Belzile, M G Fortin, and M V Strömvik. 2008. "Effect of transgenes on global gene expression in soybean is within the natural range of variation of conventional cultivars." *Journal of Agricultural and Food Chemistry* 56 (9):3057–3067.

Cross, J M, M von Korff, T Altmann, L Bartzetko, R Sulpice, Y Gibon, N Palacios, and M Stitt. 2006. "Variation of enzyme activities and metabolite levels in 24 *Arabidopsis* accessions growing in carbon-limited conditions." *Plant Physiology* 142 (4):1574–1588.

Croteau, R, R E B Ketchum, R M Long, R Kaspera, and M R Wildung. 2006. "Taxol biosynthesis and molecular genetics." *Phytochemistry Reviews* 5 (1):75–97.

D'Auria, J C, and J Gershenzon. 2005. "The secondary metabolism of *Arabidopsis thaliana*: growing like a weed." *Current Opinion in Plant Biology* 8 (3):308–316.

Daayf, F, M Ongena, R Boulanger, I El Hadrami, and R R Bélanger. 2000. "Induction of phenolic compounds in two cultivars of cucumber by treatment of healthy and powdery mildew-infected plants with extracts of *Reynoutria sachalinensis*." *Journal of Chemical Ecology* 26 (7):1579–1593.

Davies, H V, L V T Shepherd, D Stewart, T Frank, R M Röhlig, and K-H Engel. 2010. "Metabolome variability in crop plant species–When, where, how much and so what?" *Regulatory Toxicology and Pharmacology* 58 (3): S54–S61.

Diaz, C, M Kusano, R Sulpice, M Araki, H Redestig, K Saito, M Stitt, and R Shin. 2011. "Determining novel functions of *Arabidopsis* 14-3-3 proteins in central metabolic processes." *BMC Systems Biology* 5 (1):192.

Do, P T, M Prudent, R Sulpice, M Causse, and A R Fernie. 2010. "The influence of fruit load on the tomato pericarp metabolome in a *Solanum chmielewskii* introgression line population." *Plant Physiology* 154:1128–1142.

Dobson, G, T Shepherd, S R Verrall, S Conner, J W McNicol, G Ramsay, L V T Shepherd, H V Davies, and D Stewart. 2008. "Phytochemical diversity in tubers of potato cultivars and landraces using a GC-MS metabolomics approach." *Journal of Agricultural and Food Chemistry* 56 (21):10280–10291.

Dobson, G, T Shepherd, S R Verrall, W D Griffiths, G Ramsay, J W McNicol, H V Davies, and D Stewart. 2010. "A metabolomics study of cultivated potato (*Solanum tuberosum*) groups andigena, phureja, stenotomum, and tuberosum using gas chromatography-mass spectrometry." *Journal of Agricultural and Food Chemistry* 58 (2):1214–1223.

Doddaraju, P, R Gunnaiah, A C Kushalappa, and A Murphy. 2013. "Metabolomics of potato genotypes with contrasting levels of resistance to late blight." *Plant Molecular Biology Reporter* 32 (2):584–595.

Donkin, S G, M A Eiteman, and P L Williams. 1995. "Toxicity of glucosinolates and their enzymatic decomposition products to *Caenorhabditis elegans*." *Journal of Nematology* 27 (3):258.

Doughty, K J, A J R Porter, A M Morton, G Kiddle, C H Bock, and R Wallsgrove. 1991. "Variation in the glucosinolate content of oilseed rape (*Brassica napus* L.) leaves: II. Response to infection by *Alternaria brassicae* (Berk.) Sacc." *Annals of Applied Biology* 118 (2):469–477.

Eckardt, A N. 2008. "Epistasis and genetic regulation of variation in the *Arabidopsis* Metabolome." *The Plant Cell* 20 (1):1185–1186.

Farre, E M, A R Fernie, and L Willmitzer. 2008. "Analysis of subcellular metabolite levels of potato tubers (*Solanum tuberosum*) displaying alterations in cellular or extracellular sucrose metabolism." *Metabolomics* 4 (2):161–170.

Fernie, A R, Y Tadmor, and D Zamir. 2006. "Natural genetic variation for improving crop quality." *Current Opinion in Plant Biology* 9 (2):196–202.

Fiehn, O. 2001. "Combining genomics, metabolome analysis, and biochemical modelling to understand metabolic networks." *International Journal of Genomics* 2 (3):155–168.

Francki, M G, S Hayton, J P A Gummer, C Rawlinson, and R D Trengove. 2016. "Metabolomic profiling and genomic analysis of wheat aneuploid lines to identify genes controlling biochemical pathways in mature grain." *Plant Biotechnology Journal* 14 (2):649–660.

Fraser, P D, E M A Enfissi, M Goodfellow, T Eguchi, and P M Bramley. 2007. "Metabolite profiling of plant carotenoids using the matrix-assisted laser desorption ionization time-of-flight mass spectrometry." *The Plant Journal* 49 (3):552–564.

Fukusaki, E, and A Kobayashi. 2005. "Plant metabolomics: potential for practical operation." *Journal of Bioscience and Bioengineering* 100 (4):347–354.

Fukushima, A, M Kusano, N Nakamichi, M Kobayashi, N Hayashi, H Sakakibara, T Mizuno, and K Saito. 2009. "Impact of clock-associated *Arabidopsis* pseudo-response regulators in metabolic coordination." *Proceedings of the National Academy of Sciences* 106 (17):7251–7256.

Fukushima, A, M Kusano, H Redestig, M Arita, and K Saito. 2009. "Integrated omics approaches in plant systems biology." *Current Opinion in Chemical Biology* 13 (5–6):532–538.

Fukushima, A, M Kusano, H Redestig, M Arita, and K Saito. 2011. "Metabolomic correlation-network modules in *Arabidopsis* based on a graph-clustering approach." *BMC Systems Biology* 5 (1):1.

González-Lamothe, R, G Mitchell, M Gattuso, M S Diarra, F Malouin, and K Bouarab. 2009. "Plant antimicrobial agents and their effects on plant and human pathogens." *International Journal of Molecular Sciences* 10 (8):3400–3419.

Gray, G R, and D Heath. 2005. "A global reorganization of the metabolome in *Arabidopsis* during cold acclimation is revealed by metabolic fingerprinting." *Physiologia Plantarum* 124 (2):236–248.

Gregersen, P L, H Brinch-Pedersen, and P B Holm. 2005. "A microarray-based comparative analysis of gene expression profiles during grain development in transgenic and wild type wheat." *Transgenic Research* 14 (6):887–905.

Großkinsky, D K, E van der Graaff, and T Roitsch. 2012. "Phytoalexin transgenics in crop protection—Fairy tale with a happy end?" *Plant Science* 195:54–70.

Gunnaiah, R, A C Kushalappa, R Duggavathi, S Fox, and D J Somers. 2012. "Integrated metabolo-proteomic approach to decipher the mechanisms by which wheat QTL (Fhb1) contributes to resistance against *Fusarium graminearum*." *PloS One* 7 (7):e40695.

Hakeem, K R, M Öztürk, P Ahmad, and A R Memon. 2012. "Biotechnology as an aid for crop improvement to overcome food shortage." In *Crop Production for Agricultural Improvement*, edited by Ashraf et al, 239–261. Verlag, NY, USA: Springer.

Hakeem, K R, P Ahmad, and M Ozturk. 2013. *Crop Improvement: New Approaches and Modern Techniques*: Springer Science & Business Media.

Hakeem, K R, H Tombuloğlu, and G Tombuloğlu. 2016. *Plant Omics: Trends and Applications*: Springer.

Hall, R D. 2006. "Plant metabolomics: from holistic hope, to hype, to hot topic." *New Phytologist* 169 (3):453–468.

Hara, M, A Harazaki, and K Tabata. 2013. "Administration of isothiocyanates enhances heat tolerance in *Arabidopsis thaliana*." *Plant Growth Regulation* 69 (1):71–77.

Harrigan, G G, L G Stork, S G Riordan, T L Reynolds, W P Ridley, J D Masucci, S MacIsaac, S C Halls, R Orth, and R G Smith. 2007. "Impact of genetics and environment on nutritional and metabolite components of maize grain." *Journal of Agricultural and Food Chemistry* 55 (15):6177–6185.

Hediji, H, W Djebali, C Cabasson, M Maucourt, P Baldet, A Bertrand, L Boulila Zoghlami, C Deborde, A Moing, and R Brouquisse. 2010. "Effects of long-term cadmium exposure on growth and metabolomic profile of tomato plants." *Ecotoxicology and Environmental Safety* 73 (8):1965–1974.

Hegeman, A D. 2010. "Plant metabolomics—meeting the analytical challenges of comprehensive metabolite analysis." *Briefings in Functional Genomics* 9 (2):139–148.

Hendrawati, O, Q Yao, H K Kim, H J M Linthorst, C Erkelens, A W M Lefeber, Y Hae Choi, and R Verpoorte. 2006. "Metabolic differentiation of *Arabidopsis* treated with methyl jasmonate using nuclear magnetic resonance spectroscopy." *Plant Science* 170 (6):1118–1124.

Heuberger, A L, F M Robison, S Marie A Lyons, C D Broeckling, and J E Prenni. 2014. "Evaluating plant immunity using mass spectrometry-based metabolomics workflows." *Frontiers in Plant Science* 5:291.

Hinks, C F, and O Olfert. 1992. "Cultivar resistance to grasshoppers in temperate cereal crops and grasses: a review." *Journal of Orthoptera Research* (1):1–9.

Hirai, M Y, M Klein, Y Fujikawa, M Yano, D B Goodenowe, Y Yamazaki, S Kanaya, Y Nakamura, M Kitayama, and H Suzuki. 2005. "Elucidation of gene-to-gene and metabolite-to-gene networks in *Arabidopsis* by integration of metabolomics and transcriptomics." *Journal of Biological Chemistry*.

Hirai, Masami Yokota, and Kazuki Saito. 2008. "Analysis of systemic sulfur metabolism in plants using integrated '-omics' strategies." *Molecular BioSystems* 4 (10):967–973.

Hirai, M Y, K Sugiyama, Y Sawada, T Tohge, T Obayashi, A Suzuki, R Araki, N Sakurai, H Suzuki, and K Aoki. 2007. "Omics-based identification of *Arabidopsis* Myb transcription factors regulating aliphatic glucosinolate biosynthesis." *Proceedings of the National Academy of Sciences* 104 (15):6478–6483.

Hirai, M Y, M Yano, D B Goodenowe, S Kanaya, T Kimura, M Awazuhara, M Arita, T Fujiwara, and K Saito. 2004. "Integration of transcriptomics and metabolomics for understanding of global responses to nutritional stresses in *Arabidopsis thaliana*." *Proceedings of the National Academy of Sciences* 101 (27):10205–10210.

Hirochika, H, E Guiderdoni, G An, Y Hsing, M Y Eun, C-d Han, N Upadhyaya, S Ramachandran, Q Zhang, and Pereira. 2004. "Rice mutant resources for gene discovery." *Plant Molecular Biology* 54 (3):325–334.

Hong, J, L Yang, D Zhang, and J Shi. 2016. "Plant Metabolomics: An indispensable system biology tool for plant science." *International Journal of Molecular Sciences* 17 (6):767.

Hu, C, J Shi, S Quan, B Cui, S Kleessen, Z Nikoloski, T Tohge, D Alexander, L Guo, and H Lin. 2014. "Metabolic variation between japonica and indica rice cultivars as revealed by non-targeted metabolomics." *Scientific Reports* 4:5067.

Iijima, Y, and K Aoki. 2009. "Application of metabolomics to improve tomato fruit productivity and quality." *Journal of the Japanese Society for Horticultural Science* 78 (1):14–22.

Ioset, J-R, B Urbaniak, K Ndjoko-Ioset, J Wirth, F Martin, W Gruissem, K Hostettmann, and C Sautter. 2007. "Flavonoid profiling among wild type and related gm wheat varieties." *Plant Molecular Biology* 65 (5):645–654.

Itoh, T, T Tanaka, R A Barrero, C Yamasaki, Y Fujii, P B Hilton, B A Antonio, H Aono, R Apweiler, and R Bruskiewich. 2007. "Curated genome annotation of *Oryza sativa* ssp. japonica and comparative genome analysis with *Arabidopsis thaliana*." *Genome Research* 17 (2):000–000.

Jones, O A H, M L Maguire, J L Griffin, Y-H Jung, J Shibato, R Rakwal, G K Agrawal, and N-S Jwa. 2011. "Using metabolic profiling to assess plant-pathogen interactions: an example using rice (*Oryza sativa*) and the blast pathogen *Magnaporthe grisea*." *European Journal of Plant Pathology* 129 (4):539–554.

Kang, Y, and M Udvardi. 2012. "Global regulation of reactive oxygen species scavenging genes in *Alfalfa* root and shoot under gradual drought stress and recovery." *Plant Signaling and Behavior* 7 (5):539–543.

Karre, S S. 2014. "Functional elucidation of wheat qtl fhb4 for resistance against *Fusarium* head blight, based on non-targeted metabolomics." *Plant Animal Genome.*

Keurentjes, J JB, R Sulpice, Y Gibon, M-C Steinhauser, J Fu, M Koornneef, M Stitt, and D Vreugdenhil. 2008. "Integrative analyses of genetic variation in enzyme activities of primary carbohydrate metabolism reveal distinct modes of regulation in *Arabidopsis thaliana*." *Genome Biology* 9 (8):R129.

Klie, S, S Osorio, T Tohge, M F Drincovich, A Fait, J J Giovannoni, A R Fernie, and Z Nikoloski. 2014. "Conserved changes in the dynamics of metabolic processes during fruit development and ripening across species." *Plant Physiology* 164 (1):55–68.

Kliebenstein, D J, J Gershenzon, and T Mitchell-Olds. 2001. "Comparative quantitative trait loci mapping of aliphatic, indolic and benzylic glucosinolate production in *Arabidopsis thaliana* leaves and seeds." *Genetics* 159 (1):359–370.

Kliebenstein, D J. 2004. "Secondary metabolites and plant/environment interactions: a view through *Arabidopsis thaliana* tinged glasses." *Plant, Cell and Environment* 27 (6):675–684.

Kolesnik, T, I Szeverenyi, D Bachmann, C Santhosh Kumar, S Jiang, R Ramamoorthy, M Cai, Z G Ma, V Sundaresan, and S Ramachandran. 2004. "Establishing an efficient Ac/Ds tagging system in rice: large-scale analysis of Ds flanking sequences." *The Plant Journal* 37 (2):301–314.

Kumar, R, A Bohra, A K Pandey, M K Pandey, and A Kumar. 2017. "Metabolomics for plant improvement: status and prospects." *Frontiers in Plant Science* 8 (1):1302.

Kusano, M, A Fukushima, M Arita, P Jonsson, T Moritz, M Kobayashi, N Hayashi, T Tohge, and K Saito. 2007. "Unbiased characterization of genotype-dependent metabolic regulations by metabolomic approach in *Arabidopsis thaliana*." *BMC Systems Biology* 1 (1):53.

Kusano, M, A Fukushima, H Redestig, M Kobayashi, H Otsuki, H Onouchi, S Naito, M Y Hirai, and K Saito. 2010. "Comparative metabolomics charts the impact of genotype-dependent methionine accumulation in *Arabidopsis thaliana*." *Amino Acids* 39 (4):1013-1021.

Kusano, M, A Fukushima, H Redestig, and K Saito. 2011. "Metabolomic approaches toward understanding nitrogen metabolism in plants." *Journal of Experimental Botany* 62 (4):1439–1453.

Kusano, M, and K Saito. 2012. "Role of metabolomics in crop improvement." *Journal of Plant Biochemistry and Biotechnology* 21 (1):24–31.

Kusano, M, M Tabuchi, A Fukushima, K Funayama, C Diaz, M Kobayashi, N Hayashi, Y N Tsuchiya, H Takahashi, and A Kamata. 2011. "Metabolomics data reveal a crucial role of cytosolic glutamine synthetase 1; 1 in coordinating metabolic balance in rice." *The Plant Journal* 66 (3):456–466.

Kusano, M, T Tohge, A Fukushima, M Kobayashi, N Hayashi, H Otsuki, Y Kondou, H Goto, M Kawashima, and F Matsuda. 2011. "Metabolomics reveals comprehensive reprogramming involving two independent metabolic responses of *Arabidopsis* to UV-B light." *The Plant Journal* 67 (2):354–369.

Lee, E A, P F Byrne, M D McMullen, M E Snook, B R Wiseman, N W Widstrom, and E H Coe. 1998. "Genetic mechanisms underlying apimaysin and maysin synthesis and corn earworm antibiosis in maize (*Zea mays* L.)." *Genetics* 149 (4):1997–2006.

Leiss, K A, Y H Choi, I B Abdel-Farid, R Verpoorte, and P G L Klinkhamer. 2009. "NMR metabolomics of thrips (*Frankliniella occidentalis*) resistance in *Senecio* hybrids." *Journal of Chemical Ecology* 35 (2):219–229.

Leiss, K A, F Maltese, Y H Choi, R Verpoorte, and P G L Klinkhamer. 2009. "Identification of chlorogenic acid as a resistance factor for thrips in *Chrysanthemum*." *Plant Physiology* 150 (3):1567–1575.

Lemmens, M, Uwe Scholz, F Berthiller, C Dall'Asta, A Koutnik, R Schuhmacher, G Adam, H Buerstmayr, Á Mesterházy, and R Krska. 2005. "The ability to detoxify the mycotoxin deoxynivalenol colocalizes with a major quantitative trait locus for *Fusarium* head blight resistance in wheat." *Molecular Plant-Microbe Interactions* 18 (12):1318–1324.

Li, Q, S D Eigenbrode, G R Stringam, and M R Thiagarajah. 2000. "Feeding and growth of *Plutella xylostella* and *Spodoptera eridania* on *Brassica juncea* with varying glucosinolate concentrations and myrosinase activities." *Journal of Chemical Ecology* 26 (10):2401–2419.

Lopes, A S, E C Santa Cruz, A Sussulini, and A Klassen. 2017. "Metabolomic strategies involving mass spectrometry combined with liquid and gas chromatography." In *Metabolomics: From Fundamentals to Clinical Applications*, 77–98. Springer.

López-Gresa, M P, P Lisón, H K Kim, Y H Choi, R Verpoorte, I Rodrigo, V Conejero, and J M Bellés. 2012. "Metabolic fingerprinting of tomato mosaic virus infected *Solanum lycopersicum*." *Journal of Plant Physiology* 169 (16):1586–1596.

Ludwig-Müller, J, B Schubert, K Pieper, S Ihmig, and W Hilgenberg. 1997. "Glucosinolate content in susceptible and resistant Chinese cabbage varieties during development of clubroot disease." *Phytochemistry* 44 (3):407–414.

Luengwilai, K, M Saltveit, and D M Beckles. 2012. "Metabolite content of harvested Micro-Tom tomato (*Solanum lycopersicum* L.) fruit is altered by chilling and protective heat-shock treatments as shown by GC–MS metabolic profiling." *Postharvest Biology and Technology* 63 (1):116–122.

Lugan, R, M F Niogret, L Kervazo, F R Larher, J Kopka, and A Bouchereau. 2009. "Metabolome and water status phenotyping of *Arabidopsis* under abiotic stress cues reveals new insight into ESK1 function." *Plant, Cell and Environment* 32 (2):95–108.

Manici, L M, L Lazzeri, and S Palmieri. 1997. "In vitro fungitoxic activity of some glucosinolates and their enzyme-derived products toward plant pathogenic fungi." *Journal of Agricultural and Food Chemistry* 45 (7):2768–2773.

Maruyama, K, M Takeda, S Kidokoro, K Yamada, Y Sakuma, K Urano, M Fujita, K Yoshiwara, S Matsukura, and Y Morishita. 2009. "Metabolic pathways involved in cold acclimation identified by integrated analysis of metabolites and transcripts regulated by DREB1A and DREB2A." *Plant Physiology* 150 (4):1972–1980.

Matsuda, F, M Y Hirai, E Sasaki, K Akiyama, K Yonekura-Sakakibara, N J Provart, T Sakurai, Y Shimada, and K Saito. 2010. "AtMetExpress development: a phytochemical atlas of *Arabidopsis* development." *Plant Physiology* 152 (2):566–578.

Matsuda, F, Y Okazaki, A Oikawa, M Kusano, R Nakabayashi, J Kikuchi, J-I Yonemaru, K Ebana, M Yano, and K Saito. 2012. "Dissection of genotype–phenotype associations in rice grains using metabolome quantitative trait loci analysis." *The Plant Journal* 70 (4):624–636.

Matsuda, F, K Yonekura-Sakakibara, R Niida, T Kuromori, K Shinozaki, and K Saito. 2009. "MS/MS spectral tag-based annotation of non-targeted profile of plant secondary metabolites." *The Plant Journal* 57 (3):555–577.

Ménard, R, J-P Larue, D Silué, and D Thouvenot. 1999. "Glucosinolates in cauliflower as biochemical markers for resistance against downy mildew." *Phytochemistry* 52 (1):29–35.

Menéndez, C M, E Ritter, R Schäfer-Pregl, B Walkemeier, A Kalde, F Salamini, and C Gebhardt. 2002. "Cold sweetening in diploid potato: Mapping quantitative trait loci and candidate genes." *Genetics* 162 (3):1423–1434.

Meyer, R C, M Steinfath, J Lisec, M Becher, H Witucka-Wall, O Törjék, O Fiehn, Ä Eckardt, L Willmitzer, and J Selbig. 2007. "The metabolic signature related to high plant growth rate in *Arabidopsis thaliana*." *Proceedings of the National Academy of Sciences* 104 (11):4759–4764.

Morreel, K, G Goeminne, V Storme, L Sterck, J Ralph, W Coppieters, P Breyne, M Steenackers, M Georges, and E Messens. 2006. "Genetical metabolomics of flavonoid biosynthesis in *Populus*: a case study." *The Plant Journal* 47 (2):224–237.

Moudi, M, R Go, C Y S Yien, and M Nazre. 2013. "Vinca alkaloids." *International Journal of Preventive Medicine* 4 (11):1231.

Munns, R, and M Tester. 2008. "Mechanisms of salinity tolerance." *Annual Review of Plant Biology* 59:651–681.

Niemeyer, H M. 2009. "Hydroxamic acids derived from 2-hydroxy-2 H-1, 4-benzoxazin-3 (4 H)-one: key defense chemicals of cereals." *Journal of Agricultural and Food Chemistry* 57 (5):1677–1696.

Nikiforova, V J, J Kopka, V Tolstikov, O Fiehn, L Hopkins, M J Hawkesford, H Hesse, and R Hoefgen. 2005. "Systems rebalancing of metabolism in response to sulfur deprivation, as revealed by metabolome analysis of *Arabidopsis* plants." *Plant Physiology* 138 (1):304–318.

Nishizawa, A, Y Yabuta, and S Shigeoka. 2008. "Galactinol and raffinose constitute a novel function to protect plants from oxidative damage." *Plant Physiology* 147 (3):1251–1263.

Obata, T, and A R Fernie. 2012. "The use of metabolomics to dissect plant responses to abiotic stresses." *Cellular and Molecular Life Sciences* 69 (19):3225–3243.

Okazaki, Y, M Shimojima, Y Sawada, K Toyooka, T Narisawa, K Mochida, H Tanaka, F Matsuda, A Hirai, and M Y Hirai. 2009. "A chloroplastic UDP-glucose pyrophosphorylase from *Arabidopsis* is the committed enzyme for the first step of sulfolipid biosynthesis." *The Plant Cell* 21 (3):892–909.

Pan, Q, Q Wang, F Yuan, S Xing, J Zhao, Y H Choi, R Verpoorte, Y Tian, G Wang, and K Tang. 2012. "Overexpression of ORCA3 and G10H in *Catharanthus roseus* plants regulated alkaloid biosynthesis and metabolism revealed by NMR-metabolomics." *PloS One* 7 (8):e43038.

Patti, G J, O Yanes, and G Siuzdak. 2012. "Innovation: Metabolomics: the apogee of the omics trilogy." *Nature Reviews Molecular Cell Biology* 13 (4):263.

Perez de Souza, L, T Naake, T Tohge, and A R Fernie. 2017. "From chromatogram to analyte to metabolite. How to pick horses for courses from the massive web resources for mass spectral plant metabolomics." *GigaScience* 6 (7):gix037.

Qian, Y, Y Wang, R Sa, H Yan, X Pan, Y Yang, and Y Sun. 2013. "Metabolic fingerprinting of *Angelica sinensis* during growth using UPLC-TOFMS and chemometrics data analysis." *Chemistry Central Journal* 7 (1):42.

Rao, A, and L G Rao. 2007. "Carotenoids and human health." *Pharmacological Research* 55 (3):207–216.

Rasmussen, S, A J Parsons, and C S Jones. 2012. "Metabolomics of forage plants: a review." *Annals of Botany* 110 (6):1281–1290.

Resham, S, F Akhter, M Ashraf, and A Gul Kazi. 2014. "Metabolomics Role in Crop Improvement." In *Emerging Technologies and Management of Crop Stress Tolerance*, 39–55. Elsevier.

Roetschi, A, A Si-Ammour, L Belbahri, F Mauch, and B Mauch-Mani. 2001. "Characterization of an *Arabidopsis*–Phytophthora pathosystem: resistance requires a functional PAD2 gene and is independent of salicylic acid, ethylene and jasmonic acid signalling." *The Plant Journal* 28 (3):293–305.

Rostás, M, R Bennett, and M Hilker. 2002. "Comparative physiological responses in Chinese cabbage induced by herbivory and fungal infection." *Journal of Chemical Ecology* 28 (12):2449–2463.

Saito, K, M Y Hirai, and K Yonekura-Sakakibara. 2008. "Decoding genes with coexpression networks and metabolomics–'majority report by precogs'." *Trends in Plant Science* 13 (1):36–43.

Sankawa, U. 1999. *Comprehensive Natural Products Chemistry*. Edited by D H R Barton, K Nakanishi and O Meth-Cohn. Vol. 1, *Polyketides and Other Secondary Metabolites Including Fatty Acids and Their Derivatives*. Oxford, UK: Elsevier Science.

Sarma, B K, and U P Singh. 2003. "Ferulic acid may prevent infection of *Cicer arietinum* by *Sclerotium rolfsii.*" *World Journal of Microbiology and Biotechnology* 19 (2):123–127.

Sato, Y, B A Antonio, N Namiki, H Takehisa, H Minami, K Kamatsuki, K Sugimoto, Y Shimizu, H Hirochika, and Y Nagamura. 2010. "RiceXPro: a platform for monitoring gene expression in japonica rice grown under natural field conditions." *Nucleic Acids Research* 39 (suppl_1):D1141-D1148.

Sawada, Y, K Akiyama, A Sakata, A Kuwahara, H Otsuki, T Sakurai, K Saito, and M Y Hirai. 2008. "Widely targeted metabolomics based on large-scale MS/MS data for elucidating metabolite accumulation patterns in plants." *Plant and Cell Physiology* 50 (1):37–47.

Sawada, Y, and M Y Hirai. 2013. "Integrated LC-MS/MS system for plant metabolomics." *Computational and Structural Biotechnology Journal* 4 (5):e201301011.

Sawada, Y, K Toyooka, A Kuwahara, A Sakata, M Nagano, K Saito, and M Y Hirai. 2009. "*Arabidopsis* bile acid: sodium symporter family protein 5 is involved in methionine-derived glucosinolate biosynthesis." *Plant and Cell Physiology* 50 (9):1579–1586.

Schauer, N, Y Semel, I Balbo, M Steinfath, D Repsilber, J Selbig, T Pleban, D Zamir, and A R Fernie. 2008. "Mode of inheritance of primary metabolic traits in tomato." *The Plant Cell* 20 (3):509–523.

Schauer, N, Y Semel, U Roessner, A Gur, I Balbo, F Carrari, T Pleban, A Perez-Melis, C Bruedigam, and J Kopka. 2006. "Comprehensive metabolic profiling and phenotyping of interspecific introgression lines for tomato improvement." *Nature Biotechnology* 24 (4):447.

Sharp, R E, V Poroyko, L G Hejlek, W G Spollen, G K Springer, H J Bohnert, and H T Nguyen. 2004. "Root growth maintenance during water deficits: Physiology to functional genomics." *Journal of Experimental Botany* 55 (407):2343–2351.

Skot, L, J Humphreys, M O Humphreys, D Thorogood, J Gallagher, R Sanderson, I P Armstead, and I D Thomas. 2007. "Association of candidate genes with flowering time and water soluble carbohydrate content in *Lolium perenne* (L.)." *Genetics*.

Sun, H-J, H Kataoka, M Yano, and H Ezura. 2007. "Genetically stable expression of functional miraculin, a new type of alternative sweetener, in transgenic tomato plants." *Plant Biotechnology Journal* 5 (6):768–777.

Suzuki, M, M Kusano, H Takahashi, Y Nakamura, N Hayashi, M Kobayashi, T Ichikawa, M Matsui, H Hirochika, and K Saito. 2010. "Rice-*Arabidopsis* FOX line screening with FT-NIR-based fingerprinting for GC-TOF/MS-based metabolite profiling." *Metabolomics* 6 (1):137–145.

Tabuchi, M, K Sugiyama, K Ishiyama, E Inoue, T Sato, H Takahashi, and T Yamaya. 2005. "Severe reduction in growth rate and grain filling of rice mutants lacking OsGS1; 1, a cytosolic glutamine synthetase1; 1." *The Plant Journal* 42 (5):641–651.

Tan, J, P Ednarek, J Liu, B Schneider, A Svatoš, and K Hahlbrock. 2004. "Universally occurring phenylpropanoid and species-specific indolic metabolites in infected and uninfected *Arabidopsis thaliana* roots and leaves." *Phytochemistry* 65 (6):691–699.

Thevissen, K, K KA Ferket, I EJA François, and B PA Cammue. 2003. "Interactions of antifungal plant defensins with fungal membrane components." *Peptides* 24 (11):1705–1712.

Tohge, T, Y Nishiyama, M Y Hirai, M Yano, J-i Nakajima, M Awazuhara, E Inoue, H Takahashi, D B Goodenowe, and M Kitayama. 2005. "Functional genomics by integrated analysis of metabolome and transcriptome of *Arabidopsis* plants over-expressing an MYB transcription factor." *The Plant Journal* 42 (2):218–235.

Turner, M F, A L Heuberger, J S Kirkwood, C C Collins, E J Wolfrum, C D Broeckling, J E Prenni, and C E Jahn. 2016. "Non-targeted metabolomics in diverse *Sorghum* breeding lines indicates primary and secondary metabolite profiles are associated with plant biomass accumulation and photosynthesis." *Frontiers in Plant Science* 7:953.

Urano, K, K Maruyama, Y Ogata, Y Morishita, M Takeda, N Sakurai, H Suzuki, K Saito, D Shibata, and M Kobayashi. 2009. "Characterization of the ABA-regulated global responses to dehydration in *Arabidopsis* by metabolomics." *The Plant Journal* 57 (6):1065–1078.

Uri, C, Z Juhász, Z Polgár, and Z Bánfalvi. 2014. "A GC–MS-based metabolomics study on the tubers of commercial potato cultivars upon storage." *Food Chemistry* 159:287–292.

Van der Heijden, R, D I Jacobs, W Snoeijer, D Hallard, and R Verpoorte. 2004. "The *Catharanthus* alkaloids: pharmacognosy and biotechnology." *Current Medicinal Chemistry* 11 (5):1241–1253.

Wahid, A, S Gelani, M Ashraf, and M R Foolad. 2007. "Heat tolerance in plants: An overview." *Environmental and Experimental Botany* 61 (3):199–223.

Wang, C, S Hu, C Gardner, and T Lübberstedt. 2017. "Emerging avenues for utilization of exotic germplasm." *Trends in Plant Science* 22 (7):624–637.

Watanabe, M, M Kusano, A Oikawa, A Fukushima, M Noji, and K Saito. 2008. "Physiological roles of the β-substituted alanine synthase gene family in *Arabidopsis*." *Plant Physiology* 146 (1):310–320.
Watanabe, M, Keiichi M, T Kato, S Tabata, N Yoshimoto, M Noji, and K Saito. 2008. "Comparative genomics and reverse genetics analysis reveal indispensable functions of the serine acetyltransferase gene family in *Arabidopsis*." *The Plant Cell* 20 (9):2484–2496.
Wen, W, K Li, S Alseek, N Omranian, L Zhao, Y Zhou, Y Xiao, M Jin, N Yang, and H Liu. 2015. "Genetic determinants of the network of primary metabolism and their relationships to plant performance in a maize recombinant inbred line population." *The Plant Cell*:tpc. 15.00208.
Yang, W-L, and M A Bernards. 2007. "Metabolite Profiling of Potato (*Solanum tuberosum* L.) Tubers During Wound-Induced Suberization." *Metabolomics* 3 (2):147–159.
Yonekura-Sakakibara, K, and K Saito. 2009. "Functional genomics for plant natural product biosynthesis." *Natural Product Reports* 26 (11):1466–1487.
Yonekura-Sakakibara, K, T Tohge, F Matsuda, R Nakabayashi, H Takayama, R Niida, A Watanabe-Takahashi, E Inoue, and K Saito. 2008. "Comprehensive flavonol profiling and transcriptome coexpression analysis leading to decoding gene–metabolite correlations in *Arabidopsis*." *The Plant Cell* 20 (8):2160–2176.
Yonekura-Sakakibara, K, T Tohge, R Niida, and K Saito. 2007. "Identification of a flavonol 7-O-rhamnosyltransferase gene determining flavonoid pattern in *Arabidopsis* by transcriptome coexpression analysis and reverse genetics." *Journal of Biological Chemistry* 282 (20):14932–14941.
Yonekura-Sakakibara, K, A Fukushima, R Nakabayashi, K Hanada, F Matsuda, S Sugawara, E Inoue, T Kuromori, T Ito, and K Shinozaki. 2012. "Two glycosyltransferases involved in anthocyanin modification delineated by transcriptome independent component analysis in *Arabidopsis thaliana*." *The Plant Journal* 69 (1):154–167.
Zasada, I A, SLF Meyer, J M Halbrendt, and C Rice. 2005. "Activity of hydroxamic acids from Secale cereale against the plant-parasitic nematodes *Meloidogyne incognita* and *Xiphinema americanum*." *Phytopathology* 95 (10):1116–1121.
Zhen, S, K Dong, X Deng, J Zhou, X Xu, C Han, W Zhang, Y Xu, Z Wang, and Y Yan. 2016. "Dynamic metabolome profiling reveals significant metabolic changes during grain development of bread wheat (*Triticum*

aestivum L.)." *Journal of the Science of Food and Agriculture* 96 (11):3731–3740.

Zörb, C, K H Mühling, M Hasler, V Gödde, K Niehaus, D Becker, and C-M Geilfus. 2013. "Metabolomic responses in grain, ear, and straw of winter wheat under increasing sulfur treatment." *Journal of Plant Nutrition and Soil Science* 176 (6):964–970.

CHAPTER 10

MELATONIN: A POTENTIAL PLANT GROWTH REGULATOR FOR AGRICULTURAL APPLICATION

JUN-LAN XIONG,[1,2*] HUAN ZHANG,[1] CHUN-LEI ZHANG[2]

[1]School of Life Science, Lanzhou University, Lanzhou, China
[2]Oilcrops Research Institute, Chinese Academy of Agricultural Science, Wuhan, China
xiongjunlan09@163.com

Abstract

As a ubiquitously and endogenously-produced indoleamine in plants, melatonin has emerged as a bioregulator of immense importance in plant sciences. For more than twenty years, the presence of melatonin has been investigated in almost all plant species. This review addresses the significance of melatonin in terms of plant growth and development. Melatonin helps plants with regard to promoting root and shoot growth and delaying leaf senescence, as well as regulating reproduction and yield formation. Moreover, a variety of environmental stresses, such as drought, salinity, and extreme temperatures greatly inhibit plant growth and productivity. Melatonin has been confirmed to aid plants against harsh environmental conditions. Research has shown that abiotic stress can cause an enhancement in the endogenous melatonin level in plants. The transgenic plants overexpressing melatonin biosynthetic genes, exhibited increased melatonin levels, and enhanced tolerance in response to abiotic stresses. Exogenous melatonin treatment is considered effective for improving abiotic stress tolerance in numerous plant species. The mechanism of melatonin in reducing stress-induced damage is discussed, which mainly focuses on its ability of reactive oxygen species detoxification. In summary, melatonin is recognised as having great potential for promoting crop production, especially under environmental stresses.

Key words: melatonin, biosynthesis, growth, leaf senescence, abiotic stress tolerance

Introduction

Melatonin, structurally named *N*-acetyl-5-methoxytrytamine, was first detected in the bovine pineal gland by Lerner et al. (1958). It has also been reported in almost all living organisms which consist of animals, photosynthetic bacteria, red and green algae, fungi, and plants (Reiter et al. 2015). Melatonin was found in over 100 species from many plant families after its discovery in land plants by two independent research groups (Dubbels et al. 1995, Hattori et al. 1995, Hardeland 2014). Widely diverse melatonin levels ranged from a few pg/g to several mg/g in different plant species. The plant's age, harvest time, and environmental conditions (such as photoperiod, growth conditions, and abiotic and biotic stresses) can alter melatonin concentration in plants (Arnao and Hernández-Ruiz 2013).

In animals, melatonin has long been utilised in several cellular actions as an antioxidant, and contributes to the regulation of circadian rhythms, immunology, sexual behaviour, and so on (Tan et al. 2011, Hardeland 2014). In plants, the role of melatonin in physiology has been explored over the past decade. Preliminary studies have concluded that melatonin possesses a growth-promoting ability which can regulate morphogenesis, photosynthetic efficiency, leaf senescence, flowering, fruit ripening, and crop production. In addition to functioning in plant growth and development, melatonin is involved in encountering environmental stress conditions, such as drought, salinity, cold and high temperature, and heavy metal stress, due to its antioxidant ability.

Understanding the regulatory mechanism in plant growth and the environmental stress response of melatonin will be helpful in the use of melatonin in agriculture. In this review, we emphasise the promotional effect of melatonin on plant growth and development, as well as its function in alleviating abiotic stress damage.

The Biosynthesis Pathway of Melatonin in Plants

Back et al. (2016) speculated that there exist multiple (at least four) biosynthetic pathways in plants. The classic biosynthetic pathway of melatonin, which may occur under normal growth conditions, begins with tryptophan, and consists of four enzymatic steps: decarboxylation reaction catalyses tryptophan to form tryptamine by the enzyme tryptophan decarboxylase (TDC); hydroxylation reaction changes tryptamine to serotonin mediated by tryptamine 5-hydroxylase (T5H); its *N*-acetylation to form N-acetylserotonin by a serotonin N-acetyltransferase (SNAT); and finally, O-methylation to form melatonin by *N*-acetylserotonin *O*-methyltransferase (ASMT) (Hardeland 2016, Zhang, Zhang, Zhao, et al. 2014).

Another major pathway has been proposed, based on the enzyme kinetics: the tryptophan/tryptamine/serotonin/5-methoxytryptamine/melatonin pathway, which particularly occurs when plants produce large amounts of serotonin, e.g. upon senescence in plants (Tan et al. 2016). The steps from tryptophan to serotonin are identical to those shown in the classic pathway. From serotonin to melatonin is a two-step reaction, by production of a 5-methoxytryptamine from serotonin through ASMT/O-methyltransferase (COMT), and subsequent conversion of 5-methoxytryptamine to melatonin by SNAT.

In addition to the above major pathways, another two routes in plants originate from tryptophan, which are catalysed by a putative tryptophan hydroxylase (TPH) enzyme to produce 5-hydroxytrytophan (Back, Tan, and Reiter 2016, Fitzpatrick 1999). Then, 5-hydroxytryptophan is converted into serotonin by TDC; this reaction is significantly involved in melatonin biosynthesis in plants, with higher catalytic efficiency than TDC catalysing tryptophan into tryptamine (Back, Tan, and Reiter 2016, Park et al. 2008). The penultimate, biosynthesis step in the two pathways, serotonin can be catalysed to *N*-acetylserotonin and 5-methoxytrytamine by SNAT and ASMT/COMT, respectively. The last step is production of melatonin from *N*-acetylserotonin by ASMT/COMT or from 5-methoxytrytamine by SNAT, respectively. Melatonin synthesis occurs in chloroplasts where the final-step enzyme is SNAT, while ASMT/COMT is involved in the terminal reaction in the cytoplasm (Byeon and Back 2014b).

The Role of Melatonin in Plant Growth and Development

Numerous studies have described how melatonin plays several pivotal roles in higher plants, where it can regulate the growth of root and shoot, delay leaf senescence, function in reproductive processes (flower development and fruit ripening), and affect crop yield.

Growth

During the last decade, the growth effects of melatonin in aerial parts and roots have been found in both monocots and dicots, but there exist some differences among various plant species. The main effect of melatonin on growth is its auxin-like action, which was supported by observation in shoot and root growth (REFS). Melatonin has a similar structure and the same biosynthetic precursor—tryptophan—as that of auxin (Reiter et al. 2015). Melatonin can promote growth in the aerial parts in some crops, e.g. wheat, barley, oat, rice, tomato, soybean, Arabidopsis, maize, rapeseed, and sunflower (Arnao and Hernández-Ruiz 2017b, Arnao and Hernández-Ruiz 2017a). As an example, compared with control, melatonin triggered a twofold increase in the leaves of maize (Zhao et al. 2015). Arnao and Hernández-Ruiz (2014) have documented that melatonin can modify the development pattern of stems and leaves. Relative to the underground plant organs, the main conclusion of previous studies has proposed that melatonin is involved in several root morphogenetic processes which consist of lateral and adventitious root formation, primary root number formation, root biomass accumulation, and root gravitropic response (Arnao and Hernández-Ruiz 2017b, Arnao 2014, Arnao and Hernández-Ruiz 2014, Arnao and Hernández-Ruiz 2007). Seed priming with 500 μM melatonin showed increased lateral root number in cucumber (Zhang, Zhang, Zhao, et al. 2014). Melatonin at a concentration of 50 μM induced the adventitious root formation of de-rooted tomato seedlings (Wen et al. 2016).

In addition, the auxin-like action of melatonin is also proved by comparisons with auxin, such as indoleacetic acid (IAA) (Hardeland 2014, Koyama et al. 2013). The strong dose dependence of effects has also supported that melatonin has auxin-like activity in growth, i.e. high concentrations inhibit plant organ growth, whereas low concentrations have a promoting effect (Arnao and Hernández-Ruiz 2014, Hardeland 2014). Chen et al. (2009) showed that exogenous melatonin, at a concentration of 0.1 μM, stimulated root growth of etiolated *Brassica juncea* seedlings, but 100 μM had an inhibitory effect. The root parameters, including rooting percentage, number of roots per rooted explants, and root length of sweet cherry, were enhanced by melatonin at low concentrations, but decreased at high concentrations (Sarropoulou, Therios, and Dimassi-Theriou 2012).

Although the effect of melatonin is similar in IAA on plant growth, some studies have suggested that the induction of plant growth caused by melatonin is an auxin-independent manner (Arnao and Hernández-Ruiz 2014, 2017a). As melatonin and auxins shared the common biosynthetic precursor—tryptophan—melatonin presents dual effects on endogenous auxin levels (Arnao and Hernández-Ruiz 2017a, Wang, An, et al. 2016).

The transgenic tomato plants, overexpressing *oAANAT*, contained higher melatonin concentration and lower auxin levels (Wang, Zhao, et al. 2014). Although there is crosstalk between melatonin and the auxin-signalling pathway (Shi et al. 2018), the minimal expression differences of the auxin-related genes were found between melatonin and control in cucumbers (Zhang, Zhang, Zhao, et al. 2014). Melatonin stimulated lateral and adventitious root formation in *Arabidopsis thaliana*, but did not activate auxin-inducible gene (*DR5:uidA* and *BA3:uidA*) expression (Pelagio-Flores et al. 2012).

There is one model to explain the effects of melatonin as an auxin-like regulator in plants (Arnao and Hernández-Ruiz 2017b). Nitric oxide (NO) plays an important role in a series of physiological processes during plant growth and development (Lee et al. 2015, Yun et al. 2016). More importantly, previous studies have highlighted that NO can modulate auxin response in adventitious and lateral root growth, root hair formation, and root gravitropism (Correa-Aragunde, Graziano, and Lamattina 2004, Lombardo et al. 2006, Pagnussat et al. 2002). Melatonin functioned in increasing adventitious root formation, and enhanced the accumulation of endogenous NO level mediated by down-regulating *S*-nitrosoglutathione reductase (*GSNOR*) in de-rooted tomato seedlings. The action of melatonin on roots could be removed by a NO-scavenger, or in *GSNOR*-overexpressed plants (Wen et al. 2016). Arnao and Hernández-Ruiz (2017b) presented a hypothetical model where NO acted as a common signalling molecule involved in melatonin, and auxin signalling pathways in plant growth, such as rooting and tropism.

Nevertheless, another study has highlighted that melatonin acts as a positive regulator of skotomorphogenesis mediated by hormone brassinosteroids (BR) other than auxin biosynthesis (Hwang and Back 2018). The *SNAT2* (melatonin biosynthetic gene) RNAi rice lines showed a decline in the levels of melatonin and BR. Exogenous application of melatonin up-regulated BR biosynthetic genes, *DWARF4*, *D11*, and *RAL1* in rice plants.

Leaf senescence

The major pigment for photosynthesis, chlorophyll, is vital for plant growth and development. The regulatory role of melatonin in leaf senescence is common in plants. For example, protection by melatonin against chlorophyll loss and delayed leaf senescence are observed in many species, such as barley, apple, cucumber, rice, peach, ryegrass, cassava tuber, and *Arabidopsis* (Arnao and Hernández-Ruiz 2009b, Arnao and Hernández-Ruiz 2017a).

The mechanism of melatonin functioning in delaying leaf senescence is still unclear. However, two major possible ways are given: 1) the regulation of senescence-associated genes; and 2) interaction with other phytohormone signalling pathways. The reduction in chlorophyll degradation caused by melatonin has been shown to be associated with the suppression of senescence-associated genes (Reiter et al. 2015). The proteomic analysis revealed that melatonin treatment down-regulated the proteins that are usually up-regulated during leaf senescence (Wang, Sun, et al. 2014). RNA-sequence results indicated that melatonin inhibited chlorophyll degradation genes, encoding chlorophyllase and pheophorbide a oxygenase (PAO), and two chlorophyll-degrading enzymes in *Arabidopsis* (Weeda et al. 2014). Overexpression of a sheep *SNAT* in seven day-old seedlings of rice led to decreased expression of several senescence-associated genes, encoding senescence-related protein (SAG29), and polygalacturonase isoenzyme (Byeon et al. 2013). Notably, several investigations have shown that exogenously applied melatonin could exhibit preservation of the chlorophyll by a delay in the up-regulation of senescence-related genes in some species. For example, a melatonin supplement slowed leaf senescence in accordance with a reduction in the transcript levels of *PAO* (Wang, Sun, Chang, et al. 2013, Wang et al. 2012) and *SAG 12* (Wang et al. 2012) in apples. Application of melatonin dramatically inhibited the senescence-related genes, mitigated chlorophyll degradation, and delayed senescence in rice (Liang et al. 2015). Suppression of the expression of chlorophyll degradation genes and senescence marker genes (*LpSAG12.1* and *Lph36*) was found in melatonin-treated perennial ryegrass plants (Zhang et al. 2016, Zhang et al. 2017).

In addition, it is important to consider the interaction between melatonin and the pathways of other phytohormones. Of these, abscisic acid (ABA) accelerates the senescence process in plants, whereas cytokinins slow it down (Khan, Rozhon, and Poppenberger 2014). Arnao and Hernández-Ruiz (2009b) reported that the melatonin-induced inhibition of leaf senescence exhibited both in the ABA and cytokinin-treated barley. Melatonin application to perennial ryegrass delayed leaf senescence parallel to the inhibition of ABA biosynthesis and signalling pathway, whilst promoting cytokinin synthesis and signalling pathways (Zhang et al. 2017). Another study does not support the view that melatonin delays leaf senescence by triggering a senescence-inducing hormone signalling pathway, such as ABA and jasmonic acid. Weeda et al. (2014) documented that a large number of genes associated with senescence-promoting hormone pathways, such as ABA, ethylene, and jasmonic acid were up-regulated by melatonin in *Arabidopsis*. An attractive model to explain the actions of melatonin in delaying leaf senescence has been proposed. Melatonin is reported to function in reactive oxygen species (ROS) detoxification to protect plants, especially against a stress environment. ROS detoxification is mainly achieved in three ways, including the attenuation of radical formation, direct radical scavenging, and indirect actions via antioxidant systems, which could slow the up-regulation of senescence-associated genes such as *PAO* and *SAG12*. As an example, melatonin application resulted in a marked increase in activity and expression levels of SOD and CAT, but had no significant effect in the non-enzymatic substances, such as AsA and DHA, along with the delaying chlorophyll degradation in perennial ryegrass (Zhang et al. 2016). This interpretation reveals that senescence-delaying effects of melatonin may work more efficiently in plants subjected to environmental stress, where the plants generate more ROS.

Reproductive function

In plants, melatonin was detected in reproductive tissues of rice and bananas during the mid-1990s (Dubbels et al. 1995, Hattori et al. 1995). Until now, melatonin has been measured in both vegetative and reproductive tissues of almost all plant species, where productive organs such as fruits, seeds, and/or nuts contain relatively high levels of melatonin (Arnao 2014, Erland et al. 2015, Paredes et al. 2008, Park et al. 2013).

Preliminary studies indicated that the growth periods with highest melatonin level during reproduction varied greatly among plant species. In some plants, the highest melatonin level was found at maturity in fruits. During the ripening of pepper, the melatonin concentration was increased (Korkmaz, Değer, and Cuci 2014). In *Solanum lycopersicum*, the highest values of melatonin were observed in mature red fruits in three cultivars (Okazaki et al. 2009, Van Tassel et al. 2001). However, other studies reported that the highest level of melatonin existed at the early stage of fruit development. Murch et al. (2009) showed that the highest melatonin content was obtained in the first days after anthesis of *Dutura*. In *Prunus avium* and *Vitis vinifera*, highest melatonin levels were reported during early maturity (Murch et al. 2010, Zhao et al. 2013).

However, the function of high melatonin levels in flowering and reproduction has not been clearly understood. Some reports suggest that a high level of melatonin was associated with a delay in flowering (Byeon and Back 2014a, Kolář, Johnson, and Macháčková 2003, Shi et al. 2016). For example, melatonin application delayed flowering in *C. rubrum* (Kolář, Johnson, and Macháčková 2003). Other work showed that there was a positive relationship between the regenerative capacity of pollen and melatonin levels during the reproductive period of St. John's wort (*Hypericum perforatum* L.) (Murch and Saxena 2002).

Crop production

As the promotion effect of melatonin on preserving chlorophyll was found, so preliminary studies also reported that melatonin increased the photosynthetic ability in plants (REFS). Melatonin treatment had higher photosynthetic activity and end-products in some crops (Tan et al. 2011, Wang et al. 2012, Zhou et al. 2016). However, whether the elevation of photosynthesis caused by melatonin can accumulate more biomass, especially grain yield, is the major question for melatonin application. Wei et al. (2014) focused on the possible impact of melatonin on crop yield in soybean, and found that seed coating with melatonin led to a significant increase in pod numbers, seed numbers and seed yields at maturity. Similarly, seed priming with melatonin elevated corn production in the field (Posmyk and Janas 2009, Tan et al. 2011). Tomato fruits treated with melatonin had a higher water-soluble pectin and fruit softening (Sun et al. 2014). But some studies did not support the earlier results and showed that the melatonin elevation in transgenic rice overexpressing sheep *SNAT* caused low grain yield at maturity (Byeon and Back 2014a).

The Influence of Melatonin on Abiotic Stress Tolerance in Plants

The ability of melatonin to alleviate the negative effects of environmental stresses on plants has been documented in several plants: most notably, stress conditions are drought, salinity, high and cold temperature, heavy metal, and UV stress. Melatonin acts as an antioxidant that seems to aid them in resisting marked changes in harsh environments (Arnao and Hernández-Ruiz 2014). Numerous environmental stresses provoke a significant increase in endogenous melatonin level, indicating its possible role as an effector in such conditions (Arnao and Hernández-Ruiz 2015). Overexpression of the melatonin biosynthetic genes, and exogenous application of melatonin can promote the ability of plants responding to various environmental stress agents (REFS).

Drought/soil salinization:

Drought, one of the major abiotic stresses worldwide, limits plant growth and crop production. Previous work focused on exploring the relationship between drought and melatonin in some species, such as maize, lupin, bermudagrass and apple. Water shortages increased endogenous melatonin in lupin seedlings, which was fourfold higher than in well-watered plants (Arnao and Hernández-Ruiz 2013). Drought treatment elevated melatonin level (2.6-fold) in bermudagrass (*Cynodon dactylon*) (Shi, Jiang, et al. 2015). In drought-tolerant and drought-sensitive *Malus* species, transcript levels of four melatonin biosynthesis genes, *MdTDC1*, *MdT5H4*, *MdAANAT2*, and *MdASMT1*, were up-regulated by drought treatment (Li et al. 2014). Overexpression of ovine *HIOMT* increased the levels of melatonin, and improved drought tolerance in "Micro-Tom" tomato (Wang, Zhao, et al. 2014). Melatonin application functioned in attenuation of the cell damage caused by drought stress in bermudagrass, and they proposed that melatonin might act as a regulator of metabolic homeostasis involved in the photorespiratory and carbohydrate/nitrogen metabolism (Shi, Jiang, et al. 2015). An increased seed germination rate and improved root growth were found in water-stressed cucumbers subjected to melatonin (Zhang et al. 2013).

Melatonin can act as a scavenger of ROS by stimulating the enzymatic and non-enzymatic substances to protect plants against drought stress. Under drought conditions, the transgenic plants with overexpression of melatonin biosynthetic genes, such as *MzASMT1* or *MzSNAT5* resulted in an increase in melatonin level and a decrease in ROS accumulation, compared with wild type plants (Wang et al. 2017, Zuo et al. 2014). Exogenous melatonin drenched in roots resulted in a remarkable delay in leaf senescence, enhanced photosynthetic traits, and increased the potency of antioxidant enzymes in apples under prolonged drought stress (Wang, Sun, Li, et al. 2013, Wang et al. 2012). Melatonin-treated maize plants showed higher photosynthetic rates, stomatal conductance and transpiration rates, lower H_2O_2 and malonaldehyde (MDA), as well as higher enzymatic and non-enzymatic antioxidant activity than non-treated plants (Ye et al. 2016). Melatonin maintained ROS homeostasis by lowering H_2O_2 and $O_2^{\cdot-}$ in wheat under 25% polyethylene glycol 6000 (PEG) stress (Cui et al. 2018). Pretreatment with melatonin significantly increased photosynthesis ability, lowered ROS damage, as well as keeping proline homeostasis in grapes (Meng et al. 2014) and in alfalfa under drought stress (Antoniou et al. 2017). Moreover, drought triggers the accumulation of ABA leading to the closure of the stomata and regulation of the expression of drought-responsive genes in plants (Seki et al. 2007). Melatonin application reduced the expression of ABA synthetic gene 9-*cis*-epoxycarotenoid dioxygenase (*MdNCED3*) and increased its catabolic genes *MdCYP707A1* and *MdCYP707A2*, along with decreasing ABA level in apple plants under drought stress (Li et al. 2014). Fleta-Soriano et al. (2017) agreed and reported that water stress significantly enhanced the accumulation of endogenous melatonin and ABA in maize. This combined effect of ROS-scavenge and reduced-ABA accumulation, caused by melatonin, could keep the stomata open and enhance plant drought tolerance.

Another osmotic stress in plants is salinity, which disturbs plant growth because it induces physiological drought and ion toxicity. The effect of melatonin on plant salt tolerance is similar to that in drought tolerance, which consists of increasing melatonin level, decreasing ROS and regulating ABA accumulation, and so on. The accumulation of melatonin and its biosynthetic precursor, serotonin, was increased by salt stress in sunflower seedlings (Mukherjee et al. 2014). Pre-treatment with melatonin dramatically reduced growth inhibition in apples (Li et al. 2016), maize (Jiang et al. 2016) and cucumber (Wang, Liu, et al. 2016) subjected to salinity, and decreased the accumulation of ROS and MDA. ABA catabolism genes (two CYP707 monooxygenases) were up-regulated, and ABA biosynthetic gene (*NCED*) was down-regulated in melatonin-treated plants, which led to a decreased ABA level during seed germination under salt stress (Zhang, Zhang, Yang, et al. 2014). In addition, other studies indicated that exogenous melatonin maintained ion homeostasis in plants under salt stress. For example, melatonin application elevated the K^+ concentration and K^+/Na^+ in maize seedlings exposed to salinity (Jiang et al. 2016). Similarly, melatonin could increase the expression of K^+ transporters in apples under salt stress (Li et al. 2016). Compared with control plants, melatonin-treated maize seedlings exhibited increased photosynthetic activity and decreased Na^+ contents in leaves and roots under 150 mM NaCl (Chen et al. 2018). The expression levels of ion channel genes *MdNHX1* and *MdAKT1* were increased by 0.1 μM melatonin to maintain the ion homeostasis in apple under salinity condition (Li et al. 2012).

Temperature

Cold or heat stress, which occurs at any time throughout the growth period of plants, inhibits growth and development. In the case of cold stress, melatonin is effective in improving cold tolerance in plants. Cold stress affects the biosynthetic rate of melatonin, resulting in the increase of endogenous levels. The endogenous melatonin concentration in lupin plants treated with 6°C was 2.5-fold higher than the control planted at 24°C (Arnao and Hernández-Ruiz 2013). In *Arabidopsis*, the melatonin level was increased by cold treatment (4°C) (Shi and Chan 2014). Exogenous melatonin has protective actions in plants against cold stress. Melatonin pretreatment for 5 days mitigated cold-induced apoptosis in carrot suspension cells (Lei et al. 2004). Posmyk et al. (2009) confirmed that melatonin solution enhanced seed germination in cucumbers during chilling stress (10°C). Foliar application and root drench with melatonin alleviated the cold damage to leaf photosynthesis in tomato (Yang et al. 2018). The maize seedlings treated with melatonin under cold stress had a higher chlorophyll level, relative water content, activities of antioxidant enzymes, and mineral elements compared to those of the respective control (Turk and Erdal 2015). Melatonin had a positive effect in alleviating cold stress (4°C) mediating by up-regulating of cold-responsive genes (*CBFs/DREBs* and *COR15a*), and transcription factors of ROS-associated antioxidant genes (*ZAT10* and *ZAT12*) in *Arabidopsis* (Bajwa et al. 2014, Shi and Chan 2014). In addition, melatonin treatment could accumulate the plant stress-responsive hormone, such as ABA. Higher ABA levels were found in wild-type barley applied to melatonin under cold stress (Li et al. 2016).

High temperature often provokes an increase in endogenous melatonin level in plants. Melatonin contents were higher in heat-treated rice and *Arabidopsis* seedlings, compared with those in plants grown in normal conditions (Byeon and Back 2014a, Shi, Jiang, et al. 2015). Melatonin treatment has been reported to have an effect in promoting thermotolerance of plants. *Phacelia tanacetifolia* seeds treated with melatonin exhibited improved germination under high temperature treatment (Tiryaki and Keles 2012). Exogenous melatonin reversed the inhibition of heat stress on survival rates in *Arabidopsis* (Shi, Tan, et al. 2015). In this case, exogenous melatonin and high temperature also up-regulated several heat-responsive genes, like *HSFA2*, *HSA32*, *HSP90*, and *HSP101*, confirming thermotolerance in *Arabidopsis*. Xu et al (2016) reported that foliar application of 10 μM melatonin or overexpression of *ASMT* reduced the photoinhibition caused by heat stress, and improved cellular protein protection in tomato plants grown at 40°C.

Heavy metal stress

Elevated levels of melatonin may help protect plants against environmental stress induced by soil pollutants. Previous studies reported that chemical stressors, such as zinc, copper, and cadmium could enhance endogenous melatonin in crops, including barley, lupin, and rice (Arnao and Hernández-Ruiz 2009a, Byeon et al. 2015, Reiter et al. 2015). Although the expression of penultimate gene *SNAT* during melatonin synthesis was down-regulated by cadmium in rice, other melatonin biosynthetic genes *TDC*, *T5H*, and *HIOMT* were up-regulated accompanying the cadmium-induced synthesis (Byeon et al. 2015). However, Lee, Choi, and Back (2017) demonstrated that chloroplastidic SNAT functioned in persevering increased melatonin production in rice plants subjected to cadmium. In tomato seedlings, the transcription factor HsfA1a could trigger melatonin biosynthetic gene *COMT1*, and accumulate melatonin for up-regulating *HSPs* to confer cadmium tolerance (Cai et al. 2017). Melatonin supplementation reversed the inhibitory effect of copper contamination, which showed increased phytoremediative efficiency and survival ability of pea plants against copper contamination (Tan et al. 2007). Seed priming with 1 or 10 mM melatonin improved seed germination and increased fresh weight of seedlings in red cabbage subjected to copper treatment (Posmyk et al. 2008). In tobacco suspension cells, melatonin treatment had a restricting effect on lead-triggered programmed cell death by elevating the expression of Bax inhibitor-1 (BI-1) which is a well-conserved protein for the inhibition of ROS-induced cell death in plants (Kobylińska and Posmyk 2016). Melatonin was involved in the amelioration of fluoride-induced detrimental effects on growth in *Cajanus cajan* L. (Yadu et al. 2018). They showed that the exogenous application of melatonin increased biomass accumulation, the scavenging ability of ROS, and the accumulation of proline over the respective controls. However, several studies indicated that melatonin did not reduce the endogenous levels of heavy metals. Low concentration (50 μmol/kg) of melatonin was found to promote the biomass and increase the activities of antioxidant enzymes, but had no significant effect on cadmium concentration in *Cyphomandra betacea* seedlings treated with cadmium (Lin et al. 2018).

UV stress

High intensity ultraviolet (UV) light results in generating excessive free ROS in tissues of plants subjected to this radiation (Afreen, Zobayed, and Kozai 2006). Plants treated with high UV conditions probably contain high melatonin levels. Simopoulos et al. (2005) reported that compared with plants treated with low UV in their natural habitat, the same species of Alpine and Mediteranean plants, grown under high UV exposure, exhibited higher endogenous melatonin levels. UV-B treatment dramatically increased the melatonin concentration in roots of *Glycyrrhyza uralensis* (Afreen, Zobayed, and Kozai 2006). *Nicotiana sylvestris* plants with overexpression of key melatonin biosynthetic genes (*AANAT* and *HIOMT*) had a significantly higher melatonin content, and showed a reduction in DNA damage caused by UV-B (Zhang et al. 2012). Erland and Saxena (2018) reviewed that the ROS scavenging ability of melatonin may lead to the protection of plants against UV stress. Additionally, the defensive role of melatonin in high light stress has been confirmed in *Arabidopsis thaliana* treated with exogenous melatonin or overexpression of *SNAT* (Lee and Back 2018). In transgenic plants overexpresing *SNAT*, the MDA and ion leakage were decreased; exogenous melatonin reduced damage induced by high light stress through decreasing ROS production.

Conclusions and Future Perspectives

Recently, the beneficial influence of melatonin on plants may be emphasised: (1) melatonin promotes root development for better nutrient absorptive capacity from soil; (2) melatonin enhances chlorophyll preservation, delays leaf senescence, promotes photosynthetic ability, synthesises more carbohydrates, and increases their above-ground biomass accumulation; and (3) melatonin improves plant tolerance against a variety of environmental stresses, especially drought, salinity, cold and hot temperatures, and heavy metal stress. Nevertheless, several fundamental issues around melatonin urgently require exploration. Although the transcriptomic and proteomic analyses revealed that melatonin regulated gene expression in plants, the primary signalling pathway induced by melatonin is still not fully understood. There is a demand for clarification of key components, such as the possible melatonin receptor involved in the melatonin signalling pathway. In addition, melatonin biosynthesis and catabolism are important for regulation of the melatonin level, which can stimulate a series of biological processes in plants. Identification of the role of isoenzymes in the melatonin biosynthetic and catabolic route is an urgent task in order to track the melatonin metabolism.

References

Afreen, F, S M A Zobayed, and T Kozai. 2006. "Melatonin in *Glycyrrhiza uralensis*: response of plant roots to spectral quality of light and UV-B radiation." *Journal of Pineal Research* 41 (2):108–115.

Antoniou, C, G Chatzimichail, R Xenofontos, J J Pavlou, E Panagiotou, A Christou, and V Fotopoulos. 2017. "Melatonin systemically ameliorates drought stress-induced damage in *Medicago sativa* plants by modulating nitro-oxidative homeostasis and proline metabolism." *Journal of Pineal Research* 62 (4):e12401.

Arnao, M B. 2014. "Phytomelatonin: discovery, content, and role in plants." *Advances in Botany* 2014.

Arnao, M B, and J Hernández-Ruiz. 2014. "Melatonin: plant growth regulator and/or biostimulator during stress?" *Trends in Plant Science* 19 (12):789–797.

Arnao, M B, and J Hernández-Ruiz. 2017a. "Melatonin and its relationship to plant hormones." *Annals of Botany* 121 (2):195–207.

Arnao, M B, and J Hernández-Ruiz. 2007. "Melatonin promotes adventitious-and lateral root regeneration in etiolated hypocotyls of *Lupinus albus* L." *Journal of Pineal Research* 42 (2):147–152.

Arnao, M B, and J Hernández-Ruiz. 2009a. "Chemical stress by different agents affects the melatonin content of barley roots." *Journal of Pineal Research* 46 (3):295–299.

Arnao, M B, and J Hernández-Ruiz. 2013. "Growth conditions determine different melatonin levels in *Lupinus albus* L." *Journal of Pineal Research* 55 (2):149–155.

Arnao, M B, and J Hernández-Ruiz. 2015. "Functions of melatonin in plants: a review." *Journal of Pineal Research* 59 (2):133–150.

Arnao, M B, and J Hernández-Ruiz. 2017b. "Growth activity, rooting capacity, and tropism: three auxinic precepts fulfilled by melatonin." *Acta Physiologiae Plantarum* 39 (6):127.

Arnao, M B, and J Hernández-Ruiz. 2009b. "Protective effect of melatonin against chlorophyll degradation during the senescence of barley leaves." *Journal of Pineal Research* 46 (1):58–63.

Back, K, D-X Tan, and R J Reiter. 2016. "Melatonin biosynthesis in plants: multiple pathways catalyze tryptophan to melatonin in the cytoplasm or chloroplasts." *Journal of Pineal Research* 61 (4):426–437.

Bajwa, V S, M R Shukla, S M Sherif, S J Murch, and P K Saxena. 2014. "Role of melatonin in alleviating cold stress in *Arabidopsis thaliana*." *Journal of Pineal Research* 56 (3):238–245.

Byeon, Y, and K Back. 2014a. "An increase in melatonin in transgenic rice causes pleiotropic phenotypes, including enhanced seedling growth, delayed flowering, and low grain yield." *Journal of Pineal Research* 56 (4):408–414.

Byeon, Y, and K Back. 2014b. "Melatonin synthesis in rice seedlings in vivo is enhanced at high temperatures and under dark conditions due to increased serotonin N-acetyltransferase and N-acetylserotonin methyltransferase activities." *Journal of Pineal Research* 56 (2):189–195.

Byeon, Y, H Y Lee, O J Hwang, H-J Lee, K Lee, and K Back. 2015. "Coordinated regulation of melatonin synthesis and degradation genes in rice leaves in response to cadmium treatment." *Journal of Pineal Research* 58 (4):470–478.

Byeon, Y, S Park, Y S Kim, and K Back. 2013. "Microarray analysis of genes differentially expressed in melatonin-rich transgenic rice expressing a sheep serotonin N-acetyltransferase." *Journal of Pineal*

Research 55 (4):357–363.
Chen, Q, W-b Qi, R J Reiter, W Wei, and B-m Wang. 2009. "Exogenously applied melatonin stimulates root growth and raises endogenous indoleacetic acid in roots of etiolated seedlings of *Brassica juncea*." *Journal of Plant Physiology* 166 (3):324–328.
Chen, Y-E, J-J Mao, L-Q Sun, B Huang, C-B Ding, Y Gu, J-Q Liao, C Hu, Z-W Zhang, and S Yuan. 2018. "Exogenous melatonin enhances salt stress tolerance in maize seedlings by improving antioxidant and photosynthetic capacity." *Physiologia Plantarum*.
Correa-Aragunde, N, M Graziano, and L Lamattina. 2004. "Nitric oxide plays a central role in determining lateral root development in tomato." *Planta* 218 (6):900–905.
Cui, G, F Sun, X Gao, K Xie, C Zhang, S Liu, and Y Xi. 2018. "Proteomic analysis of melatonin-mediated osmotic tolerance by improving energy metabolism and autophagy in wheat (*Triticum aestivum* L.)." *Planta* 248 (1):69–87.
Dubbels, R, R J Reiter, E Klenke, A Goebel, E Schnakenberg, C Ehlers, H W Schiwara, and W Schloot. 1995. "Melatonin in edible plants identified by radioimmunoassay and by high performance liquid chromatography-mass spectrometry." *Journal of Pineal Research* 18 (1):28–31.
Erland, L AE, S J Murch, R J Reiter, and P K Saxena. 2015. "A new balancing act: the many roles of melatonin and serotonin in plant growth and development." *Plant Signaling and Behavior* 10 (11):e1096469.
Erland, L AE, and P K Saxena. 2018. "Melatonin in plant morphogenesis." *In Vitro Cellular and Developmental Biology-Plant* 54 (1):3–24.
Fitzpatrick, P F. 1999. "Tetrahydropterin-dependent amino acid hydroxylases." *Annual Review of Biochemistry* 68 (1):355–381.
Fleta-Soriano, E, L Díaz, E Bonet, and S Munné-Bosch. 2017. "Melatonin may exert a protective role against drought stress in maize." *Journal of Agronomy and Crop Science* 203 (4):286–294.
Hardeland, R. 2014. "Melatonin in plants and other phototrophs: advances and gaps concerning the diversity of functions." *Journal of Experimental Botany* 66 (3):627–646.
Hardeland, R. 2016. "Melatonin in plants–diversity of levels and multiplicity of functions." *Frontiers in Plant Science* 7:198.
Hattori, A, H Migitaka, M Iigo, M Itoh, K Yamamoto, R Ohtani-Kaneko, M Hara, T Suzuki, and R J Reiter. 1995. "Identification of melatonin in plants and its effects on plasma melatonin levels and binding to melatonin receptors in vertebrates." *Biochemistry and Molecular Biology International* 35 (3):627–634.

Hwang, O J, and K Back. 2018. "Melatonin is involved in skotomorphogenesis by regulating brassinosteroid biosynthesis in rice plants." *Journal of Pineal Research*: e12495.

Jiang, C, Q Cui, K Feng, D Xu, C Li, and Q Zheng. 2016. "Melatonin improves antioxidant capacity and ion homeostasis and enhances salt tolerance in maize seedlings." *Acta Physiologiae Plantarum* 38 (4):82.

Khan, M, W Rozhon, and B Poppenberger. 2014. "The role of hormones in the aging of plants-a mini-review." *Gerontology* 60 (1):49–55.

Kobylińska, A, and M M Posmyk. 2016. "Melatonin restricts Pb-induced PCD by enhancing BI-1 expression in tobacco suspension cells." *Biometals* 29 (6):1059–1074.

Kolář, J, C H Johnson, and I Macháčková. 2003. "Exogenously applied melatonin (N-acetyl-5-methoxytryptamine) affects flowering of the short-day plant *Chenopodium rubrum*." *Physiologia Plantarum* 118 (4):605–612.

Korkmaz, A, Ö Değer, and Y Cuci. 2014. "Profiling the melatonin content in organs of the pepper plant during different growth stages." *Scientia Horticulturae* 172:242–247.

Koyama, F C, T LG Carvalho, E Alves, H B da Silva, M F De Azevedo, A S Hemerly, and C RS Garcia. 2013. "The structurally related auxin and melatonin tryptophan-derivatives and their roles in *Arabidopsis thaliana* and in the human malaria parasite *Plasmodium falciparum*." *Journal of Eukaryotic Microbiology* 60 (6):646–651.

Lee, D I, G Zhu, T Sasaki, G-S Cho, N Hamdani, R Holewinski, S-H Jo, T Danner, M Zhang, and P P Rainer. 2015. "Phosphodiesterase 9A controls nitric-oxide-independent cGMP and hypertrophic heart disease." *Nature* 519 (7544):472.

Lee, Hyoung Yool, and Kyoungwhan Back. 2018. "Melatonin induction and its role in high light stress tolerance in *Arabidopsis thaliana*." *Journal of Pineal Research*: e12504.

Lee, K, G-H Choi, and K Back. 2017. "Cadmium-induced melatonin synthesis in rice requires light, hydrogen peroxide, and nitric oxide: key regulatory roles for tryptophan decarboxylase and caffeic acid O-methyltransferase." *Journal of Pineal Research* 63 (4): e12441.

Lei, X-Y, R-Y Zhu, G-Y Zhang, and Y-R Dai. 2004. "Attenuation of cold-induced apoptosis by exogenous melatonin in carrot suspension cells: the possible involvement of polyamines." *Journal of Pineal Research* 36 (2):126–131.

Lerner, A B, J D Case, Y Takahashi, T H Lee, and W Mori. 1958. "Isolation of melatonin, the pineal gland factor that lightens melanocyteS1." *Journal of the American Chemical Society* 80 (10):2587–2587.

Li, C, D-X Tan, D Liang, C Chang, D Jia, and F Ma. 2014. "Melatonin mediates the regulation of ABA metabolism, free-radical scavenging, and stomatal behaviour in two *Malus* species under drought stress." *Journal of Experimental Botany* 66 (3):669–680.

Li, C, P Wang, Z Wei, D Liang, C Liu, L Yin, D Jia, M Fu, and F Ma. 2012. "The mitigation effects of exogenous melatonin on salinity-induced stress in *Malus hupehensis*." *Journal of Pineal Research* 53 (3):298–306.

Li, X, D-X Tan, D Jiang, and F Liu. 2016. "Melatonin enhances cold tolerance in drought-primed wild-type and abscisic acid-deficient mutant barley." *Journal of Pineal Research* 61 (3):328–339.

Liang, C, G Zheng, W Li, Y Wang, B Hu, H Wang, H Wu, Y Qian, X-G Zhu, and D-X Tan. 2015. "Melatonin delays leaf senescence and enhances salt stress tolerance in rice." *Journal of Pineal Research* 59 (1):91–101.

Lin, L, J Li, F Chen, M Liao, Y Tang, D Liang, H Xia, Y Lai, X Wang, and C Chen. 2018. "Effects of melatonin on the growth and cadmium characteristics of *Cyphomandra betacea* seedlings." *Environmental Monitoring and Assessment* 190 (3):119.

Lombardo, M C, M Graziano, J C Polacco, and L Lamattina. 2006. "Nitric oxide functions as a positive regulator of root hair development." *Plant Signaling and Behavior* 1 (1):28–33.

Meng, J-F, T-F Xu, Z-Z Wang, Y-L Fang, Z-M Xi, and Z-W Zhang. 2014. "The ameliorative effects of exogenous melatonin on grape cuttings under water-deficient stress: antioxidant metabolites, leaf anatomy, and chloroplast morphology." *Journal of Pineal Research* 57 (2):200–212.

Mukherjee, S, A David, S Yadav, F Baluška, and S C Bhatla. 2014. "Salt stress-induced seedling growth inhibition coincides with differential distribution of serotonin and melatonin in sunflower seedling roots and cotyledons." *Physiologia Plantarum* 152 (4):714–728.

Murch, S J, A R Alan, J Cao, and P K Saxena. 2009. "Melatonin and serotonin in flowers and fruits of *Datura metel* L." *Journal of Pineal Research* 47 (3):277–283.

Murch, S J, B A Hall, C H Le, and P K Saxena. 2010. "Changes in the levels of indoleamine phytochemicals during véraison and ripening of wine grapes." *Journal of Pineal Research* 49 (1):95–100.

Murch, S J, and P K Saxena. 2002. "Mammalian neurohormones: potential significance in reproductive physiology of St. John's wort (*Hypericum perforatum* L.)?" *Naturwissenschaften* 89 (12):555–560.

Okazaki, M, K Higuchi, Y Hanawa, Y Shiraiwa, and H Ezura. 2009. "Cloning and characterization of a *Chlamydomonas reinhardtii* cDNA arylalkylamine N-acetyltransferase and its use in the genetic engineering of melatonin content in the Micro-Tom tomato." *Journal of Pineal Research* 46 (4):373–382.

Pagnussat, G C, M Simontacchi, S Puntarulo, and L Lamattina. 2002. "Nitric oxide is required for root organogenesis." *Plant Physiology* 129 (3):954–956.

Paredes, S D, A Korkmaz, L C Manchester, D-X Tan, and R J Reiter. 2008. "Phytomelatonin: a review." *Journal of Experimental Botany* 60 (1):57–69.

Park, M, K Kang, S Park, and K Back. 2008. "Conversion of 5-hydroxytryptophan into serotonin by tryptophan decarboxylase in plants, *Escherichia coli*, and yeast." *Bioscience, Biotechnology, and Biochemistry* 72 (9):2456–2458.

Park, S, T-N Nguyen Le, Y Byeon, Y S Kim, and K Back. 2013. "Transient induction of melatonin biosynthesis in rice (*Oryza sativa* L.) during the reproductive stage." *Journal of Pineal Research* 55 (1):40–45.

Pelagio-Flores, R, E Muñoz-Parra, R Ortiz-Castro, and J López-Bucio. 2012. "Melatonin regulates *Arabidopsis* root system architecture likely acting independently of auxin signaling." *Journal of Pineal Research* 53 (3):279–288.

Posmyk, M M, M Bałabusta, M Wieczorek, E Sliwinska, and K M Janas. 2009. "Melatonin applied to cucumber (*Cucumis sativus* L.) seeds improves germination during chilling stress." *Journal of Pineal Research* 46 (2):214–223.

Posmyk, M M, and K M Janas. 2009. "Melatonin in plants." *Acta Physiologiae Plantarum* 31 (1):1.

Posmyk, M M, H Kuran, K Marciniak, and K M Janas. 2008. "Presowing seed treatment with melatonin protects red cabbage seedlings against toxic copper ion concentrations." *Journal of Pineal Research* 45 (1):24–31.

Reiter, R J, D-X Tan, Z Zhou, M H Coelho Cruz, L Fuentes-Broto, and A Galano. 2015. "Phytomelatonin: assisting plants to survive and thrive." *Molecules* 20 (4):7396–7437.

Sarropoulou, V N, I N Therios, and K N Dimassi-Theriou. 2012. "Melatonin promotes adventitious root regeneration in in vitro shoot tip explants of the commercial sweet cherry rootstocks CAB-6P (*Prunus cerasus* L.), Gisela 6 (P. *cerasus*× P. *canescens*), and MxM 60 (P. *avium*× P. *mahaleb*)." *Journal of Pineal Research* 52 (1):38–46.

Seki, M, T Umezawa, K Urano, and K Shinozaki. 2007. "Regulatory metabolic networks in drought stress responses." *Current Opinion in Plant Biology* 10 (3):296–302.

Shi, H, C Jiang, T Ye, D Tan, R J Reiter, H Zhang, R Liu, and Z Chen. 2015. "Comparative physiological, metabolomics, and transcriptomic analysis reveal mechanisms of improved abiotic stress resistance in bermudagrass [*Cynodon dactylon* (L). Pers.] by exogenous melatonin." *Journal of Experimental Botany* 66 (3):681–694.

Shi, H, and Z Chan. 2014. "The cysteine2/histidine2-type transcription factor zinc finger of *Arabidopsis thaliana* 6-activated C-Repeat-binding factor pathway is essential for melatonin-mediated freezing stress resistance in *Arabidopsis*." *Journal of Pineal Research* 57 (2):185–191.

Shi, H, D-X Tan, R J Reiter, T Ye, F Yang, and Z Chan. 2015. "Melatonin induces class A1 heat-shock factors (HSFA 1s) and their possible involvement of thermotolerance in *Arabidopsis*." *Journal of Pineal Research* 58 (3):335–342.

Shi, H, Y W, Q Wang, R J Reiter, and C He. 2016. "Melatonin mediates the stabilization of DELLA proteins to repress the floral transition in *Arabidopsis*." *Journal of Pineal Research* 60 (3):373–379.

Shi, H, S Zhang, D Lin, Y Wei, Y Yan, G Liu, R J Reiter, and Z Chan. 2018. "Zinc finger of *Arabidopsis thaliana* 6 is involved in melatonin-mediated auxin signaling through interacting Indeterminate domain 15 and Indole-3-Acetic Acid 17." *Journal of Pineal Research*: e12494.

Simopoulos, A P, D-X Tan, L C Manchester, and R J Reiter. 2005. "Purslane: a plant source of omega-3 fatty acids and melatonin." *Journal of Pineal Research* 39 (3):331–332.

Sun, Q, N Zhang, J Wang, H Zhang, D Li, J Shi, R Li, S Weeda, B Zhao, and S Ren. 2014. "Melatonin promotes ripening and improves quality of tomato fruit during postharvest life." *Journal of Experimental Botany* 66 (3):657–668.

Tan, D-X, R Hardeland, L C Manchester, A Korkmaz, S Ma, S Rosales-Corral, and R J Reiter. 2011. "Functional roles of melatonin in plants, and perspectives in nutritional and agricultural science." *Journal of Experimental Botany* 63 (2):577–597.

Tan, D-X, L C Manchester, P Helton, and R J Reiter. 2007. "Phytoremediative capacity of plants enriched with melatonin." *Plant Signaling and Behavior* 2 (6):514–516.

Tan, D-X, R Hardeland, K Back, L C Manchester, M A Alatorre-Jimenez, and R J Reiter. 2016. "On the significance of an alternate pathway of melatonin synthesis via 5-methoxytryptamine: comparisons across species." *Journal of Pineal Research* 61 (1):27–40.

Tiryaki, I, and H Keles. 2012. "Reversal of the inhibitory effect of light and high temperature on germination of *Phacelia tanacetifolia* seeds by melatonin." *Journal of Pineal Research* 52 (3):332–339.

Turk, H, and S Erdal. 2015. "Melatonin alleviates cold-induced oxidative damage in maize seedlings by up-regulating mineral elements and enhancing antioxidant activity." *Journal of Plant Nutrition and Soil Science* 178 (3):433–439.

Van Tassel, D L, N Roberts, A Lewy, and S D O'Neill. 2001. "Melatonin in plant organs." *Journal of Pineal Research* 31 (1):8–15.

Wang, L, C Feng, X Zheng, Y Guo, F Zhou, D Shan, X Liu, and J Kong. 2017. "Plant mitochondria synthesize melatonin and enhance the tolerance of plants to drought stress." *Journal of Pineal Research* 63 (3):e12429.

Wang, L, Y Zhao, R J Reiter, C He, G Liu, Q Lei, B Zuo, X Dong Zheng, Q Li, and J Kong. 2014. "Changes in melatonin levels in transgenic 'Micro-Tom'tomato overexpressing ovine AANAT and ovine HIOMT genes." *Journal of Pineal Research* 56 (2):134–142.

Wang, L Y, J L Liu, W X Wang, and Y Sun. 2016. "Exogenous melatonin improves growth and photosynthetic capacity of cucumber under salinity-induced stress." *Photosynthetica* 54 (1):19–27.

Wang, P, X Sun, C Chang, F Feng, D Liang, L Cheng, and F Ma. 2013. "Delay in leaf senescence of *Malus hupehensis* by long-term melatonin application is associated with its regulation of metabolic status and protein degradation." *Journal of Pineal Research* 55 (4):424–434.

Wang, P, X Sun, C Li, Z Wei, D Liang, and F Ma. 2013. "Long-term exogenous application of melatonin delays drought-induced leaf senescence in apple." *Journal of Pineal Research* 54 (3):292–302.

Wang, P, X Sun, Y Xie, M Li, W Chen, S Zhang, D Liang, and F Ma. 2014. "Melatonin regulates proteomic changes during leaf senescence in *Malus hupehensis*." *Journal of Pineal Research* 57 (3):291–307.

Wang, P, L Yin, D Liang, C Li, F Ma, and Z Yue. 2012. "Delayed senescence of apple leaves by exogenous melatonin treatment: toward regulating the ascorbate–glutathione cycle." *Journal of Pineal Research* 53 (1):11–20.

Wang, Q, B An, Y Wei, R J Reiter, H Shi, H Luo, and C He. 2016. "Melatonin regulates root meristem by repressing auxin synthesis and polar auxin transport in *Arabidopsis*." *Frontiers in Plant Science* 7:1882.

Weeda, S, N Zhang, X Zhao, G Ndip, Y Guo, G.A Buck, C Fu, and S Ren. 2014. "Arabidopsis transcriptome analysis reveals key roles of melatonin in plant defense systems." *PLoS One* 9 (3):e93462.

Wei, W, Q-T Li, Y-N Chu, R J Reiter, X-M Yu, D-H Zhu, W-K Zhang, B Ma, Q Lin, and J-S Zhang. 2014. "Melatonin enhances plant growth and abiotic stress tolerance in soybean plants." *Journal of Experimental*

Botany 66 (3):695–707.

Wen, D, B Gong, S Sun, S Liu, X Wang, M Wei, F Yang, Y Li, and Q Shi. 2016. "Promoting roles of melatonin in adventitious root development of *Solanum lycopersicum* L. by regulating auxin and nitric oxide signaling." *Frontiers in Plant Science* 7:718.

Yadu, B, V Chandrakar, R K Meena, A Poddar, and S Keshavkant. 2018. "Spermidine and melatonin attenuate fluoride toxicity by regulating gene expression of antioxidants in *Cajanus cajan* l." *Journal of Plant Growth Regulation*:1–14.

Yang, XL, H Xu, D Li, X Gao, TL Li, and R Wang. 2018. "Effect of melatonin priming on photosynthetic capacity of tomato leaves under low-temperature stress." *Photosynthetica* 56 (3):884–892.

Ye, J, S W Wang, X P Deng, N YanL, and B L Xiong. 2016. "Melatonin increased maize (*Zea mays* L.) seedling drought tolerance by alleviating drought-induced photosynthetic inhibition and oxidative damage." *Acta Physiologiae Plantarum* 38 (2):48.

Yun, B-W, M J Skelly, M Yin, M Yu, B-G Mun, S-U Lee, A Hussain, S H Spoel, and G J Loake. 2016. "Nitric oxide and S-nitrosoglutathione function additively during plant immunity." *New phytologist* 211 (2):516–526.

Zhang, H-J, N A Zhang, R-C Yang, L Wang, Q-Q Sun, D-B Li, Y-Y Cao, S Weeda, B Zhao, and S Ren. 2014. "Melatonin promotes seed germination under high salinity by regulating antioxidant systems, ABA and GA 4 interaction in cucumber (*Cucumis sativus* L.)." *Journal of Pineal Research* 57 (3):269–279.

Zhang, J, H Li, B Xu, J Li, and B Huang. 2016. "Exogenous melatonin suppresses dark-induced leaf senescence by activating the superoxide dismutase-catalase antioxidant pathway and down-regulating chlorophyll degradation in excised leaves of perennial ryegrass (*Lolium perenne* L.)." *Frontiers in Plant Science* 7:1500.

Zhang, J, Y Shi, X Zhang, H Du, B Xu, and B Huang. 2017. "Melatonin suppression of heat-induced leaf senescence involves changes in abscisic acid and cytokinin biosynthesis and signaling pathways in perennial ryegrass (*Lolium perenne* L.)." *Environmental and Experimental Botany* 138:36–45.

Zhang, LJ, JF Jia, Y Xu, Y Wang, JG Hao, and TK Li. 2012. "Production of transgenic *Nicotiana sylvestris* plants expressing melatonin synthetase genes and their effect on UV-B-induced DNA damage." *In Vitro Cellular and Developmental Biology-Plant* 48 (3):275–282.

Zhang, N, H-J Zhang, B Zhao, Q-Q Sun, Y-Y Cao, R Li, X-X Wu, S Weeda, L Li, and S Ren. 2014. "The RNA-seq approach to discriminate gene

expression profiles in response to melatonin on cucumber lateral root formation." *Journal of Pineal Research* 56 (1):39–50.
Zhang, N, B Zhao, H-J Zhang, S Weeda, C Yang, Z-C Yang, S Ren, and Y-D Guo. 2013. "Melatonin promotes water-stress tolerance, lateral root formation, and seed germination in cucumber (*Cucumis sativus* L.)." *Journal of Pineal Research* 54 (1):15–23.
Zhao, H, T Su, L Huo, H Wei, Y Jiang, L Xu, and F Ma. 2015. "Unveiling the mechanism of melatonin impacts on maize seedling growth: sugar metabolism as a case." *Journal of Pineal Research* 59 (2):255–266.
Zhao, Y, D-X Tan, Q Lei, H Chen, L Wang, Q-T Li, Y Gao, and J Kong. 2013. "Melatonin and its potential biological functions in the fruits of sweet cherry." *Journal of Pineal Research* 55 (1):79–88.
Zhou, X, H Zhao, K Cao, L Hu, T Du, F Baluška, and Z Zou. 2016. "Beneficial roles of melatonin on redox regulation of photosynthetic electron transport and synthesis of D1 protein in tomato seedlings under salt stress." *Frontiers in Plant Science* 7:1823.
Zuo, B, X Zheng, P He, L Wang, Q Lei, C Feng, J Zhou, Q Li, Z Han, and J Kong. 2014. "Overexpression of MzASMT improves melatonin production and enhances drought tolerance in transgenic *Arabidopsis thaliana* plants." *Journal of Pineal Research* 57 (4):408–417.

CHAPTER 11

ORGANIC AGRICULTURE – A CASE STUDY FROM TURKEY

HAKKI ZAFER CAN, TANSEL ASCIOGUL, DURSUN ESIYOK

Ege University Faculty of Agriculture Horticulture Department Bornova-İzmir.
dursun.esiyok@ege.edu.tr

Abstract

Organic agriculture, which is closely related to rural sustainable development, must encompass food production alongside the protection of natural resources and the local environment. In the last few decades, organic agriculture has developed rapidly across the world due to the increased awareness of food safety and environmental concerns among consumers, and because of the important role it plays in reducing surpluses, promoting high-quality products, and incorporating environmental conservation practices and animal welfare.

Beginning in the early 1970s in Europe with the initiatives of pioneers from different geographical areas, organic food production experienced a huge increase in the 1990s. In Turkey, such activities started around the 1980s with the involvement of certain European companies because of an increasing demand in Europe for Turkish products, such as dried grapes and figs. Organic agricultural production, in terms of production area, number of farmers, product diversity, and consumption, increased rapidly with this push. The reason for this was that the majority of organic products were exported mainly to European countries. The organic production pattern in Turkey has, therefore, been shaped by the demand from the exporter countries for Turkish organics, resulting from its own specific structure and climatic conditions. Due to its distinctive ecological and phytogeographical characteristics, the historical development of organic agriculture and implementation strategies in Turkey can provide a model for other countries.

Our main objective here is to present the historical development and current status of Turkish organic agriculture, regional distribution of organic farms, land use, state legislation, regulatory and institutional aspects, inspection and certification system bodies, support policies, producer organisations, education, action plans, research, and market issues.

Since organic agriculture is a sustainable farming system, its main goal is to maintain a healthy and clean environment and create a prosperous society. The global climate change and socio-economic aspects are key factors for developing organic agriculture, which is a key system in alleviating the negative effects of global climate change and economic fluctuations. Here, an attempt is also made to evaluate the interactions between organic agriculture and global climate change, with major suggestions for future developments.

Keywords: Turkey, organic agriculture, history, system, global climate change

Introduction

Organic agriculture, which is highly related to rural sustainable development, must encompass food production alongside protection of natural resources and the native environment (Hakeem, Ahmad, and Ozturk 2013, Ozturk 2018). From this point of view, organic agriculture has developed rapidly due to increased consumer awareness of food safety and environmental concerns because such practices could play an important role in reducing surpluses, promoting quality goods, and integrating environmental conservation practices and animal welfare (Ozturk et al. 2018). The organic production of crops began in the 1970s, in Europe, with pioneers from different areas. It received a great impetus in the 1990s. In contrast, beginning in the 1980s in Turkey, organic agriculture quickly developed through increasing demand by companies in Europe and the USA, which sought dried grapes and figs, not grown in some European countries. Developments in this direction rapidly increased in terms of production area, the number of farmers, product diversity, and consumption. Today, the majority of Turkish organic products are exported, primarily to Europe. The organic production pattern in Turkey has thus been mainly shaped by demands from Europe. Due to its distinctive characteristics, historical development of organic agriculture, and implementation strategies in the country, Turkey is expected to serve as a model for other countries.

Over the last few decades, organic agriculture and environmentally-friendly agricultural practices have spread rapidly, not only in developed countries, but also in several fast-industrialising countries, such as Turkey. Since food insecurity is becoming a pervasive issue on a global scale, increasing consumer demand for certified products originating from organic agriculture is on the rise, and this type of production is becoming more prevalent. According to the latest survey by the Research Institute of Organic Agriculture (FiBL) (Anonymous 2018) on certified organic agriculture worldwide, the area of organic agricultural land, including converted areas, reached 57.8 million hectares in 178 countries by the end of 2016. The countries with the highest organic agricultural land cultivation are Australia (27.4 million hectares), Argentina (3 million hectares), and China (2.3 million hectares). Nearly 1.2% of the world's agricultural land is currently organic, and some countries have gained far higher shares in this area, e.g. Liechtenstein (37.7%) and French Polynesia (31.3%). Furthermore, in 15 countries including Austria, Samoa and Estonia, 10% or more of total agricultural land is organic.

From 1999 to 2016, organic farmland in Australia increased by 15%. However, important increases have also been reported from many other countries, such as China, Uruguay, India, and Italy. While the number of organic producers was 2.34 million in 2015, it reached 2.7 million in 2016. A quarter of the world's organic agricultural producers are now emerging as markets from countries like Turkey.

In parallel with Europe, organic agriculture grew rapidly in Turkey during the last decade. This increase is mainly due to the growing awareness of Turkish people regarding developments in health and environmental concerns (Öztürk et al. 2015, Ozturk et al. 2013). However, the demand for organic products that cannot be grown in Europe and North America has resulted in the development of an international market for organic food products (Hameed et al. 2018), and has received an impetus in countries like Turkey with appropriate ecological and climatic conditions. Turkey is now emerging as a model among industrialising countries due to its establishment of organic agriculture based on export potential, and nearly 85% of its current organic production being exported to Europe (Demiryürek, Stopes, and Güzel 2008).

In this paper, we present the historical development and current status of Turkish organic agriculture, state legislation, regulatory and institutional aspects, support policies, related NGOs, and research activities undertaken in Turkey. Since organic agriculture is a sustainable farming system, its main goal is to maintain a healthy and clean environment, and create a society with a high level of prosperity. The global climate change and socio-economic aspects are key factors for developing organic agriculture, which, in turn, is a significant system for alleviating the negative effects of global climate change and economic fluctuations (Ashraf et al. 2012). In view of this, the interactions between organic agriculture and global climate change are also examined here, with major suggestions for future developments.

Some Important Milestones

Organic agriculture first attracted the attention of European countries and later expanded to other countries. It has experienced dramatic growth during the last few decades. From the the mid-1980s to 2014, agricultural area as farmland under organic production increased to 10.3 million hectares (Anonymous 2016a, 2016b). This great interest has been triggered mainly by concerns over the environment and health, as well as social and economic conditions. Consumer demand for organic food has led to the development of international trade.

In Europe, organic production has received an acceptable support per unit area in the transitional process, which has facilitated its rapid spread. The first organic agricultural activity in Turkey started with the involvement of some European companies, not with local input (Aksoy and Altindisli 1999). Its development in the country was realised from the ground up, with raisins and dried figs in line with the requests from European buyers. With such parallel demands from foreign countries, particularly for organic dried fruits, international certification bodies, such as BCS, IMO, Ecocert, and SKAL became active in Turkey.

Sun-dried fruits and nuts, for example, raisins, figs, apricots and hazelnuts are seen as core Turkish exports because the country is like a homeland for these products (Aksoy and Engiz 2007). These products, not being available in Europe, pushed organic traders to enter the Turkish market in the 1980s. These traders cooperated with Turkish processors and exporters, and because of their reputation, organic projects were developed at the level of farmers, including the initiation of organic cotton production projects for the first time in the world.

After the European Economic Community 2092/91 regulation, a similar regulation was adopted by Turkey in 1994, and renewed in 2010. The last amendment came into practice on January 10, 2018. Before the legal regulation issued in 1994, production in Turkey followed the private rules of certification bodies according to the IFOAM rules, and later, EU legislation. The National Regulation of 1994 was based on EU regulation 2092/91, and an amendment to this regulation was drawn up in 1995. The National Ecological Farming and Ecological Farming Guidance Committees were established in 1995. Since then, amendments to the EU basic regulation have been followed, and revised regulations were prepared accordingly and published in 2002. To improve coordination of activities, a special law on organic agriculture was prepared in 2004. The National Trade and National Research and Projects Committees were established in the field, and the term "organic" was used instead of "ecological" from 2002 onwards (Aksoy and Engiz 2007). Organic agriculture was characterised by a product range consisting of eight traditional export products during 1985–1990. After 1990, the product range increased, because foreign demand became the determinant of product quantity and type, and thus organic products were intended more for the foreign market until 2000 (Aksoy 2001). All activities relating to organic production were foreign-based until the domestic market started to develop. Work on both legal regulations and scientific studies gained a considerable momentum with the establishment of the Association for Ecological Agriculture Organisations (ETO), an NGO that acted as the major source for dissemination activities related to organic agriculture. ETO is the umbrella organisation in Turkey, established with the participation of all volunteers involved in various fields, such as producers, consumers, processors, traders, controllers, researchers, and technical staff, to ensure fast and healthy development of organic agriculture in the country. Another important association for the organic movement was Buğday, which began to operate officially in 2002. Besides these root organisations, a unit and organic research group were formed at the then-called Ministry of Agriculture. Parallel to the rise in organic agriculture and trade, the Organic Product Producers and Industrialists Association was established in 2004 to support development of organic production and industrialisation in the country. In 2016, the Association of Control and Certification Bodies of Agricultural Products was founded to support the development and promotion of testing, examination and certification of conformity with technical regulations or standards for the agriculture and food sector, in order to ensure that all stakeholders and consumers were informed about the certification of agricultural products.

Throughout this process, the flow of scientific information provided by Ege University, Faculty of Agriculture (EUFA) has been extremely important. ETO, the Ministry of Agriculture and the EUFA Department of Horticulture jointly organised the Third Agribiomediterraneo and the First National Symposium on Organic Agriculture in 1992 and 1999, respectively. Particularly in 2000, organic agriculture developed rapidly and spread widely. At this time, the state supported all certified organic farms based on areas particularly in horticultural crops, animal husbandry, aquaculture, and beekeeping. Additionally, an organic agriculture unit was established in each province of Turkey by the Ministry of Agriculture. The ministry started to collect data on organic food and farming through an electronic data collection system called the "Organic Farming Information System". A strategic agenda and an action plan were also devised together with stakeholders, by focusing on organic agriculture and the responsibilities of the ministry. The ministry and the Organic agriculture National Guidance Committee prepared the Turkish organic food and farming action plan based on the National Organic Strategy. The aim of this action plan was to support the sustainable development of organic agriculture and follow the market for organic products in Turkey and abroad, as a way of achieving rural development and protecting the environment and human health in Turkey (Redman 2007). The organic food and farming action plan had important objectives, such as ensuring the implementation of policies, provide practical support for organic food and farming, create a strong stakeholder network, support development of organic production, and develop domestic markets and export facilities.

Another important attempt was the organisation of the 18th Organic World Congress by IFOAM and the Buğday Association for Supporting Ecological Living in 2014, with the participation of thousands of people from 81 countries. The Organic World Congress brought together farmers, scientists and organic leaders to share developments from their fields, laboratories and libraries, and to inspire new ways to practice and promote organic agriculture.

In 2017, the EKO-TIP working group was established under the leadership of Izmir Governorate Chamber of Medical Doctors and ETO to share knowledge about the importance of organic products in human health, and provide continuous information for the public on scientific and ethical values. This working group included scientists from the faculties of Medicine, Agriculture and Science from Ege University, Izmir. This group made great contributions to the development of people's understanding of organic agriculture, especially because the information was provided by medical doctors as reliable sources for the public.

Current Situation

Organic production in Turkey is increasing, mostly in plant production. The total area of organic agriculture has reached 523,777 hectares, and the organic share of total agricultural land is around 2%; together with wild collection areas, it has reached 661,210 hectares. Of the total area of 523,777 hectares, 76.4% consists of arable crops, and 16.5% permanent crops, while another 4.7% is left over as permanent grasslands and grazing areas. The key arable crops are cereals (197,877 hectares), green fodder from arable land (174,136 hectares), and crops for industrial uses (14,315 hectares). The key permanent crops are olives (36,261 hectares), nuts (19,320 hectares), and apricots (4,946 hectares). On the other hand, total production has increased only slightly, but there have been several fluctuations over recent years in terms of both the number and location of producers, as well as production area. In terms of the numbers of organic producers, Turkey is among the top 10 countries with 67,879 producers, 1,422 processors, 61 importers, 46 exporters, and 235 types of products. Turkey is also one of the prominent organic cotton producers in the world. In 2016, certified organic cotton was grown in 18 countries, although 97% of global supply came from just seven countries: India, China, Kyrgyzstan, Turkey, Tajikistan, The United States, and Tanzania. Turkey has attracted importers not only with organic temperate fruit, olive, cotton, and dried fruit but also by organic beehives, as well as wild-collected herbs.

According to the statistical database of the ministry, there are more than 1 million hectares of land where organic agriculture can potentially be performed, and organic agriculture has been included in the scope of support programme since 2015. For this reason, a regular increase is expected in the area of organic land in the future.

The executive authority responsible for the implementation of the national organic legislation is the Plant Production General Directorate operating under the recently renamed Ministry of Food, Agriculture and Livestock (MoFAL); and organic agriculture is defined by both the Organic Agriculture Law and the Turkish Organic Regulation. Currently, 35 control and certification companies are active, and organic products inspected and certified by an approved Turkish control body have to be labelled with the national organic logos. According to the Turkish Regulation on Principles and Application of Organic agriculture, manufacturers, wholesalers and retailers of organic agricultural products and substances must use one of the official organic agriculture logos on their packaging. (Fig. 1). The Turkish organic logos cannot be used on products imported in their final packaging; however, if the imported product is semi-processed and the final processing or packaging is undertaken in Turkey, then these logos are placed on the products. Products that do not include this logo cannot be sold or advertised on domestic markets and do not qualify for patent eligibility as organic products.

Figure 1. Examples of the official organic logo

The process of organic control and certification is also undertaken by private control bodies approved and supervised by the Department of Good Agricultural Practices and Organic Agriculture, a subunit of the Plant Production General Directorate of MoFAL. Besides the legal regulations, following the results of studies performed by the ministry such as the strategic plan for organic agriculture, the action plan to improve and extend organic agriculture, and the organic farming information system, there has been a 39% increase in the number of products, an 83.7% increase in production area, and 43.0% increase in the amount of organic production realised within the last decade.

Agricultural policies in Turkey have been providing financial support for organic agriculture via area payments. Farmers engaging in organic plant production and organic bee farming are currently financially supported by the state. Besides area payments, interest-reduced agricultural loans are provided, and the programme for the Conservation of Agricultural Resources for the Environment is implemented to support organic farmers. The supporting policy is very important in terms of both economy and the environment due to the social and financial barriers facing organic farmers. In Turkey, there was no supporting payment for organic agriculture until 2004, and it must be said that the level of financial supports is still limited. Currently, the highest support payment made to farmers engaged in fruit and vegetable production is around $25 per decare. Additional support is available for the purchase of certified propagation material, biological control, organic fertilisers, soil analysis, environmentally fragile areas, and receiving agricultural consultancy.

The organic agricultural system in Turkey is based on a form of the contract model. According to this agreement, the farmers must implement all instructions and in return companies are to purchase the products, as agreed, at a premium price. Inspection and certification are carried out by independent and accredited inspection companies authorised by the agricultural ministry and the EU. The major market is still exports, with

European countries being the main destinations. Although organic products have lately been promoted by some big supermarket chains, production is still mainly dependent on the demand from importers because of the relatively higher prices than domestic markets, and lack of awareness of customers, especially within the country. One of the most important obstructions is the weak domestic market. For this reason, organic farmers' markets have been established and supported in many cities to meet the demand for fresh and healthy local food. The Buğday Association founded the first ecological farmers' market in 2006, in Istanbul. Starting with 48 stalls, the market expanded to 250 stalls within four years. Following the success of this first market, more open farmers' markets have been established in other cities by Buğday and other ETO organisations. These markets provide opportunities for local farmers to directly market their products at retail prices. As a result, local farmers have gained greater control over production and marketing decisions. At the same time, these markets offer a host of social, economic and health benefits for consumers. Besides access to fresh food, they provide an opportunity to develop relationships with local growers, and have also become places to meet, socialise, and experience alternative shopping. Approximately 20 local marketplaces for organic products are operational throughout the country, and their number is gradually increasing (Fig. 2).

The range of organic products includes dried fruit and nuts, herbs, aromatic plants, pulses, and fruit juice, providing an impetus towards upgrading the domestic market for plant and animal products. Fresh vegetables and fruit are the main products in open markets. Cotton is still an important crop and is processed into yarn and textile for the export market. In the domestic market, the main channels are retail chains and open markets in major cities; however, specialised shops, internet sales and direct sales are also increasing.

Ecology Izmir, which is the only fair that represents certified Turkish organic products, both nationally and internationally, has been organised uninterruptedly for the last nine years (Fig. 3). It has a wide range of product categories, including organic certified food, textiles, personal care and cleaning products. It also provides information on organic inputs, ecotourism, renewable energy, control and certification companies, analysis laboratories, and alternative education programmes.

Figure 2. An open organic bazaar supported by ETO and the local administration. *Photos obtained from the ETO web page.*

Figure 3. General views from the Ecology Izmir certified organic products fair. *Photos obtained from the IFAŞ web page.*

Education Projects

The first four-year organic agricultural management undergraduate programme was started at Pamukkale University. Currently, there are a total of eight ongoing programmes, run in five universities. Besides these undergraduate programmes, 48 pre-university organic agriculture programmes (two years) are implemented by various universities in Turkey. Some universities also offer several MSc and PhD programmes.

Agricultural research and scientific projects in Turkey are conducted mainly by the related subunits of the Scientific and Technological Research Council, the MoFAL General Directorate of Agricultural Research and Policies (TAGEM), and universities. Research related to organic agriculture was started by TAGEM in 2001, and by 2016, 67 projects had been completed. TAGEM also supports research in organic agriculture through 30 national research institutes located throughout the country, and a research group that meets annually. The main research topics are: growing techniques, plant nutrition, disease and pest management, sustainable soil and environment management, beekeeping, and animal fodder and husbandry. Currently, a total of 28 national projects are being conducted. Furthermore, the project on the expansion and control of organic agriculture has been implemented by MoFAL since 1997, with the aim of expanding organic agriculture and allowing control of organic agriculture products without polluting soil, water resources, or air. This project is presently being carried out in cooperation with 51 provincial directorates and 11 research institutes. Another comprehensive project is the Gokceada & Bozcaada Agricultural Development and Settlement Project, which aims to extend organic agricultural production by utilising available natural resources, increase agricultural activities by providing producers with agricultural supports, raise the income levels of the people involved, make agriculture a sustainable source of income and, finally, create an active and settled population in Gokceada and Bozcaada.

The project, Organic Agriculture for Turkey, funded by the EU and implemented by MoFAL, is related to policy development, capacity building, and training. It aims to create sustainable development for the organic sector in accordance with the EU regulations. The most important task of this project is to establish farmer schools in several pilot areas to create a model for the future. The organic agriculture cluster project implemented by the Southeast Anatolia Project Regional Development Administration under the technical assistance of UNDP targets the improvement of competitiveness of the organic agriculture sector in the Southeast Anatolia Region of Turkey. In parallel with this overall target, one of the important aspects of the project has been its contribution to the development of the region in a sustainable manner and utilisation of organic agriculture potential. This project has also been supported by the European Secretariat for Cluster Analysis-ESCA. As a consequence of the development of a national clustering policy, the Izmir Organic Food Cluster was established with the partnership of the Association of Ecologic Producers, ETO, Ege University, Izmir Chamber of Commerce, Izmir Metropolitan Municipality, and Izmir Development Agency. An EU-funded technical assistance project was also launched by the Under Secretariat of Foreign Trade in 2007. A roadmap for the Izmir Organic Food Cluster was developed with the direct involvement of a broad range of stakeholders, such as enterprises, governmental organisations and NGOs, and academic and research institutions. Some outputs of the cluster workshops have been realised, and some are still ongoing, e.g. the Farm to Fork Food Traceability Infrastructure, the Izmir Certified Open Organic Market, the Ecology Izmir Fair, the Organic Agriculture in Karaburun Peninsula, the Organic Karaağaç Village Project, and the Acting Organic by Thinking Organic Project. Many national and regional projects, supported by several related ministries and NGOs, are conducted in Turkey. Besides state-supported projects, many important projects are undertaken by NGOs, such as ETO and Buğday. The Acting Organic by Thinking Organic Project, funded by the EU and implemented by ETO, was completed in 2009. The aim of this project was to enrich the network, both nationally and internationally, in order to provide a better environment for the organic sector, and build a bridge between organisations from Turkey and the EU that represent professionals, farmers and academics involved in promoting organic agriculture and environmental consciousness.

Another completed project, German-Turkish Bilateral Cooperation Organic Agriculture, was conducted by ETO and FiBL with the aim of supporting the development of the organic agriculture sector in Turkey, and promoting the export of such products into the European market. The cooperation of all market partners from the organic agricultural sector was strengthened, and an educational training programme was initiated. The action fields of this project can be summarised as follows: improving the know-how of all stakeholders; organising knowledge transfer activities; setting up a network of demonstration farms; the organisation of technical workshops; offering training courses for agricultural extension; strengthening cooperation between the Turkish and German market partners; and facilitating multi-stakeholder dialogue. Within the scope of this project, an e-learning platform and organic input list were created, and research was conducted to seek solutions to the contamination problem. Additionally, a pilot project called Europe Fit & Go was implemented to improve the knowledge and experience of organic product exporters for entering international markets.

A civil society dialogue project, Easy Access to Organic Products, co-funded by the EU and the Republic of Turkey, was implemented by ETO, Ökomarkt e.V. (Germany), and Avalon Foundation (The Netherlands). The target of this project was to develop channels to facilitate consumer access to organic products and to promote a sustainable lifestyle within society. The outcomes of this project include: increased consumer awareness, guidelines for educational institutions to introduce organic catering to universities and schools, quality guidelines for organic open farmers' markets for municipalities, policy recommendations for authorities, and an internet portal for accessing organic products and services in Turkey. Another significant project output has been the creation of Turkey's organic product search engine (http://www.bio123.net/).

One of the non-profit NGOs, the Buğday Association for Supporting Ecological Living, has also initiated many great projects, such as nature-friendly city gardens; 100% ecological markets; eco-agro tourism; voluntary exchanges of knowledge and skills on organic farms, "the community goes farming"; and "Turkey composts" projects. The "community goes farming" project focusses on community-supported agricultural models and examples, organic agriculture, and self-sufficient and sustainable communities, as well as sharing knowledge and experience concerning these issues. This project was conducted with the contribution of Heeley City Farm (England), Kulturverein Hof Medewege (Germany), ALEJA inzeniring d.o.o (Slovenia), and ZIARNO Ecological and Cultural Association (Poland) within the scope of the EU Lifelong Learning Program Leonardo da Vinci.

The Buğday Association undertook another project, known as Supporting the Development of a National Agri- Environment Program for Turkey, which was funded by the Netherlands Ministry of Agriculture. The project provided high nature value farmland potential in Turkey and agri-environmental policy making. The Çamtepe Ecological Centre, opened in 2010, is the rural centre of the Buğday Association, designated to operate as a self-sufficient rural model, built with local, nature-friendly materials, combining both traditional and modern knowledge, such as a solar-powered water system. This centre, and ecological farming visits, are managed by the European Voluntary Service (EVS), a youth programme with diverse training, exchange and volunteer platforms to support young people's international activities. EVS allows young people to work in different areas in EU projects and share experiences with different cultures.

Eco-Agro Tourism and Voluntary Exchange (TaTuTa) is another project that was initiated by the Buğday Association, in 2003. It allows visitors and volunteers to visit ecological farms located throughout Turkey. In this project, the visitor develops an understanding of the value of food by helping out and spending time with the farmer, and the farmer benefits from the contribution of the visitor. The TaTuTa project also received funding from the UNDP GEF Small Grants Programme, with the Buğday Association being an official member of the European Centre for Ecological and Agricultural Tourism (ECEAT), a national, accredited EVS organisation with global involvement, linking volunteers with organic farmers and growers to promote cultural and educational experiences (Fig. 4). The TaTuTa project provides extra income and support for smallholding organic farmers, connects consumers living in the cities with the soil and nature, and contributes to rural development. Currently, there are many farms that host hundreds of visitors and volunteers from all around the world every year.

Figure 4. The red markers on the map show the approximate locations of the EVS-accredited farms

Last but not least, the organic agriculture project was funded and implemented by the Istanbul Municipality as a social responsibility project. Within this project, a bread factory (Istanbul Halk Ekmek) was set up by the municipality, and small farmers were contracted to produce organic wheat for making organic bread. The bread produced in the factory was sold to consumers in Istanbul. Another important aim of this project was to adopt organic agricultural practices for small farmers with the market guarantee and premium prices.

Conclusions and Suggestions

Although organic agriculture in Turkey has grown rapidly over the past few decades, there are still challenges and weakness in the programmes. A rapid growth in organic agriculture is due mainly to an increase in the awareness of health and environmental concerns in Turkey, as in several other countries. However, the improvements in the organic sector remain dependent on exports, with the demand in the domestic market being low due to relatively higher prices. Turkey has important advantages in terms of organic agriculture, such as diverse climatic features; rich biodiversity; the variety of agricultural products; nearness and accessibility to European markets; a robust, contracted production system; familiarity with traditional growing methods; low input use due to relatively high prices; and availability of labour. The country also has great potential in this area due to its geographically diverse ecological conditions being suitable for the cultivation of various crops. Some organic products, particularly dried fruit, nuts, olive oil, and medicinal plants are important exports, and have great implications for the economy. Moreover, there is no heavy chemical pollution due to relatively low agrochemical use in comparison with developed countries. In addition, rich biodiversity results in a considerable number of species that are resistant to diseases. The relatively low input use appears to be one of the main advantages for organic agriculture, and the transition from conventional to organic agriculture. In both conventional and organic agriculture systems, a substantial amount of input is imported. However, organic inputs are relatively expensive compared with artificial inputs. For this reason, local input production and use are very important for this type of farming.

Despite these advantages of organic agriculture, one of the main obstacles remains the lack of awareness among both producers and consumers. In view of this, for organic agriculture to continue to develop, further research, good education and training programmes, and consultation services are needed. With the support of extension and dissemination activities, and with the collaboration of stakeholders, organic agriculture is expected to extend to higher levels. Increased awareness of research into organic agriculture, especially in national research institutions and NGOs, and organic agriculture being among research priorities, are also mentioned in the action plan; however, organic research suffers from a lack of funding, and only a few individual researchers in many research institutes and universities are involved in organic agriculture. The absence of a multidisciplinary and integrated approach to assess research needs at the local level, and lack of interest in many agriculture and related faculties, are serious barriers for the development of research and project activities.

Organic agriculture systems are confronted with a lack of multidisciplinary research projects. There is an urgent need, not only in Turkey, but also in all the countries in the Mediterranean Basin, to cooperate to cope with the predicted changes in the region. Elevated CO_2 is expected to facilitate forest expansion and greater carbon storage if precipitation increases; however, for Europe, only minor biome-level shifts are projected for the Mediterranean vegetation types, in contrast with the projection that between 60% and 80% of current species will not exist in the southern European Mediterranean region (Bakkenes et al. 2002). Since Mediterranean countries are among those most impacted by global change, it is essential that organic agriculture studies should be well supported (Sala et al. 2000).

Turkey is located at the intersection of three continents and three climate zones: The Mediterranean—in an arid and semi-arid region—the Continental, and Eurosiberian. It is among the countries heavily affected by global warming and climate change, which have been increasing during the last century. All these problems constitute serious threats to organic agriculture. However, organic agriculture can also be used as an effective measure to reduce the effects of such unfavourable changes. For this purpose, a series of studies should be initiated and shifted toward mitigating risks and damages due to degradation and drought (Anonymous 2016a, 2016b).

Organic agriculture is practiced in all regions in Turkey. As a partly-Mediterranean country, it is expected to be severely affected by global climate change, which will mostly affect cultivated crops and fruit species. Since agriculture is both a source of greenhouse gas emissions and a reservoir for greenhouse gases, organic agriculture practices can also be important to alleviate the negative impacts of climate change. Organic practices are generally well-suited to enhancing productivity and efficiency, particularly if practiced by domestic inputs. The implementation of organic agriculture will increase the ability of farming systems to function, even when faced with the adverse effects of climate change, by increasing resilience within the agro-ecosystem to temperature extremes, drought and soil degradation (Niggli et al. 2009). Long-established and locally-optimised organic agriculture practices are needed to mitigate the impact of climate change. Rich biological diversity, local species and varieties, integrated crop management and production, multifunctionality, local input use and local consumption must be the target of future research in Turkish agricultural systems. Some basic organic methods, such as crop rotation and efficient farming design, manure management, livestock management, pasture and fodder supply improvement, fertile soil maintenance, and restoration of degraded land can reduce the environmental stress; however, further scientific studies are needed to clarify these issues (Aksoy and Can 2012).

Low soil organic matter content is a problem in many regions of the country because of conventional practices and marginal climatic conditions. Thus, biological activity in soil, soil fertility and organic matter must be enhanced by adopting practices such as composting, cover crop use, and mulching. Furthermore, drought, the most crucial abiotic stress factor, should be investigated in detail in relation to the developing organic agriculture practices in Turkey to facilitate farming in water deficit conditions. Drought resistant and salt-tolerant species and varieties must be developed to overcome reduced efficiency in water use. There is also a need for precise scientific findings based upon long-term evaluations under organic management (Aksoy and Can 2012).

Many products including olives, figs, dried fruit, and endemic plants are produced by traditional methods, a practice that has been adopted for thousands of years. However, traditional agronomic methods are gradually being replaced with modern techniques, by new generations. The same situation is valid for fertilisation and plant nutrition. The fact that organic agriculture uses traditional farming methods is of great importance in this respect. Furthermore, domestic consumption should have attached to it as much importance as is placed on the export of products. The domestic market for organic products is still very limited in Turkey due to the lack of awareness, promotion, and relatively high prices. Consumer awareness must be increased by educational and promotional activities, and local organic unions, cooperatives and marketing companies should be supported by the government. Organic agriculture activities should also be integrated with other sectors, such as health, wellness, catering, local markets and restaurants, and particularly tourism. It is not difficult to integrate all these sectors with organic agriculture to create added value to the resulting products. Finally, the rapid and strong growth of organic agriculture depends on the contribution of women to economic activities, particularly in Turkey. To this end, extensive education programmes should be organised for women, as they are very important in the adoption of organic agriculture practices.

References

Aksoy, U., 2001. "Ecological agriculture: an overview." Second National Symposium on Ecological Agriculture, Antalya, Turkey.

Aksoy, U., and M. Engiz, 2007. Country report on organic farming in Turkey Ankara: General Directorate of Agricultural Production Development (TÜGEM).

Aksoy, U, and A. Altindisli, 1999. The Production, Export and Development Possibilities of Ecological Agricultural Products in the World and Turkey. Publication.

Aksoy, U., and H. Z. Can, 2012. "Expected effects of climate change on organic agriculture in Turkey." *New Medit* 11 (4 SI):S3–S3.

Anonymous, 2016*a*. Organic in Europe. Prospects and Developments.

Anonymous, 2016*b*. "Land Degradation Neutrality. Turkey Report. United Nations Convention to Combat Desertification."

Anonymous, 2018. "Survey on certified organic agriculture worldwide. FIBL."

Ashraf, M, M Öztürk, M S A Ahmad, and A Aksoy, 2012. *Crop production for agricultural improvement*: Springer Science & Business Media.

Bakkenes, M., J R M. Alkemade, F Ihle, R Leemans and J B Latour, 2002. "Assessing effects of forecasted climate change on the diversity and distribution of European higher plants for 2050." *Global change biology* 8 (4):390–407.

Demiryürek, K., C Stopes and A Güzel, 2008. "Organic agriculture: the case of Turkey." *Outlook on Agriculture* 37 (4):261–267.

Hakeem, K R, P Ahmad, and M Ozturk, 2013. *Crop improvement: new approaches and modern techniques*: Springer Science & Business Media.

Hameed, M, M S A Ahmad, M Ashraf, M Ozturk and S Fatima, 2018. "Plant Genetic Resources of Major and Minor Crops: Origin, Sustainable Use, and Conservation." In *Crop Production Technologies for Sustainable Use and Conservation*, edited by Ozturk et al, 541. Apple Academic Press, USA-CRC Press, a Taylor & Francis Group.

Niggli, U, A Fließbach, P Hepperly and N Scialabba, 2009. Low greenhouse gas agriculture: mitigation and adaptation potential of sustainable farming systems. Food and Agriculture Organization of the United Nations, Rome.

Ozturk, M, S Gucel, S Sakcali and S Baslar, 2013. "Nitrate and edible plants in the Mediterranean Region of Turkey: an overview." *Nitrate in leafy vegetables—toxicity and safety measures. IK Intern. Publ. House Pvt. Ltd, New Delhi-Bangalore (India)*:17–51.

Ozturk, M, K R Hakeem, M Ashraf, M S A. Ahmad, 2018. *Global Perspectives on Over-Utilized Crops*. USA: Apple Academic Press, CRC Press and Taylor & Francis Group.

Oztürk, M, M Ashraf, A Aksoy, M S A Ahmad and K R Hakeem, 2015. *Plants, pollutants and remediation*: Springer.

Redman, M. 2007. Opportunities for Organic Food and Farming in the Turkey Rural Development Plan. Organic Agriculture for Turkey, Technical Assistance Project to Ministry of Agriculture and Rural Affairs for the alignment of organic agriculture legislation to the EU acquis and the development of organic agriculture in Turkey.

Sala, O, E F S Chapin, J J Armesto, E Berlow, J Bloomfield, R Dirzo, E Huber-Sanwald, L F Huenneke, R B Jackson and A Kinzig, 2000. "Global biodiversity scenarios for the year 2100." *Science* 287(5459):1770–1774.

CHAPTER 12

PRINCIPLES OF FERTILIZER USE IN ORGANIC AGRICULTURE

SINAN UZUNLU

Alanya Alaaddin Keykubat University, Rafet Kayıs Engineering Faculty, Food Engineering Department, Alanya-ANTALYA/TURKEY.
sinan.uzunlu@alanya.edu.tr

Abstract

To maintain food security, we need to have a well-managed system of agriculture for sustainable living. The soil undoubtedly plays the main role in agriculture, which not only ensures the yield of our produce but also adds value to the biodiversity of its habitat. Organic farming is a highly managed system based on scientific methods to improve soil fertility. The role of soil in supplying nutrients to a crop is key to the broad term of soil fertility. The quantities of nutrients to be supplied is managed in the soil and affected by its chemical, physical and biological properties. An inadequate supply of nutrients to the crops is a limiting factor in agriculture. Nutrients that are essential for the growth of plants generally include seventeen elements. Three of them—carbon, hydrogen and oxygen—come from the air, while the rest are derived from the soil. Each element is essential and has equal importance for the metabolism and growth of plants. Fertilisers, specifically organics, are naturally-derived materials of biological or mineral origin and have low nutrient concentrations or solubility features. The main difference that leads to both advantages and disadvantages of organic fertilisers from synthetic fertilisers is the slow release of nutrients. Slow release is generated by environmental factors in terms of moisture, temperature and microbial activity respondent with the soil. Organic farming systems usually don't use chemicals as fertilisers, as care for the diversity of life in the soil is of great importance.

Key words: ecological agriculture, manure, nutrient cycling, plant nutrition

The Need for Organic Agriculture

Fertiliser, whether it is obtained organically or synthetically, provides the nutrients required by plants. For proper usage of fertilisers, the basic elements for providing nutrient sources for crops need to be identified. However, before focusing on relationships that occur in the soil among organisms, plants and fertilisers, it is necessary to discuss the need for organic agriculture.

Food for human nutrition is provided by agricultural production. Plant-based foods, meat and meat products, and aquaculture products are the main foods of human beings. Since the foundations of agriculture ten thousand years ago, agricultural production has continuously evolved, with many different methods being developed.

Justus von Liebig proposed using elements in crop growing that formed the basis for the use of inorganic fertilisers in conventional agriculture. It was the early start of using synthetically produced materials in agriculture that probably led to the use of synthetic chemicals for pest management. However, pests, diseases and weeds are becoming increasingly resistant to agro-chemicals. Apart from the question of whether conventional agriculture has succeeded in producing enough food for the whole world, the improper use of agro-chemicals has resulted in an unsustainable ecosystem. The quantity of produced foods is enough for the world population, but more than 15% still go hungry annually. Securing food for those groups, which are mainly located in African and Asian countries, is of great concern (Halberg & Muller 2013).

Using toxic pesticides and fertilisers and loading animals with antibiotics are obviously incompatible with a sustainable farming system. Organic agriculture therefore promises to prevent the negative effects of current intensive agriculture by harmonising thousands of years of traditional knowledge with innovative practices (Bavec & Bavec 2006).

Crop Nutrition and Nutrient Cycling

Crops must be supplied by nutrients, and pests must be controlled independently of the applied methods in agriculture. Crops need seventeen out of the ninety-two elements that are identified in the periodic table. Carbon, hydrogen and oxygen are derived from the air, while the rest of them must be deposited in the soil, if we do not consider water-based (soil-less) farming or hydroponic farming. None of them is superior; all of them are essential elements. There is a clear differentiation between macronutrients and micronutrients, which is determined by the required amounts of the related nutrient. Whereas macronutrients vary from about 0.3% to 5% or higher in plants, the presence of micronutrients is one to several hundred parts per million in the dry basis of plants. Nitrogen, phosphorous, potassium, calcium, magnesium and sulphur are classified as macronutrients, while iron, zinc, copper, manganese, molybdenum, boron, chlorine and nickel are classified as micronutrients. These elements constitute 14 out of the 17 elements essential for plants (Barker 2010).

In order to be classified as a plant nutrient, an element should meet several criteria, in particular: i) the element must be required for its life-cycle—deficiency of an element can lead to abnormal growth or premature death; ii) the need for the element must be universal among plants; and iii) full substitution of an element is not covered by the term "nutrient", but partial substitution might be among some elements. The role of each element in plant nutrition is unique, however. If an element does not meet the above criteria, it can still be evaluated as being a beneficial element. Elements such as sodium, silicon, selenium and cobalt do not meet the requirements for essential nutrient sources, but they stimulate the growth of plants (Barker 2010).

To a chemist, materials containing carbon are organic, whilst to a biologist, materials that are obtained from living organisms, are organic. However, neither of these definitions reflects the term "organic" in organic agriculture practices. For instance, calcium carbonate is derived from living organisms and contains carbon atoms, but is an inorganic compound, whereas rock and dusts are non-living and lack carbon, but are considered organic fertilisers (Barker 2010). Thus, a farmer who wishes to be certified for an organic product should consult with a local certification organisation regarding allowable practices in the fertilisation of crops.

Carbon is the basic element of living organisms, and carbon atoms can bond with up to four other atoms, including other carbon atoms. As a result, there are a vast number of carbon-based compounds, such as carbohydrates, proteins and lipids, found in the dry basis of foods. Hydrogen is the most common element in the universe. Hydrogen is used to make ammonia through the Haber-Bosch process, and ammonia is used for many industrial products, such as plastic, nitric acid and inorganic fertilisers. Last but not least, oxygen is the most abundant element on Earth and is vital for life (Stockley et al. 2007).

The ability to supply nutrients to a crop is described by the wider term "soil fertility". This is closely related to the quantity of nutrients, and the physical, chemical and biological properties of the soil. Acidity of soil, temperature, depth of soil, nutrient-holding capacity, water-holding capacity, aeration, tilth, drainage, and presence of both harmful and beneficial organisms are interrelated and affect total soil fertility or soil health. Plant nutrition is therefore governed by the uptake of nutrients, as well as the transportation and functioning of nutrients in plants. The success of the operation of an organic system can be evaluated on the basis of nutrient recycling. Maximum recycling and minimum losses of nutrients from a closed organic production system are therefore essential. The closed system must be applied, not only at an individual farm level, but also at the regional level. This entails healthy and living soil by using the organic matter of harvested crops (Stinner 2007; Barker 2010).

The best approach to fertiliser and manure usage basics is to have knowledge of which crops remove nutrients in ways that minimise losses and maximise returns, recycling the ability of the soil, and to understand the effects of cultivations and rotations in the long term. The removal of nutrients depends on harvesting, leaching, volatilisation and erosion. The quantity of nutrients removed by crops is determined for many crops and these values will be beneficial for the required amounts of nutrient sources in farming practice. The level of maximising returns to the system and the availability of nutrients to crops are therefore determined by the rate at which nutrients are recycled within the system. In other words, an organic farmer should consider the law of return when managing agricultural practices. The law of return is a great opportunity to build up soil fertility. All non-harvested natural and organic materials will return to the soil according to the law of return. Nevertheless, care should be taken because contamination of heavy metals and hazardous substances should be discarded. The waste materials must be properly composted to destroy pathogenic organisms, and farmers should check and balance the adequacy of minerals to build up fertile soil, as recycled wastes often do not provide nutrients themselves. Suppression of soil-borne pathogenic micro-organisms by using mulches and organic amendments is also possible; where organic materials are spread on the soil surface, attributes for the term "mulches", and incorporated organic materials, are considered to be soil amendments (Thurston 1997; Stinner 2007; Heckman et al. 2009). Slash/mulching systems are ubiquitous for soil fertility, accumulation of soil organic matter and effective use of fertiliser in the field. For instance, the direct use of phosphorous fertiliser results in a decreased yield of avocado rather than incorporating the fertiliser into the mulches. Moreover, mulching protects erosion, controls weed growth, increases infiltration of soil, and promotes biological activity of the soil. Application of organic amendments for long periods increases the carbon rate up to 90% more than unfertilised soil, and up to 100% more than inorganic fertiliser treatments (Chianu et al. 2012).

Nitrogen recycling for an organic grower is an opportunity, since nitrogen is an essential building block of living tissue, and a component of soil organic matter. Nitrogen constitutes 78% of atmospheric air and nitrogen-fixing bacteria turn it into a usable form by using legumes. A widely known micro-organism is the *Rhizobium* bacteria species. Conversely, conventional farmers take nitrogen from the air by means of the Haber-Bosch process and use large amounts in fertilising. Well-integrated organic farms can be self-sufficient when incorporating legumes and livestock production into crop rotations. Legumes are, therefore, one of the recommended mulching practices in organic agriculture (Thurston 1997; Heckman et al. 2009).

When there is a nutrient deficiency occurring in crops, amending practices should be assessed through the physical and biological conditions of the soil, the balance of the rotation or intercropping and the proper practices of farming before using fertilisers, whether their source is organic or inorganic. Combining inorganic and organic fertilisers, therefore, improves the fertility of soil and increases crop yields, according to researchers (Oad et al. 2004). Specifically, soils which have a low level (<1%) of soil organic matter (SOM) might be combined with inorganic and organic fertilisers to increase the rate of SOM during the three years of conversion from conventional to organic farming (Barker 2010).

In short, soil must be biologically active, and nutrients should be given regularly, in a controlled manner, because organic matter accumulation in the soil, which might take a decade, is crucial for sustaining the soil's biological activity. It will be achieved by not only supplying nutrients, but also retaining and enriching the plants and organisms in the field (Heckman et al. 2009; Lampkin et al. 2017). The uptake of crop nutrients will be summarised in the following section: physical and chemical parameters of the soil.

Physical and Chemical Parameters of Soil

The role of soil in supplying nutrients to plants is governed by releasing nutrients in a suitable form and at an optimum rate that plants can use. Nutrients are taken up by plants, primarily in the form of electrically charged ions, such as NO_3^-, $H_2PO_4^-$, K^+, Mg^{2+}, Ca^{2+} and SO_4^{2-}. These ions should be deposited in the soil, and soil should hold them safely, and make them available for use, as needed, by crops. Soil colloids, which are negatively charged absorptive sites of soil, hold positive ions (cations), such as Ca^{2+}, Mg^{2+}, K^+, Na^+, H^+, Al^{3+}, Fe^{2+}, Mn^{2+}, Zn^{2+} and Cu^{2+} through electrostatic forces (negative soil particles attract positive cations). This ability of the soil is measured by the cation exchange capacity (CEC). The largest amounts of cations are calcium, magnesium and potassium used by plants. Soils have CEC because clay particles and organic matter in the soil tend to be negatively charged (Ketterings et al. 2007; Barker 2010; Brown & Lemon 2017).

Clay particles and humus (decomposed organic matter) contribute to soil cation exchange capacity. When compared, organic matter has 4–50 times higher CEC than clay particles per given weight, owing to having carboxyl groups that dissociate negatively charged sites. This makes a significant contribution to the cation exchange capacity, even at low levels of organic matter. A study found that the total soil carbon and nitrogen, as well as CEC, were higher in compost treatment (total N input was 1648 kg N ha^{-1}) than high manure (total N input was 2690 kg N ha^{-1}) treatment. Compost was prepared by using greenhouse waste (discarded plants with potting media) and yard waste, while spring incorporation of broiler litter was used as high manure. Organic acid dissociation depends on the soil's pH, because net negative charge caused by organic acids is balanced by cations in the soil. Therefore, CEC is very dependent on the soil's pH. Neutral soil (pH ~7) has a higher CEC than acid soil (pH 5). Sandy soils that have a low amount of organic matter have a very low CEC, while heavier clay soils or soils that have a high amount of organic matter have much greater CEC, which means they have higher water- and nutrient-holding capacity than do low CEC soils. Soil pH is very important for crop physiology and compost treatment resulted in a greater increase in CEC than high manure treatment. pH might decrease to undesirable levels (<5.5) for raspberries, due to some other parameters such as high rainfall, low CEC, ammonium-based fertilisers and low organic matter in the soil. The compost treatment was also beneficial for many root diseases by increasing the calcium availability in the soil. Moreover, soil electrical conductivity (EC) was found to be high (0.37 $mScm^{-1}$) in compost treatment, directly affecting suppression of salt accumulation in early growth of crops (Forge et al. 2016).

Nutrient-enriched fertilisers result in increased ratios of micronutrients in plants, which prevents soil deficiencies, increases yields, and adds value to human health. Selenium and zinc intakes are increased in the populations of Finland and Turkey, for instance (Green et al. 2016).

Increasing organic matter in soils located in tropical regions, such as Bangladesh, to a high level ($>1\%$) can be carried out by applying integrated nutrient management. For this purpose, inorganic fertilisers and organic amendments are proposed (Saleque et al. 2004). The combined use of organic matter and inorganic fertilisers provides increased yields in the short term, and enhances soil organic matter in the long term (Wei et al. 2016). Growers might prefer to use organic fertiliser to increase the soil's organic matter over time, where inorganic fertilizers have been used as a source of nutrients, until after three years of not using inorganic fertilisers they can be certified as an organic grower by the certifying authorities (Barker 2010). Organic agricultural practices have resulted in higher soil organic matter and provided more aggregates than conventionally cultivated soil. Moreover, the porosity of the soils is greater at the meso- and micro-scale, comprising smaller pores, than at the macro-scale because of the fewer large pores in conventionally farmed soil (Papadopoulos et al. 2006).

Soil organic matter is a heterogeneous mixture, and degradability and turnover rates vary, and need to be determined for nutrient cycling, such as for nitrogen (Tian et al. 2016). Soil organic matter is found in various situations and performs different functions in the soil. It can be grouped into three types: i) fresh and incompletely decomposed plant and animal residues that are still active; ii) organic residues that are in an advanced state of decomposed products and products that are re-synthesised by micro-organisms; and iii) humic substances that have high molecular weight. Humus, which is a widely-cited substance in organic agriculture, contributes to the latter two groups. Humus contributes to soil structure by giving rise to high cation exchange capacity, and by having the ability to complex with clays containing amino, hydroxyl and carboxyl humic compounds. Humic substances act as sources of nutrients, whilst resisting pathogenic organisms. Plant roots have the role of moving nutrients through the soil to crops. They need to be stimulated by minerals, notably nitrogen and phosphorous, to branch out and explore new areas. Organic agriculture proposes the movement of nutrients to the deep areas of roots in the soil, where surface fertilising leaves crops vulnerable to harmful conditions owing to plant leaves growing on the soil surface in an attempt to reach nutrients. The role of nutrients in plant growth is not limited to serving as a source of growth, but also puts up resistance to other plants or stimulating the growth of nematodes and micro-organisms. The movement of nutrients is part of the activity of organisms as well as physical and chemical reactions. Metabolites of microbial growth ease the availability of nutrients, such as phosphate for plants (Barker 2010).

Fertilisers

Fertilisers compensate for the lack of nutrient sources in the soil, because soil does not have an unlimited capability to nourish crops. Organic fertilisers are naturally-derived materials of biological or mineral origin (geological resources), and have low nutrient concentrations or solubility properties. Permitted minerals in organic agriculture include lime, gypsum, rock phosphate, guano, sulphur (S), dolomite and various kinds of ground silicate minerals. Green manure, animal manure and compost are farm-made organic materials, whereas fish, blood and bone meal, seaweed extracts and microbial products are mostly off-farm organic material sources. While inorganic fertilisers are synthetically (purification, chemical treatment) derived materials from various sources, such as air (N source) and rock dust (phosphorous source) (Kristiansen et al. 2006; Barker 2010), inorganic fertilisers are readily available in the market, and sold more cheaply than organic fertilisers, where the grower must use organic fertilisers in bulk amounts. For instance, a tonne of farm manure or compost contains equal or smaller amounts of nutrients than an 80lb bag of inorganic fertiliser. The price is also varied among the organic fertilisers, where growers pay USD2.40 per lb total nutrients of dried cow manure; USD1.89 per lb total nutrients of seed meals, and USD6.67 per lb total nutrients of dried blood (Barker 2010).

However, the main difference between organic and inorganic fertilisers is the release rate of the fertilisers. Synthetically-produced inorganic fertilisers dissolve rapidly in water, while organic ones require biological activity, or need strict physical and chemical reactions for the release of elements in ionic forms. In other words, organic fertilisers are converted slowly into the ionic forms of nutrients that plants absorb from the soil solution. Therefore, the release rate of the nutrients is much more important than the amount of nutrients added to the soil. The carbon and nitrogen ratio affects the rate of release of nutrients into the soil. For instance, the C/N ratio should be 25:1 to 40:1 for composting raw materials. And the temperature must be between 55°C and 77°C for three days for in-vessel/static aerated pile systems, or for fifteen days in the case of windrow-composting systems. Organic material having a high ratio of carbon to nitrogen (C/N) must be carefully added to the soil (Heckman et al. 2009; Barker 2010). For example, paper has 5000/1, whereas sawdust has 500/1, and grass hay has 33/1. These are called carbonaceous materials, not fertilisers, because they lack sufficient nitrogen to be mineralised or decomposed, whereas micro-organisms need available nitrogen to mineralise a given material (Barker 2010).

Consumption or decomposition of nitrogen by micro-organisms is called immobilisation, and immobilised nutrients are not available for nutrition of plants until the micro-organisms die. The use of newly-seeded crops or transplants can be supplemented with a high-quality and low C/N ratio compost during the initial five weeks. Therefore, carbon-rich materials should be composted in order to stabilise organic matter and reduce nitrogen pollution (Heckman et al. 2009; Barker 2010).

The quantities of nutrients are affected by the texture and organic content of the soil because of the turnover of nutrients, energy supplied by organic matter and chemical weathering processes taking place in the soil. It is a requirement for organic fertilisers that biological activity occurs in the living soil to provide nutrients through a release process called mineralisation (Barker 2010). Micro-organisms play an active role in the recycling of elements, such as sulphur, ferrous and nitrogen. Nitrogen is a key element for microbial cells and is subjected to three processes: nitrification, denitrification and nitrogen fixation (Madigan & Martinko 2012).

Microbial cells supply the nitrogen source from nitrogen fixation (N2+8H → 2NH3+H2) from air, which can be called a farmer-friendly process. Certain prokaryotes, and some aerobic and anaerobic micro-organisms, can fix nitrogen. Some of these bacteria need a host whilst some of them do not. Both legume (soybean, pea, clover) and non-legume plants can serve as hosts for symbiotic bacteria, such as *Rhizobium*, *Bradyrhizobium*, *Sinorhizobium*, *Mesorhizobium*, *Azorhizobium* and *Photorhizobium*. Species of these bacteria, when colonised on a legume plant, stimulate the formation of root nodules, which are used in the fixation of nitrogen. Therefore, the symbiotic relationship between the legume plant and bacteria fix nitrogen into an available form (ammonium or nitrate), which results in an increased nitrogen rate in the soil. In other words, bacteria need the leghemoglobin protein present in the nodules for immobilisation of O_2 because the enzyme used in nitrogen fixation, nitrogenase, must survive under oxygen (protect from inactivation). The protection of a cell by a protein complex is called conformational protection. These bacteria must be genetically eligible to stimulate the formation of nodules by plants. Nitrogen fixation is simply and effectively measured by observing the reduction of acetylene to ethylene by nitrogenase activity in gas chromatography (Madigan & Martinko 2012).

In nitrogen fixation, nitrogen is reduced to ammonium, and ammonium is converted into an organic compound. The release of nitrogen from a nitrogenous organic content (amino acids, nucleotides) is termed nitrogen mineralisation or ammonification and can be carried out by several micro-organisms. Ammonia (NH_3) is the first compound released from an organic source and is oxidised by *Nitrosomonas* genius bacteria into nitrite (NO_2^-); nitrite is oxidised by *Nitrobacter* genius bacteria into nitrate (NO_3^-), which is called the nitrification process. Nitrate is released by nitrification, whereas it is consumed in denitrification (NO3- → N2) by *Bacillus*, *Paracoccus*, *Pseudomonas* bacteria. These obligate aerobic bacteria are widely found in the soil and water, and the nitrification process occurs under aerobic conditions. However, when these bacteria are grown on the reduced nitrogen substrates, ammonia can be oxidised under anoxic conditions. This is called anammox (anoxic ammonia oxidation), which is carried out by *Brocadia*, and is beneficial for removing anoxic nitrogen in polluted waters, which is not desirable in soils with the denitrification process (Heckman et al. 2009; Barker 2010; Madigan & Martinko 2012).

Silicate minerals, which have varied contents of elements and dissolution rates, are permitted by organic certification standards as organic fertiliser and soil ameliorants (Kristiansen et al. 2006). Silicates are the major rock-forming minerals of almost all igneous and metamorphic rocks. The term "rock-forming minerals" is attributed to the major and essential minerals associated with each type of rock (Harley & Gilkes 2000). Silicate minerals are also a source of micronutrients for plant growth. When taking into account that organic fertilisers have low amounts of P and K, sources of silicate minerals, such as rock phosphate and guano, might be used in cases of having inadequate quantities of manure to replace the P loss, in organic farming systems. Moreover, highly weathered soils, acidic tropical soils, nutrient deficient soils and leaching soils (e.g. sands) are the proper fields for silicate minerals to be used as fertiliser (Kristiansen et al. 2006).

The dissolution of silicate minerals is determined by size of the material and surface area. Dissolution is increased in the soils that have low pH, high moisture and temperature, and have no equilibrium with mineral surfaces. Heavy rainfall events also favour the dissolution of silicate minerals in the soil. It is not only physical and chemical processes in the soil that affect the dissolution, but also plants and micro-organisms increase silicate mineral dissolution. Organic ligands are released by those organisms that attack mineral surfaces and form complexes, and result in a decreased pH in the soil by released H^+ ions and organic acids in the soil. Therefore, it is a result of silicate minerals having a slow release rate, which is documented in the literature (Kristiansen et al. 2006).

The interaction between silicate minerals and micro-organisms is well-documented. Fungal species of *Aspergillus*, mycorrhizal or non-mycorrhizal growth on the silicate minerals (e.g. quartz, apatite, potassium feldspar and marble) results in the increased mineralisation rate of K, for instance, (Manning 2010) because the slow release of minerals results in decreased yields owing to being less readily available in the plants. Therefore, *Bacillus megaterium* var. *phosphaticum* might be used as a phosphate-solubilising bacteria to convert insoluble phosphate in rocks to soluble forms to enhance the availability of plant uptake (Supanjani et al. 2006).

Over 80% of plants form some type of symbiotic association with mycorrhizal fungi. Mycorrhizal fungi colonise plant roots, absorb nutrients and water from the soil, and usually obtain plant photosynthates in return. About six thousand fungal species may form ectomycorrhizal associations with the plants where the plants growing in symbiosis with an ectomycorrhizal fungus (e.g. *Hysterangium* spp., *Hydnellum* spp., *Gautieria* spp.) are defined as ectomycorrhizal plants. The expanding mycorrhizal mycelium exploits a larger soil volume than the plant roots have. These micro-organisms enable the uptake of plant nutrients in two ways: enzyme production and organic acid excretion. Available forms of nitrogen (NH_4^+) and phosphorous ($H_2PO_4^-$) are provided by degradation of high molecular weight substances, such as proteins, carbohydrates, nucleic acids by extracellular enzymes of fungi; and the mobilisation of P, K, Ca, and Mg from solid mineral substrates is provided by organic acid excretion. Ectomycorrhizal mycelium is used to transport the dissolved nutrient elements to the host plant (Landeweert et al. 2001).

To sum up, the condition of the soil must be properly maintained for microbial life. For instance, the pH of the soil should provide the optimum condition of organisms, as well as availability of nutrients. Low pH results in reduced biological activity, the release of nutrients and slower turnover of organic matter. Biologically active soil is dependent on micro-organisms, insects, mites, various invertebrates and earthworms. Besides biological resources, careful management of soil, water and nutrients is essential for building a bio-diversified soil. Cultivated species of major crops should be increased—no more than seventy plant species are cultivated over a total area of approximately 1440 million hectares, globally. Crop rotations, cover crops, intercropping, agroforestry and crop/livestock mixtures will all result in an increase in diversity. Moreover, correctly practised biodiversity will bring about natural pest regulation for controlling insect pests, diseases and nematodes. And preserving the activity of soil biota will provide optimal nutrient recycling and soil conservation (Bavec & Bavec 2006).

Thorough management of soil facilitates the productive use of fertiliser. A high porosity rate (>48%) and increased water- and nutrient-holding capacity of the soil will enhance the prevention of fertiliser leaching with the downward movement of water. This is considerably reduced when using organic fertilizers, owing to their slow release. Hence, slow release might be a problem for nutrient-deficient crops. Besides this disadvantage, having low nutrient concentrations of organic fertilisers forces farmers to use bulk amounts in the field as compared to synthetic fertilisers. Before choosing an organic fertiliser, its rate of release (mineralisation) of nutrients should be taken into consideration. For example, cow manure releases nitrogen more slowly than do seed meals (cottonseed meal, linseed meal, soybean meal) and dried blood. Precautions such as one to two weeks between fertilising and seedlings should be taken against ammonium accumulation in the soil to avoid ammonium toxicity in a crop. The release rate also affects the humus content of the soil in one growing season. Cow manure contributes to humus content in the soil, whilst dried blood makes no significant contribution (Heckman et al. 2009; Barker 2010), whereas humus is relatively stable to mineralisation among soil organic matters. For instance, the mineralisation rate of nitrogen in humus is about 1 or 2% in a six-inch depth of soil (in the plough layer). In a fertile mineral soil in the plough layer that has about 4000 lb of nitrogen, the humus would therefore be mineralised to 40–80lb of available nitrogen per acre in a year (Chianu et al. 2012).

Fertiliser use, which has been the subject of recent studies, will now be summarised. In the field, there are many applications for organic fertiliser use. Biochar is one of the C-rich products that has recently been studied. It is derived from plants by pyrolysis to enhance soil quality. It has much slower rates of decomposition than other organic amendments, but lacks a significant source of plant nutrients such as inorganic N. Having an aromatic structure provides its mineralisation to CO_2 at a low rate (approx. 6% decayed during the first 8.5 years). Therefore, combining organic fertilisers with biochar seems a promising process strategy to enhance soil quality and crop yields towards sustainable agricultural management (Plaza et al. 2016). The addition of Biochar to the soil can differentially affect individual soil organic matter fractions. For instance, it results in an increase in dissolved organic C content, microbial biomass C and light fraction organic C, when compared to unamended soil (Tian et al. 2016).

Nutrient management in soil depends on environmental conditions, such as crop rotation, fertilisation and soil biology characteristics. Single nutrient application might result in deficiencies in plants, which can be seen primarily when inorganic fertilisers are used rather than organic fertilisers. The effect of using combined chicken and cattle manure has been demonstrated through the increased height of maize, for instance (Oad et al. 2004). Feeding livestock with poor quality materials results in an N content below 2%, whereas the nutrient content of animal manures is periodically controlled in North America and Europe through legislation. The availability coefficients of nutrients are present in the Manure Management Planner software, which is developed by Purdue University. The coefficients provide better application rates for crop production. Therefore, they are of paramount importance as the knowledge of fertiliser equivalency values of organic materials minimises the risk of excessive and insufficient nutrition of crops (Chianu et al. 2012).

Combining organic fertiliser (composed of amino acid fertilizer and pig manure composts) with antagonistic micro-organisms, such as *Bacillus subtilis*, *Paenibacillus polymyxa* and *Trichoderma harzianum*, to control *Fusarium* wilt of cucumber plants, was found to be useful for effective biocontrol. A disease caused by *Fusarium* was decreased 83% by means of bio-fertiliser application (Qiu et al. 2012). Additionally, another type of bio-fertiliser has been used to increase the availability of plant nutrients. Phosphorous-Potassium (PK) rocks are inoculated with PK-solubilising bacteria (*Bacillus megaterium* var. *phosphaticum* and *Bacillus mucilaginosus* strain KCTC3870) to enhance the nutrient uptake, growth and production of hot pepper plants (*Capsicum annuum* L.). In low fertility soils this method is proposed for sustainable fertiliser use as an alternative to inorganic fertilisers for hot pepper production (Supanjani et al. 2006).

Composts made from manures have lower total nitrogen contents, and less readily available forms of nitrogen than raw manures and, as a result, may not lead to nitrate leaching. A compost made from a mixture of poultry manure and other feedstocks does not pose the same risk of nitrate leaching as does manure, a study concluded (Forge et al., 2016). Overuse and improper use of fertilisers result in water contamination, soil acidification and greenhouse gas emissions (Green et al. 2016).

The rice-milling process results in an inedible waste product known as rice husk, which, by using an earthworm, *Eudrillus eugeniae,* is bio-transformed into an organic fertiliser, combined with the use of market-rejected fruits (banana, honeydew, or papaya). The results of one study concluded that bio-transforming of rice husk through vermicomposting is a promising strategy for organic fertiliser use (Lim et al., 2012).

Sewage sludge is used as a beneficial source of organic matter in crop planting. However, it carries risks, such as heavy metals (Cd, Pb, Hg, Zn), organic pollutants, pathogenic micro-organisms, malodours, ammonia and genetic damage to different crops. Therefore, taking into account the European Commission (EC) regulation, 1907/2006, composting sewage sludge supplies waste management and helps to turn it into a soil conditioner (Rossini-Oliva et al. 2017).

Acknowledgements: I would like to thank Scribendi editing and proofreading service (Canada) for proofreading this chapter.

References

Barker, A V. 2010. “Science and Technology of Organic Farming.” CRC Press, Boca Raton FL.

Bavec, F, and M Bavec. 2006. “Organic Production and Use of Alternative Crops.” CRC Press, Boca Raton FL.

Brown, K, and J Lemon. 2007. “Cations and Cation Exchange Capacity.” *Soilquality.org.au*. http://www.soilquality.org.au/factsheets/cation-exchange-capacity.

Chianu, J N, J N Chianu, and F Mairura. 2012. “Organic Fertilizers in Sub-Saharan Farming Systems.” In: Organic Fertilisation, Soil Quality and Human Health. Ed: E. Lichtfouse. Sustainable Agriculture Reviews 9, doi: 10.1007/978-94-007-4113-3_3. Springer Science+Business Media B. V.

Forge, T, E Kenney, N Hashimoto, D Neilsen, and B Zebarth. 2016. “Compost and Poultry Manure as Preplant Soil Amendments for Red Raspberry: Comparative Effects on Root Lesion Nematodes, Soil Quality and Risk of Nitrate Leaching.” *Agriculture, Ecosystems and Environment* 223. Elsevier B.V.: 48–58. doi:10.1016/j.agee.2016.02.024.

Green, H, P Broun, I Cakmak, L Condon, N Fedoroff, J Gonzalez-Valero, I Graham, et al. 2016. “Planting Seeds for the Future of Food.” *Journal of the Science of Food and Agriculture* 96 (5): 1409–14. doi:10.1002/jsfa.7554.

Halberg, N, and A Muller. 2013. “Organic Agriculture for Sustainable Livelihoods.” Published by Routledge, Abingdon, OX.

Harley, A D, and R J Gilkes. 2000. “Factors Influencing the Release of Plant Nutrient Elements from Silicate Rock Powders: A Geochemical Overview.” *Nutrient Cycling in Agroecosystems* 56: 11–36.

Heckman, J R, R Weil, and F Magdoff. 2009. “Practical Steps to Soil Fertility for Organic Agriculture.” In: Organic Farming: The Ecological System. Agronomy Monograph 54. Ed: Charles Francis. Published by American Society of Agronomy, Crop Science Society of America, Soil Science Society of America. Madison, WI.

Ketterings, Q M, S Reid, and R Rao. 2007. “Cation Exchange Capacity (CEC).” *Cornell University Cooperative Extension*, 1–2. doi:10.1080/01431160305010.

Kristiansen, P, A Taji, and J Reganold. 2006. “Organic Agriculture: A Global Perspective. CSIRO Publishing, Collingwood VIC, Australia.

Lampkin, N, M Measures, and S Padel. 2017. "2017 Organic Farm Management Handbook." Published by Organic Research Centre, Newbury, UK.

Landeweert, R, E Hoffland, R D. Finlay, T W. Kuyper, and N van Breemen. 2001. "Linking Plants to Rocks: Ectomycorrhizal fungi mobilize nutrients from minerals. *Trends in Ecology and Evolution* 16 (5): 248–54.

Lim, Ta Yeong Wu, Edwin Yih Shyang Sim, Pei Nie Lim, and Charles Clarke. 2012. "Biotransformation of Rice Husk into Organic Fertilizer through Vermicomposting." *Ecological Engineering* 41. Elsevier B.V.: 60–64. doi:10.1016/j.ecoleng.2012.01.011.

Madigan, M T, and J M Martinko. 2012. "Brock Biology of Microorganisms." Eleventh Ed. Pearson Education, Inc. Publishing as Prentice Hall. Indiana, USA.

Manning, D A C. 2010. "Mineral Sources of Potassium for Plant Nutrition. A Review." Agronomy for Sustainable Development, Springer Verlag/EDP Sciences/INRA, 2010, 30 (2), <10.1051/agro/2009023>.<hal-00886529>.

Oad, F C, U A Buriro, and S K Agha. 2004. "Effect of Organic and Inorganic Fertilizer Application on Maize Fodder Production." *Asian Journal of Plant Science* 3 (3): 375–7.

Papadopoulos, A, N R A Bird, A P Whitmore, and S J Mooney. 2006. "The Effects of Organic Farming on the Soil Physical Environment." *Aspects of Applied Biology* 79: 263–7.

Plaza, C, B Giannetta, J M Fernández, E G López-de-Sá, A Polo, G Gascó, A Méndez, and C Zaccone. 2016. "Response of Different Soil Organic Matter Pools to Biochar and Organic Fertilizers." *Agriculture, Ecosystems and Environment* 225: 150–59. doi: 10.1016/j.agee.2016.04.014.

Qiu, M, R Zhang, C Xue, S Zhang, S Li, N Zhang, and Q Shen. 2012. "Application of Bio-Organic Fertilizer Can Control Fusarium Wilt of Cucumber Plants by Regulating Microbial Community of Rhizosphere Soil." *Biology and Fertility of Soils* 48 (7): 807–16. doi:10.1007/s00374-012-0675-4.

Rossini-Oliva, S, M D Mingorance, and A Peña. 2017. "Effect of Two Different Composts on Soil Quality and on the Growth of Various Plant Species in a Polymetallic Acidic Mine Soil." *Chemosphere* 168: 183–90. doi:10.1016/j.chemosphere.2016.10.040.

Saleque, M A, M J Abedin, N I Bhuiyan, S K Zaman, and G M Panaullah. 2004. "Long-Term Effects of Inorganic and Organic Fertilizer Sources on Yield and Nutrient Accumulation of Lowland Rice." *Field Crops Research* 86 (1): 53–65. doi:10.1016/S0378-4290(03)00119-9.

Stinner, D H. 2007. "The Science of Organic Farming." In: Organic Farming An International History. Ed: W Lockeretz. CABI Publishing. Oxfordshire, UK.

Stockley, C, C Oxlade, and J Wertheim. 2007. "The Usborne Illustrated Dictionary of Science." Published by Usborne Publishing Ltd, London, England.

Supanjani, H S, H S Han, J S Jung, and K D Lee. 2006. "Rock-Phosphate Potassium and Rock-Solubilising Bacteria as Alternative, Sustainable Fertilisers. *Agronomy for Sustainable Development* 26: 233–40. doi:10.1051/agro:2006020.

Thurston, H. D. 1997. "Slash/Mulch Systems: Sustainable Methods for Tropical Agriculture." Westview Press. Boulder, CO.

Tian, J, J Wang, M Dippold, Y Gao, E Blagodatskaya, and Y Kuzyakov. 2016. "Biochar Affects Soil Organic Matter Cycling and Microbial Functions but Does Not Alter Microbial Community Structure in a Paddy Soil." *Science of the Total Environment* 556. Elsevier B.V.: 89–97. doi:10.1016/j.scitotenv.2016.03.010.

Wei, W, Y Yan, J Cao, P Christie, F Zhang, and M Fan. 2016. "Effects of Combined Application of Organic Amendments and Fertilizers on Crop Yield and Soil Organic Matter: An Integrated Analysis of Long-Term Experiments." *Agriculture, Ecosystems and Environment* 225. Elsevier B.V.: 86–92. doi:10.1016/j.agee.2016.04.004.

CHAPTER 13

PHENOMICS

SAIMA SIDDIQUE,[1] MADEEHA HAMID,[1] ALVINA GUL,[1] NAMRA HAQ,[1] RABIA AMIR,[1] VOLKAN ALTAY,[2] MUNIR OZTURK[3*]

[1] Atta-ur-Rahman School of Applied Biosciences, National University of Sciences and Technology, Islamabad, Pakistan
[2] Biology Department, Faculty of Science & Arts, Hatay Mustafa Kemal University, Hatay, Turkey
[3] Botany Department & Centre for Environmental Studies, Ege University, Izmir, Turkey
munirozturk@gmail.com

Abstract

Highly productive genotypes of new farming crops adapted to climate change are required to cope with issues like food security. A perfect goal of phenomics is to plug the gap between plant function, genomics and agronomic characters. Acknowledgment of the restrictions of existing methods in phenomics has motivated interest in high-throughput phenotyping procedures that can be used to precisely portray huge numbers of appearances. High-throughput phenotyping of ideal plants, like *Arabidopsis*, utilising non-invasive imaging tools, is a fast-evolving field. Field-based phenotyping (FBP) is being progressively recognised as the only approach adept at delivering the requisite throughput in periods of figures of plants or populations, as well as an unquestionable description of trait signs in real-world cropping schemes. The aim in FBP is at simultaneous proximal feeling for spectral reflectance, canopy warmth, and vegetation architecture. Large numbers of imaging techniques, like three-dimensional (3D), far infrared (FIR), near infrared (NIR), fluorescence and magnetic resonance imaging (MRI) are being used to screen large numbers of plants for traits or phenotypes of interest. With innovative global-scale technologies evolving for the toolbox of plant functional genomics, accepted mutant screening is now on the verge of transitioning from small-scale to non-biased, large-scale phenotype recognising mutant collections. On the other hand, plant phenomics is proposing the solution for speeding up vegetation breeding of crops in order to increase the yield of crops.

Keywords: phonemics, plants, genomics, climate, crops, stress

Introduction

Global agricultural production is facing a huge challenge regarding the provision of food for continuously growing populations (Ray et al. 2015). An unexpected increase in the prices of primary food commodities in 2008 raised questions about securing the food supply for a growing world population, which is estimated to reach nine billion by 2050 (Godfray and Garnett 2014). Worldwide security of food, the identification of applicable and effective plant-based biofuel raw materials, and managing with climatic changes are at the forefront of the minds of politicians, scientists and local people. Worldwide farming and the plant biology supporting it are facing major challenges and need novel methods for practical genomics and herbal propagation (Tester and Langridge 2010). On the other hand, the Food and Agriculture Organization (FAO) has estimated that in order to feed such a growing population, developing countries need to increase production of food by 77% (Ye et al. 2016). Hence, yields of fuel and food need to be improved using limited resources (Fahlgren, Gehan, and Baxter 2015). In order to enable these matters, we need highly productive genotypes of new farming crops adapted to climate change. Farming crops which satisfy future fuel and food requirements must show both extraordinary essential yield and yield constancy under abiotic pressures. Yearly growths in yield achieved from conventional global breeding programmes are no longer adequate to meet the expected demand for all three major cereal crops (Fig. 1): maize (*Zea mays*), wheat (*Triticum aestivum*) and rice (*Oryza sativa*) (Tester and Langridge 2010).

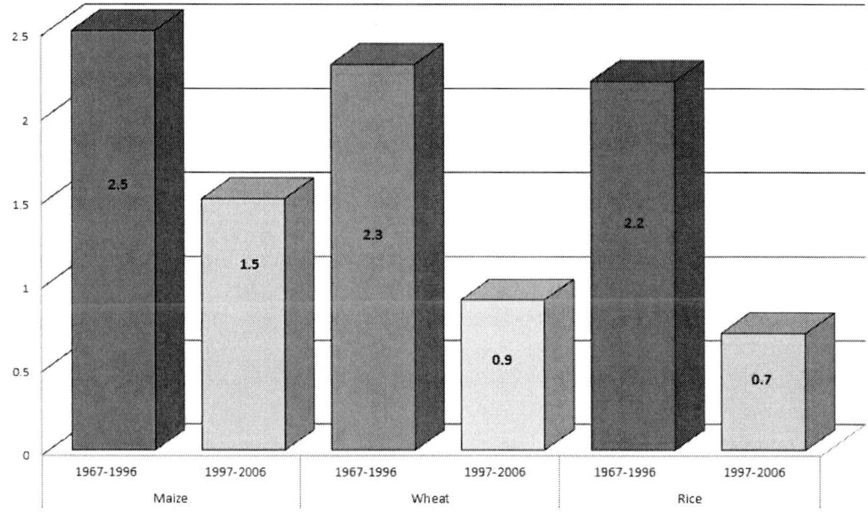

Fig. 1. Annual increase (%) in grain produced

Study to improve phenotyping methods is termed "phenomics"; the term "-omics" is from the plant sciences (Furbank 2009).

Phenome = Gene x Environment or the expression of the genome as characters in a specified environment (Furbank and Tester 2011).

Plant phenomics is the study of plant composition performance and growth (Furbank and Tester 2011). Phenomics involves high-throughput phenotyping for the improvement of crops with respect to different demographic and climate scenarios (Yang et al. 2013). The phenome of a plant can be defined by three general groups: the biological, chemical and physical phenotypes (Kuromori et al. 2006). Phenomics studies involve data related to physiological traits (photosynthetic activity and water efficiency), biochemical traits (anthocyanin content) and morphological characteristics (leaf area, fruit number, and plant height) (Gago et al. 2014, Paulus et al. 2014, Rahaman et al. 2015). Genomic data is increasing at a fast pace due to faster and cheaper sequencing methods. Genomic breeding, genotyping by sequencing, and high-throughput genotyping have become standard in plant breeding (Kuijken et al. 2015b).

Field-Based Phenomics for Plant Genetics Research

The agronomic yield needs to be in accordance with population demands—the population is estimated to exceed nine billion by 2050, which will exceed yield. This leaves plant scientists and crop specialists with the intensive task of making phenotypic forecasts using genetic organisation of cultivars or lines (White et al. 2012). By linking phenotype to genotype, highly productive, stress-resistant plants can be chosen, both for celerity and proficiency, more than is currently possible. Remarkable developments in "next generation" DNA sequencing are fast reducing genotyping rates (Jackson et al. 2011). On the other hand, plant phenotyping has become a fashion over the last three decades; nonetheless, attaining appropriate, usable and pertinent phenotypic knowledge, based on a particular plant, remains an exigent task (White et al. 2012). This is apt with regards to composite characters such as from yield perspective and abiotic stress resistance, which are important as regards the improvement of crops and, eventually, viable production (Myles et al. 2009). The practical and functional use via genome collection (Jannink, Lorenz, and Iwata 2010) or by employing genome-wide linked studies (Myles et al. 2009) will similarly involve phenotyping of thousands of lines that are heritably distinct (either indicated or linked populations) being developed as facsimilia across various different environmental conditions. This is practiced in order to both assess and evaluate differences in expression patterns of various different genes (i.e. expression of a genotype when it comes in contact with its ambience) (Furbank et al. 2009, White et al. 2012).

Forward and Reverse Phenomics

Forward phenomics is a method of discovering the plant species that are the "best of the best". Reverse phenomics is separating the "best" species to find out why they are the best (Weirman 2010). Widespread research on reverse genetics approaches were arrayed for allocating biological function to a particular gene or genetic product (Ben-Amar et al. 2016). Forward phenomics practices phenotyping tools to "sieve" groups of germplasms for valued characteristics. The sieve could be high-throughput, completely mechanical and of low resolution, trailed by lower-throughput, higher-resolution measurements. Screens might contain biotic or abiotic pressure tests and must be replicable and of biological significance (Furbank and Tester 2011). Forward phenomics quickens up plant breeding by screening a huge number of plants at the seedling period using mechanical imaging skill. This makes it quicker to classify remarkable characters, as the plants are skipped from growing into an adult. A large number of plants can be separated in pots moving along a conveyor belt, and drifting through a room comprising mechanical imaging methods, such as 3D and infrared cameras. The pots are tagged with radio tags or barcodes, so that the method can find those pots containing plants with remarkable characters. The particular plants can then be grown to produce seed for future breeding and analysis (Weirman 2010).

In reverse phenomics, the chosen characteristic—such as drought resistance—is first identified (Weirman 2010). Reverse phenomics is the thorough partition of characters shown to be valuable to expose systematic understanding and permit novel manipulation of this tool. This can include decrease of a biological characteristic to biophysical or biochemical procedures and eventually a genetic factor or genes (Fig. 2) (Furbank and Tester 2011). Scientists then practice the methods that regulate the characters or genes that control them. Once the gene has been recognised, plants with the specific genes can be separated, or the gene can be reproduced into a new species. Phenomics quickens the breeding of plants via automatic systems, which can be used to separate several plant species in less time (Weirman 2010).

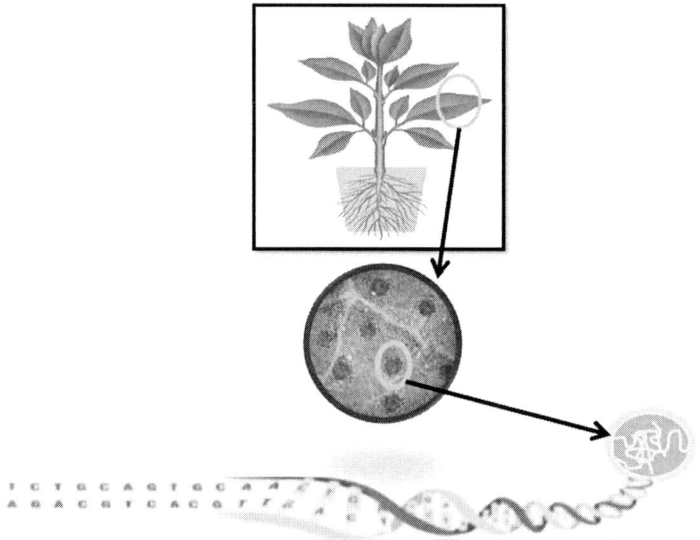

Fig. 2. Plant phenomics is the study of plant growth, performance and composition (Furbank and Tester 2011).

Acknowledgment of the restrictions of existing methods in phenomics has motivated interest in high-throughput phenotyping procedures that can be used to portray huge numbers of appearances or single plants precisely, and that involve cost, labour and time of existing methods (Montes, Melchinger, and Reif 2007).

High-Throughput Phenomics of Model Systems: The Phenomics-Genomics Pipeline

A perfect goal of phenomics is to link the gap between plant function, genomics and agronomic characters. Mainly in the perspective of ideal methods, where accessibility of genomic structure is increasing, there is an urgent requirement for a phenotypic database that is searchable, and that links plant structure, composition, performance and development to gene sequence, all measured in an openly definite environment (Furbank and Tester 2011, Ilic et al. 2007).

Worldwide, significant reduction occurs in agricultural productivity due to plant diseases. Such plant diseases affect growth and development of plants, making them unfit for consumption and reducing their yield at the same time. Knowledge, regarding the physiological mechanisms linking the infection caused by a plant pathogen and the disease symptoms seen in the host, is lacking for many plant-pathogen systems. For this purpose, large numbers of quantitative high-throughput, image-based methods are being developed, which will help in studying the quantitative effect of pathogen infection on host physiology (Mutka and Bart 2015).

Nowadays, multidisciplinary technologies, consisting of software and hardware, are being adopted in phenomics (Li, Zhang, and Huang 2014, Lin 2015) in countries like America, Belgium, Germany, Australia, France, China, Japan, England. Korea and India are making huge investments in phenomics, and have already set up large facilities to analyse various phenotypic traits of fruits, vegetables and grains. The most widely analysed traits include biotic-abiotic stress responses (Balachandran et al. 1997); yield-related traits (Duan et al. 2011); photosynthetic efficiency (Bauriegel, Giebel, and Herppich 2011); root morphology (Flavel et al. 2012, Kumar et al. 2014); leaf characteristics (Arvidsson, Pérez-Rodríguez, and Mueller-Roeber 2011); fruit characteristics (Monforte et al. 2014); and biomass (Golzarian et al. 2011).

Plant phenotypic data obtained from various phenomics platforms is growing exponentially, and it needs to be managed well before, during and after the research cycle (Yang et al. 2013). Management of phenotypic data obtained from various studies is a complex task. It is also the most critical process in crop improvement programmes (Kim et al. 2017). Data management basically involves management, storage and sharing of research data (Brown et al. 2014, Klukas, Chen, and Pape 2014). Proper data management is crucial for analysis of raw and metadata (Yang et al. 2013), and also for its optimal application. The outcome of research in part depends upon how well data is managed which, in turn, depends upon collection, storage and use of data during the research period (Li et al. 2013).

To create a significant database of phenotypic traits in ideal varieties, the first condition must be to have data accurately defined, possibly in a calculated, easily searchable and digitised format (Ilic et al. 2007). Secondly, the information on how the test was done, the plant type used, and the development conditions used must be noted in a consistent format (the "metadata"). Thirdly, proper reporting of the phenotyping methods used is needed to permit assessment of data collections. The insufficiency of these reports and normalisation of tentative management in the literature, and in existing databases, have severely restricted the value of both metabolic and transcriptional records (Hannemann et al. 2009).

Researchers have been working on plant phenotyping for decades. A huge effort has been required in the form of workforce and time in the case of traditional phenotyping. Recently, however, algorithm analysis and high-throughput imaging technology are being used to overcome the limitations of manual visual techniques (Li, Zhang, and Huang 2014).

High-throughput phenotyping of ideal plants, like *Arabidopsis*, utilising non-invasive imaging tools, is a fast-evolving field (Finkel 2009). Image examination and scientific management of imaging records to extract morphological traits, growth dynamics and spatially defined photosynthetic factors are thought-provoking, and need refined linkage and storage of calculated data and primary images. Use of numerical imaging and image examination to evaluate developmental changes from probable leaf part is benign, and is useful with regards to rosette plants, such as *Arabidopsis.* It can be achieved using the high-throughput technique in which, usually, a tray comprising twenty or more plants is employed (Granier et al. 2006, Walter et al. 2007). This developmental-imaging technique has already been employed in order to scrutinise drought tolerance in *Arabidopsis* accessions (Furbank and Tester 2011, Granier et al. 2006). This technique has recently been furthered by employing pulse-modulated chlorophyll fluorescence imaging in order to study photosynthetic reactions in response to drought stress in *Arabidopsis*; in addition to developmental change being brought to light (Furbank and Tester 2011, Jansen et al. 2009). High-throughput image study for programmed phenotyping is used to reveal numerous phenotypic factors, like development, production and pressure tolerance of the plants.

In this way, the laborious manual study of a variety of phenotypes can be reduced (Clark et al. 2013).

Phenomics Ontology Driven Data Management (PODD)

Phenomics studies utilise both low and high-throughput imaging and calculating devices to capture records, which are subsequently used for study. As a result, large volumes of records are made on a daily basis, making annotation, storage, distribution and management a taxing job. Adequate related information, the metadata, must also be retained to enable the distribution of these records. The task is further complicated by the necessity of maintaining evolving tools and procedures in phenomics studies. The objective of PODD is to capture, manage, annotate and dispense the records created by phenotyping stages. It maintains both Australian and international genetic research groups by providing source and information publication facilities. PODD is envisioned to be an open platform that permits any consumer to access records that are either printed or openly shared with them by the record holders. Furthermore, PODD is intended to offer information management facilities for a huge variety of phenotyping platforms and study projects (Li et al. 2010).

PODD is also envisioned to facilitate both animal and plant phenomics and, even though related to Xemlab in concentrating on ontologies and metadata linked to a web-based, graphical user interface, it employs diverse software keys to attain these objectives. For plant uses, it is envisioned to provide a tool for recovery and archiving of phenotypic records created from spectral, imaging analysis and a huge collection of biological (phenomics) data, created from the Australian Plant Phenomics Facility (Furbank and Tester 2011).

Phenomics is a fast developing, record-intensive discipline in biology with novel skills and procedures quickly evolving and emerging. As a consequence, its information management systems and domain model must also be able to advance to handle the dynamics, scale and complexity of the data (Li et al. 2013).

- In phenomics, dossiers comprise knowledge being garnered via both low and high-throughput techniques. The scale of study begins at the cellular level (micro level), and extends to the point where we consider one particular individual at the highest level, i.e. ground level (macro level) (Li et al. 2013).
- The scaling, imaging and scrutinising of plants on such a massive scale provides us with a huge sum of information. As for plants, the

intention is to focus on developmental expertise involving phenotypes that start from analysing the genome of a particular plant and then extending further, underpinning its developmental stage, ambience and disease basics. Novel imaging techniques employed at the High Resolution Plant Phenomics Centre (HRPPC), a section of Australian Plant Phenomics Facility (APPF), makes it practically possible to conduct extensive analysis and imaging of a plant's functional role and morphological comprehension (Li et al. 2013).

- By graphical imaging plant traits, like leaf biomass, the structure, damage and area can be measured and known. Distant infrared imaging reveals the shapes of leaves and can measure the definite content of water in soil substrates and plant tissue. Hyper-spectral imaging and FTIR explain the existence and spreading of chemical materials at the cellular level.
- Collectively, these stages can provide us with an overall summary regarding the phenome of plants, especially when studied over several growing ranges or time slots. HRPPC facilitates the study of various different plant shapes comprising distinct genotypes, which can further lead us towards their phenotypes (Li et al. 2013).
- Facts and figures based on ecological settings and management of the plant shapes are recorded as well, and are made accessible in the form of continuous metadata, as well as the data to preserve the key phenotyping data and analysis being conducted. This includes data obtained from plant growth sections and cabinets; meticulous environment; mechanical watering and treatment systems; and sensor records from field trials (Li et al. 2013).
- These wide-ranging records stacking for an individual plant source are further typified in the form of PODD as an amalgamation of data streams and metadata. The PODD region model explains the plant genotypes (genotype); records generation platforms (platform); the single plants that are phenotype (material); their growth ambience (environment) and their related systems (including both treatment material and treatment).
- Subsequently, PODD not only directs but also discharges the design of data entities (information) in order to manage data record groups generated by platforms, and also in order to bring it to a subsidiary level with the plant ingredients. Lastly, PODD transmits the objects of examination (analysis) in order to capture secondary data garnered via the study conducted on the main data archives (Li et al. 2013).

Phenomics Research

The High Resolution Plant Phenomics Centre (HRPPC) has applied a workflow organisation for observing its metadata, phenomics records, and testing procedures (Fig. 3). This composite task was completed by forming a multidisciplinary collection of IT experts, and embedding managers with phenomics skills into it. The outcome of this method is a state-of-the-art records-mining and art-computational environment, enhanced for data sharing, data discovery and data access, which also delivers the tractability for relating genomic information by using RDF triples. CSIRO Data Access Portal (DAP) has a part to play in interpreting and storing processed and raw datasets. DAP also delivers extensive secure storage for records collections, and the capability to examine, govern access to, and mention them via Digital Object Identifiers. PODD accomplishes the recording of collections found in DAP to PODD projects, providing for the storage of huge documents and images unsuitable for RDF databases (Ansell et al. 2013). Lately, the Leibniz Institute of Plant Genetics and Crop Plant Research and the German Plant Phenotyping Network have mutually introduced the Plant Genomics and Phenomics Research Data Repository (PGP) as a setup to broadly distribute the plant study information (Arend et al. 2016).

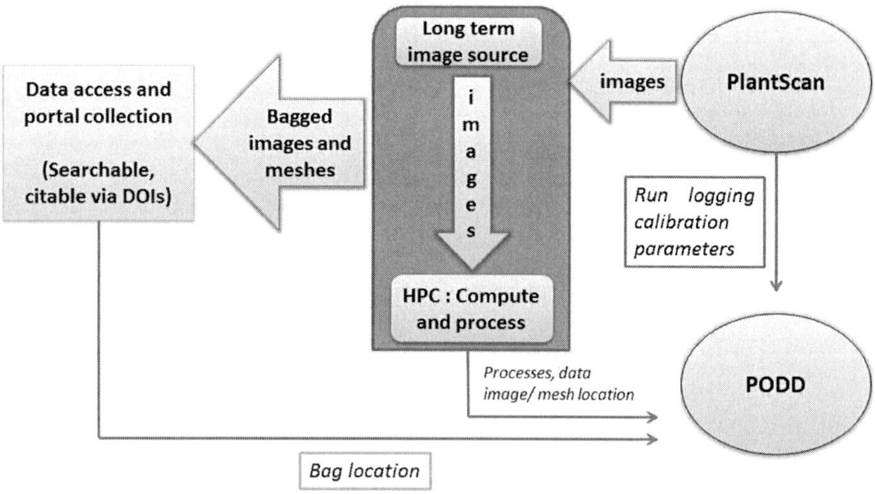

Fig. 3. HRPCC components relationships and data flow (Ansell et al. 2013)

In Australia, some phenomics projects are under way at the High Resolution Vegetation Phenomics Centre (http://www.plantphenomics.org.au/HRPPC); and a project to fast-track gene breakthrough in wheat by utilising *Brachypodium* as a form for origin architecture and infection traits is taking place at the Commonwealth Scientific and Developed Study Association in Canberra (Brkljacic et al. 2011).

Field-Based Phenotyping

Mapping populations for foremost crops may comprise 20 or more families, each comprising as many as 200 lines, necessitating field tests with over 20,000 plots in a single place. Buying the assets and labour needed to quantify even a couple of agronomic traits for linkage with genetic markers in such huge populations is actually impractical for most breeding programmes (White et al. 2012) Field-based phenotyping (FBP) is being progressively recognised as the only approach adept at delivering the requisite throughput in periods of figures of plants or populations, as well as an unquestionable description of trait sign in real-world cropping schemes (Babar et al. 2006). The aims in FBP are centred on simultaneous proximal feeling for spectral reflectance, canopy warmth, and vegetation architecture where a vehicle, bearing replicated sets of sensors, notes facts and figures on multiple plots with the potential to record data all through the crop life cycle. The potential to consider traits, such as adaptations to water deficits or acute heat tension, at times during a single diurnal cycle is particularly precious for quantifying stress recovery. Simulation modelling and related devices can help estimate physiological traits, such as canopy conductance and rooting (White et al. 2012).

Root exudation research is lacking a reliable and fast phenotyping method (Kuijken et al. 2015a). Inaccessibility of soil matrix (Kuijken et al. 2015b), binding of exudates to soil matrix (Radersma and Grierson 2004), microbial degradation of exudates around the root (Kuijken et al. 2015a), and the complexity of root traits hampered the genetic studies of root traits. Recently, however, scientists and breeders have overcome this limitation by making use of automated phenotyping techniques (Kuijken et al. 2015b).

Imaging Techniques Used in Phenomics

Phenomics scrounges imaging methods from surgery in order to permit those conducting studies to understand the inside functioning activities of leaves, roots, or even the entire plant. Various different imaging techniques can be employed in order to screen a vast variety of plants for desirable traits or phenotypes, such as drought or saline tolerance, to name a few. As plants are not laid to waste, they can be further used more than once at various different stages of growth. Phenomics imaging techniques include:

- To examine the structure and form of vegetation, three-dimensional (3D) imaging is used.
- To estimate the temperature (heat being produced) of leaf, far infrared imaging (FIR) is used.
- To estimate the amounts of water of other parts (departs) and soil (roots), near infrared imaging (NIR) is used.
- To analyse photosynthesis and overall health of vegetation fluorescence imaging is used.
- To analyse the development of root in dirt magnetic resonance imaging (MRI) is used.
- To examine the chemical make-up of vegetation, spectral reflectance is used (Weirman 2010).

Employing Three-Dimensional Imaging (3D) Technique

The digital photos of both the apex and ends of plants are taken using a line of cameras employing prominent light weight. These images are joined together in a 3D format of the vegetation using a computer programme. Once a 3D format is obtained, various measurements can be conveniently devised, without going back and looking at the vegetation (Weirman 2010). These include the following attributes:

- Leaf hue.
- Leaf number, shape and bend.
- Shoot mass.
- Leaf wellbeing—if it is living, dying or dead (Weirman 2010).

3D imaging, along with other software, can be used to analyse the root structure. CT scan, X-ray and MRI are some of the techniques which can be used to study the structural heterogeneity of soil and nutrient uptake phenomenon in roots (Janodia et al. 2013).

Infrared Imaging

Infrared thermography as a surrogate for measurements of transpiration and photosynthetic response to tension shows a large pledge in high-throughput vegetation phenomics, and can be used both in controlled environments (Qiu, Omasa, and Sase 2009) and in the field (Jones et al. 2009). The conductance of leaf stomata in a variety of genes can mainly be studied through infrared imaging. It can also be used to screen the distorted populations as it covers large canopy areas (Prashar et al. 2013). The strength of this approach is that it allows high-throughput screening at the seedling stage to be validated at the canopy level in a field environment, utilising the identical genotypes and devices. Due to its large chilling effect, evaporation is commonly the main determinant of leaf temperature, and there is a direct connection between leaf warmth, transpiration rate, and stomatal conductance (Jones et al. 2009). Dehydration-tolerant genotypes, and genotypes with a greater capability for water uptake can sustain a higher stomatal conductance and also a higher rate of photosynthesis (Medrano et al. 2002). These genotypes could, therefore, be recognised as being the plants which have the cooler departs. Canopy temperature has been used effectively as an assortment criterion in breeding programmes for drought prone environments (Balota et al. 2008). Colour images have been used effectively to recognise sunlit and shaded departs in a canopy, and to work out the respective leaf warmth (Leinonen and Jones 2004). The combination of hue and thermal imaging thus promises to boost the data and precision of warmth measurements compared with thermal imaging solely, making this approach apt for discussions on the issues of leaf orientation and canopy structure (Berger, Parent, and Tester 2010).

a. Far Infrared (FIR) Imaging

This technique is employed to study the temperature (warmth) of the plant part. It makes use of the far IR region of the spectrum. This technique is used to compute variations either in internal parts of vegetation or between different parts of plants. The variations in warmth can be used to examine salinity (salt) tolerance. For instance, it has been noted that the temperature (warmth) of wheat and barley leaf enhances as saline levels increase. Furthermore, it was found that the leaves that are able to maintain low temperatures in the saline environment are better able to endure salt tolerance, and thus reveal better prospects in salted soils. It has been made possible to estimate the salt stress in young seedlings rather than of other plants using FIR, thus enabling us to conduct phenotyping in a comparatively limited time frame (Weirman 2010).

Furthermore, FIR can also be employed to estimate those plants with cooling canopies in the fields. The plants that depict cooling properties are the ones that typically absorb large amounts of water compared with their counterparts (Weirman 2010).

Additionally, FIR cameras have the capacity to measure differences in stomatal conductance, which provides us with an estimated value at which water transpires through the stomata, leaves, or pores of plants. Thus, stomatal conductance can provide us with an estimation at which pace photosynthesis is occurring in plants under study (Weirman 2010).

b. Near Infrared (NIR) Imaging

This technique is used to estimate water content and their mode of action in plant parts. NIR cameras typically work by throwing light in the NIR region of the spectrum (900–1550 nanometres). The investigations are made possible by growing plants in unusually diaphanous pots so that plant roots can be imaged during the period of plant growth. The NIR measurements of soil reveal how much water is being absorbed from the source, and how much is being consumed by the vegetation. Normal digital images of the sections, termed origins, can be superimposed on NIR images in order to estimate how these origins have been used. Additionally, the NIR imaging technique can also be used to estimate the carbohydrate contents of plant leaves, thus can also be used to estimate the rate at which plants alter carbon dioxide to synthesise carbohydrate during photosynthesis (Weirman 2010).

In the near infrared (NIR) region of the spectrum, physically sound plants depict marked spectral images compared with their weak and fragile counterparts; nevertheless, in clear regions of the spectrum, both cannot be very well differentiated. Thus, this technique can be employed by researchers to gain insight into whether the plant is encountering some stress well before it is visible to the naked eye and the damage is irreparable. Thus, this can be employed at the seedling stage of plants in order to identify if the plant is facing some unfavourable conditions (Weirman 2010). This NIR technique is pollution-free, safe, fast-paced and cost-effective. Currently, in the field of agriculture, it has been successfully employed to pinpoint crop infection, production stress, adulterations and the vegetation mapping of various different types of plants (Dale et al. 2013).

Fluorescence Imaging Technique

Fluorescence imaging is employed to estimate both plants health and photosynthesis. Fluorescence occurs when any object absorbs lightweight of one particular wavelength and shows off lightweight of a typical wavelength. In plant phenomics, fluorescence imaging dusts azure lightweight wavelength (500 nanometres) on the plants in focus which, in turn, show in the lightweight zone (600–750 nanometres). A snapshot of this fluorescence can be taken and can be changed into colours other than true using a computer programme, so that researchers can easily notice differences in fluorescence; whilst merely about 1% of the lightweight dropping on plant fluoresces, this is enough to examine transport of water in the vegetation in order to evaluate the photosynthesis and tension rates (Weirman 2010).

Chlorophyll fluorescence is routinely utilised in phenomics to see the effect of different genes or ecological situation on the effectiveness of photosynthesis. It involves the photosystem II (PSII) of plants and does not invade this part. This technique is used to detect how a plant responds to biotic and abiotic changes in the environment (Murchie and Lawson 2013). Chlorophyll fluorescence is influenced by changes in photosynthesis. This can be observed before any other measurable change can be noticed in plants, thus it makes it extraordinarily convenient to evaluate whether plants are stressed or growing normally at the very early stage of plant growth and development (Weirman 2010).

Chlorophyll fluorescence has become comparatively reasonably priced, particularly with reference to leaf grade, as it has been one of the pieces of equipment routinely used; now, however, it has been supplemented with 2D digital imaging in order to conduct screening based on the phenotype of the complete plant so that the abiotic stress tolerance of it can be evaluated (Table 1) (Furbank 2009, Jansen et al. 2009). If 3D imaging (Hosoi and Omasa 2009) is brought into practice, the effects of herbicides on plants can be monitored (Furbank 2009, Konishi et al. 2009). Furthermore, chlorophyll fluorescence imaging has been employed to pinpoint the mutants in photorespiration (Badger et al. 2009), and the same technique can be used to examine foliar disease symptoms that can lead towards biotic stress resistance screening (Scholes and Rolfe 2009). Quite recently, in selections of mutant populations, particularly the ones being prepared using activation tagging (Ayliffe and Pryor 2009) to serve the purpose of examining biotic stress resistance, tension has typically been found to be gruelling, and thus the application of such modern approaches can actually be an alternative (Furbank 2009).

Table 1. Summary of non-destructive imaging methods to study physiological changes occurring during drought (Image source: New Scientist, June 2008)

Physiological changes occurring under drought	Non-destructive measurement method	Technical challenges
Decreased stomatal conductance	Thermal infrared	Influence of environmental conditions (air temp, wind speed, etc.)
		Orientation of leaves to light source and camera
Decreasing leaf water content	Near infrared	Low sensitivity
		Absolute measurements not possible at this stage, only monitoring of changes over time
Early onset of senescence	RGB visible or fluorescence if available	
Decreased growth rate and biomass accumulation	RGB / visible	
Decreased photosynthetic activity only during stress	Fluorescence	Pulsed imaging setups currently limited to about 100 cm^2

Magnetic Resonance Imaging (MRI)

Magnetic resonance imaging (MRI) is typically used to examine roots of vegetation with reference to plants. MRI takes into account the magnetic zone and wireless waves to snapshot images of origins in exactly the same way as it snapshots images of other parts, for example, organs and supply tissues, in health-linked projects; nevertheless, as the roots are not visible, unlike other parts, this makes them more difficult to examine, which is why, in contrast to other parts, we have limited knowledge about the facts pertaining to them, such as how they coalesce with surrounding soil and how the origin organisations occur. An attempt to break the origins somewhere deep actually ruins them, and makes it literally impossible to study how the origin system actually governs inside the soil or how it actually copes with filth around. Thus, in order to cast light on how origins work inside soil, these need to be grown in clear agar and, additionally, high-throughput can be employed; nevertheless, it does not provide the exact activities inside the soil, but can provide us with an inkling of them. MRI allows the 3D geometric patterns of origins to be examined in plants grown in the tubes filled with sand or soil; however, it is not a high-throughput method, unlike other methods that are usually used in phenomics, but is quite useful in corroborating the ones growing in the agar containers (Weirman 2010). Nevertheless, of those being grown in filth, screening for optimum structural development and growth has proven to be an exhaustive challenge (Gregory et al. 2009).

Reaching water levels deep down into the earth, to the point where certain cereal species are developing on accumulated filth moisture, has been given particular consideration, especially with regards to drought tolerance and the process of identifying species in order to throw light on genes that are responsible for origin-linked attributes that are under the process of development. Development of cereal crops with a shorter life span that are more sustainable are becoming a scientific fashion these days, typically with regards to phenomics screening (Watt et al. 2009). Advances, ranging from imaging in low amounts of filth or artificial media to MRI and X-ray CT-scanning, comes under the same umbrella being discussed (Faget et al. 2009, Furbank 2009, Gregory et al. 2009, Nagel et al. 2009, Yazdanbakhsh and Fisahn 2009).

Spectral Reflectance

Spectral reflectance is employed in order to determine the chemical composition of various different plant parts. Spectral reflectance makes use of only a fraction of light that is made to echo by an opaque outer surface, nevertheless, it is dependent on the part of the electromagnetic spectrum being used in order to examine reflectance as different types of materials produce different patterns of reflectance. This type of spectral pattern being obtained is known as the "spectral signature", which is dependent on the material-type in particular. Spectral reflectance can rapidly assess different plant traits under controlled conditions (Hackl et al. 2013).

Phenomics can use spectral reflectance technology to supervise several plant properties in the area at the identical time. These encompass the grades of water-soluble carbohydrates and nitrogen in leaves, stems and kernels

(Weirman 2010). Through this modern technique, both the function and chemical composition of various different parts of vegetation, ranging from the tissue to canopy can be examined non-invasively (Furbank 2009).

Major Applications of Phenomics
Mutant collection-Phenotypic profiling

Phenomics is that branch of contemporary biology which encompasses the huge diversity of plants with regard to their complex morphology. Classical genetics mainly focussed on the identification of mutant plants and then the identification of the gene causing that mutation. Thus, the scientific community laid emphasis on finding out the phenotype which revealed mutations and, ultimately, the gene being mutated in the phenotype under study. Nevertheless, with scientific advancement, it is now considered important to conduct scientific research in a holistic context to have an overall idea of the existing crisis, thus shifting the trend from merely analysing the single gene mutation, to adopting an all-embracing approach where phenotypic analysis of mutant collections is conducted. This type of phenotypic profiling not only allows for study of the chemical attributes at one particular stage of plant development but it also allows for study of different stages of plant development (Holtorf, Guitton, and Reski 2002). Assembling facts and figures on a certain set of morphological and developmental traits permits investigators to create a comprehensive picture of a mutant's phenome simultaneously (Boyes et al. 2001). Therefore, the reasonable approach is to not only have the mutant's phenotypic profile but also to have the one of the wild type noted down at the same time so that useful comparisons can be made. These contemporary approaches will definitely provide us with the complete picture, where both mutant and wild type species of plant will be taken into account. Almost a decade ago this approach was attempted on 141 different accessions of *Arabidopsis*, and the results obtained were appreciable; clear distinctive, characteristic patterns were obtained in response to differing ambiences set by the researchers (Maloof et al. 2001). Nevertheless, apart from these moderate scales attempted thus far by the scientific community, phenomics on a larger scale has been amongst one of the many foci of the business community, and ventured by them (Boyes et al. 2001). It has been shown that once a useful mutant has been identified, then the data generated, based on phenomics, can be interpreted in the light of the data being generated based on earlier approaches, such as transcriptomics, proteomics and metabolomics to get a complete picture of exactly how the mutant gene performs its job (Holtorf, Guitton, and Reski 2002).

Adapting to climate change

It has been observed that, in Australia, people have long been battling with drought, but in doing so have increased the salinity of the land. As the world's climate is constantly changing and is becoming more and more rigid and static towards crop plantation, it has been suggested by the scientific community that the solution to such an outcome is to develop crops that are able to cope with such environmental conditions. Therefore, it has been thought that, as phenomics is a holistic approach, it can duly serve the purpose of crop development that will be robust enough to thrive in such corrosive environments (Weirman 2010). The tasks include development of the following types of crops that will be better able to cope up with both drought and salinity stress:

- Drought-tolerant wheat.
- Salt-tolerant wheat and barley.

Drought-tolerant Wheat

Water consumption by plants is a multifaceted trait. The consumption of water depends not only on the growth and developmental stages of a plant, but also on its ambience. Wheat plant can reasonably compete with dry environmental conditions; nevertheless, it is important to examine how varieties of it thrive in the fields during the entire course of a season, characterised by plant growth (Weirman 2010).

The process of interpreting the data, received using the various techniques discussed thus far, can provide researchers with substantial information regarding the subject of exactly how plants cope with stress conditions when stumbled upon during the entire season. This study will provide us with the foundation to detect the species of wheat amongst the multitude that will be better able to thrive in the conditions of extreme water deficit-drought (Weirman 2010).

In Australia, various different research groups are working and collaborating with other leading research organisations around the world in order to effectively come up with the wheat varieties that will thrive during water shortage-drought (Weirman 2010).

Salt-tolerant Wheat and Barley

In Australia, nearly two thirds of the land is affected by salt excess-salinity. There, one of the world's leading institutes, CSIRO, is working on projects pertaining to the subject of salt overload-salinity. Researchers are looking for those varieties of wheat and barley tolerant to salt excess-salinity. The contemporary techniques mentioned above are being used extensively to achieve the desired goals (Weirman 2010).

Drought-resistant Bananas

Bananas are one of the important food crops that give us a potential amount of energy. Production of bananas on a large scale under limited water conditions is to be prioritised. Worldwide agricultural systems have shown that phenotyping drought-resistant bananas in germplasm can be used to improve the production of this staple food crop on a larger scale. Banana roots also play a very important role in signalling drought conditions, and study of all the root parameters through different phenomics techniques, may be able to recognise drought tolerance in relation to the banana plant. Moreover, the *Musa* genome (considered to have more drought tolerating activity) can be sequenced with the help of good phenotyping techniques (Ravi et al. 2013).

Phenotyping for Biotic and Abiotic Stress Tolerance

Agriculturalists are facing the dilemma of both yield and yield stability. The solution to this actually lies in screening of the varieties that are able to thrive in both biotic and abiotic stress conditions. To achieve these goals, application of modern phenomics can essentially serve the purpose (Munns et al. 2010, Munns and Tester 2008, Weirman 2010).

Phenomics-based Screening for Biotic Stress Resistance

Recently, modern techniques, such as digital imaging in the visible spectrum and chlorophyll fluorescence imaging have been employed to study the development of disease symptoms in plant parts (Scholes and Rolfe 2009). The chlorophyll fluorescence imaging, in particular, helps us in diagnosing the earlier symptoms of disease development (even before the symptoms can be traced with the naked eye); provides us with an estimation of how much of a region is affected; and also helps in the identification of the susceptible and resistant zones, for example, mildew on barley leaves at a very basic level (Chaerle et al. 2009, Swarbrick, Schulze-Lefert, and Scholes 2006). Digital imaging, on the other hand, does not increase its potential to be identified over the naked eye, but it does help us in practically quantifying the region being affected, in terms of lesion or chlorosis.

Imaging techniques, such as image arrest, image investigation, and hue classification, can be collectively applied to study the development of a lesion over the course of its development (Furbank and Tester 2011). Reduced transpiration, stomatal closure, probably making canopies hot, can be used as the indicators of some disturbance in xylem tissue. These indicators typically support non-invasive screening techniques, such as thermography computer display for measuring differences in transpiration rates in lieu of other, more severe, invasive methods (Furbank and Tester 2011, Sirault, James, and Furbank 2009).

Phenomics-based screening for Abiotic Stress Tolerance

The two main abiotic stress traits one is in search of are drought and stress tolerance (Munns et al. 2010, Munns and Tester 2008, Weirman 2010). There are times when these become similar to phenotypic effects, and thus the use of phenomics techniques for screening exhibits a significant crossover (Fig. 4). For example, stomatal closure is one of the very first signs of a plant having been exposed to salinity stress and, later, the ability of the roots to absorb water from the soil in the aftermath of such events (Furbank and Tester 2011, Munns and Tester 2008). This type of osmotic stress, in the aftermath of both drought and salinity stress, has been regarded as "chemical drought", nonetheless, the screenings, based on either photosynthetic parameters or stomatal conductance, are slow-paced and often found to be less reproducible (Munns et al. 2010). Similar to various plant phenomics devices, with regard to the situation under discussion, surrogate approximation can be employed to a computer display to study stomatal or photosynthetic responses to osmotic stresses. One example where the phenomics approach is used, in conjunction with surrogate approximation, is with regard to carbon isotope discrimination (referred to as carbon isotope discrimination (CID) in plant breeding), that indicates transpiration efficiency in both crop physiology and plant breeding (Richards et al. 2010, Weirman 2010). This technique is based on an observation made almost two decades ago that plants can distinguish between the heavier isotope of carbon (13C) and the one occurring naturally in atmospheric carbon dioxide, both with regard to the diffusion of CO_2 in the leaves of plants and in the photosynthesis-linked metabolic processes (Condon et al. 2004). This type of isotopic differentiation can be noticed with reference to the dry matter of plants and in C3 plantings. CID standards are considered to be the major players with reference to stomatal conductance and transpiration ability for photosynthesis (Condon et al. 2004, Richards et al. 2010). Besides the phenomenal help that this can provide, it has additionally been employed to locate genetic variation in wheat with reference to transpiration efficiency and exploitation of commercial varieties with far greater water-use efficiency and gross production (Condon et al. 2004). The prominent advantages of using this approach are that the samples can be garnered at the end of the growing period, and the isotopic composition reveals the integrated effect of the entire growing period, eliminating the need for evaluating the leaves and other parts at basic physiological stages. Nonetheless, the technique is exorbitant, which makes it a difficult option. Moreover, it is not known whether CID estimations can suffice to the readings being jotted down, typically at the seedling stage in a controlled environment. In recent times, infrared thermography has been successfully employed at the young seedling stage in wheat and barley in order to locate genotypes able to maintain stomatal conductance under osmotic stress (Weirman 2010).

Therefore, there are pros and cons associated with both pot and field trials as far as the traits evaluation is concerned. There will be some that will be more vigorous in the controlled environment whilst others will not exhibit the same energy. It will depend on the type of trait under study (Munns et al. 2010, Weirman 2010).

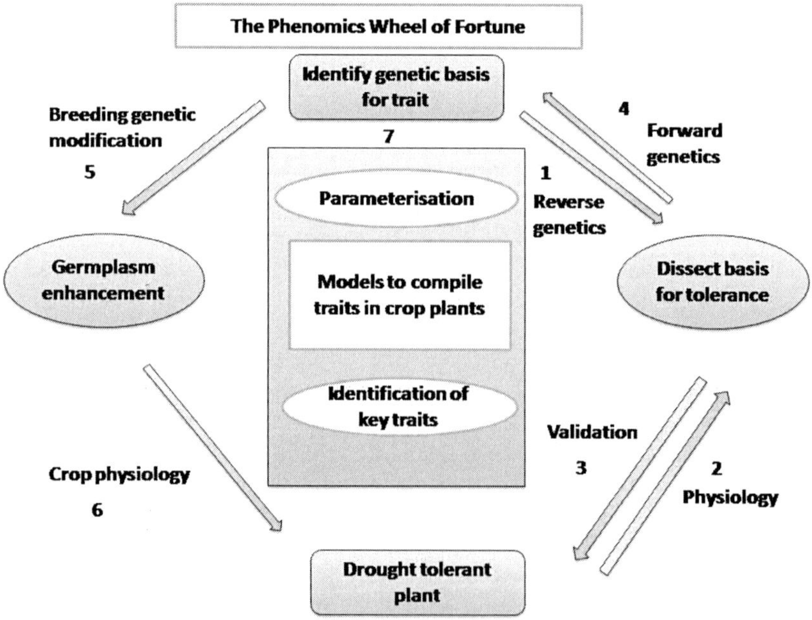

Fig. 4. Closing the gene to genotype loop with phenomics (Furbank and Tester 2011)

Crop yield improvement and enhancement

The FAO estimates that, due to the expected population explosion, by 2050 the world will need 70% more food in order to deal with the issues pertaining to food insecurity. Since the 1960s, scientists have fought a constant battle to achieve the goal of increased yield to meet the demands; as a consequence of their steadfast approach, a nearly 2.5% increase in overall crop yield has been observed annually; now, however, we are facing the issue of yield being static. Thus, in order to cope with the existing situation, phenomics warrants serious consideration. The following phenomics-based approaches are currently in fashion (Weirman 2010):

a. "Supercharging" photosynthesis.
b. Increasing wheat yield.
c. *Brachypodium*—the cereal "lab rat".

"Supercharging" Photosynthesis

In recent times, this has been attempted by rice experts to enhance rice yields in order to enhance their photosynthetic activity. There are two patterns of photosynthesis, namely, C3 and C4. Rice, wheat and other crop plants depict a C3 pattern of photosynthesis, whilst maize, sorghum and sugarcane depict a C4 pattern of photosynthesis. It has been observed that C4 plants perform better under stress conditions and are better able to cope with extreme temperatures, water shortages and limited nitrogen availability. It has been shown that, when provided with same amount of water and nitrogen, C4 maize increases two-fold compared to C3 rice. The reason behind this is the intricate cellular structure in C4 plants which makes them better able to concentrate CO_2 in their leaf interior than their counterpart. Thus, an option could be to substitute the C3 pattern with the more efficient C4 pattern by altering their genetic composition. The rice consortium has thus been minded to use phenomics in order to achieve its current goals (Weirman 2010).

Improving Wheat Yield

Researchers employing phenomics for isolation and development of better yielding crops have taken the lead in recent years. The activity of an enzyme, Rubisco, holds a very important place as far as the photosynthetic activity of crops is concerned. Some crop plants have better performing forms of Rubisco than others. Researchers are, therefore, in search of wheat varieties, using the phenomics approach, that possess better forms of Rubisco, and are also better able to cope with the usual plant stresses being discussed (Weirman 2010).

Phenomics researchers based in Australia are collaborating with the International Maize and Wheat Improvement Center in this research. As well as looking for wheat with improved photosynthesis, researchers are searching for varieties that will adapt best to land affected by environmental stresses, such as nutrient deficiency, drought and salinity (Weirman 2010).

Brachypodium—The cereal "lab rat"

Brachypodium distachyon is regarded as a "lab rat" amongst the scientific community due to its several genomic and phenotypic features. Thus, it can serve as model plant that has the potential and can greatly accelerate wheat research (Weirman 2010). Some of the advantages of it serving as a model plant are that its genome has been sequenced thus far; it depicts rapid growth rate; and some of its genetic composition are in line with that of wheat, at least in outline (Weirman 2010).

Application of Plant Phenomics to Trait-based Physiological Breeding

Recently, it has been considered more reasonable by the scientific community that, in order to study and quantify traits intended for physiological breeding, non-invasive options should be opted for, owing to their being more efficient (Furbank and Tester 2011). Reverse phenomics—where either only once being a useful trait or else yield components lying at the heart of fine crops are located then it is required to pinpoint the genomic region responsible for this trait by MAS in breeding—or else, with regards to multigene traits, a powerful phenotypic marker is indispensable (Richards et al. 2010, Weirman 2010), such as Ppd, Vrn, and Rht in wheat (Reynolds and Langridge 2016).

Application of Phenomics in Production of Biofuels

Biofuels are plant-based fuels that are mainly produced from plant materials—feedstocks. Biofuels from plants are the richest source of renewable energy, contributing to the production of butanol and ethanol, and long-chain hydrocarbons, from cellulose, starch oils and hemicellulose. Biofuels are produced mostly from crop plants, mainly corn and soybeans

(Weirman 2010)), but they can also be produced from other plants, such as rapeseed and sugar beet (Aro 2016). There is growing concern around the world that edible crops, when used as biofuel, will thus contend with the land meant for human food resources. Therefore, attempts have been made to use non-food plants as feedstocks for biofuel that can grow on lands that are comparatively less fertile, marginal lands; nevertheless, these plants need to be strong and robust enough to grow in unfavourable, hostile environments.

Switchgrass (*Panicum virgatum*), when exploited, has yielded promising results in this regard. It has the potential to be grown in lands that are limited in both water and nutrient resources; nevertheless, the downside is its lengthy life cycle and extensively intricate genome that make its breeding slow-paced.

Phenomics researchers are using *Brachypodium distachyon* as a model plant to attain celerity as its genome is somewhat similar to other related plants but less intricate, and its life cycle is brief compared with other crop plants. These attempts will serve as the ones underlying those looking towards the development of, not only cost-effective, but efficient biofuels (Weirman 2010).

Future Prospects

Current plant-breeding methods make use of large number of technologies to identify simple and multifaceted characteristics with increasing potential to accelerate the integration of useful genetic traits into parental germplasm. A large number of activities are being practiced in different areas of plant breeding to increase crop adaption to salinity, drought and water logging. There is a need for an ongoing evaluation of the potential for pathogens and pests to cause a shift in their geographic distribution, an increase in population, and to evolve into such forms that could present serious threats to production. In the case of major crop plants, there are prospects for genetic variations in temperature tolerance, plant phenology, and the effect of elevated CO_2, all of which can lead to an increase in yield, especially in localities where the temperature is predicted to rise (Chapman et al. 2012). In addition to the development of advanced phenotyping tools, there are also some other challenges in plant phenomics, like acquiring large quantity of data and images, data storage and its handling, and analysis (Yang et al. 2013).

Conclusion

Conventional techniques for plant phenotyping were destructive as well as laborious, and required removal of the biomass of a plant in order to carry out analysis. Recent advances in high-throughput imaging methods have made it possible to obtain quantitative data on plant characteristics. These recent technologies are non-destructive since they allow plant researchers to acquire a large number of images of the same plant at different wavelengths (Berger, Parent, and Tester 2010). Plant phenomics is a rapidly evolving area of plant research. It combines rapidly advancing imaging technologies with genomic and physiological data of plants. Hence, work is required in these scientific areas in order to incorporate expertise in plant physiology. In addition, high-throughput quantification of visible plant phenotypes has turned out to be a valuable tool for characterising the function of genes, for improvements in plant biomass production and growth performance (Yazdanbakhsh and Fisahn 2009). A large number of imaging and non-destructive spectroscopic techniques are accessible to evaluate plant function, chemical composition and photosynthetic performance (Furbank 2009, Furbank et al. 2009). Nevertheless, for breeding, selection of the parent should be carried out only after the results obtained in controlled settings are validated in real environments and real soil (Munns et al. 2010).

References

Ansell, P, R Furbank, K Gunasekera, J Guo, D Benn, G Williams, and X Sirault. 2013. "Flexible Scientific Data Management for Plant Phenomics Research." CEUR Workshop Proceedings.

Arend, D, A Junker, U Scholz, D Schüler, J Wylie, and M Lange. 2016. "PGP repository: a plant phenomics and genomics data publication infrastructure." *Database* 2016.

Aro, E-M. 2016. "From first generation biofuels to advanced solar biofuels." *Ambio* 45 (1):24–31.

Arvidsson, S, P Pérez-Rodríguez, and B Mueller-Roeber. 2011. "A growth phenotyping pipeline for *Arabidopsis thaliana* integrating image analysis and rosette area modeling for robust quantification of genotype effects." *New Phytologist* 191 (3):895–907.

Ayliffe, M A, and A J Pryor. 2009. "Transposon-based activation tagging in cereals." *Functional Plant Biology* 36 (11):915–921.

Babar, M A, M P Reynolds, M Van Ginkel, A R Klatt, W R Raun, and M L Stone. 2006. "Spectral reflectance indices as a potential indirect selection criteria for wheat yield under irrigation." *Crop Science* 46 (2):578–588.

Badger, M R, H Fallahi, S Kaines, and S Takahashi. 2009. "Chlorophyll fluorescence screening of *Arabidopsis thaliana* for CO_2 sensitive photorespiration and photoinhibition mutants." *Functional Plant Biology* 36 (11):867–873.

Balachandran, S, V M Hurry, S E Kelley, C B Osmond, S A Robinson, J Rohozinski, G G R Seaton, and D A Sims. 1997. "Concepts of plant biotic stress. Some insights into the stress physiology of virus-infected plants, from the perspective of photosynthesis." *Physiologia Plantarum* 100 (2):203–213.

Balota, M, W A Payne, S R Evett, and T R Peters. 2008. "Morphological and physiological traits associated with canopy temperature depression in three closely related wheat lines." *Crop Science* 48 (5):1897–1910.

Bauriegel, E, A Giebel, and W B Herppich. 2011. "Hyperspectral and chlorophyll fluorescence imaging to analyse the impact of *Fusarium culmorum* on the photosynthetic integrity of infected wheat ears." *Sensors* 11 (4):3765–3779.

Ben-Amar, A, S Daldoul, G M Reustle, G Krczal, and A Mliki. 2016. "Reverse genetics and high throughput sequencing methodologies for plant functional genomics." *Current Genomics* 17 (6):460–475.

Berger, B, B Parent, and M Tester. 2010. "High-throughput shoot imaging to study drought responses." *Journal of Experimental Botany* 61 (13):3519–3528.

Boyes, D C, A M Zayed, R Ascenzi, A J McCaskill, N E Hoffman, K R Davis, and J Görlach. 2001. "Growth stage–based phenotypic analysis of *Arabidopsis*: a model for high throughput functional genomics in plants." *The Plant Cell* 13 (7):1499–1510.

Brkljacic, J, E Grotewold, R Scholl, T Mockler, D F Garvin, P Vain, T Brutnell, R Sibout, M Bevan, and H Budak. 2011. "*Brachypodium* as a model for the grasses: today and the future." *Plant Physiology* 157 (1):3–13.

Brown, T B, R Cheng, X R R Sirault, T Rungrat, K D Murray, M Trtilek, R T Furbank, M R Badger, B J Pogson, and J O Borevitz. 2014. "TraitCapture: genomic and environment modelling of plant phenomic data." *Current Opinion in Plant Biology* 18:73–79.

Chaerle, L, S Lenk, I Leinonen, H G Jones, D Van Der Straeten, and C Buschmann. 2009. "Multi-sensor plant imaging: Towards the development of a stress-catalogue." *Biotechnology Journal: Healthcare Nutrition Technology* 4 (8):1152–1167.

Chapman, S C, S Chakraborty, M F Dreccer, and S M Howden. 2012. "Plant adaptation to climate change—opportunities and priorities in breeding." *Crop and Pasture Science* 63 (3):251–268.

Clark, R T, A N Famoso, K Zhao, J E Shaff, E J Craft, C D Bustamante, S R Mccouch, D J Aneshansley, and L V Kochian. 2013. "High-throughput two-dimensional root system phenotyping platform facilitates genetic analysis of root growth and development." *Plant, Cell and Environment* 36 (2):454–466.

Condon, A G, R A Richards, G J Rebetzke, and G D Farquhar. 2004. "Breeding for high water-use efficiency." *Journal of Experimental Botany* 55 (407):2447–2460.

Dale, L M, A Thewis, C Boudry, I Rotar, P Dardenne, V Baeten, and J A F Pierna. 2013. "Hyperspectral imaging applications in agriculture and agro-food product quality and safety control: A review." *Applied Spectroscopy Reviews* 48 (2):142–159.

Duan, L, W Yang, C Huang, and Q Liu. 2011. "A novel machine-vision-based facility for the automatic evaluation of yield-related traits in rice." *Plant Methods* 7 (1):44.

Faget, M, J M Herrera, P Stamp, I Aulinger-Leipner, E Frossard, and M Liedgens. 2009. "The use of green fluorescent protein as a tool to identify roots in mixed plant stands." *Functional Plant Biology* 36 (11):930–937.

Fahlgren, N, M A Gehan, and I Baxter. 2015. "Lights, camera, action: high-throughput plant phenotyping is ready for a close-up." *Current Opinion in Plant Biology* 24:93–99.

Finkel, E. 2009. "With 'phenomics,' plant scientists hope to shift breeding into overdrive." *Science* 325 (5939):380–381. doi: 1126/science.325_380.

Flavel, R J, C N Guppy, M Tighe, M Watt, A McNeill, and I M Young. 2012. "Non-destructive quantification of cereal roots in soil using high-resolution X-ray tomography." *Journal of Experimental Botany* 63 (7):2503–2511.

Furbank, R T. 2009. "Plant phenomics: from gene to form and function." *Functional Plant Biology* 36 (10):5–6.

Furbank, R T, and M Tester. 2011. "Phenomics–technologies to relieve the phenotyping bottleneck." *Trends in Plant Science* 16 (12):635–644.

Furbank, R T, S von Caemmerer, J Sheehy, and G Edwards. 2009. "C4 rice: a challenge for plant phenomics." *Functional Plant Biology* 36 (11):845–856.

Gago, J, C Douthe, I Florez-Sarasa, J M Escalona, J Galmes, A R Fernie, J Flexas, and H Medrano. 2014. "Opportunities for improving leaf water use efficiency under climate change conditions." *Plant Science* 226:108–119.

Godfray, H C J, and T Garnett. 2014. "Food security and sustainable intensification." *Philosophical Transactions of the Royal Society B: Biological Sciences* 369 (1639):20120273.

Golzarian, M R, Ross A Frick, K Rajendran, B Berger, S Roy, M Tester, and D S Lun. 2011. "Accurate inference of shoot biomass from high-throughput images of cereal plants." *Plant Methods* 7 (1):2.

Granier, C, L Aguirrezabal, K Chenu, S J Cookson, M Dauzat, P Hamard, J-J Thioux, G Rolland, S Bouchier-Combaud, and A Lebaudy. 2006. "PHENOPSIS, an automated platform for reproducible phenotyping of plant responses to soil water deficit in *Arabidopsis thaliana* permitted the identification of an accession with low sensitivity to soil water deficit." *New Phytologist* 169 (3):623–635.

Gregory, P J, A G Bengough, D Grinev, S Schmidt, W B TB Thomas, T Wojciechowski, and I M Young. 2009. "Root phenomics of crops: opportunities and challenges." *Functional Plant Biology* 36 (11):922–929.

Hackl, H, B Mistele, Y Hu, and U Schmidhalter. 2013. "Spectral assessments of wheat plants grown in pots and containers under saline conditions." *Functional Plant Biology* 40 (4):409–424.

Hannemann, J, H Poorter, B Usadel, O E Blaesing, A Finck, F Tardieu, O K Atkin, T Pons, M Stitt, and Yves Gibon. 2009. "Xeml Lab: a tool that supports the design of experiments at a graphical interface and generates computer-readable metadata files, which capture information about genotypes, growth conditions, environmental perturbations and sampling strategy." *Plant, Cell and Environment* 32 (9):1185–1200.

Holtorf, H, M-C Guitton, and R Reski. 2002. "Plant functional genomics." *Naturwissenschaften* 89 (6):235–249.

Hosoi, F, and K Omasa. 2009. "Detecting seasonal change of broad-leaved woody canopy leaf area density profile using 3D portable LIDAR imaging." *Functional Plant Biology* 36 (11):998–1005.

Ilic, K, E A Kellogg, P Jaiswal, F Zapata, P F Stevens, L P Vincent, S Avraham, L Reiser, A Pujar, and M M Sachs. 2007. "The plant structure ontology, a unified vocabulary of anatomy and morphology of a flowering plant." *Plant Physiology* 143 (2):587–599.

Jackson, S A, A Iwata, S-H Lee, J Schmutz, and R Shoemaker. 2011. "Sequencing crop genomes: approaches and applications." *New Phytologist* 191 (4):915–925.

Jannink, J, A J Lorenz, and H Iwata. 2010. "Genomic selection in plant breeding: from theory to practice." *Briefings in Functional Genomics* 9 (2):166–177.

Janodia, M D, S Dharmagadda, V Ligade, P Muragundi, P Musmade, and N Udupa. 2013. "High throughput plant phenotyping in agriculture." *Current Science* 105 (4):434.

Jansen, M, F Gilmer, B Biskup, K A Nagel, U Rascher, A Fischbach, S Briem, G Dreissen, S Tittmann, and S Braun. 2009. "Simultaneous phenotyping of leaf growth and chlorophyll fluorescence via GROWSCREEN FLUORO allows detection of stress tolerance in Arabidopsis thaliana and other rosette plants." *Functional Plant Biology* 36 (11):902–914.

Jones, H G, R Serraj, B R Loveys, L Xiong, A Wheaton, and A H Price. 2009. "Thermal infrared imaging of crop canopies for the remote diagnosis and quantification of plant responses to water stress in the field." *Functional Plant Biology* 36 (11):978–989.

Kim, S-L, N Solehati, I-C Choi, K-H Kim, and T-R Kwon. 2017. "Data management for plant phenomics." *Journal of Plant Biology* 60 (4):285–297.

Klukas, C, D Chen, and J-M Pape. 2014. "Integrated analysis platform: an open-source information system for high-throughput plant phenotyping." *Plant Physiology*:506–518.

Konishi, A, A Eguchi, F Hosoi, and K Omasa. 2009. "3D monitoring spatio–temporal effects of herbicide on a whole plant using combined range and chlorophyll a fluorescence imaging." *Functional Plant Biology* 36 (11):874–879.

Kuijken, R C P, J F H Snel, M M Heddes, H J Bouwmeester, and L F M Marcelis. 2015a. "The importance of a sterile rhizosphere when phenotyping for root exudation." *Plant and Soil* 387 (1–2):131–142.

Kuijken, R C P, F A van Eeuwijk, L F M Marcelis, and H J Bouwmeester. 2015b. "Root phenotyping: from component trait in the lab to breeding." *Journal of Experimental Botany* 66 (18):5389–5401.

Kumar, P, C Huang, J Cai, and S J Miklavcic. 2014. "Root phenotyping by root tip detection and classification through statistical learning." *Plant and Soil* 380 (1–2):193–209.

Kuromori, T, T Wada, A Kamiya, M Yuguchi, T Yokouchi, Y Imura, H Takabe, T Sakurai, K Akiyama, and T Hirayama. 2006. "A trial of phenome analysis using 4000 Ds-insertional mutants in gene-coding regions of *Arabidopsis*." *The Plant Journal* 47 (4):640–651.

Leinonen, I, and H G Jones. 2004. "Combining thermal and visible imagery for estimating canopy temperature and identifying plant stress." *Journal of Experimental Botany* 55 (401):1423–1431.

Li, L, Q Zhang, and D Huang. 2014. "A review of imaging techniques for plant phenotyping." *Sensors* 14 (11):20078–20111.

Li, Y-F, G Kennedy, F Davies, and J Hunter. 2010. *PODD: An ontology-driven data repository for collaborative phenomics research.* Verlag Berlin Heidelberg: Springer.

Li, Y-F, G Kennedy, F Ngoran, P Wu, and J Hunter. 2013. "An ontology-centric architecture for extensible scientific data management systems." *Future Generation Computer Systems* 29 (2):641–653.

Lin, Y. 2015. "LiDAR: An important tool for next-generation phenotyping technology of high potential for plant phenomics?" *Computers and Electronics in Agriculture* 119:61–73.

Maloof, J N, J O Borevitz, T Dabi, J Lutes, R B Nehring, J L Redfern, G T Trainer, J M Wilson, T Asami, and C C Berry. 2001. "Natural variation in light sensitivity of *Arabidopsis*." *Nature Genetics* 29 (4):441.

Medrano, H, J M Escalona, J Bota, J Gulías, and J Flexas. 2002. "Regulation of photosynthesis of C3 plants in response to progressive drought: stomatal conductance as a reference parameter." *Annals of Botany* 89 (7):895–905.

Monforte, A J, A Diaz, A Caño-Delgado, and E Van Der Knaap. 2014. "The genetic basis of fruit morphology in horticultural crops: lessons from tomato and melon." *Journal of Experimental Botany* 65 (16):4625–4637.

Montes, J M, A E Melchinger, and J C Reif. 2007. "Novel throughput phenotyping platforms in plant genetic studies." *Trends in Plant Science* 12 (10):433–436.

Munns, R, R A James, X R R Sirault, R T Furbank, and H G Jones. 2010. "New phenotyping methods for screening wheat and barley for beneficial responses to water deficit." *Journal of Experimental Botany* 61 (13):3499–3507.

Munns, R, and M Tester. 2008. "Mechanisms of salinity tolerance." *Annual Review of Plant Biology* 59:651–681.

Murchie, E H, and T Lawson. 2013. "Chlorophyll fluorescence analysis: a guide to good practice and understanding some new applications." *Journal of Experimental Botany* 64 (13):3983–3998.

Mutka, A M, and R S Bart. 2015. "Image-based phenotyping of plant disease symptoms." *Frontiers in Plant Science* 5:734.

Myles, S, J Peiffer, P J Brown, E S Ersoz, Z Zhang, D E Costich, and E S Buckler. 2009. "Association mapping: critical considerations shift from genotyping to experimental design." *The Plant Cell* 21 (8):2194–2202.

Nagel, K A, B Kastenholz, S Jahnke, D Van Dusschoten, T Aach, M Mühlich, D Truhn, H Scharr, S Terjung, and A Walter. 2009. "Temperature responses of roots: impact on growth, root system architecture and implications for phenotyping." *Functional Plant Biology* 36 (11):947–959.

Paulus, S, H Schumann, H Kuhlmann, and J Léon. 2014. "High-precision laser scanning system for capturing 3D plant architecture and analysing growth of cereal plants." *Biosystems Engineering* 121 (1):1–11.

Prashar, A, J Yildiz, J W McNicol, G J Bryan, and H G Jones. 2013. "Infra-red thermography for high throughput field phenotyping in *Solanum tuberosum*." *PLoS One* 8 (6):e65816.

Qiu, G Y, K Omasa, and S Sase. 2009. "An infrared-based coefficient to screen plant environmental stress: concept, test and applications." *Functional Plant Biology* 36 (11):990–997.

Radersma, S, and P F Grierson. 2004. "Phosphorus mobilization in agroforestry: Organic anions, phosphatase activity and phosphorus fractions in the rhizosphere." *Plant and Soil* 259 (1–2):209–219.

Rahaman, M, D Chen, Z Gillani, C Klukas, and M Chen. 2015. "Advanced phenotyping and phenotype data analysis for the study of plant growth and development." *Frontiers in Plant Science* 6:619.

Ravi, I, S Uma, M M Vaganan, and M M Mustaffa. 2013. "Phenotyping bananas for drought resistance." *Frontiers in Physiology* 4:9.

Ray, D K, J S Gerber, G K MacDonald, and P C West. 2015. "Climate variation explains a third of global crop yield variability." *Nature Communications* 6:5989.

Reynolds, M, and P Langridge. 2016. "Physiological breeding." *Current Opinion in Plant Biology* 31:162–171.

Richards, R A, G J Rebetzke, M Watt, A G T Condon, W Spielmeyer, and R Dolferus. 2010. "Breeding for improved water productivity in temperate cereals: phenotyping, quantitative trait loci, markers and the selection environment." *Functional Plant Biology* 37 (2):85–97.

Scholes, J D, and S A Rolfe. 2009. "Chlorophyll fluorescence imaging as tool for understanding the impact of fungal diseases on plant performance: a phenomics perspective." *Functional Plant Biology* 36 (11):880–892.

Sirault, X R R, R A James, and R T Furbank. 2009. "A new screening method for osmotic component of salinity tolerance in cereals using infrared thermography." *Functional Plant Biology* 36 (11):970–977.

Swarbrick, P J, P Schulze-Lefert, and J D Scholes. 2006. "Metabolic consequences of susceptibility and resistance (race-specific and broad-spectrum) in barley leaves challenged with powdery mildew." *Plant, Cell and Environment* 29 (6):1061–1076.

Tester, M, and P Langridge. 2010. "Breeding technologies to increase crop production in a changing world." *Science* 327 (5967):818–822.

Walter, A, H Scharr, F Gilmer, R Zierer, K A Nagel, M Ernst, A Wiese, O Virnich, M M Christ, and B Uhlig. 2007. "Dynamics of seedling growth acclimation towards altered light conditions can be quantified via GROWSCREEN: a setup and procedure designed for rapid optical phenotyping of different plant species." *New Phytologist* 174 (2):447–455.

Watt, M, K Schneebeli, P Dong, and I W Wilson. 2009. "The shoot and root growth of *Brachypodium* and its potential as a model for wheat and other cereal crops." *Functional Plant Biology* 36 (11):960–969.

Weirman, A. 2010. "Australian Plant Phenomics Facility. From Plant phenomics teacher resource." accessed 7 May 2013. http://www.plantphenomics.org.au/files/teacher/Final_Phenomics_for_word_with_images.doc.

White, J W, P Andrade-Sanchez, M A Gore, K F Bronson, T A Coffelt, M M Conley, K A Feldmann, A N French, J T Heun, and D J Hunsaker. 2012. "Field-based phenomics for plant genetics research." *Field Crops Research* 133:101–112.

Yang, W, L Duan, G Chen, L Xiong, and Q Liu. 2013. "Plant phenomics and high-throughput phenotyping: accelerating rice functional genomics using multidisciplinary technologies." *Current Opinion in Plant Biology* 16 (2):180–187.

Yazdanbakhsh, N, and J Fisahn. 2009. "High throughput phenotyping of root growth dynamics, lateral root formation, root architecture and root hair development enabled by PlaRoM." *Functional Plant Biology* 36 (11):938–946.

Ye, L-M, J-P Malingreau, H-J Tang, and E Van Ranst. 2016. "The breakfast imperative: The changing context of global food security." *Journal of Integrative Agriculture* 15 (6):1179–1185.

CHAPTER 14

PRANIC AGRICULTURE FOR ENHANCED CROP PRODUCTION

K. NAGENDRA PRASAD, N. JOIS SRIKANTH*, N.S PAPANNA

World Pranic Healing Foundation, India, Research Centre, Mysuru–570009, Karnataka, India
srikanth@pranichealing.co.in

Abstract

Pranic agriculture is an ancient science that uses prana to enhance plant growth. Prana is the energy needed by a plant for its growth. It involves the projection of prana to plants to promote faster development. The current study aims to understand the application of pranic agriculture practices on tomatoes, pole beans, cucumber, and European cucumber plants. Physical parameters, including the length of the plant, stem diameter of the plant, flower initiation, the time taken for fruit development and fruit yield were studied in pranic-treated plants against the control group. An increase in plant length of 18.5% and 49% was noticed in tomato and cucumber when compared with control groups. Also, the pranic group had 12–24% more stem diameter than the control. The pranic group had 31.7% additional flowers per plant when compared to the control in tomato. When, group-wise, crop yield was verified, the pranic yielded significantly more in tomato (31%), cucumber (17%), in European cucumber (18%), and pole beans (10.7%) than the respective control groups. This clearly shows that pranic agriculture could play a significant role in improving the quality and quantity of vegetables.

Keywords: prana, plant growth, flowering, vegetables, yield

Introduction

Pranic agriculture utilises prana to treat plants and is considered an ancient science. Prana is that vital energy that is essential to keep the body alive and in a healthy condition (Sui 1992). Prana is also referred to as Ki or bioplasmic energy. Sui (1992) specified that there are three major sources of prana, namely, ground prana, sun prana, and air prana. Water absorbs prana from the ground, sunlight, and air. Trees and plants absorb prana from the ground, sun, and air. Using Kirlian photography, investigators have been able to take photographs, observe and study pictures of prana or small bioplasmic objects, including bioplasmic leaves (Kirlian 1949).

Master Choa Kok Sui, the founder of Pranic Healing, suggested that we can feel or scan the prana and aura and use it to diagnose the problem areas. Clinically, Pranic Healing is useful in treating diseases, such as depression (Rajagopal et al. 2018), COPD (Mahesh et al. 2017), asthma (John 1995), musculoskeletal pain (Jain et al. 1999), and increases wellbeing by manipulating the energy fields of subjects (Tsuchiya and Motoyama 2009).

Pranic Healing is also practiced by well-educated followers to achieve personal and professional growth (Jauregui et al. 2012). It is gaining popularity in physics, biology, health, and agriculture.

Using Pranic Healing in agriculture is referred to as Pranic Agriculture. It cannot be used as a substitute for farming; rather it is complementary to it. It enhances the yield, improves the shelf-life of products and their nutritional quality. Recently, Asna et al. (2016) have reported that cucumber storage qualities increased in the pranic group significantly, against the control. Polyphenol content and antioxidant activities were also higher in pranic-treated cucumber. Seedling vigour test and germination of green gram were also higher in pranic treated seeds, against the control, and the results were significant (Jois, Prasad, and Roohie 2016a).

Method

The agricultural land and the seeds were given pranic treatment for a few minutes, whilst the remaining land and seeds, which did not receive any treatment, were referred to as the control. The plants were grown by conventional agricultural practices. The evaluation was done by measuring seed germination, plant height, stem diameter, and the number of flowers, the time taken to flowering, the number of flowers per truss, and vegetable yield.

Effect of Pranic Agriculture on Tomato Plants

The height of pranic-treated tomato plants was 114.2 cm when compared with the control (92.9 cm), and the results were found to be significant ($p < .001$) (Table 1). Plants treated with pranic energy had 18.5% more height than the control. The stem diameter of pranic group plants was 16 cm compared with the control (14 cm). The number of flowers per truss, and the number of flowers per plant, were more in the pranic group against the control. The pranic group had 31.7% additional flowers per plant when compared to the control. When the yield per row was analysed, the pranic group obtained a yield of 117.07 kg, whilst the control yielded only 80.58 kg. Plants treated with pranic energy yielded significantly more, with an increase of 31% against the control (Jois et al. 2016b).

Table 1. Effect of Pranic Agriculture on Tomato Plants

Growth Parameters	Treatment	Mean	*p*
Plant height (cm)	Control	92.90	< .001
	Pranic	114.20	
Diameter of stem (cm)	Control	14.00	< .001
	Pranic	16.00	
Number of flowers/plant	Control	10.00	< .001
	Pranic	24.40	
Number of flowers/truss	Control	1.70	< .001
	Pranic	3.87	
Overall yield/row (kg)	Control	80.50	< .001
	Pranic	117.07	

Effect of Pranic Agriculture on European Cucumber Plants

Pranic cucumber seeds had higher seed germination (94.8%) than the control (Table 2). Pranic-treated cucumber plants needed 54.3 days to get 50% flowering in cucumber plants, whereas the control needed 59.1 days, and the results were found to be significant ($p < .001$). Pranic-treated plants needed fewer days to flower than the control group. The control plants had more flowers (24.1) than the pranic group (23.1), but the results were insignificant. Pranic treated cucumber plants had 22.7 fruits per plant, whereas the control had only 19.6 fruits per plant ($p < .001$). Plants given pranic treatment had 14% additional fruits per plant than the control. Pranic-treated cucumber plants needed only 10.7 fruits to get 1kg, whereas the control group needed 11.3 fruits. However, the results were found to be insignificant. Pranic-treated fruits had more weight than the control. When yield per plant was verified, the pranic group yield was 1.9 kg/plant when compared to the control (1.58 kg), and the results were found to be statistically significant ($p < .001$). Plants receiving pranic treatment had 18% more yield than the control (Yathindra et al. 2017a).

Table 2. Effect of Pranic Agriculture on European Cucumber plants

Growth Parameters	Treatment	Mean	*p*
Plant germination (%)	Control	94.7	< .003
	Pranic	94.8	
Days to 50% flowering	Control	59.1	< .001
	Pranic	54.3	
Number of flowers/plant	Control	24.1	< .130
	Pranic	23.1	
Number of fruits/plant	Control	19.6	< .001
	Pranic	22.7	
Number of fruits/kg	Control	11.3	< .148
	Pranic	10.7	
Total yield/plant (kg)	Control	1.5	< .001
	Pranic	1.9	

Effect of Pranic Agriculture on Pole Beans Plants

Pole beans treated by pranic energy had a germination rate of 90.2% as against the control group (87.1%). This shows that plants treated with pranic energy had higher germination than control, and the results were significant. Similarly, the time taken to flower in the pranic group was also less (30.4 days), compared to the control (31.1 days). The pranic group took 31.2 days to the fruit set, whereas the control took more days (32.2) to the fruit set. Finally, total fruit yield was also higher in the pranic as against the control. There was an increase in the yield of 10.7% in the pranic-treated plants as against the control group (Table 3) (Yathindra, Srikanth, and D'Souza 2017b).

Table 3. Effect of Pranic Agriculture on Pole Beans Plants

Stages	Group	Mean	p
Germination (%)	Pranic	90.2±1.8	< .001
	Control	87.1±1.7	
Flowering (days)	Pranic	30.4±0.4	< .001
	Control	31.1±0.30	
Flowering to fruit set (days)	Pranic	31.2±0.30	< .002
	Control	32.2±0.90	
Yield (kg)	Pranic	0.49±0.03	< .002
	Control	0.44±0.03	

Effect of Pranic Agriculture on Cucumber Plants

The pranic group achieved a longer plant length of 72.8 cm compared to the control (37 cm). Plants that received pranic treatment had 49% more plant length than the control. Pranic plants' stem diameter was 18.7 cm, whilst control plants had 14.2 cm (Table 4). Plants that received pranic treatment had significantly more stem diameter (24.1%) than the control. We found the pranic group's total yield to be 63.8 kg compared to the control (52.8 kg). Pranic treatment yielded significantly more, with an increase of 17% as compared to the control (Jois, Prasad, and Lancy 2017).

Discussion

Seeds of peas and wheat, when treated with chi energy, germinated faster, since chi energy could impact the structure of a germination correlated gene. Chi energy can also alter the structure of cellular organs, like mitochondria, and influence the metabolism of the cell. Chi energy can alter the molecular structure of treated cells, affect gene expression, enzyme activity and nucleotide polymerisation (Bai et al. 2000). Research on mitomycin, using Kirlian photography, was conducted. Mitomycin has anti-tumor activity. The effect of cleansing the mitomycin, followed by its application, has shown great results, which were evident in the Kirlian photography of the cells treated with mitomycin (Sui 1992). The images showed that the energy level almost doubled and the length of the energy streamers also increased (Sui 1992). Similarly, when the pranic agricultural protocol is applied to seeds, the pranic energy level in the seeds increases, resulting in a greater germination percentage. Application of energy would increase the rate of cellular growth and division, and increase ATPase activity during germination of the dividing cells (Bai et al. 2000). These result in more energy supplied to the cells during cell division, growth and differentiation, probably resulting in enhanced growth of plants (Bai et al. 2000). However, the exact mechanism needs to be investigated further.

Table 4. Effect of Pranic Agriculture on Cucumber Plants

Stages	Group	Mean	*P*
Plant length (cm)	Pranic	72.8±7.50	<0.001
	Control	37±6.90	
Stem Diameter (cm)	Pranic	18.7±0.69	<0.001
	Control	14.2±1.80	
Yield (kg)	Pranic	63.80	<0.001
	Control	52.80	

Conclusions

The results show that Pranic Healing can have a significant impact on agriculture to increase the yield, seed germination, plant growth, stem diameter, the number of flowers and fruits, and decrease the time for flowering. The application of Pranic Healing in agriculture for these benefits does not require specialised laboratory conditions, special equipment or the use of special chemicals. Any interested person can be trained in the art of Pranic Healing for application in agriculture. This would have a positive impact on the environment (air and water) and human health. Further studies are needed to compare the nutritional and sensory qualities of the vegetables obtained, and the molecular mechanism to determine the faster growth of plants when treated with Pranic Agriculture.

Acknowledgments Master Choa Kok Sui, the founder of Modern Pranic Healing for imparting the precise knowledge of Pranic Healing. Acharya Daniel Gorgonia, Mr. Sriram Rajagopal of World Pranic Healing Foundation, India for encouragement and support and World Pranic Healing Foundation, India for funding the study. Dean of the College of Horticulture, Mysore, and Head of the Department of Food Science and Nutrition, the University of Mysore for their generous support to conduct the study.

References

Asna, U, T Keerthika, Devaki C S, and Florence S. 2016. "Studies on nutritional and quality characteristics of Cucumis sativus varieties." *Agriculture Science Research Journal* 6:79–85.

Bai, F, C Sun, Z Liu, J Shen, Y Shen, R Ge, C Bei, J Zhang, X Shi, and Y Liu. 2000. "Seeds induced to germinate rapidly by mentally projected'qi energy'are apparently genetically altered." *The American journal of Chinese medicine* 28 (01):3–8.

Jain, R, R Nagarathna, H R Nagendra, and S Telles. 1999. "Effect of 'pranic'healing in chronic musculoskeletal pain–a single blind control study." *International Journal of Alternative and Complementary Medicine* 17 (8):14–17.

Jauregui, M, T L Schuster, M D Clark, and J P Jones. 2012. "Pranic Healing: Documenting Use, Expectations, and Perceived Benefits of a Little-Known Therapy in the United States." *Journal of Scientific Exploration* 26 (3):569–588.

John, R P. 1995. "Pranic Healing Suggestion in the Treatment of Asthma." *Philippine Journal of Psychology* 28:50–81.

Jois Srikanth, N, K Nagendra Prasad, and K R Roohie. 2016a. "Traditional pranic healing therapy for enhanced germination and growth of green gram." *Agriculture in Sanskrit Literature*:337–340.

Jois Srikanth, N, K Nagendra Prasad, and L D'Souza. 2017. "Morphology of Cucumber Plants as Influenced by Pranic Agriculture." *Asian Journal of Agricultural Research* 11:33–35.

Jois Srikanth N, K Roohie, L D'Souza, F Suma, C S Devaki, A Urooj, R Krell, and K Nagendra Prasad. 2016b. "Physico-chemical qualities of tomato fruits as influenced by pranic treatment-an ancient technique for enhanced crop development." *Indian Journal of Science and Technology* 9 (46):1–5.

Kirlian, S D. 1949. "Method for receiving photographic pictures of different types of objects." *USSR Patent* N106401.

Mahesh, P A, J N Srikanth, M S Ananthakrishna, G Parthasarathi, S K Chaya, R Rajgopal, M R Devi, L D'Souza, and D H Amrutha. 2017. "Amelioration of quality of life and lung function of chronic obstructive pulmonary disease by pranic healing as adjuvant therapy: A randomised double blind placebo controlled pilot study." *Australasian Medical Journal* 10 (8):665–673.

Rajagopal, R, N Srikanth Jois, S Mallikarjuna Majgi, M N Anil Kumar, and H B Shashidhar. 2018. "Amelioration of mild and moderate depression through Pranic Healing as adjuvant therapy: randomised double-blind controlled trial." *Australasian Psychiatry* 26 (1):82–87.

Sui, C K. 1992. *The Ancient Science and Art of Pranic Healing*. Vol. 17. Bangalore: India: Institute of Inner Studies Publishing Foundation India Private Ltd.

Tsuchiya, K, and H Motoyama. 2009. "Study of body's energy changes in non-touch energy healing 1. pranic healing protocol applied for a breast cancer subject." *Subtle Energies and Energy Medicine Journal Archives* 20 (2):15–29.

Yathindra, H A, N J Srikanth, and L D'Souza. 2017b. "Enhancement of seed germination, fruit set and yield of Pole Beans (*Phaseolus vulgaris* L.) by Pranic Healing." *Ecology, Environment and Conservation* 23 (2):813–814.

Yathindra, H A, N J Srikanth, K Nagendra Prasad, and L D'Souza. 2017a. "Influence of pranic agriculture on germination, flowering and yield of European cucumber." *Ecology, Environment and Conservation* 23 (3):545–538.

CHAPTER 15

ENVIRONMENTAL EFFECTS OF ORGANIC FERTILISER USE IN ORGANIC AGRICULTURE

BULENT OKUR,[1] BULENT YAGMUR,[1] NUR OKUR,[1] DURSUN ESIYOK,[2] TANSEL KAYGISIZ ASCIOGUL,[2] OMAR MOHAMMED[2]

[1] Ege University Faculty of Agriculture Department of Soil Science and Plant Nutrition Bornova-Izmir
[2] Ege University Faculty of Agriculture Department of Horticulture Bornova-Izmir
dursun.esiyok@ege.edu.tr

Abstract

Chemical pollution is the most pressing environmental problem relating to the soil ecosystem. Irrigation with polluted waters, inorganic fertilisers and their compounds, such as sodium, potassium, phosphorus, calcium, magnesium, iron, zinc, copper, and manganese, and use of various heavy metals and pesticides are the major soil-polluting factors. Organic agriculture is considered a more environmentally-friendly agricultural practice than conventional agriculture since it does not involve the use of synthetic chemicals. Organic fertilisers are obtained from nature as well as from animal and plant waste. This not only reduces the leaching of plant nutrients and erosion but also increases carbon storage. Organic fertilisers positively affect the physical and chemical productivity characteristics and microbial properties of soil and add important plant nutrients as well as a significant amount of carbon to the soil. This type of production is also quite important for preventing carbon pollution, which has gained importance in the last decade from the point of view of environmental pollution.

Fertilisers used in organic production include compost, humic acids, fulvic acids, leonardite, farmyard manure, green manures, seaweeds, and mycorrhizas. Some fertilisers contain enzymes, amino acids, seaweed extracts, and microorganisms. However, excessive use of fertilisers, permitted in organic production, also cause accumulations of phosphorus, salt, and nitrogen in soil and groundwater. Thus, in interlinked ecosystems, eutrophication begins and results in a significant decline of living organisms, microbial populations, and water oxygen content. Although environmentally friendly fertilisers are used in organic agriculture, their uninformed or excessive use can lead to damage of the environment. For this reason, fertilisers of organic origin, permitted in this production system, must also be used properly, at the specified times and in the permitted doses. This study aims to provide further information on the damage caused by conventional farming and to contribute to the appropriate and knowledgeable use of organic fertilisers in organic agriculture.

Keywords: organic agriculture, manure, pollution, fertiliser, soil

Introduction

Environmental pollution can be defined as the deterioration of the natural balance caused by the depletion of resources of air, water and soil, which can occur naturally or as a result of human activities (Hakeem et al. 2013, Hakeem et al. 2014, Öztürk et al. 2015a, Öztürk et al. 2015b). Agriculture, based on natural systems using soil, water and living organisms, has been defined as sustainable agriculture, ecological agriculture, organic agriculture, biological farming, and good agricultural practices (Ozturk et al. 2008). All these systems mainly focus on increased quality and environmental friendliness. Agriculture refers to activities that aim to restore the natural balance but is slowly disappearing due to improper practices in our ecological system. Besides being a human and environmentally-friendly production system, organic farming is an alternative type of production, which targets increased quality more than quantity, based on the principles of using organic materials, green manure, and crop rotation, and taking advantage of parasites and predators in improving plant resistance and soil conservation.

Organic agriculture is much more environmentally friendly than conventional agriculture as it not only avoids using many inorganic chemicals but also utilises organic fertilisers obtained from nature (plant residues and animal manures). As a result, plant nutrient loss is reduced, erosion is controlled, and there is more carbon storage. Organic fertilisers mostly enhance microbial diversity and improve the physical and chemical properties of soil. In addition, organic fertilisers play an important role in increasing the emission of carbon dioxide, which is one of the current pollutants causing global environmental problems.

In recent years, the fertilisers used in organic production have been diversified, e.g. compost, humic and fulvic acids, leonardite, farmyard manure, green manure, seaweed, and mycorrhiza (Fig.1). Those containing various microorganisms, enzymes and seaweed extracts, in addition to organic materials, have also begun to be produced on a commercial scale. The use of organic fertilisers has particularly spread in Turkey in recent years due to concerns about human health. As ecologically friendly products, organic fertilisers do not contain any additives and undergo chemical processes of plant or animal origin. When compared to conventional agriculture, organic agriculture has fewer negative effects on the environment (Ozturk et al. 2008, de Magistris and Gracia 2008, Aldanondo Ochoa and Almansa-Sáez 2009). Furthermore, organic agriculture practices are regulated to international standards, prepared by the Food and Agricultural Organisation, the World Health Organisation and the International Federation of Organic Agriculture Movements.

Organic agriculture intends to produce high-quality food without using mineral fertilisers, synthetic pesticides, animal drugs, and food additives that may have adverse health effects. Organic agriculture should fit the cycles and balances in nature without exploiting it by utilising local resources and efficient management of materials and energy. It should provide a good quality of life, contribute to food sovereignty, and take into account and enhance the wellbeing of humans. Before adopting any new technology for organic agriculture, necessary precautions should be taken, and significant risks should be prevented by rejecting unpredictable technology, such as genetic engineering (Tuomistoa et al. 2012).

Even using fertilisers allowed in organic production involves certain risks, e.g. although farmyard manure enriches the soil with nutrients, such as nitrogen, phosphor, potassium, and sulphur, its excessive use can cause the accumulation of salt, and fractions of nitrogen and phosphor in the soil, and contaminate groundwater. Since ecosystems are interlinked, eutrophication can also occur, due to excessive concentrations of nitrogen and phosphorus in still water, which leads to a significant decrease in organisms, microbial populations, and water and oxygen content in the environment. Even though the fertilisers allowed in organic agriculture are environmentally friendly, their thoughtless and excessive use can prove detrimental to the environment. Therefore, fertilisers of organic origin, permitted in the production system, must be used properly in terms of application times and doses, otherwise, they will first affect the environment, then all living organisms under and above ground through the food chain and, ultimately, people who consume produce grown in the contaminated environments.

Fig 1. Organic Fertilizers

Effect of Organic Matter on Soil Characteristics, and Quality and Quantity of Yield

Organic and mineral matter constitute the solid part of the soil. Although the organic matter is represented by a low percentage (5–7%) in soil, it has very significant and versatile positive effects on all aspects of soil properties. Organic matter in the agricultural lands of Turkey shows very low levels due to ecological conditions, climate, topography, exploitation of plant nutrients due to monoculture practices, misuse of agricultural lands, destruction of forests and pasture, intensive tillage, lack of rotation systems, and erosion. In Turkey, nearly 64% of the soils contain less than 1% organic matter, 22% contain 1–2%, and 14% contain over 2% of organic matter. Good quality land should contain at least 5% organic matter. This indicates that the soil in Turkey is poor in organic matter (Tenth Development Plan 2014), which is an important material, positively affecting the physical, chemical and biological characteristics of the soil.

Effects of Organic Matter on the Physical Properties of Soil

One of the greatest effects of soil organic matter is observed in the physical properties of soil, which can be summarised as follows (Altınbas et al. 2016):

1- Low cohesion and plasticity features of humus make it possible to obtain a loose structure in clay soils, balancing the high cohesion and plasticity of the minerals in the clay.
2- Improving the structure of sandy soil by connecting soil particles, thus increasing the low water holding capacity.
3- Loosening the soil and porous structures, thus increasing its aeration capacity.
4- Leading to faster soil heating and early germination of seeds through the dark colour of the organic matter, highly dispersed and well mixed into the soil.

Effects of Organic Matter on the Chemical Properties of Soil

1- Good source of nutrients for plants, providing more than 90% of the nitrogen in the soil.
2- Regulates the sudden changes in soil pH by creating a high buffering effect via its high adsorbing power and, accordingly, alkali (Na, K, Rb, etc.) and alkaline (Ca, Mg, etc.) elements.
3- Adsorbs cations with negative (-) electrical charges, preventing them from being washed away from the soil, and keeping them ready to be utilised as plant nutrients.
4- Increases soil fertility and the cation exchange capacity of the soil (the cation exchange capacity of organic material is 2–3 times higher than that of clay minerals).

Effects of Organic Matter on the Biological Activity of Soil

1- Provides a good living environment for soil organisms, and, in sufficient amounts, leads to high biological activity.
2- With its organic carbon content, serves as the main source of energy and nutrients for most soil organisms. Easily decomposable carbon sources are of great importance in the fixation and enrichment of air nitrogen, which requires particularly high energy expenditure.
3- Facilitates mineralisation and humification, which are important processes for soil organisms, ensuring the continuity of soil fertility and sufficient plant nutrients.

Effects of Organic Matter on Soil Erosion

Soil organic matter plays a major role against water and wind erosion, which threatens agricultural land and causes soil losses. The organic matter layer on the soil surface:

1- Reduces the impact of rain droplets on the soil, thus preventing the formation of the duff layer caused by droplets (The infiltration of soil remains high for a long time, and water is prevented from passing to the surface stream, as such, sediment transport from the soil is reduced to the lowest level).
2- Prevents the transport and loss of soil caused by wind.
3- Together with mulch (i.e. plant residues) covering the soil surface, prevents loss of soil water by evaporation.

Effects of Organic Matter on the Product Quality and Quantity

Soil organic matter regulates almost all soil properties and improves fertility and productivity. These are reflected in plant development, as well as crop yield. For these reasons, it is recommended that farmers use farmyards and green manure to solve problems related to soil fertility. With the use of organic matter in agricultural areas, the crop production period can be extended. Although both the quality and quantity of a crop are significant in agricultural production, the former is the most important preference for dealers and consumers. Currently, a broad range of chemicals is used in an unlimited and uncontrolled manner in agricultural practices. Many countries are taking serious measures on the import of agricultural products subject to the application of chemicals to offer healthy products to their citizens and keep the quality criterion at the forefront. This brings to the fore the role of organic agriculture, which is not a new phenomenon. Traditional ways of agriculture followed by our ancestors are now being adapted to present-day conditions with certain modifications. Many studies show that if the soil contains enough organic matter, the quality of fruit and vegetable increases because it has a positive effect on the dry matter content, fruit weight, aromatic features, sugar content, and pH value. The spillage and cracking features also decrease in some fruit varieties, preventing production loss and increasing resistance to diseases and pests.

Organic Production and Farmyard Manure

Among the most commonly used inputs in organic crop production are organic materials and fertilisers of animal origin. The latter is farmyard manure, which is an essential source of plant nutrients in the soil. It improves many physical and chemical properties of the soil. It is most important that manure should be allowed to mature before spreading on or ploughing into the soil. How the manure should be used is determined after analysis of the soil and the manure. Farmyard manure is especially rich in nitrogen and also provides macro and micronutrients, such as phosphorus, potassium, and sulphur.

A common method for the use of farmyard manure in organic agriculture is composting these materials. However, this process cannot be achieved at the desired level due to the producers' insufficient knowledge. Plant nutrients, especially nitrogen, are reduced during the unconscious implementation of the composting process. On the other hand, lack of required conditions or exposure to raw farmyard manure leads to environmental pollution.

The nutrient content of farmyard manure shows great variation depending on animal species and age, amount of feed and its value, work done by animals, the type of the shed used, manure content of solid stool and urine, barn conditions, and storage method. Application of animal manure without the desired characteristics or without performing necessary soil and manure analyses will have detrimental effects on the soil and lead to the deterioration of soil properties, ultimately polluting the environment. The use of animal manure, without composting and in excessive quantities, causes problems in plant growth and contaminates microbial populations. The quantity and quality of the product, soil structure, and underground and surface water are adversely affected by excessive application of manure and urine (liquid faeces). Several reports have stressed that due to the overuse of this material, there is a heavy smell in the environment, and disease-causing factors have emerged in soil, plants, water, animal, and people. Organic farmyard manure mustn't contain antibiotic bacteria, which results from farm animals being given antibiotics in their feed.

Since commercial and chemical fertilisers are inappropriate for organic agriculture systems, farmyard manure constitutes the most important fertiliser material for organic crop production. The use of organic fertilisers depends on the climatic conditions of the region where it is to be applied. When these factors are taken into consideration, ideal application seasons appear to be spring and autumn, taking into account the climate and soil factors. Organic fertiliser applied to the field with a plough is buried to decrease the loss of the nutritive value of the manure. The depth at which the manure will be buried varies, depending on the soil structure. Generally, for fine-textured soils, farmyard manure will be laid on the surface, and then ploughed in deeply; however, for heavy-textured soils, it will be laid on the surface and mixed in gently. Organic fertiliser practices have shown that low quantities and frequent fertilisation programmes are more effective than using large quantities over longer periods. Determining the amount of organic fertiliser to be applied to the soil is also affected by the soil structure, organic matter quantity, climate factors, and cultivated plant variety. Organic fertilisers play a more important role in enhancing soil properties than the nutrients they provide for the plants. Organic fertilisers, applied to the field, increase the soil permeability values and moisture capacity, as well as produce a positive effect on the soil's physical properties, and a more suitable soil structure for crop cultivation.

Proper Use of Organic Fertilisers is Very Important for the Prevention of Environmental Pollution

It is very important to set aside a place where organic fertilisers can be stored. The barn should be constructed in such a way that no water can leach in. The litter must be spread over the barn floor to absorb the animal urine and prevent losses from washing, as well as the loss of ammonia. The organic matter and plant nutrients present in the litter should be supplied as a fertiliser and its use facilitated. To prevent the loss of available nutrients in the farmyard manure during maturation, materials, such as gypsum, calcium chloride, and calcium nitrate can be added. If phosphorus fertiliser is required, 7–8 kg phosphorus (P_2O_5) can be added to each tonne of farmyard manure. Loss of nitrogen should be prevented: in this way manure fertiliser will become enriched with phosphorus. The manure from the barn, and especially fresh manure, should be kept in an open area as a uniform pile. During the storage of the farmyard manures, one must be sure that it is composted under anaerobic conditions and compressed as much as possible. Thus, the loss of nitrogen and organic matter will be minimised. The farmyard manure must be kept covered to prevent nutrient losses, particularly in areas where rainfall is excessive. Manure should be composted on concrete or a compacted surface to prevent leakage and losses of the manure. Also, loss of the liquid part, due to rain, should be prevented, and in the case of such losses, if possible, liquids should be returned to the manure pile. However, in areas where there is low rainfall, manure should be kept in a well-compacted pile in a fairly high place sheltered from the wind. The manure pile should be covered with a thin layer of soil or organic matter. The variations in organic fertilisers should be differently determined, taking into account many factors, such as the amount of usage, the form of application, application time, irrigation system or method to be applied, soil characteristics, and cultivar for each fertiliser.

Careful Application of Organic Fertilisers

Organic fertilisers must be well composted, otherwise immature forms can harm vegetation when applied to the soil, because the maturing process in the soil will continue to decay and the bacteria will use the nutrients in the soil. Organic waste, mineralised materials and partly decomposed and fermented compost are stabilised products. One of the methods used to increase the organic matter content of soils, especially in organic agriculture, is the use of compost. Eventually, organic materials disappear from the soil in different ways; using compost returns the organic material to the soil, and thus decreases the loss of nutrients to the plants. Compost provides nutrients, allows air circulation, and increases soil temperature. Agricultural and industrial enterprises produce raw waste rich in organic matter and containing plant nutrients, but it cannot be used directly as a fertiliser. However, it is possible to use these organic compounds as fertilisers after composting. To ensure maximum benefit from the organic fertilisers to be applied, the ratio of $C/N < 1/20$ should be considered. The contents of the nutrients before application must be analysed, and the amount to be applied should be determined accordingly.

Green manure, used in organic agriculture, has no negative effects on the environment. While using green manure, it is important to consider the appropriate time for the green manure crops to be ploughed into the soil, and sufficient time should be allowed for decomposition, before planting the intended main crop.

Green manure crops can be grown in three different systems: main crop plantation, secondary crop plantation, and sowing into stubble plantation. There must be enough heat and rainfall to cultivate the green manure crop and ensure the composting process when returned to the soil. Therefore, not every crop may be suitable for green manuring, but the following are considered suitable: legumes, such as clover and clover species, vetch, soybeans and forage peas, as well as non-legumes like barley, oats, rye, millet, wheat, grass, mustard, rapeseed, Sudan grass, and buckwheat plants. In the selection of species, the plant to be sown should be deeply rooted, resistant to diseases and pests, rapidly developing, able to seed easily, and compatible with climate and soil conditions.

Animal Fertilisers and Soil and Environmental Pollution

The adverse effects of animal waste on the environment, and removal of such wastes from barns have become major problems. In countries where there is intensive livestock agriculture, such as the Netherlands, Taiwan, and Korea—ie a high human population and scarce land—it has become an environmental problem. Animal waste causes serious problems by polluting soil and water resources. One of the major problems in this type of pollution is the nitrogen and phosphor fertilisers (Hakeem et al. 2013). According to Edmeades (2003), due to the long-term application of farmyard manure, which has a high nitrogen and phosphorus content, when these nutrients are not removed from the soil through cultivation, an excessive amount of nitrogen and phosphorus is accumulated in it. The author reported that there might be a higher amount of organic phosphorus washed from surface runoff in the soil with more P and N losses, especially under low P-holding capacity. The use of farmyard manure, compared to inorganic fertilisers, can cause poor water quality by increasing oxygen demand. As such, long-term use of farmyard manure has been estimated to increase soil quality in terms of productivity, but which adversely affects the water quality more than chemical fertiliser applications with the same nutrient content (Edmeades 2003).

An important part of the phosphorus element in poultry feed mixes is phytin phosphorus, and the poultry metabolism cannot absorb this form of phosphorus; therefore, a considerable amount of phosphorus is excreted in the faeces. Poultry manure contains more phosphorus than that of bovine animals (Cromwell 1991). This phosphorus can be adsorbed by soil colloids by turning it into insoluble forms that cannot be taken up by plants from the soil (Konca and Uzun 2012, Okur and Ortas 2012). The phosphorus in acidic soils forms complexes, which are insoluble in iron (Fe) and aluminium.

Aluminium (Al) in the earth, whilst complexes which are not soluble in Ca and Mg in the soil are seen in alkaline soil. P transforms into insoluble forms, remains relatively immobile and then enters into the state of being a potential contaminant in the soil. It is accumulated in the soil and transported to rivers and lakes by rainwater or due to erosion of such lands. This element is restricted by the ground movement; therefore, groundwater contamination by P is rare. However, eutrophication does occur due to the enrichment of nutrients, especially nitrogen and phosphorus in surface water (Walker 2000). Phosphor is transported in significant quantities to lakes and rivers by precipitation or erosion from the fertilised soil or manure pile and is regarded as a nutrient that mostly restricts the growth of aquatic plants. In phosphorus-contaminated surface water, the production of algae increases and reaches a level that prevents the growth of other aquatic plants and animals. Some of the dense algal biomasses on the surface move towards the bottom and start being decomposed by aerobic bacteria, which ultimately reduces the amount of oxygen in such aquatic ecosystems, deteriorates the quality of freshwater resources, affects fish survival, and results in the emergence of unfavourable surroundings for other wildlife. This situation is particularly dangerous for cold-water fish, and in the lakes suffering from eutrophication, carp is replaced by pike and other fish with low O_2 requirements (Roland, Gordon, and Rao 1993, Walker 2000).

Eutrophication in freshwater systems is a serious problem and the culprit is phosphorus. Several restrictions have now been placed on nitrogen and phosphorus element contents in fertilisers being applied to agricultural land in Western Europe. The phosphate level allowed for use in agricultural land in the Netherlands has been reduced from 350 kg/ha to 250 kg/ha at the first level, followed by 175 kg P_2O_5 ha^{-1} for later levels (Cole 1991). In Turkey, the amount of nitrogen fertiliser to be applied to the soil in organic agricultural practices is recommended to be 17 kg per decare (Official newspaper 2010).

Soil Contamination by Farmyard Manures Containing Antibiotics

Antibiotics are highly effective bioactive compounds, which continue to be active in the soil and environment for a long time. Residue ratio levels in the soil vary between µg and gr per kg (Thiele-Bruhn 2003). Some of the medicine given to animals for treatment or to encourage growth is excreted in faeces and urine. Carballo et al. (2006) reported significant antibiotic residue in pigs, chickens and turkey fertiliser, and in the soil fertilised with these products. Kemper (2008) reported that antibiotic residues sometimes dissolved very quickly in the soil, but others might remain in the soil for a very long time, and some bacteria are resistant to one or more antibiotics being formed in these residues. According to Séveno et al. (2002), the genetic structure of some bacteria interacts with antibiotics, and they undergo a change. Fründ et al. (2000) and Sengeløv et al. (2003) found that tetracycline-resistant bacteria increased in the soil and groundwater in areas fertilised with antibiotic-applied animal manure. Kumar et al. (2005) stated that antibiotics were transferred to the soil with the fertiliser—this residue was then transferred to the plants growing in such soil, and antibiotic-resistant bacteria were transmitted to humans who consume the plants. Furthermore, antibiotics and medicines transferred to the soil with animal fertilisers change the profile of microorganisms living there, which leads to an increase in antibiotic resistance in some pathogenic microorganisms. In animal manure deposits and storage areas, urine rapidly turns into ammonia, becoming a gas, and hurting the environment. Another danger in farm fertilisers is the pollution of the environment due to their excessive nitrate content (Hakeem, et al. 2013). In the past three years, more than 2,500 publications have focused on researching nitrate washing in agricultural areas. Legumes used as green manure provide the highest nitrogen gains in the soil of most organic fertilisers. For an average hectare, this nitrogen gain is around 400 kg/year—about 50% of this can be released within the first two months. The transport of nitrate from farms to groundwater, and the increasing acceptable nitrate levels are the most important issues. Thus, recently, methods have been developed, such as Life Cycle Assessment or Life Cycle Impact Assessment to determine the effects of inputs used in the production on the agriculture sector, especially on humans and the environment. These methods calculate the results of all activities carried out in the production process, directly or indirectly affecting living systems, and they are regarded as ecological indicators. Agricultural production comprises not just those undertaken in the field, but also the environmental impacts of production, transport, and storage. In organic agriculture, most of the fertilisers that plants need include organic materials; however, in some cases, the nutrient needs of plants are provided by commercial fertilisers.

Major Commercial Organic Fertilisers Used in Organic Agriculture

Guano is a dry fertiliser obtained especially from marine and other birds. The amount of plant nutrients in the composition, which is completely natural, varies depending on the age of the plant beds and the climatic conditions of the area. Bird fertilisers are quite rich in potassium and the nitrogen contained in guano can be easily obtained by the plant.

Bone flour is a commercial fertiliser obtained by converting different animal bones into powder. It is very rich in phosphorus.

Horn and nail flour, like bone flour, is obtained by milling the remaining nails and horns from the slaughtered animals. It is used as a source of nitrogen and phosphorus in organic agriculture. The nitrogen in horn and nail flour is in plant-soluble form.

Water or sea algae, in both freshwater and marine forms, are very rich in macro and micronutrients. Algae contain nitrogen, phosphorus, potassium, calcium, sulphur, boron, copper and zinc.

Blood flour is obtained from the blood of slaughtered animals. The blood obtained is dried and ground and is quite rich in nitrogen and protein.

Blood flour can be absorbed more quickly by the plant than other commercial fertilisers.

Fish flour comes from fish waste, which is converted into flour, and treated as commercial organic fertiliser. Fish remains must be sterilised and their bad odour removed. Fish flour contains a high ratio of nitrogen, many microelements, amino acids, and several vitamins.

Tea residue serves as a commercial organic fertiliser obtained from the waste produced by tea factories. This residue is rich in nitrogen and regarded as good compost.

Oil coal comes from petroleum refiners. As a commercial organic fertiliser, it is used as a source of nitrogen and phosphorus in organic agriculture.

In addition to the commercial organic fertilisers listed above, residual or non-hazardous waste materials obtained from alcohol factories, textile workshops and various food processing plants are also used as commercial fertilisers in organic agriculture:

Liquid organic fertilisers are usually prepared from animal fertilisers or plant materials. These substances are left in water; the resulting filtrate is diluted with pure water at certain ratios, which is applied to the soil or plant in determined concentrations. Liquid organic fertilisers with different properties and contents are commercially available and sold in the agricultural markets. Some of the liquid organic fertilisers are obtained from guano, sea algae, and fish waste. Organic fertilisers obtained from these substances must be diluted because of their dense structure. Liquid organic fertilisers can be used at lower doses than similar products. They are also safer because they do not cause clogging in irrigation systems.

Other Uses of Environmentally Friendly and Human-Friendly Fertilisers in the Organic Production System with no Chemicals or Bio-fertilisers

Biofertilisers contain viable or latent microorganisms that are capable of nitrogen fixation, solubilise phosphates and other elements, promoting plant growth through the production of secondary metabolites, or protecting them against certain diseases. Such fertilisers can be applied to seeds, soil, leaves and compost, to accelerate microbial involvement, which, in turn, increases the usefulness of nutrients a plant can easily ingest and increases the useful microorganism population (Mahdi et al. 2010). The benefits of biofertilisers can be summarised as follows:

- Being environmentally friendly, organic agricultural inputs are less costly than chemical fertilisers.
- Enriching the soil with nitrogen and increasing the nitrogen feeding of legumes.
- Increasing the solubility of phosphate compounds found in the soil in an insoluble form.
- Improving the phosphorus feeding of plants by providing for the mobility of the surrounding phosphorus.
- Releasing substances that stimulate some plant growth.
- Protecting the plant from pathogenic fungi by showing antifungal activity under certain conditions.
- Increasing plant production by 20–30%.
- Protecting the environment by reducing the use of sulphur and phosphorus from chemical fertilisers by about 25%.

The major biofertilisers used in sustainable agriculture systems are:

1. Those containing micro-organisms capable of N2 fixation: *Rhizobium*, *Azospirillum*, *Azotobacter*, and *Cyanobacteria* (blue-green algae).
2. Those dissolving phosphates in the soil: *Bacillus* and *Pseudomonas* strains of bacteria, *Aspergillus,* and *Penicillum* strains of fungi.
3. Those increasing plant uptake of phosphorus and other plant nutrients in the soil: *Mycorrhizal* fungi.
4. Those containing rhizobacteria (PGPR) that promote plant growth: *Bacillus, Burkholderia, Clostridium, Enterobacter, Erwinia, Flavobacterium, Klebsiella, Microccocus, Pseudomonas, Rhizobium, and Serratia*.

Of the phosphorus fertilisers applied to the soil, 75–90% become useless since they are held by iron, aluminium and calcium elements, depending on soil properties, such as pH and lime content (Gyaneshwar et al. 2002). In such soils, inoculation of plant seeds with bacteria dissolving unavailable phosphorus will increase the availability of phosphorus (Jones and Darrah 1994). Phosphorus also becomes available in a more beneficial form in the soil with the help of mycorrhiza (Okur and Ortas 2012), which has an important place among the biological fertilisers as they increase phosphorus solubility in systems in which plant roots and certain fungi live a joint life.

This lifestyle is one of the most common forms of symbiosis between microorganisms and plants. In this symbiotic relationship, the plant provides the fungus with carbohydrates and some nutrients, whilst the fungus mainly increases the uptake of phosphorus and nutrients and water by the plant (Demir 1998). These organisms, extending from their environment to the root or from the roots to the soil, facilitate the uptake of phosphorus, zinc and copper. According to Karacal and Tufenkci (2006), in chickpea, uptake of P in the mycorrhiza-applied plots increases by 10.3% compared to control plots.

Biofertilisers are a pollution-free source, and the use of these fertilisers reduces environmental pollution by decreasing the use of chemical fertilisers, which tend to be imported, thus causing an economic burden. Biofertilisers are produced from microorganisms, located and adapted to the soil of a territory or country. They are very important in terms of preventing foreign microorganisms from entering agricultural lands, as well as preventing microorganisms from transferring to another country.

Although there are many types of biological fertilisers, vermicompost has begun to be used in recent years. These worms can live in suitable soil conditions, producing stools known as "biohumus", which are rich in nutrients and create a biological fertiliser. Vermicompost is particularly based on the California Worm veneta types —*Humbricus rubellis* and *Eisenia foetida* and *Dendrobena*—which are cultivated and fed under appropriate moisture, pH and salt conditions. They are produced in large quantities by the private sector. These obtained fertilisers also contain plant nutrients that are dissolved and are beneficial for plants, e.g., many organic compounds, amino acids, enzymes, humic acid, fulvic acid, bacteria, nitrogen-fixing bacteria, fungus, and micro-organisms.

Importance of humic substances in organic production

These products, formed after the humification of the organic matter, are also important because they are very effective in binding plant nutrients in the soil; increasing nitrogen, phosphorus and sulphur uptake; accelerating the soil microbial activity; increasing the water-retention capacity of the soil and the solubility of plant nutrients; and improving soil structure (Altınbas et al. 2016). These materials are mainly obtained from lignite coal beds or leonardite and can be used safely in organic agriculture. Organic substances, such as humic acid, fulvic acid and amino acid, are included among the approved plant growth regulators in Turkey, but not in other countries. They are used in organic agriculture applications and affect plants indirectly, mainly by stimulating enzymes, supporting root growth, increasing permeability of stem cell membranes, encouraging plants to take more plant nutrients from the soil, and facilitating the formation of chlorophyll, sugar and amino acids in plants, thereby increasing the content of vitamins and minerals in the plant, and promoting photosynthesis, leading to good quality products.

The application of the permissible type, and amount of farm fertilisers in organic agriculture, improves the physical, biological and chemical properties of soils, helps in the recycling of organic waste, reduces the use of synthetic or artificial chemicals, increases the quality of animal and vegetable production, controls landslides, erosion, salinity and alkalinity in soils, and contributes to the beneficial effects of ecology.

Some negative aspects are originating from the presence of organic materials in the environment. These include the increased acidity of soil, eutrophication, damage due to nitrate washing, and the serious environmental impacts of cadmium, nickel, chromium, copper, lead, and zinc from the use of treated waste sludge. Soil acidity also increases as a negative factor due to long-term application of green manure and organic substances. Similarly, nitrogen gas emissions and nitrogen wash from organic fertilisers negatively affect the climate. Nitrification in soils occurs in oxygenic conditions and denitrification under anaerobic conditions or those in the absence of oxygen. The gas released enters the atmosphere directly. Ammonia gas, originating directly from farm fertilisers, is regarded as an acidic contaminant, and it adversely affects bacteria and fungi by interfering with their life cycle. However, all these negative effects are not permanent or large-scale when compared to the harm from chemical fertilisers applied to agricultural land. As a result, organic production is nowadays being adopted as a way of thinking and lifestyle, allowing for healthy food production for humans.

Conclusion

The most important issue in organic agriculture is to ensure that the organic matter in the soil is considered very valuable in the long term and that it supports the sustainability of the soil. Many factors must be considered, from the appropriate tillage techniques to the selection of irrigation systems and methods. In short, animal or farmyard manure of different origins improves almost all soil characteristics, provides nutrients, and accelerates the soil microbiological activity. It is an indispensable input for organic production, and farmyard manures are essential in this context. They should be applied to the soil after they are completely mature or composted. In Turkey, farmyard manure is, in general, environmentally friendly and more beneficial than commercial fertilisers. However, to make full use of its benefits, it is necessary to take great care in how it is stored, and pay strict attention to how it is applied to the soil and in what quantities.

Organic matter, despite being highly beneficial for agricultural soils, is underused in Turkey. The important and valuable sources of organic matter, such as farmyard manure, become destructive instead of beneficial and lose environmentally-friendly features, even becoming harmful, if left unused. Raw or excessive use of manure in the soil can have negative impacts, such as causing both microbial and environmental contamination, thus leading to contamination of underground and surface water, and spreading infections and diseases in humans and animals.

References

Aldanondo-Ochoa, A M, and C Almansa-Sáez. 2009. "The private provision of public environment: Consumer preferences for organic production systems." *Land Use Policy* 26 (3):669–682.

Altınbas, U M, H Cengel, B Uysal, N Okur, Y Kurucu, and S Delibacak. 2016. *Soil Science., Ege University, Faculty of Agriculture Publications No: 156*. Bornova-Izmir, Turkey: E.Ü. Rectorate Printing House.

Carballo, J L, E Avila, S Enríquez, and L Camacho. 2006. "Phenotypic plasticity in a mutualistic association between the sponge *Haliclona caerulea* and the calcareous macroalga *Jania adherens* induced by transplanting experiments. I: morphological responses of the sponge." *Marine Biology* 148 (3):467. 10

Cole, J A. 1991. "The role of nutritionist in designing feed for the future. Biotechnology in the feed industry." Proceedings of Alltech's Seventh Annual Symp.

Cromwell, G L. 1991. "Phosphorus-a key essential nutrient, yet possible major pollutant It's central role in animal nutrition, Biotechnology in the Feed Industry." Proceedings of Alltech's 7th Annual Symp. Alltech Technical Publications, Kentucky p.

de Magistris, T, and A Gracia. 2008. "The decision to buy organic food products in Southern Italy." *British Food Journal* 110 (9):929–947.

Demir, S. 1998. "Studies on the Formation of Vesicular Arbuscular Mycorrhiza (VAM) in Some Cultural Plants and Its Role in Plant Development and Durability." Doctorate, Institute of Science and Technology, Ege University.

Edmeades, D C. 2003. "The long-term effects of manures and fertilisers on soil productivity and quality: a review." *Nutrient cycling in Agroecosystems* 66 (2):165–180.

Fründ, H-C, A Schlösser, and H Westendarp. 2000. "Effects of tetracycline on the soil microflora determined with microtiter plates and respiration measurement." *Mitteilgn. Dtsch. Bodenkundl. Gesellsch* 93:244–247.

Gyaneshwar, P, G Naresh Kumar, L J Parekh, and P S Poole. 2002. "Role of soil microorganisms in improving P nutrition of plants." *Plant and Soil* 245 (1):83–93.

Hakeem, K R, P Ahmad, and M Ozturk. 2013. *Crop improvement: new approaches and modern techniques*: Springer Science and Business Media.

Hakeem, K, M Sabir, M Ozturk, and A R Mermut. 2014. *Soil remediation and plants: prospects and challenges*: Academic Press.

Jones, D L, and P R Darrah. 1994. "Role of root derived organic acids in the mobilization of nutrients from the rhizosphere." *Plant and soil* 166 (2):247–257.

Karacal, I, and S Tufenkci. 2006. "New Approaches to Plant Nutrition and Fertilizer-Environment Relationships." TMMOB VII. Technical Congress, Ankara, Turkey.

Kemper, N. 2008. "Veterinary antibiotics in the aquatic and terrestrial environment." *Ecological indicators* 8 (1):1–13.

Konca, Y, and O Uzun. 2012. "Effect Of Animal Waste On Soil And Environment." 4th Congress of Soil Scientists of Azerbaijan, Bakü, Azerbaijan

Kumar, K, S C Gupta, S K Baidoo, Y Chander, and C J Rosen. 2005. "Antibiotic uptake by plants from soil fertilized with animal manure." *Journal of environmental quality* 34 (6):2082–2085.

Mahdi, S S, G I Hassan, S A Samoon, H A Rather, S A Dar, and B Zehra. 2010. "Bio-fertilizers in organic agriculture." *Journal of Phytology* 2 (10).

Official newspaper. 2010. Principles of Organic Agriculture and Directing on Implementation. edited by Issue: 27676. Ankara.

Okur, N, and I. Ortas. 2012. "Microbiological Fertilization and Mycorrhiza in Agriculture." In *Plant nutrition*, edited by M.R. Karaman. Ankara: Gübretaş Guide Books Series: 2.

Ozturk, M, A Guvensen, S Sakcali, and S Celik. 2008. "Agriculture and Ecological Interactions." In *Ecological / Organic Agriculture and Environment* 14–26. Bursa, Turkey: Özsan Press, EKODER.

Öztürk, M, M Ashraf, A Aksoy, and M S A Ahmad. 2015a. *Phytoremediation for green energy*: Springer.

Öztürk, M, M Ashraf, A Aksoy, M S A Ahmad, and K R Hakeem. 2015b. *Plants, Pollutants and Remediation*: Springer.

Roland, D A, S K Gordon, and Rao. 1993. Phosphorus solubilization and its affects on the enviroment with examples of common misformulation problems associated with calcium and phosphorus. Muti-State Poult.

Feeding and Nutr. Conf., Indianorpolis, Indiana., Indiana, USA, 22–25 May.

Sengeløv, G, Y Agersø, B Halling-Sørensen, S B Baloda, J S Andersen, and L B Jensen. 2003. "Bacterial antibiotic resistance levels in Danish farmland as a result of treatment with pig manure slurry." *Environment International* 28 (7):587–595.

Séveno, N A, D Kallifidas, K Smalla, J D van Elsas, J-M Collard, A D Karagouni, and E M H Wellington. 2002. "Occurrence and reservoirs of antibiotic resistance genes in the environment." *Reviews in Medical Microbiology* 13 (1):15–27.

Tenth Development Plan. 2014. Report on the Sustainable Use of Agricultural Land. edited by Ministry of Development. Ankara.

Thiele-Bruhn, S. 2003. "Pharmaceutical antibiotic compounds in soils–a review." *Journal of Plant Nutrition and Soil Science* 166 (2):145–167.

Tuomistoa, H L, I D Hodgeb, P Riordana, and D W Macdonalda. 2012. "Does organic farming reduce environmental impacts?".

Walker, F. 2000. "Best management practices for phosphorus in the environment." *Agricultural Extension Service, University of Tennessee, PB* 1645.

CHAPTER 16

PLANT NUTRITION IN DROUGHT STRESSED CROPS UNDER THE CURRENT SCENARIO OF CLIMATE CHANGE

CENGIZ KAYA,[1]* MEHMET SENBAYRAM,[1] SALIH AYDEMIR,[1] MUHAMMAD ASHRAF[2]

[1]Harran University, Faculty of Agriculture, Plant Nutrition and Soil Science Department, Sanliurfa/TURKEY.
[2] Pakistan Academy of Sciences, Islamabad, Pakistan
c_kaya70@yahoo.com

Abstract

Over the past century, there has been a marked increase in atmospheric carbon dioxide (CO_2), which has, in turn, elevated the ambient temperature of the globe. However, the impact of elevated CO_2 can be offset by other environmental variables, such as elevated temperatures and changed patterns of precipitation. Predictions of many climate models show that the effects of occurrence and duration of heat and drought stress on major crops will increase in many regions, negatively affecting food security. To overcome this problem, breeders, botanists, and molecular geneticists are trying to identify genes involved in drought tolerance to enhance the genetic potential of crops for attaining high productivity under scarce water conditions. However, improved genetic potential against drought stress may increase the plant's nutrient demand. This happens because the processes which are related to drought defence of crops, such as stomatal regulation, rate of photosynthesis, assimilate transport from source to sink organs, oxidative defence system, morphological changes in leaves to prevent high-light intensities or reduction in cuticular water loss, are regulated by the availability of key plant nutrients. Furthermore, many environmental, soil and plant factors under water scarcity often limit nutrient availability and uptake, despite the high nutrient contents of the soil. In this review, we aim to discuss the complex interrelation between plant nutrients (nitrogen, phosphorus, potassium and magnesium), water scarcity, heat stress, and elevated CO_2 concentration.

Keywords: plant, drought, nutrients, nitrogen, magnesium, potassium, phosphorus

CO_2 as a Greenhouse Gas and a Plant Nutrient

It has been reported that over the last 0.8 million years, carbon dioxide (CO_2) concentration in the atmosphere has risen from 180 ppm (glacial periods) to 280 ppm (interglacial periods) (Ahn et al. 2012, Ehlers et al. 2015). On the other hand, only since the beginning of the industrial development ca. 150 years ago, have CO_2 levels increased from 280 ppm to 400 ppm, and the mean temperature has risen to 0.8°C during the same period. The rise in global atmospheric CO_2 level can be primarily attributed to the use of fossil fuels, land-use change related to population growth, and industrial expansion (Houghton, Callander, and Varney 1990). There is now a consensus that an increase in the atmospheric CO_2 along with other greenhouse gases (e.g. N_2O, and CH_4), primarily causes global warming (Pachauri et al. 2014). Many climate models suggest that by the end of this century, CO_2 concentration in the atmosphere will increase to more than 700 ppm, leading to an increased global temperature of 1.8–4.0 °C, subject to the extent of greenhouse gas emission (Pachauri et al. 2014). The growing evidence suggests that a broad range of crops, in particular C_3 crops, may be positively affected by the increased CO_2 levels in the atmosphere, particularly when cultivated under non-stressful environmental conditions (Amthor 1997, Bishop, Leakey, and Ainsworth 2014, Leakey et al. 2009, Long et al. 2004). However, the positive impact of elevated CO_2 on plant growth and development could be curtailed by other climatic variables, such as elevated temperatures, and, specifically, varied patterns of rainfall (Easterling et al. 2007).

Despite being a greenhouse gas, CO_2 is an essential plant nutrient that plays a key role in photosynthesis. Elevated CO_2 is believed to stimulate several physiological processes, such as a decrease in the stomatal opening, especially in C_3 plants (higher leaf-water use-efficiency), and increased carbon uptake (higher rate of photosynthesis (Marble et al. 2011). It has been reported that elevated CO_2 is more effective in C_3 than C_4 plants in terms of improving plant growth (Poorter and Navas 2003, Prior et al. 2003, Wang and Wu 2015). In addition to enhanced photosynthesis, elevated CO_2 also affects C partitioning and enhances root growth and/or tuber development

(Barnaby et al. 2015, Bindi et al. 1999). This is especially important for crops like potato, sugar beet, and other crop species, where the main harvest organs are below ground. Furthermore, enhanced below-ground growth under elevated CO_2 may also increase plant root development whereby plants may acquire more water from deeper soil layers. As mentioned above, elevated CO_2 can also affect plant growth via enhanced plant water relations (Kimball et al. 2002, Rogers and Roger 1993). Elevated CO_2 minimises transpiration by inducing higher internal CO_2 concentration in the leaf, causing the partial closure of the stomata (Jones and Mansfield 1970, Leakey et al. 2009, Sreeharsha, Sekhar, and Reddy 2015). This decrease in water loss and enhanced photosynthesis under elevated CO_2 concentration can significantly increase plant water use efficiency (WUE).

One way to sustain yield under drought conditions is to enhance the efficiency of water use by crops. Water-use efficiency does not have a single precise definition. Its definition depends upon the period over which efficiency is appraised, and accurate measures of efficiency should relate to carbon gain in terms of biomass production and crop productivity. Intrinsic water use efficiency (WUE_i) can generally be defined as the ratio of CO_2 assimilation into the photosynthetic products to the amount of water lost by transpiration through the stomata (Brueck and Senbayram 2009, Flexas et al. 2010, Gago et al. 2014). As elevated CO_2 concentration affects both transpiration and the rate of photosynthesis, its effect on water use efficiency is very clear in water-limited environments. Furthermore, WUE_i alone misses several other aspects (i.e. respiration during the night, unproductive water loss during the night, cuticular water loss, C allocation in the roots) that may significantly influence total C and water budgets. The term "agronomic water use efficiency" (biomass water use efficiency, WUE_T), which was defined earlier by Farquhar, Ehleringer, and Hubick (1989), can give a better estimation of the efficiency of water use. The term WUE_T, considers the plant as a whole and estimates losses of carbon and water that occur during the day and night from both photosynthetically active and non-active organs. Here, the proportion of C allocation in the roots and construction costs, which include day and night respiration, may reduce the productivity of crops significantly. However, current knowledge, especially on construction costs of the plant, is still very limited. Similarly, transpiration during the day and unproductive water loss during the night may drastically increase the water consumption of crops. In this context, the effect of elevated CO_2 on unproductive water loss, root or canopy respiration, nighttime respiration and their effect on WUE_T is still not being properly studied.

Plant Nutrient Requirements under Elevated CO_2

Given the fact that higher atmospheric CO_2 may mitigate water loss in crop plants, several authors have suggested that this may ameliorate the negative effects of water limitation (Bazzaz 1990, De Boer et al. 2012, Franks et al. 2012, Medlyn et al. 2011, Swann et al. 2016, Wullschleger, Tschaplinski, and Norby 2002), and may allow plants to maintain a high rate of photosynthesis and growth. Plant mechanisms, such as the maintenance of high photosynthetic rate, effective stomatal regulation, efficient translocation of assimilates, and enhanced root growth play a key role under drought situations. The processes relating to drought defence of crops, such as stomatal regulation; rate of photosynthesis; reactive oxygen species scavenging processes; assimilate transport from source to sink organs; and morphological modifications in plant parts, particularly in leaves to avoid high light-intensities or reduced cuticular water evaporation, are regulated if the appropriate amounts of essential nutrients are available in the plant. It is well documented that under stress conditions, higher rates of key nutrients are required (such as nitrogen, magnesium, potassium and phosphorus).

Drought and Magnesium

Magnesium (Mg) is one of the major components of many soils, comprising 2% of the Earth's crust. Soil Mg at 90–98% is not directly taken up by crop plants due to its incorporation into the crystal lattice structure of minerals. It has been reported that the total Mg content of soils varies greatly between 0.05 and 0.5% (Gransee and Führs 2013). Soil Mg has been classified into four fractions which are slowly exchangeable, rapidly exchangeable, structural forms, and organic complexes (Senbayram et al. 2015a). Magnesium in the soil solution, which is readily available and slowly refilled by the soil reserves, can readily be absorbed by the plants. In addition to soil texture, soil moisture is another key variable affecting plant Mg uptake. Magnesium is an important plant macro-nutrient, which plays a key role in several plant

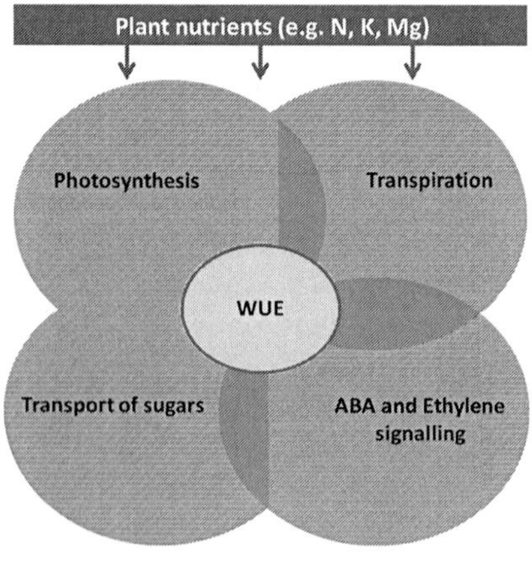

physiological processes, for example, harvesting solar energy by being a central chlorophyll constituent; as a cofactor and allosteric modulator for more than 300 enzymes, such as ATPases, kinases, RNA polymerases; in the Calvin cycle; and in chelation to nucleotidyl phosphate forms (Barker and Pilbeam 2015, Bruns 2016, Cowan 2002, Shaul 2002, Verbruggen and Hermans 2013). Thus, it is abundantly clear that severe Mg deficiency can decrease biomass yield by diminished photosynthetic processes. Typical Mg deficiency symptoms are leaf interveinal chlorosis and sugar accumulation in the leaves. As under-elevated CO_2-enhanced dry-matter partitioning is one of the key mechanisms, an adequate Mg supply may specifically be needed to maintain growth. Development of chlorosis due to degradation of chlorophyll by over-production of reactive oxygen species (ROS) occurs under severe deficiency, whilst sugar and starch accumulations in the leaves are affected in much earlier stages, which is called hidden deficiency (Hermans et al. 2005). There is strong evidence to show that optimised Mg nutrition may enhance WUE due to its role in several important metabolic processes (Tränkner et al. 2016). Anomalies in physiological processes due to hidden Mg deficiency (in the early stages of the deficiency) and their effects on drought tolerance can be partly explained by the above-mentioned functions of Mg in plants. However, the direct effect of Mg deficiency on WUE needs further validation as there is a lack of research on this subject.

Magnesium deficiency leads to an accumulation of carbohydrates in leaves, which impairs sucrose synthesis and starch accumulation (Araya, Noguchi, and Terashima 2006). It is known that starch accumulation in the chloroplasts due to diminished sugar transport causes deformation of the chloroplasts, thereby decreasing CO_2 diffusion from the intercellular spaces to the catalytic site of Rubisco, and hence the photosynthetic activity (Farhat et al. 2014, Keenan, Sabate, and Gracia 2010, Nafziger and Koller 1976). Here, lower demand in electron acceptors (due to lack of CO_2 in the catalytic site of Rubisco) may trigger ROS production and start cell damage. Alternatively, a decrease in the rate of CO_2 diffusion may also force the stomata to open further to enhance the CO_2 gradient to feed chloroplasts and increase transpiration that may significantly lower WUE (Tränkner et al. 2016).

Water shortage in soil reduces nutrient uptake (e.g. especially Mg, K or other cations) from the soil, as in dry soils, and diffusion and mass flow of nutrients to the roots are obstructed (Engels and Kirkby 2001, Hepworth et al. 2015). Also, root growth generally decreases in drier soils, and the latter may lead to a further decrease in nutrient uptake. Plant roots may increase root hairs and enhance the secretion of gelatinous substances to the soil that may increase root elasticity under drought (Carminati and Vetterlein 2012). As suboptimal Mg content of crop plants specifically affects phloem loading, this may also hamper plants' ability to develop the above-mentioned fine root cover as it necessitates high rates of root exudates and assimilates. In conclusion, Mg deficiency is expected to reduce tolerance to drought in the plants because root elongation and root exudation require a huge amount of assimilate transport from the source organs (e.g. photosynthesising leaves) (Senbayram et al. 2015a). Application of foliar Mg or granular fertilisers can ameliorate the nutritional status of Mg deficient plants. The latter has been shown to increase the concentration of chlorophyll and vegetative parts of the plants (Neuhaus, Geilfus, and Mühling 2014). In agricultural systems under changing climates, Mg fertilisation can enhance crop WUE, root elasticity, dry matter partitioning and, ultimately, plant growth (Römheld and Kirkby 2007, Senbayram et al. 2015a, Tränkner et al. 2016).

Drought and Potassium

Being one of the most important osmotica in crop plants, potassium (K) is the main determinant of the cell turgor (Wegner 2013, Zörb, Senbayram, and Peiter 2014). Furthermore, K plays a key role in a series of fundamental metabolic and physiological processes in plants such as enzymatic processes, ATP synthesis and, especially, stomatal regulation mediated by the guard cells' osmotic potential (Cakmak 2005). It has also been reported that K deficiency also reduces the plant's tolerance to water-limited environments, and those plants fed with an appropriate amount of K had a sufficient amount of water in their tissues (Mengel et al. 2001, Römheld and Kirkby 2010, Shabala and Pottosin 2014, Zain and Ismail 2016). Crop K concentration varies greatly with site, year, crop species and input of fertilisers, and, therefore, K concentration of crops, in the range between 0.4–4.3%, has been widely reported (Askegaard, Eriksen, and Johnston 2003, Whitehead 2000). Oborn et al. (2005) reported that the concentrations of K in the crops are generally lower than the conventional standard values of 2.5–3.5%. The critical concentration of K for many plants varies between the range of 0.5 to 2% on a dry matter basis (Leigh and Wyn Jones 1984). High K concentration in plants (especially in vacuole) is called luxury consumption; however, it has been reported that accumulation of K in plants at a higher rate can improve plant tolerance to abiotic stresses (Kafkafi 1990).

Plant species may vary significantly in terms of K requirement and uptake. Monocotyledons, like cereals, are known to be more efficient in K uptake than dicotyledons, such as potato and legumes (Mengel et al. 2001). The differences among various plant species in terms of K absorption are reported to be due to differences in root density and depth, as well as root hair length (Rengel and Damon 2008, Zörb, Senbayram, and Peiter 2014). A positive correlation has been found between root hair length and density and K uptake efficiency in different crops, such as tomato, oilseed rape, and maize (Jungk 2001), barley, rye, red clover, pea, and perennial ryegrass (Høgh Jensen and Pedersen 2003) in K depleted soils. Mengel and Diedrich (1985) hypothesised that ryegrass takes up K more efficiently than red clover due to possessing a denser root system and longer root hairs. Both the denser root system and longer root hairs may also deplete K in the soil solution in high amounts, and this causes the release of non-exchangeable K (Springob and Richter 1998).

The mechanism of plant K uptake is another important parameter that plays a key role in K use efficiency. Early investigations have shown that higher plants show a characteristic dual-affinity (low and high) uptake mechanisms, which perform at varying external K levels (Epstein, Rains, and Elzam 1963). Under conditions of external K concentration below 0.2 mM, high-affinity K uptake mechanism is mediated primarily by K transporters (Maathuis and Sanders 1994). These transporters differ in their energetic coupling, affinity, and selectivity for K, therefore, plant species and varieties show distinct differences in high-affinity K uptake mechanisms. Additionally, higher plants seem to develop K-deficiency-sensing structures that help them to activate high-affinity K uptake mechanisms to survive under K deficient regimes (Wang and Wu 2013). Specifically, under dry conditions, soil nutrient availability is low; effective high-affinity K uptake mechanisms may therefore play a key role in plant adaptation to water-limited conditions.

Potassium plays a critical role in cellular osmoregulation, which leads to improved water uptake, and hence plant growth (Britto and Kronzucker 2008, Shabala and Pottosin 2014, Wyn Jones, Brady, and Speirs 1979). Potassium accumulation in guard cells and root xylem vessels can effectively improve water uptake (Ahmad and Maathuis 2014, Lynch and Läuchli 1984) and stomatal regulation of the controlling water loss (Hsiao and Lauchli 1986, Shabala and Pottosin 2014). It is well accepted that K deficiency in crop plants decreases the rate of photosynthesis and plant susceptibility to drought (Gupta, Berkowitz, and Pier 1989). We may conclude that application of the adequate amount of K fertiliser to K-deficit soils may improve plant response to water deficit conditions by two different means: i) enhanced adaptive/resistance mechanisms of crop plants due to the above-mentioned physiological activities; ii) increased available water content of soils under water-limited conditions by enhancing water-holding-capacity (improvement of soil physical properties). Thus, increased levels of plant K status are known to enhance WUE of crop plants (Anderson et al. 1992, Jákli et al. 2017).

Drought and Nitrogen

Nitrogen (N), being an important constituent of DNA, amino acids, proteins and chlorophyll, is one of the key plant nutrients involved in several key plant physiological processes including photosynthesis, synthesis of amino acids, and phytohormone signalling (Criado et al. 2017, Glibert et al. 2016, Marschner 2011). Plants take up nitrogen mainly in two forms: nitrate (NO_3^-) or ammonium (NH_4^+). Movement of N is predominantly regulated by mass flow, thus water availability in the soil directly controls the plant's availability of soil N. In soils with limited water, N availability is specifically reduced, particularly due to impaired N uptake (Glibert et al. 2016, Mengel et al. 2001). Balanced N fertilisation under dry environments may alleviate the negative influence of water stress by enhancing osmoregulation, preventing damage to biological membranes and enhancing intrinsic and agronomic WUE (Brueck and Senbayram 2009, Senbayram et al. 2015b). In contrast, nitrogen nutrition also enhances photosynthetic capacity and leaf area index, so plants with a high rate of photosynthesis and high leaf area index may consume more soil-available water for biomass production, which may enhance plant susceptibility to the water supply. Therefore, excess N fertilisation needs to be avoided in dry conditions. As mentioned in the earlier sections, elevated atmospheric CO_2 may promote biomass production and dry matter partitioning, so it can also decrease biomass N concentrations in many plant species (Buysse, Smolders, and Merckx 1996, Leakey et al. 2009, Long et al. 2004, Ruiz-Vera et al. 2017). A decrease in specifically N concentration in crop plants may negatively affect yield and quality, specifically under optimal water supply conditions. The occurrence of low tissue N content under elevated CO_2 may be due to the dilution effect as a higher rate of photosynthesis increases the levels of non-structural carbohydrates. The effect of elevated CO_2 on tissue N content can be ameliorated by N fertilisation (Leakey et al. 2009, Long et al. 2004, Ruiz-Vera et al. 2017, Taub and Wang 2008). The latter suggests that plants grown under elevated CO_2 have higher nutrient needs, especially nitrogen. Thus, nutrition programmes need to be optimised accordingly.

Drought and Phosphorus

Phosphorus (P) is well known as one of the essential mineral nutrients required in large amounts to sustain plant growth and development. It also plays a crucial role in both the maintenance and transfer of energy in cell metabolism (Amtmann and Blatt 2009, Artuso et al. 2018, Jin et al. 2006). Phosphorus is available in both organic and inorganic forms in the soil; the organic form of P is combined with non-decomposed plant and animal residual materials whilst its inorganic form is often associated with calcium in alkaline soils or with aluminium in acidic soils (Celi and Barberis 2005, Gama-Rodrigues et al. 2014, Hou et al. 2016, Vitousek et al. 2010). The inorganic form of P in soil is mostly available in the form of orthophosphate and it is absorbed by the plants as $H_2PO_4^-$ in acidic soils or HPO_4^{2-} in alkaline soils (Mengel et al. 2001, Vitousek and Chadwick 2013, Vitousek et al. 2010). However, in some cases, phosphorus can also be taken up in the form of organic phosphate, such as nucleic acids by plants. Absorbed inorganic phosphorus can form organic compounds rapidly after accessing the roots and moving to the shoots of the plant (Möller et al. 2018, Sanchez 2007).

The addition of phosphorus (P) to the soil having low P improves the growth of plants exposed to elevated CO_2 (eCO_2), especially for legumes (Edwards, McCaffery, and Evans 2005, Jin et al. 2013, Jin et al. 2012), and has also been reported to improve drought tolerance in crops (Graciano, Guiamét, and Goya 2005, Jin et al. 2015, Jun et al. 2017). It has also been reported that application of P improves drought tolerance under the conditions of eCO_2 in white clover and soybean (Jin et al. 2006, Singh et al. 2000) by increasing hydraulic conductivity of the root, thereby sustaining leaf water potential (Radin and Eidenbock 1984, Singh, Sale, and McKenzie 1997) and increasing root volume to uptake more water in deep soil layers (Jin et al. 2005). Furthermore, increasing levels of P could enhance the production of osmotically active carbohydrates in the leaf cells which leads to sustained leaf water potential in plants exposed to water deficit conditions, mainly due to the crucial role of inorganic P (Pi) in moving triose sugars out of chloroplasts (Abel, Ticconi, and Delatorre 2002, Ahmad et al. 2017, Lambers et al. 2006, Rychter and Rao 2005). This seems to be a likely mechanism of how P supply regulates the effect of eCO_2 on plant drought tolerance (Jin et al. 2015).

Overall, in the future, under eCO_2 environment conditions, the addition of supplementary P fertiliser may mitigate the deleterious effects of drought on plant growth, because drought is expected to be predominant in some parts of the world (Allen et al. 2010). Current reports have indicated that alterations in CO_2 or P may change tolerance to drought stress for a broad range of crops, such as white clover, cotton, soybean, alfalfa, and barley by increasing water use efficiency (Erice et al. 2006, Jin et al. 2005, Li et al. 2013, Radin and Eidenbock 1984, Robredo et al. 2007, Singh, Sale, and McKenzie 1997, Zhang et al. 2018). It has also been reported by Shen et al. (2013) that the application of P can enhance tolerance to water stress by increasing nutrient uptake and plant growth.

Conclusion and Future Prospects

Under changing climate conditions, balanced nutrient management may ameliorate the detrimental effects of climate change (e.g. water stress) on plant growth by improving key growth parameters, such as water use efficiency, photosynthetic activity, sink-source partitioning, and leaf/water relations. However, excess application of nitrogenous and P fertilisers needs to be avoided as, specifically, an increase in reactive nitrogen in the environment is known to contribute to global climate change (responsible for the increase in non-CO_2 greenhouse gases, such as N_2O, and NH_4), as well as pollution of groundwaters. In this context, smart nutrient management techniques need to be developed for increasing the efficiency of fertilisers. The 4R Nutrient Stewardship technology developed by the International Plant Nutrition Institute is one good example of nutrient management with site- and crop-specific applications. The method is called 4R because it includes decisions relying on the right source of nutrient, applied at the right time, and at the right rate and right technique that maximise the nutrient penetration, uptake, and utilisation efficiency. In this review, advanced knowledge of the significance of N, P, K, and Mg in plant development and quality yield production under a changing climate is well described. However, this chapter provides three take-home messages for readers and researchers:

1) Enhancing the water use efficiency of crop plants is one way to sustain yield under a changing climate. In this context, plant K and Mg, as well as N content need to be adequate. The causation of the decrease in WUE under various nutrients differs, therefore site-specific trials need to be done to understand the limitations of effective nutrient management.
2) Supply of K to K-deficit soils may improve response by plants to water-deficit conditions by two different means: i) enhanced adaptive/resistance mechanisms of crop plants due to the above-mentioned physiological activities, and ii) increased available water

content of soils under water-limited conditions by enhancing water-holding capacity (improvement of soil physical properties).

3) As suboptimal content of K and Mg in crop plants specifically affects phloem loading, this may also hamper a plant's ability to develop belowground biomass necessary for seeking water in deeper soil layers, and it may affect its mechanisms to develop gelatinous root cover under drier environmental conditions. Thus, Mg and K deficiency is expected to reduce tolerance to drought in plants because root elongation and root exudation require a huge amount of assimilate transport from the source organs.

References

Abel, S, C A Ticconi, and C A Delatorre. 2002. "Phosphate sensing in higher plants."*Physiologia Plantarum* 115 (1):1–8.

Ahmad, I, and F J M Maathuis. 2014. "Cellular and tissue distribution of potassium: physiological relevance, mechanisms and regulation." *Journal of Plant Physiology* 171 (9):708–714.

Ahmad, Z, E A Waraich, R Ahmad, and M Shahbaz. 2017. "Modulation in water relations, chlorophyll contents and antioxidants activity of maize by foliar phosphorus application under drought stress." *Pakistan Journal of Botany* 49 (1):11–19.

Ahn, J, E J Brook, L Mitchell, J Rosen, J R McConnell, K Taylor, D Etheridge, and M Rubino. 2012. "Atmospheric CO_2 over the last 1000 years: A high-resolution record from the West Antarctic Ice Sheet (WAIS) Divide ice core." *Global Biogeochemical Cycles* 26 (2). doi: 10.1029/2011GB004247.

Allen, C D, A K Macalady, H Chenchouni, D Bachelet, N McDowell, M Vennetier, T Kitzberger, A Rigling, D D Breshears, E H Ted Hogg, P Gonzalez, R Fensham, Z Zhang, J Castro, N Demidova, J-H Lim, G Allard, S W Running, A Semerci, and N Cobb. 2010. "A global overview of drought and heat-induced tree mortality reveals emerging climate change risks for forests." *Forest Ecology and Management* 259 (4):660–684.

Amthor, J S. 1997. "Plant Respiratory Responses to Elevated Carbon Dioxide Partial Pressure." In *Advances In Carbon Dioxide Effects Research: As a Special Publication No. 61*, edited by L H Jr Allen, M B Kirkham, D MOlszyk and C E Whitman, 35–77. Madison, ABD: American Society of Agronomy, Crop Science Society of America, Soil Science Society of America.

Amtmann, A, and M R Blatt. 2009. “Regulation of macronutrient transport.” *New Phytologist* 181 (1):35–52. doi: 10.1111/j.1469-8137.2008.02666.x.

Anderson, J A, S S Huprikar, L V Kochian, W J Lucas, and R F Gaber. 1992. “Functional expression of a probable *Arabidopsis thaliana* potassium channel in *Saccharomyces cerevisiae*.” *Proceedings of the National Academy of Sciences* 89 (9):3736–3740.

Araya, T, K Noguchi, and I Terashima. 2006. “Effects of carbohydrate accumulation on photosynthesis differ between sink and source leaves of *Phaseolus vulgaris* L.” *Plant and Cell Physiology* 47 (5):644–652.

Artuso, F R, A Galatro, A Buet, G E Santa María, and M Simontacchi. 2018. “Key acclimation responses to phosphorus deficiency in maize plants are influenced by exogenous nitric oxide.” *Journal of Plant Physiology* 222:51–58.

Askegaard, M, J Eriksen, and A E Johnston. 2003. “Sustainable Management of Potassium.” In *Managing Soil Quality: Challenges in Modern Agriculture*, edited by Per Schjønning, Susanne Elmholt and Bent Tolstrup Christensen, 85–97. UK: CABI Publishing.

Barker, A V, and D J Pilbeam. 2015. *Handbook of plant nutrition*. Second ed. New York, USA: CRC press.

Barnaby, J Y, D Fleisher, V Reddy, and R Sicher. 2015. “Combined effects of CO_2 enrichment, diurnal light levels and water stress on foliar metabolites of potato plants grown in naturally sunlit controlled environment chambers.” *Physiologia Plantarum* 153 (2):243–252. doi: 10.1111/ppl.12238.

Bazzaz, F A. 1990. “The response of natural ecosystems to the rising global CO_2 levels.” *Annual Review of Ecology and Systematics* 21 (1):167–196.

Bindi, M, L Fibbi, A Frabotta, M Chiesi, G Selvaggi, and V Magliulo. 1999. Free air CO2 enrichment of potato (*Solanum tuberosum* L.). In: CHIP (ed) Changing climate and potential impacts on potato yield and quality: CHIP project, final report; contract ENV4- CT97-0489. Commission of the European Union. Brussels.

Bishop, K A, A D B Leakey, and E A Ainsworth. 2014. “How seasonal temperature or water inputs affect the relative response of C3 crops to elevated [CO_2]: a global analysis of open top chamber and free air CO_2 enrichment studies.” *Food and Energy Security* 3 (1):33–45.

Britto, DT, and H J Kronzucker. 2008. “Cellular mechanisms of potassium transport in plants.” *Physiologia Plantarum* 133 (4):637–650.

Brueck, H, and M Senbayram. 2009. “Low nitrogen supply decreases water-use efficiency of oriental tobacco.” *Journal of Plant Nutrition and Soil Science* 172 (2):216–223.

Bruns, H A. 2016. "Macro-nutrient concentration and content of irrigated soybean grown in the early production system of the Mid-south." *Communications in Soil Science and Plant Analysis* 47 (17):2008–2016.
Buysse, J, E Smolders, and R Merckx. 1996. "Modelling the uptake of nitrate by a growing plant with an adjustable root nitrate uptake capacity." *Plant and Soil* 181 (1):19–23.
Cakmak, I. 2005. "The role of potassium in alleviating detrimental effects of abiotic stresses in plants." *Journal of Plant Nutrition and Soil Science* 168 (4):521–530.
Carminati, A, and D Vetterlein. 2012. "Plasticity of rhizosphere hydraulic properties as a key for efficient utilization of scarce resources." *Annals of Botany* 112 (2):277–290.
Celi, L, and E Barberis. 2005. "Abiotic Stabilization of Organic Phosphorus in the Environment." In *Organic Phosphorus in the Environment*, edited by B L Turner, E Frossard and D SBaldwin, 113–132. Oxfordshire: CAI International.
Cowan, J A. 2002. "Structural and catalytic chemistry of magnesium-dependent enzymes." *Biometals* 15 (3):225–235.
Criado, M V, C G Veliz, I N Roberts, and C Caputo. 2017. "Phloem transport of amino acids is differentially altered by phosphorus deficiency according to the nitrogen availability in young barley plants." *Plant Growth Regulation* 82 (1):151–160. doi: 10.1007/s10725-017-0247-6.
De B, H Jan, M B Eppinga, M J Wassen, and S C Dekker. 2012. "A critical transition in leaf evolution facilitated the Cretaceous angiosperm revolution." *Nature Communications* 3:1221.
Easterling, W E, P K Aggarwal, P Batima, K M Brander, L Erda, S M Howden, A Kirilenko, J Morton, J-François Soussana, and J Schmidhuber. 2007. "Food, fibre and forest products." In *Adaptation and Vulnerability. Contribution of Working Group II to the Fourth Assessment Report of the Intergovernmental Panel on Climate Change*, edited by M L Parry, O F Canziani, J P Palutikof, P J Linden and C E Hanson, 273–313. Cambridge, United Kingdom and New York, NY, USA: Cambridge University Press.
Edwards, E J, S McCaffery, and J R Evans. 2005. "Phosphorus status determines biomass response to elevated CO_2 in a legume: C4 grass community."*Global Change Biology* 11 (11):1968–1981.
Ehlers, I, A Augusti, T R Betson, M B Nilsson, J D Marshall, and J Schleucher. 2015. "Detecting long-term metabolic shifts using isotopomers: CO_2-driven suppression of photorespiration in C3 plants

over the 20th century." *Proceedings of the National Academy of Sciences* 112 (51):15585–15590.
Engels, C, and E A Kirkby. 2001. "Cycling of nitrogen and potassium between shoot and roots in maize as affected by shoot and root growth." *Journal of Plant Nutrition and Soil Science* 164 (2):183–191.
Epstein, E, D W Rains, and O E Elzam. 1963. "Resolution of dual mechanisms of potassium absorption by barley roots." *Proceedings of the National Academy of Sciences of the United States of America* 49 (5):684–692.
Erice, G, J J Irigoyen, P Pérez, R Martínez-Carrasco, and M Sánchez-Díaz. 2006. "Effect of elevated CO_2, temperature and drought on photosynthesis of nodulated alfalfa during a cutting regrowth cycle." *Physiologia Plantarum* 126 (3):458–468.
Farhat, N, M Rabhi, M Krol, Z Barhoumi, A G Ivanov, A McCarthy, C Abdelly, A Smaoui, and N PA Hüner. 2014. "Starch and sugar accumulation in *Sulla carnosa* leaves upon Mg^{2+} starvation." *Acta Physiologiae Plantarum* 36 (8):2157–2165.
Farquhar, G D, J R Ehleringer, and K T Hubick. 1989. "Carbon isotope discrimination and photosynthesis." *Annual Review of Plant Physiology and Plant Molecular Biology* 40 (1):503–537.
Flexas, J, J Galmés, A Gallé, J Gulías, A Pou, M Ribas-Carbo, M Tomàs, and H Medrano. 2010. "Improving water use efficiency in grapevines: potential physiological targets for biotechnological improvement." *Australian Journal of Grape and Wine Research* 16 (1):106–121.
Franks, P J, I J Leitch, E M Ruszala, A M Hetherington, and D J Beerling. 2012. "Physiological framework for adaptation of stomata to CO_2 from glacial to future concentrations." *Philosophical Transactions of the Royal Society of London B: Biological Sciences* 367 (1588):537–546.
Gago, J, C Douthe, I Florez-Sarasa, J M Escalona, J Galmes, A R Fernie, J Flexas, and H Medrano. 2014. "Opportunities for improving leaf water use efficiency under climate change conditions." *Plant Science* 226:108–119.
Gama-Rodrigues, A C, M V S Sales, P S D Silva, N B Comerford, W P Cropper, and E Gama-Rodrigues. 2014. "An exploratory analysis of phosphorus transformations in tropical soils using structural equation modeling." *Biogeochemistry* 118 (1–3):453–469.
Glibert, P M, F P Wilkerson, R C Dugdale, J A Raven, C L Dupont, P R Leavitt, A E Parker, JA M Burkholder, and T M Kana. 2016. "Pluses and minuses of ammonium and nitrate uptake and assimilation by phytoplankton and implications for productivity and community

composition, with emphasis on nitrogen-enriched conditions." *Limnology and Oceanography* 61 (1):165–197. doi: 10.1002/lno.10203.

Graciano, C, J J Guiamét, and J F Goya. 2005. "Impact of nitrogen and phosphorus fertilization on drought responses in *Eucalyptus grandis* seedlings." *Forest Ecology and Management* 212 (1–3):40–49.

Gransee, A, and H Führs. 2013. "Magnesium mobility in soils as a challenge for soil and plant analysis, magnesium fertilization and root uptake under adverse growth conditions." *Plant and Soil* 368 (1–2):5–21.

Gupta, A S, G A Berkowitz, and P A Pier. 1989. "Maintenance of photosynthesis at low leaf water potential in wheat: role of potassium status and irrigation history." *Plant Physiology* 89 (4):1358–1365.

Hepworth, C, T Doheny-Adams, L Hunt, D D Cameron, and J E Gray. 2015. "Manipulating stomatal density enhances drought tolerance without deleterious effect on nutrient uptake." *New Phytologist* 208 (2):336–341.

Hermans, C, F Bourgis, M Faucher, R J Strasser, S Delrot, and N Verbruggen. 2005. "Magnesium deficiency in sugar beets alters sugar partitioning and phloem loading in young mature leaves." *Planta* 220 (4):541–549.

Høgh Jensen, H, and M B Pedersen. 2003. "Morphological plasticity by crop plants and their potassium use efficiency." *Journal of Plant Nutrition* 26 (5):969–984.

Hou, Enqing, C Chen, Y Kuang, Y Zhang, M Heenan, and D Wen. 2016. "A structural equation model analysis of phosphorus transformations in global unfertilized and uncultivated soils." *Global Biogeochemical Cycles* 30 (9):1300–1309.

Houghton, J Theodore, B A Callander, and S K Varney. 1990. *Climate Change: The IPCC Scientific Assessmen*. Cambridge,UK: Cambridge University Press.

Hsiao, T C, and A Lauchli. 1986. "Role of Potassium in Plant-Water Relations." In *Advances in Plant Nutrition (USA)*, edited by B Tinker and A Lauchli, 281–312. New York: Praeger Scientific.

Jákli, B, E Tavakol, M Tränkner, M Senbayram, and K Dittert. 2017. "Quantitative limitations to photosynthesis in K deficient sunflower and their implications on water-use efficiency." *Journal of Plant Physiology* 209:20–30.

Jin, J, D Lauricella, R Armstrong, P Sale, and C Tang. 2015. "Phosphorus application and elevated CO_2 enhance drought tolerance in field pea grown in a phosphorus-deficient Vertisol." *Annals of Botany* 116 (6):975–985.

Jin, J, C Tang, R Armstrong, C Butterly, and P Sale. 2013. "Elevated CO_2 temporally enhances phosphorus immobilization in the rhizosphere of wheat and chickpea." *Plant and Soil* 368 (1–2):315–328.

Jin, J, C Tang, R Armstrong, and P Sale. 2012. "Phosphorus supply enhances the response of legumes to elevated CO_2 (FACE) in a phosphorus-deficient Vertisol." *Plant and Soil* 358 (1–2):86–99.

Jin, J, G Wang, X Liu, X Pan, S J Herbert, and C Tang. 2006. "Interaction between phosphorus nutrition and drought on grain yield, and assimilation of phosphorus and nitrogen in two soybean cultivars differing in protein concentration in grains." *Journal of Plant Nutrition* 29 (8):1433–1449.

Jin, J, G Wang, X Liu, X Pan, and S J Herbert. 2005. "Phosphorus application affects the soybean root response to water deficit at the initial flowering and full pod stages." *Soil Science and Plant Nutrition* 51 (7):953–960.

Jones, R J, and T A Mansfield. 1970. "Suppression of stomatal opening in leaves treated with abscisic acid." *Journal of Experimental Botany* 21 (3):714–719.

Jun, W, L Ping, L Zhiyong, W Zhansheng, L Yongshen, and G Xinyuan. 2017. "Dry matter accumulation and phosphorus efficiency response of cotton cultivars to phosphorus and drought." *Journal of Plant Nutrition* 40 (16):2349–2357.

Jungk, A. 2001. "Root hairs and the acquisition of plant nutrients from soil." *Journal of Plant Nutrition and Soil Science* 164 (2):121–129.

Kafkafi, U. 1990. "The functions of plant K in overcoming environmental stress situations." 22nd Colloquium, International Potash Institute, Switzerland.

Keenan, T, S Sabate, and C Gracia. 2010. "The importance of mesophyll conductance in regulating forest ecosystem productivity during drought periods." *Global Change Biology* 16 (3):1019–1034.

Kimball, B A, J Zhu, L Cheng, K Kobayashi, and M Bindi. 2002. "Responses of agricultural crops to free-air CO_2 enrichment." In *Advances in Agronomy*, 293–368. Elsevier.

Lambers, H, W S Michael, D C Michael, J P Stuart, and E J Veneklaas. 2006. "Root structure and functioning for efficient acquisition of phosphorus: matching morphological and physiological traits." *Annals of Botany* 98 (4):693–713.

Leakey, A D B, E A Ainsworth, C J Bernacchi, A Rogers, S P Long, and D R Ort. 2009. "Elevated CO_2 effects on plant carbon, nitrogen, and water relations: six important lessons from FACE." *Journal of Experimental Botany* 60 (10):2859–2876.

Leigh, R A, and R G Wyn Jones. 1984. "A hypothesis relating critical potassium concentrations for growth to the distribution and functions of this ion in the plant cell." *New Phytologist* 97 (1):1–13.

Li, D, H Liu, Y Qiao, Y Wang, Z Cai, B Dong, C Shi, Y Liu, X Li, and M Liu. 2013. "Effects of elevated CO_2 on the growth, seed yield, and water use efficiency of soybean (*Glycine max* (L.) Merr.) under drought stress." *Agricultural Water Management* 129:105–112.

Long, S P, E A Ainsworth, A Rogers, and D R Ort. 2004. "Rising atmospheric carbon dioxide: plants FACE the future." *Annual Review of Plant Biology* 55:591–628.

Lynch, J, and A Läuchli. 1984. "Potassium transport in salt-stressed barley roots." *Planta* 161 (4):295–301.

Maathuis, F J, and D Sanders. 1994. "Mechanism of high-affinity potassium uptake in roots of *Arabidopsis thaliana*." *Proceedings of the National Academy of Sciences* 91 (20):9272–9276.

Marble, S C, S A Prior, G B Runion, H A Torbert, C H Gilliam, and G B Fain. 2011. "The importance of determining carbon sequestration and greenhouse gas mitigation potential in ornamental horticulture." *HortScience* 46 (2):240–244.

Marschner, H. 2011. *Marschner's Mineral Nutrition of Higher Plants*: Academic press.

Medlyn, B E, R A Duursma, D Eamus, D S Ellsworth, I C Prentice, C VM Barton, K Y Crous, P De Angelis, M Freeman, and L Wingate. 2011. "Reconciling the optimal and empirical approaches to modelling stomatal conductance." *Global Change Biology* 17 (6):2134–2144.

Mengel, K, and S Diedrich. 1985. "Potassium uptake of ryegrass (*Lolium perenne*) and red clover (*Trifolium pratense*) as related to root parameters." *Biology of Fertility of Soils* 1:53–58.

Mengel, K, E A Kirkby, H Kosegarten, and T Appel. 2001. "Potassium." In *Principles of Plant Nutrition*, 481–511. Dordrecht: Springer.

Möller, K, A Oberson, E K Bünemann, J Cooper, J K Friedel, N Glaesner, S Hörtenhuber, A-K Løes, P Mäder, and G Meyer. 2018. "Improved phosphorus recycling in organic farming: navigating between constraints." In *Advances in Agronomy*, 159–237. Elsevier.

Nafziger, E D, and H R Koller. 1976. "Influence of leaf starch concentration on CO_2 assimilation in soybean." *Plant Physiology* 57 (4):560–563.

Neuhaus, C, C-M Geilfus, and K-H Mühling. 2014. "Increasing root and leaf growth and yield in Mg-deficient faba beans (*Vicia faba*) by MgSO4 foliar fertilization." *Journal of Plant Nutrition and Soil Science* 177 (5):741–747.

Oborn, I, Y A-Rangel, M Askekaard, C A Grant, C A Watson, and A C Edwards. 2005. "Critical aspects of potassium management in agricultural systems." *Soil Use and Management* 21 (1):102–112.

Pachauri, R K, M R Allen, V R Barros, J Broome, W Cramer, R Christ, J A Church, L Clarke, Q Dahe, and P Dasgupta. 2014. *Climate change 2014: Synthesis Report. Contribution of Working Groups I, II and III to the Fifth Assessment Report of the Intergovernmental Panel on Climate Change (Original text)*: IPCC.

Poorter, H, and M-L Navas. 2003. "Plant growth and competition at elevated CO_2: on winners, losers and functional groups." *New Phytologist* 157 (2):175–198.

Prior, S A, H A Torbert, G B Runion, and H H Rogers. 2003. "Elevated atmospheric CO_2 in agroecosystems: residue decomposition in the field." *Environmental Management* 33 (1):S344–S354.

Radin, J W, and M P Eidenbock. 1984. "Hydraulic conductance as a factor limiting leaf expansion of phosphorus-deficient cotton plants." *Plant Physiology* 75 (2):372–377.

Rengel, Z, and P M Damon. 2008. "Crops and genotypes differ in efficiency of potassium uptake and use." *Physiologia Plantarum* 133 (4):624–636.

Robredo, A, U Pérez-López, H S de la Maza, B González-Moro, M Lacuesta, A Mena-Petite, and A Munoz-Rueda. 2007. "Elevated CO2 alleviates the impact of drought on barley improving water status by lowering stomatal conductance and delaying its effects on photosynthesis." *Environmental and Experimental Botany* 59 (3):252–263.

Rogers, H, and C D Roger. 1993. "Crop responses to CO_2 enrichment." *Vegetatio* 104 (1):177–131.

Römheld, V, and E A Kirkby. 2007. "Magnesium functions in crop nutrition and yield." Cambridge.

Römheld, V, and E A Kirkby. 2010. "Research on potassium in agriculture: needs and prospects." *Plant and Soil* 335 (1–2):155–180.

Ruiz-Vera, U M, A P De Souza, S P Long, and D R Ort. 2017. "The Role of Sink Strength and Nitrogen Availability in the Down-Regulation of Photosynthetic Capacity in Field-Grown *Nicotiana tabacum* L. at Elevated CO_2 Concentration." *Frontiers in Plant Science* 8:998.

Rychter, A M, and I M Rao. 2005. "Role of phosphorus in photosynthetic carbon metabolism." In *Handbook of photosynthesis*, edited by M Pessarakli, 123–148. Tucson, AZ: Taylor & Francis Group, LLC.

Sanchez, C A. 2007. "Phosphorus." In, edited by A V Barker and David J Pilbeam. New York: Handbook of Plant Nutrition. CRC Press, Taylor and Francis Group.

Senbayram, M, A Gransee, V Wahle, and H Thiel. 2015a. "Role of magnesium fertilisers in agriculture: plant–soil continuum." *Crop and Pasture Science* 66 (12):1219–1229.

Senbayram, M, M Tränkner, K Dittert, and H Brück. 2015b. "Daytime leaf water use efficiency does not explain the relationship between plant *N status* and biomass water-use efficiency of tobacco under non-limiting water supply." *Journal of Plant Nutrition and Soil Science* 178 (4):682–692.

Shabala, S, and I Pottosin. 2014. "Regulation of potassium transport in plants under hostile conditions: implications for abiotic and biotic stress tolerance." *Physiologia Plantarum* 151 (3):257–279.

Shaul, O. 2002. "Magnesium transport and function in plants: the tip of the iceberg." *Biometals* 15 (3):307–321.

Shen, J, C Li, G Mi, L Li, L Yuan, R Jiang, and F Zhang. 2013. "Maximizing root/rhizosphere efficiency to improve crop productivity and nutrient use efficiency in intensive agriculture of China." *Journal of Experimental Botany* 64 (5):1181–1192.

Singh, DJ K, P WG Sale, and B M McKenzie. 1997. "Water relations of white clover (*Trifolium repens* L.) in a drying soil, as a function of phosphorus supply and defoliation frequency." *Australian Journal of Agricultural Research* 48 (5):675–682.

Singh, DK, P WG Sale, C K Pallaghy, and B M McKenzie. 2000. "Phosphorus concentrations in the leaves of defoliated white clover affect abscisic acid formation and transpiration in drying soil." *New Phytologist* 146 (2):249–259.

Springob, G, and J Richter. 1998. "Measuring interlayer potassium release rates from soil materials. II. A percolation procedure to study the influence of the variable 'solute K'in the< 1–10 μM range." *Journal of Plant Nutrition and Soil Science* 161 (3):315–322.

Sreeharsha, R V, K M Sekhar, and A R Reddy. 2015. "Delayed flowering is associated with lack of photosynthetic acclimation in Pigeon pea (*Cajanus cajan* L.) grown under elevated CO_2." *Plant Science* 231:82–93.

Swann, A LS, F M Hoffman, C D Koven, and J T Randerson. 2016. "Plant responses to increasing CO_2 reduce estimates of climate impacts on drought severity." *Proceedings of the National Academy of Sciences* 113 (36):10019–10024.

Taub, D R, and X Wang. 2008. "Why are nitrogen concentrations in plant tissues lower under elevated CO_2? A critical examination of the hypotheses." *Journal of Integrative Plant Biology* 50 (11):1365–1374.

Tränkner, M, B Jákli, E Tavakol, C-M Geilfus, I Cakmak, K Dittert, and M Senbayram. 2016. "Magnesium deficiency decreases biomass water-use efficiency and increases leaf water-use efficiency and oxidative stress in barley plants." *Plant and Soil* 406 (1–2):409–423.

Verbruggen, N, and C Hermans. 2013. "Physiological and molecular responses to magnesium nutritional imbalance in plants." *Plant and Soil* 368 (1–2):87–99.

Vitousek, P M, and O A Chadwick. 2013. "Pedogenic thresholds and soil process domains in basalt-derived soils." *Ecosystems* 16 (8):1379–1395.

Vitousek, P M, S Porder, B Z Houlton, and O A Chadwick. 2010. "Terrestrial phosphorus limitation: mechanisms, implications, and nitrogen–phosphorus interactions." *Ecological Applications* 20 (1):5–15.

Wang, Y, and W-H Wu. 2013. "Potassium transport and signaling in higher plants." *Annual Review of Plant Biology* 64:451–476.

Wang, Y, and W-H Wu. 2015. "Genetic approaches for improvement of the crop potassium acquisition and utilization efficiency." *Current opinion in plant biology* 25:46–52.

Wegner, L H. 2013. "Root pressure and beyond: energetically uphill water transport into xylem vessels?" *Journal of Experimental Botany* 65 (2):381–393.

Whitehead, David Charles. 2000. *Nutrient Elements In Grassland: Soil-Plant-Animal Relationships*. Wallingford, UK: CABI Publisher.

Wullschleger, S D, T J Tschaplinski, and R J Norby. 2002. "Plant water relations at elevated CO_2–implications for water-limited environments." *Plant, Cell and Environment* 25 (2):319–331.

Wyn Jones, R G, C J Brady, and J Speirs. 1979. "Lonic and osmotic relations in plant cells." In *Recent Advances in the Biochemistry of Cereals*, edited by DL Laidman and RG Wyn Jones, 63–103. London: Academic Press.

Zain, N A M, and M R Ismail. 2016. "Effects of potassium rates and types on growth, leaf gas exchange and biochemical changes in rice (*Oryza sativa*) planted under cyclic water stress." *Agricultural Water Management* 164:83–90.

Zhang, B, H Zhang, H Wang, P Wang, Y Wu, and M Wang. 2018. "Effect of Phosphorus Additions and Arbuscular Mycorrhizal Fungal Inoculation on the Growth, Physiology, and Phosphorus Uptake of Wheat Under Two Water Regimes." *Communications in Soil Science and Plant Analysis* 49 (7):862–874.

Zörb, C, M Senbayram, and E Peiter. 2014. "Potassium in agriculture–status and perspectives." *Journal of Plant Physiology* 171 (9):656–669.

CHAPTER 17

STRUCTURAL AND FUNCTIONAL GENOMICS OF RICE

AYESHA SAJID,[1] AQSA ABBASI,[1] ALVINA GUL,[1*] MUNIR OZTURK,[2] VOLKAN ALTAY,[3] MIRZA HASANUZZAMAN[4]

[1]Atta-ur-Rahman School of Applied Biosciences, National University of Sciences and Technology, Islamabad, Pakistan
[2]Botany Department & Centre for Environmental Sciences, Ege University, Izmir, Turkey
[3]Biology Department, Faculty of Science and Arts, Hatay Mustafa Kemal University, Hatay, Turkey
[4]Departmet of Agronomy, Sher-e-Bangla Agricultural University, Dhaka-1207, Bangladesh
alvina_gul@yahoo.com; munirozturk@gmail.com

Abstract

Rice has been effectively used as a model plant for the genomic study of other crops. Its genome size is 430 Mb, having 12 chromosomes and >50,000 genes. There were four independent projects initiated for the sequencing of the rice genome. A high-value map-based draft sequence was released by IRGSP in 2002. Different techniques, including cytogenetics, molecular genetics and physical mapping have been used for constructing a rice genome map, thus providing information regarding general features of the rice genome. The knowledge obtained from structural genomics laid the foundation for functional genomics of rice by identifying the structure and function of genes.

Keywords: structural genomics, functional genomics, sequencing strategies, cytogenetic mapping, physical mapping, a microarray

Introduction

Rice, a staple food in most countries is an important grain crop. It is widely consumed, especially in Asia (*Oryzasativa*; Asian rice) and Africa (*O.glaberrima*; African rice) (Kennedy 2002). Both these rices are closely related species, and later species have better yield growth and tolerance to abiotic stress (Chen et al. 2018, Farooq, Saqib, and Akhtar 2015, Kikuta et al. 2017, Nada and Abogadallah 2018, Sikirou et al. 2018). Rice grain has one of the highest productions and has great economic value (Lu et al. 2018). It is estimated that 25% of the caloric requirements of half of the population around the globe are met by the consumption of rice (Kusano et al. 2015, Lu et al. 2018). Besides being the principal food consumed by half of the world (Sasaki and Burr 2000), rice is an ideal, model plant for studying other plant species, like wheat, maize and other cereals (Yuan et al. 2005). Its small-sized genome, elaborate genetic maps, dual nature, and relatively easy transformation and relationship with other crops (Moore et al. 1995) make it a suitable candidate for genomic studies (Wu et al. 2002, Yamamoto and Sasaki 1997).

The aim of the genome sequencing of rice was to obtain information that can be used to identify the structure and function of its genes. After genome sequencing, it was proposed in a project, "Rice 2020", that the function of rice genes must be determined by the year 2020. This will help in improving the traits of this crop and lead to increased production of rice (Zhang and Wang 2015).

Structural Genomics of Rice

Structural genomics deals with the structural determination and three-dimensional organisations of different proteins encoded by the organism's genes. Manipulations of rice plant characters, such as high product yield and stress-resistant abilities require complete knowledge of its structural and functional features, thus making it essential to identify and discover rice genes and proteins.

Strategies to the sequence rice genome

The DNA to be sequenced is fragmented, cloned, and then sequenced. The overlapping sequences are then arranged into contigs using computer programmes and software. The sequence quality put forward in the database is classified into four phases. Phase 0 includes the production of the preliminary raw sequence which, after assembly, is converted into phase I.

The phase II sequence is generated after orienting and ordering of the contigs. Until the last and final stage, inconsistent quality and incomplete draft sequences are generated which, after completion, are classified as phase III (Buell 2002).

To sequence a complete genome of rice, two approaches are used. The first is the clone-by-clone shotgun approach, which uses bacterial artificial chromosome (BAC) vectors. The DNA insert sequence present in the vectors is physically attached to the genome using DNA markers, leading to the development of a minimum tilling path (MTP), which is created by different techniques. The bacterial-derived artificial chromosome present in the MTP is broken down and sequenced, and the sequence of the genome can be obtained by joining the individual bacterial artificial chromosome sequences (Buell 2002, Green 2001). This cost-effective technique has provided us with a high-quality rice genome sequence, in addition to the wheat genome sequence (Sasaki 2005).

The second approach is a whole shotgun approach (WGS), in which the genome is directly sequenced after fragmentation and cloning, and these sequences are then assembled for the reconstruction of the whole genome (Green 2001). This process has been used for human genome sequencing and also for sequencing different subspecies of rice (*indica*, *japonica*) (Goff et al. 2002). A great disadvantage with WGS is the disassembly of the sequences due to the presence of a large number of repetitive elements, making this process cumbersome and difficult.

The genome sequencing of rice was followed by four groups (Buell 2002). This task was initiated by the International Rice Genome Sequencing Project (IRGSP) (Sasaki and Burr 2000), and the first draft was released by the Monsanto Company, in 2000. The draft was produced by using the clone-by-clone approach containing 52,202 contigs, representing 259Mb, which cover 60% of the rice genome (Barry 2001). Other independent sequencing programmes were launched by Syngenta (Torrey Mesa Research Institute, San Diego, USA) and Beijing Genomics Institute (BGI), China (Goff et al. 2002).

The IRGSP united the resources of different sequencing groups to sequence the rice genome. The resulting sequence had less than a single error in more than 10,000 nucleotides, thus removing all kinds of ambiguities. A high-value draft sequence, which was map-based, was released by IRGSP in 2002. Three chromosomes and two centromeres that had been completely sequenced were published. This allowed the identification of many traits and duplications in the genome, which facilitated comparative genomics between rice and other crops, such as sorghum or maize.

Mapping of the rice genome

A genetic map is a significant way of connecting the nucleotide information with the phenotypic attributes in the rice genome. The primary and important step in comprehending the rice genome at DNA and nucleotide levels is the construction of a linkage map on the basis of polymorphisms, such as simple sequence repeats and restriction fragment length polymorphism (RFLP), etc. Harushima et al. (1998) has published one of the optimum genetic maps of rice, in which the genetic markers used are crucial for the assembly and determining the position of the chromosomes on the large fragments of the DNA that are obtained from the libraries (Harushima et al. 1998). The second foremost step is the fabrication of the physical map.

Cytogenetic Mapping

Different structures and features of the genome and their relative positions can be determined by genome mapping. This can be achieved either through experiments used for analysing the genome or by visually scrutinising the chromosome. Cytogenetic mapping of the rice chromosome focuses on studying it photomicroscopically. *In situ* hybridisation for the analysis of structural constituents of the genome is highly efficient, and the development of techniques, including genomic and fluorescent *in situ* hybridisation (FISH) has made the next generation of high-quality genome maps promising. Genomic *in situ* hybridisation helps in the identification of chromosome complements of the genome (Heng, Spyropoulos, and Moens 1997).

Molecular genetic mapping

The study of linkages based on the physiological and structural characteristics of genes is fundamental in plant genetics. By the utilisation of the secondary and tertiary trisomics, the position of the centromere and the linkage group orientation has been detected, thus providing the exact rice linkage map orientation (Singh et al. 1996). In addition, by combining molecular linkage and classical cytological markers, an extensive molecular linkage map can be developed (Mohan et al. 1997).

Physical mapping

Physical mapping has been proved significant in rice genome mapping. The development of many insert clone-containing libraries (YAC/BAC), and the presence of DNA markers have made feasible the construction of the physical map. The YAC (yeast artificial chromosome) library of rice, composed of almost 7000 clones of YAC, having a 350 Kb size insert (Umehara et al. 1995), was required for constructing the rice genome's physical map. The integrated YAC genetic and physical map has been utilised for the site-mediated cloning of important rice genes, such as pest-resistant genes (Yoshimura et al. 1998) and the assignment of ESTs for the generation of a transcript map of rice covering almost 81% of its genome (Wu et al. 2002). Aside from its many advantages, the YAC physical map is unstable and difficult to purify; therefore, BAC/PAC is more widely used to construct the physical maps.

The physical map, based on BAC, was constructed by using 21.078 clones with 130.147 and a 150 kb insert size (Zhang et al. 1996). These were mapped onto the rice chromosomes with the help of DNA markers, and the rice genome project (Harushima et al. 1998) from electrophoresis-based fingerprinting. This map is a source of rice genomic research studies and comparative genomics. The rice chromosome 4 physical map has been generated recently using the data from the physical map.

Apart from these techniques and methods, the analysis of structural genomics can be achieved by gene tagging, microarray, and use of interfering RNA (iRNA).

General features of rice genome sequence

Rice has a genome size of approximately ~430 Mb and an estimated >50,000 genes (Tyagi et al. 2004). Its genome comprises 12 chromosomes and two circular DNA of mitochondria and chloroplast. Nearly 37,544 genes have been predicted in the complete genome sequence, having 2,699 bp of average gene length. Rice comprises a genetic map of 3,267 markers (Harushima et al. 1998); and sequences for 10,000 unique expressed sequence tags (Yamamoto and Sasaki 1997); and the presence of 6,591 unique expressed sequence tags forming a transcriptional map (Wu *et al.* 2002). Repetitive DNA constitutes almost 50% of the rice genome (Schoof and Karlowski 2003). Transposable elements were identified owing to their similarity with gag and pol genes that are responsible for encoding polyproteins linked to their retroelement function (Sasaki 2001). rRNA units are detected by determining the patterns of consensus sequence patterns (Goff et al. 2002). Gene density of chromosomes 1 and 3 is the highest and lowest of chromosomes 11 and 12 (Sasaki 2005). It consists of 10–25% of repeat elements. The transposable element was large in number in chromosomes 8 and 12—approximately 38% and 38.3%, respectively—whilst fewer in number in chromosomes 1, 2 and 3 (Turcotte, Srinivasan, and Bureau 2001). Three classes of duplications have been determined in the rice genome, including segmental duplication (Paterson, Bowers, and Chapman 2004, Wang et al. 2000), tandem and background duplications (Yu et al. 2005). The genome of rice has an estimated content of GC to be 43.6% (introns: 38.3% and exons: 54.2%) (Sasaki 2005), which is higher than *Arabidopsis*, most importantly in the coding region. The centromere is an important component of the chromosome binding microtubules and is associated with cell division. Rice centromeres are huge and pose difficulties in sequencing; however, some of the rice centromeres of chromosomes 4, 5 and 8 have been completely sequenced (Wu et al. 2004). Tos17, an endogenous retrotransposon, is activated in tissue culture, resulting in transposition, and can be stably transferred during plant regeneration. Two copies of this retrotransposon have been identified in Nippon bare rice cultivar. Tos17 is considered an effective tool for the analysis of functional characteristics of the rice genome because of its insertion into the gene-rich regions.

The chloroplast genome of rice contains four genes for rRNA and 100 different protein-encoding genes. The transcriptional map of the rice chloroplast DNA has been constructed by northern hybridisation of the complete RNA. Many of the chloroplast DNA genes occur in cluster form and are transcribed as long primary transcripts, among which 16 are polycistronic transcripts. The genes responsible for components of photosystem I and II, RuBisCo, ATPases, and cytochrome complex have been known to be expressed in the highest amounts (Kanno and Hirai 1993).

Approximately 564 and 519 tRNA have been found in *indica* and *japonica* rice species, respectively. It is hypothesised that nearly 46 types of tRNA, responsible for decoding all 61 possible anticodons, are present in the rice genome. Many of these tRNA share sequence identity in both species of *O.sativa,* i.e. *indica* and *japonica*, and almost 18 of the rice tRNA have similar counterparts in *Arabidopsis*. The lack of information regarding rRNA is mostly due to their existence in the form of tandem arrays in the heterochromatic area, posing a difficulty in sequencing by the whole genome shotgun method (Wang, Shi, and Hao 2004).

MicroRNAs are a class of non-coding eukaryotic RNA involved in regulating gene expression through interaction with the target mRNA. 158miRNA have been mapped on the rice pseudomolecules. 215 small nucleolar RNA and almost 93 spliceosomal RNA genes have been identified, all of which are non-coding RNAs.

For the annotation of the rice genes, two basic steps are required (Adams et al. 2000):

- Computer-based programmes for the prediction of genes.
- Validation of the predicted genes on the basis of information regarding the gene function present in the databases.

Gene prediction from computer programmes is assisted by the availability of different databases. Some of these gene annotating programmes include Genscan, FGENESH, GeneMark, etc. Multiple websites providing information regarding the annotation of the rice gene are also offered (Rice Genome Automated Annotation System, The Institute for Genomic Research, etc.). TIGR rice gene annotation Osa1 (named after the rice species *O.sativa* 1) is the most extensively used database (Yuan et al. 2005). The rice genes identified from the IRGSP sequence can be categorised into different domains, among which five are the protein kinases. Half of the proteins predicted are associated with biological processes and functioning. Almost 71% homologues between rice gene products and *Arabidopsis* have been found.

Sequences of chromosomes 1, 4 and 10 from the 12 of rice chromosomes have been reported and are discussed as follows.

Chromosome 1

Bacterial and (P1)-derived artificial chromosome vectors have been used to determine the sequence of chromosome 1 in an assembly of nine contigs (Sasaki, Matsumoto, et al. 2002). Its size is 45.7 Mb and it has the least heterochromatic content (Cheng et al. 2001). BLASTP computer programme is used to classify the predicted proteins on chromosome 1. The most common gene family belonged to the serine/threonine receptor kinase family, the majority of which are distributed along the chromosome. Genetic analysis of chromosome 1 revealed many biologically significant genes, including genes controlling dwarfism and reproductive ability (Sasaki, Ashikari, et al. 2002).

Chromosome 4

The sequence of chromosome 4 spans 34.6Mb and contributes to 97.3% representation of the chromosome (Feng et al. 2002). The centromere of chromosome 4 is the longest (124kb), with 379 CentO and 19 centromeric retro-elements, among which LTR retrotransposons is the most important family (Zhang et al. 2004). 4,658 genes encoding proteins and 70 tRNA genes have been predicted (Sasaki, Ashikari, et al. 2002).

Chromosome 10

It is considered the smallest chromosome in the rice genome, having a size of 22.4 Mbp and predicted 3,471 genes (Sasaki 2005). Heterochromatin material, rich in repetitive elements and expressed genes on 10S and 10L, respectively, is present in this chromosome. Various insertions from organellar genomes have been determined (Consortium 2003).

Chromosome 11 and 12

Chromosome 12 is a little shorter compared to chromosome 11, having lengths of 27.5Mb and 28.4Mb, respectively. Genes identified on these chromosomes are considered to be related to non-transposable elements (Rice Chromosomes 2005).

The attainment of good quality, and the completely sequenced genome of the rice crop, have led to the determination of gene function. Genes responsible for multiple agronomic traits, having economic importance, can be identified. The knowledge of the sequences of the rice genome has provided us with great insight regarding the gene content, structure, length and different characteristics of the rice chromosomes. The studies on rice structural genomics have provided opportunities to find the function of genes.

The Functional Genomics of Rice

Functional genomics involves the assessment of the function of different genes by utilising information available through structural genomics (Hieter and Boguski 1997). The function of a gene can be determined through mutagenesis, and many studies have been carried out to develop mutant resources of rice genes for the exploration of the function, along with their expression (Wang et al. 2013, Wei et al. 2013, Yang, Li, and Wu 2013).

Map-based cloning

Rice is an important plant system used in the genetic analysis of various mutants (such as disease resistance, size of seed and time required for flowering) because of its easy handling and simple, redundant genes (Delseny et al. 2001, Vo et al. 2018). Map-based (positional) cloning involves the isolation of genes on the basis of their position in the map. This process includes the isolation of the genes after mutations are mapped from BAC/YAC libraries. This method has been used for the identification of essential rice genes by chromosome walking and adding many markers (RFLP, EST), etc., in the desired gene of interest.

Xa21 was the first gene isolated by this method, responsible for providing resistance for rice against blight disease (Song et al. 1995). Similarly, Xa1, Sp17 and sd1, involved in bacterial blight resistance, heat resistance and biosynthesis of gibberellins (Monna et al. 2002, Yamanouchi et al. 2002, Yang et al. 2003, Yoshimura et al. 1998), have also been isolated with the help of positional cloning. This method was used to characterise twenty-eight genes conferring resistance to brown planthopper (BPH), which is considered a major pest of the rice crop (Wu et al. 2014), and all of them are located on six out of 12 chromosomes of *O.Sativa* (Cheng, Zhu, and He 2013). Another resistant gene identified is *BPH29* comprising B3 DNA-binding domain, and is expressed in the vascular tissue, which is the target site of BPH (Wang et al. 2015).

Gene tagging by insertional mutagenesis

In insertional mutations, a small sequence of DNA is inserted into the gene of interest, which destroys or blocks the gene expression resulting in mutations leading to visible phenotype variation, thus making screening easy. The gene of interest is also tagged by the added DNA sequence and is isolated by different hybridisation probes. This technique is beneficial and easy as it does not require any previous information about the gene. Genes are effectively isolated when T-DNA is used as tags, and this system of tagging has been discussed for rice (*japonica*) (Sallaud et al. 2003).

Transposon tags

A transposon is an efficient tool for isolating genes. The exploitation of mutagenic ability and the potential for tagging the mutated sequence of the mobile elements can be used for isolating genes for cloning purposes (Hamer et al. 2001). Transposons have a preference for gene-encoding proteins. Gene tagging in rice has been successfully achieved by the maize Ac-Ds system (Greco et al. 2001).

Retrotransposons transposed by RNA intermediate are also potential gene tags. Tobacco retrotransposon, *Tto1*, has been successfully introduced by reverse transcription into rice (Hirochika et al. 1996). Rice has its own retrotransposon families that can be activated to carry specific insertional mutations during regeneration or tissue culture, such as Tos17. Genes causing dwarfism, semi-dwarfism and narrow leaf phenotypes have been detected (Hirochika et al. 2001).

Microarrays

The expression of genes is rapidly analysed using microarrays technology. It also helps in the identification of genes activated against abiotic (Hossain et al. 2016) and biotic stress responses (Wu et al. 2016). The gene chip microarray of the rice plant has been created that was used to analyse the gene expression and identified genes involved in carbohydrates, fatty acids, proteins synthesis and transportation (Zhu et al. 2003). Recently, OsGHd2 and SALP1genes have been identified, implicated in the enhanced rice grains yield and salt tolerance, respectively (Liu et al. 2016, Yuan et al. 2016). The function of p68, a DEAD-box protein family member, was analysed through microarray, which revealed that Psp68 protein is involved in the regulation of signalling pathways, such as reactive oxygen species stress response (Banu et al. 2015).

Genome-wide association study (GWAS)

This approach is used for the identification of genetic variations associated with a particular trait. This would assist in analysing the relationship between genotypes and phenotypes, thus contributing significantly to plant breeding. This method is used to locate single nucleotide polymorphisms in the genome to specifically locate genetic markers (Si et al. 2015). GWAS has contributed to determining the genetic makeup of complex agronomic traits in many plants, including maize, *Arabidopsis*, rice, barley, and wheat, etc. (Chen et al. 2016, McCouch et al. 2016, Pasam et al. 2012). Moreover, it assists in the identification of genes associated with the domestication of rice, rice grain quality, yield and genes implicated in biotic and abiotic stress responses (Crowell et al. 2016, Kang et al. 2016, Ueda et al. 2014, Zhao et al. 2011). It helps in the identification of three candidate genes that show differential expression patterns in cold, susceptible and resistant varieties of rice. This would be of great importance for producing rice varieties tolerant to cold stress in various breeding programmes (Zhang et al. 2018).

Serial analysis of gene expression

This approach permits a swift and detailed analysis of many transcripts in one attempt. This technique can provide quantitative information of the already known transcripts, and can recognise the newly expressed genes (Velculescu et al. 1995). Long Sage is the advanced version of this technique, which is also accessible (Saha et al. 2002). Analysis of the expressed genes showed that a large number of genes expressed in abundance are related to the housekeeping genes that are responsible for coding proteins of the ribosomes, cell structure, and metabolism, and almost 2.7% of gene expression was of metallothionein. Microarrays, in combination with SAGE, can provide an effective tool for the identification of functional genes.

Micro RNA based gene regulation

MicroRNAs are small RNA molecules regulating the expression of genes at the post-transcriptional level, either by degrading target mRNA or repressing translation. This is an efficient method for studying the functional genomics of plants and regulating their major physiological traits to improve quality and yield (Xie, Zhang, and Yu 2015, Zhang 2015, Zhang and Wang 2015). It is demonstrated that the increased expression of miR319 in rice results in wider-sized leaves and also mediates tolerance to cold (Yang et al. 2013). The phytohormone, Auxin, plays an important part in the growth of roots in plants (Bellini, Pacurar, and Perrone 2014, Gutierrez et al. 2012), so microRNA-mediated modification of the auxin pathway regulates root development, as indicated by various studies (Curaba, Singh, and Bhalla 2014, Jin et al. 2013). One such study shows that the over-expression of two microRNAs of rice, such as OsmiR393a and OsmiR393b control elongation of primary roots and several adventitious roots by altering auxin-mediated signalling pathways (Bian et al. 2012).

CRISPR/Cas9 Genome Editing System

The clustered, regularly interspaced, short palindromic repeat (CRISPR)-associated protein9 (Cas9) system is employed for editing of the genome, and it enables mutation at a highly specific and targeted position for genetic enhancement in plants (Doudna and Charpentier 2014, Georges and Ray 2017). It involves endonucleases, such as sequence-specific nuclease (SSN), transcription activator-like effector nucleases, zinc finger nucleases, and CRISPR-linked nucleases. They introduce double-strand breaks at specific sites in DNA molecules and activate DNA repair pathways and homologous recombination (Alamar et al. 2017, Čermák et al. 2015, Palmgren et al. 2015). Furthermore, plants produced after gene editing are free from transgenes because the marker gene along with the transgene is removed by segregation (Lawrenson et al. 2015, Zhou et al. 2014).

In *O.sativa*, resistance to bispyribac sodium (herbicide) was achieved by incorporating two mutations in the acetolactate synthase gene using CRISPR technology (Sun et al. 2016). In addition, several genes in rice have been targeted using this approach, such as *OsSWEET11*, *OsSWEET14*, *OsSPP*, *OsYSA*, *Oso2g23823*, *OsMPK2* and *OsBEL* genes. Recent reports show the construction of the rice mutant library using the CRISPR/Cas9 system, and its potential benefits for identifying and improving a gene's functions in rice (Meng et al. 2017).

Functional identification of genes having agronomic importance

A gene responsible for tillering in rice

Tillering is an important trait for the development of grains of many crops, including rice (Sreenivasulu and Schnurbusch 2012). A tiller is an independently growing branch developed on the basal internodes and bear grains. The MONOCULM 1 (MOC1) gene, involved in tillering in rice, has been identified and characterised. MOC1 gene encodes a family of nuclear proteins belonging to the GRAS family that are expressed and initiate the function of axillary buds responsible for tillering (Li et al. 2003). A WUS (WUSCHEL) protein, involved in the development of shoot apical meristem, also regulates the growth of tillers, thus controlling the rice production (Tanaka et al. 2015, Wang et al. 2014).

Major and minor loci for grain length, width, weight, and thickness

The analysis of grain size at genetic and molecular levels is essential for the advancement in the production of the rice crop. Multiple quantitative trait loci have been detected, associated with grain size. The most important is Grain Size 3 (GS3) on chromosome 3, which has been isolated through map-based cloning (Fan et al. 2006); and it has a major effect on the size of the rice grain (Tan et al. 2000, Xing et al. 2002). Almost 80%–90% of the changes in rice grain weight and length are associated with *GS3*. These loci act as a minor QTL for grain width as well as thickness (Han et al. 2007). Another such locus, *GL3.1*, is also characterised as controlling rice grain size and yield by modulating phosphorylation of cell cycle regulation proteins (Qi et al. 2012). The over-expression of QTL, such as *GS5* and *GW8,* results in increased grain width and production of rice. *GS5* encodes a serine carboxypeptidase, and is considered as a major QTL controlling grain size (Li et al. 2011), whilst *GW8*, also known as *OsSPL16*, regulates proliferation of cell cycle (Wang et al. 2012). Research findings have revealed that the CYP78A13 gene controls the size of rice grain through cell cycle regulatory pathways. The expression of CYP78A13 was observed in the big grain 2 (bg2-D) rice mutant, and its activation resulted in the greater-sized rice grain (Xu et al. 2015).

Tolerance to phosphate starvation

The tolerance to phosphate deficiency is provided by a transcription factor OSPTF1 (Yi et al. 2005). It is expressed in lateral roots and only in phloem cells in primary roots. It is also involved in effective phosphate recycling in case of deficiency of phosphate. Another study showed that the over-expression of the *PSTOL1* gene is associated with the increased production of the modern varieties of rice which are not tolerant to phosphorous deficiency. Moreover, this gene also enhances the early growth of roots, thus facilitating plants to acquire phosphorous efficiently from soil (Gamuyao et al. 2012).

Disease resistant genes

The most common and serious rice disease is bacterial blight. By using map-based cloning, a gene Xa26 and another gene Xa4 were isolated, conferring resistance to this disease (Han et al. 2007, Sun et al. 2004). Almost 289 resistance genes, and 28 genes associated with defence responses, have been discovered on chromosomes 11 and 12. The XA21 receptors have an important role in innate immunity in rice. These receptors identify RaxX, a peptide from bacterium *Xanthomonas oryzae* pv. *oryzae* (Xoo), which is responsible for blight disease in rice, and activates immune responses. Another protein, XB21, was identified conferring increased resistance to Xoo in rice. Its enhanced expression results in the upregulation of vesicular transport and cell-death-associated genes (Park et al. 2017). There are several studies that have described the role of transcription factor MYC2 in regulating immunity in rice by providing resistance against blight disease. Its overexpression results in the activation of genes linked with the jasmonic acid signalling pathway, thus conferring resistance against the pathogen. It also controls the expression of genes related to senescence and regulates development of spikelet in *O.sativa* (Giri et al. 2017, Ogawa, Kawahara-Miki, et al. 2017, Ogawa, Miyamoto, et al. 2017, Uji, Akimitsu, and Gomi 2017, Uji et al. 2016).

The fungus *Magnaportheoryzae* causes a destructive blast disease in rice that affects quality and yield. Recently, a new gene was found in the rice genome that confers resistance to variants of *M.oryzae*, thus resistance at a broader level can be achieved with the expression of this gene that induces changes in the single base sequence in the promoter region of the bsr-d1 gene. As a result, the expression of the bsrd1 gene is suppressed due to the binding of transcription factor MYB. This ultimately leads to reduced expression of peroxidases and increased immunity against the fungal pathogen. Among 3000 varieties of rice, the SNP33-G allele is found in 10% of varieties (Li et al. 2017, Takatsuji and Hayashi 2017). Furthermore, the *WRKY45* transcription factor has an important role in providing resistance to the above-mentioned bacterial and fungal pathogen. It activates genes associated with defence responses (Shimono et al. 2012). Similarly, *OsWRKY30* gene also mediates resistance against *M.oryzae* (Peng et al. 2012).

Floral organ identity genes

By studying the loss of function strategy, the function of the two MADS-box genes has been determined. The Os-MADS4 gene is specific for the identity of stamen and petals, whilst OsMADS3 controls the formation of the carpel and stamen (inner whorls). The mutant rice plants showed changes in their whorls (Kang et al. 1998). The identification and characterisation of polycombgroup (PcG) genes, such as *OsFIE1*, *OsFIE2* and *OsEMF2b* have demonstrated their role in the development of floral organs in rice (Nallamilli et al. 2013, Yang, Li, and Wu 2013, Zhang et al. 2012). The role of epigenetic factors in the development of floral parts in rice was determined in a study. It showed that the DEFORMED FLORAL ORGAN 1 (OsDFO1) gene epigenetically represses the expression of floral identity gene *OsMADS58*, resulting in the floral organ development in rice (Zheng et al. 2015).

Genes for improvement of the root system

The root structure of plants has a significant role in the uptake of water and mineral resources from the soil to ensure the survival of plants. A rice mutant that contains ARL1 genes lacks adventitious roots during the period of development (Liu et al. 2005) due to the impaired formation of the adventitious primordium, and forms an arl1 (adventitious rootless1) mutant phenotype. It was demonstrated that the anatomical characteristic of roots improves the yield of crops in water and nutrient-deficient environments (Lynch, Chimungu, and Brown 2014, York, Nord, and Lynch 2013). Hence, the research community takes on the responsibility for identifying such traits having a role in the improvement of crop quality through advanced phenotypic methods (Clark et al. 2013, Topp et al. 2013). During the past few years, QTLs have been identified for structural and functional traits of roots in rice varieties, such as DRO1, which increases the rice yield by conferring drought resistance (Uga et al. 2013). Other QTLs have also been characterised, associated with increased production of rice during drought conditions (Kumar, Dixit, and Henry 2013). Moreover, efforts are ongoing to characterise the QTL of roots for producing better quality rice (Ahmadi et al. 2014).

Abiotic Stress resistance genes

Abiotic stress, including salinity, cold, drought and heat constitutes a major challenge in rice production. It is reported that more than one hundred stress-resistant genes, along with five QTLs, have been recognised in rice (Jiang et al. 2012). Genome modification in rice using the CRISPR system helps in the identification of the OsAnn3 gene responsible for providing tolerance to cold. It is demonstrated that mutants having the OsAnn3 gene knocked out were less tolerant to cold stress (Shen et al. 2017). Another newly-discovered gene is *qCTS-9,* conferring resistance to cold at the stage of seedling development in rice (Zhao et al. 2017). In rice plants, most of the mitogen-associated protein kinases, such as OsMSRMK2, have been activated in response to abiotic stresses, including salt, drought and abscisic acid signalling (Danquah et al. 2014). Recent research findings revealed that cytochrome P450 enzymes play an important part in mediating tolerance to drought stress as well as regulation of developmental processes (Wei and Chen 2018).

In plants, members of G- protein signalling cascade have a role in mediating responses against abiotic stress. The Gα subunit plays an important part in regulating salt stress responses in many plants, including maize, rice (Urano, Colaneri, and Jones 2014) and *Arabidopsis* (Colaneri et al. 2014, Jangam, Pathak, and Raghuram 2016). In addition, genes associated with calcium signalling pathways also regulate stress signalling in *O.sativa* (Batistič and Kudla 2012). Findings have shown that modification in stress-responsive genes through transgenic methods improve tolerance to abiotic stresses in rice (Campo et al. 2014). Furthermore, the role of genes associated with nucleosides modified on tRNA molecules has been reported in modulating stress responses in plants, such as *Arabidopsis* and rice. A study demonstrated that salt stress results in the activation of 2'-O-methyladenosine (Am) nucleoside in different plant species, including Poplar, Rice, *Brachypodium*, and *Arabidopsis*. The tolerance to salt stress is enhanced owing to the high expression of the *OsTRM13* gene associated with modification of nucleoside in rice (Wang, Li, et al. 2017, Wang, Pang, et al. 2017).

The developments of different tagging tools and approaches have been crucial in the research of functional genomics in rice. These researches have a great impact on sustainable agriculture and development. Knowledge of the function of all rice genes will lead to the improvement of the rice plant, thus protecting it from different diseases and increasing its yield and productivity.

Concluding remarks and future perspectives

The high demand for rice is fulfilled by its increased productivity, owing to this plant having been extensively studied and researched. The rice genome sequencing was a scientific milestone that led to the discoveries of the structure and function of different genes. The functional and structural genomic studies have broadened our knowledge, and thus opened up ways for exploitation of the rice genes and their alleles for the development of new strains for crop improvement and sustainable agriculture systems. Moreover, the rapid development of molecular markers and advanced genomic tools would further help in understanding genomics of rice.

Acknowledgments The authors extend their sincere thanks to the Atta-ur-Rahman School of Applied Biosciences, NUST, Islamabad from Pakistan; Sher-e-Bangla Agricultural University in Bangladesh, and Ege University, as well as Hatay Mustafa Kemal University in Turkey for their full support in this and ongoing project collaborations.

References

Adams, M D, S E Celniker, R A Holt, C A Evans, J D Gocayne, P G Amanatides, S E Scherer, P W Li, R A Hoskins, and R F Galle. 2000. "The genome sequence of *Drosophila melanogaster*." *Science* 287 (5461):2185–2195.

Ahmadi, N, A Audebert, M J Bennett, A Bishopp, A Costa de Oliveira, B Courtois, A Diedhiou, A Diévart, P Gantet, and A Ghesquière. 2014. "The roots of future rice harvests." *Rice* 7 (1):29.

Alamar, M C, R Tosetti, S Landahl, A Bermejo, and L A Terry. 2017. "Assuring potato tuber quality during storage: A future perspective." *Frontiers in Plant Science* 8:2034.

Banu, M S A, K M K Huda, R K Sahoo, B Garg, S Tula, S M S Islam, R Tuteja, and N Tuteja. 2015. "Pea p68 imparts salinity stress tolerance in rice by scavenging of ROS-mediated H 2 O 2 and interacts with argonaute." *Plant Molecular Biology Reporter* 33 (2):221–238.

Barry, G F. 2001. "The use of the Monsanto draft rice genome sequence in research." *Plant Physiology* 125 (3):1164–1165.

Batistič, O, and J Kudla. 2012. "Analysis of calcium signaling pathways in plants." *Biochimica et Biophysica Acta (BBA)-General Subjects* 1820 (8):1283–1293.

Bellini, C, D I Pacurar, and I Perrone. 2014. "Adventitious roots and lateral roots: similarities and differences." *Annual Review of Plant Biology* 65:639–666.

Bian, H, Y Xie, F Guo, N Han, S Ma, Z Zeng, J Wang, Y Yang, and M Zhu. 2012. "Distinctive expression patterns and roles of the miRNA393/TIR1 homolog module in regulating flag leaf inclination and primary and crown root growth in rice (*Oryza sativa*)." *New Phytologist* 196 (1):149–161.

Buell, C R. 2002. "Obtaining the sequence of the rice genome and lessons learned along the way." *Trends in Plant Science* 7 (12):538–542.

Campo, S, P Baldrich, J Messeguer, E Lalanne, M Coca, and B S Segundo. 2014. "Overexpression of a calcium-dependent protein kinase confers salt and drought tolerance in rice by preventing membrane lipid peroxidation." *Plant Physiology*:pp. 113.230268.

Čermák, T, N J Baltes, R Čegan, Y Zhang, and D F Voytas. 2015. "High-frequency, precise modification of the tomato genome." *Genome Biology* 16 (1):232.

Chen, G, C Liu, Z Gao, Y Zhang, A Zhang, L Zhu, J Hu, D Ren, L Yu, and G Xu. 2018. "Variation in the abundance of OsHAK1 transcript underlies the differential salinity tolerance of an indica and a japonica rice cultivar." *Frontiers in Plant Science* 8:2216.

Chen, G, H Zhang, Z Deng, R Wu, D Li, M Wang, and J Tian. 2016. "Genome-wide association study for kernel weight-related traits using SNPs in a Chinese winter wheat population." *Euphytica* 212 (2):173–185.

Cheng, X, L Zhu, and G He. 2013. "Towards understanding of molecular interactions between rice and the brown planthopper." *Molecular Plant* 6 (3):621–634.

Cheng, Z, C R Buell, R A Wing, M Gu, and J Jiang. 2001. "Toward a cytological characterization of the rice genome." *Genome Research* 11 (12):2133–2141.

Clark, R T, A N Famoso, K Zhao, J E Shaff, E J Craft, C D Bustamante, S R Mccouch, D J Aneshansley, and L V Kochian. 2013. "High-throughput two-dimensional root system phenotyping platform facilitates genetic analysis of root growth and development." *Plant, Cell and Environment* 36 (2):454–466.

Colaneri, A C, M Tunc-Ozdemir, J P Huang, and A M Jones. 2014. "Growth attenuation under saline stress is mediated by the heterotrimeric G protein complex." *BMC Plant Biology* 14 (1):129.

Consortium, Rice Chromosome 10 Sequencing. 2003. "In-depth view of structure, activity, and evolution of rice chromosome 10." *Science* 300 (5625):1566–1569.

Crowell, S, P Korniliev, A Falcao, A Ismail, G Gregorio, J Mezey, and S McCouch. 2016. "Genome-wide association and high-resolution phenotyping link *Oryza sativa* panicle traits to numerous trait-specific QTL clusters." *Nature Communications* 7:10527.

Curaba, J, M B Singh, and P L Bhalla. 2014. "miRNAs in the crosstalk between phytohormone signalling pathways." *Journal of Experimental Botany* 65 (6):1425–1438.

Danquah, A, A de Zelicourt, J Colcombet, and H Hirt. 2014. "The role of ABA and MAPK signaling pathways in plant abiotic stress responses." *Biotechnology Advances* 32 (1):40–52.

Delseny, M, J Salses, R Cooke, C Sallaud, F Regad, P Lagoda, E Guiderdoni, M Ventelon, C Brugidou, and A Ghesquière. 2001. "Rice genomics: present and future." *Plant Physiology and Biochemistry* 39 (3–4):323–334.

Doudna, J A, and E Charpentier. 2014. "The new frontier of genome engineering with CRISPR-Cas9." *Science* 346 (6213):1258096.

Fan, C, Y Xing, H Mao, T Lu, B Han, C Xu, X Li, and Q Zhang. 2006. "GS3, a major QTL for grain length and weight and minor QTL for grain

width and thickness in rice, encodes a putative transmembrane protein." *Theoretical and Applied Genetics* 112 (6):1164–1171.
Farooq, M A, Z A Saqib, and J Akhtar. 2015. "Silicon-mediated oxidative stress tolerance and genetic variability in rice (*Oryza sativa* L.) grown under combined stress of salinity and boron toxicity." *Turkish Journal of Agriculture and Forestry* 39 (5):718–729.
Feng, Q, Y Zhang, P Hao, S Wang, G Fu, Y Huang, Y Li, J Zhu, Y Liu, and X Hu. 2002. "Sequence and analysis of rice chromosome 4." *Nature* 420 (6913):316.
Gamuyao, R, J H Chin, J Pariasca-Tanaka, P Pesaresi, S Catausan, C Dalid, I Slamet-Loedin, E M Tecson-Mendoza, M Wissuwa, and S Heuer. 2012. "The protein kinase Pstol1 from traditional rice confers tolerance of phosphorus deficiency." *Nature* 488 (7412):535.
Georges, F, and H Ray. 2017. "Genome editing of crops: a renewed opportunity for food security." *GM Crops and Food* 8 (1):1–12.
Giri, M K, J K Gautam, V B R Prasad, S Chattopadhyay, and A K Nandi. 2017. "Rice MYC2 (OsMYC2) modulates light-dependent seedling phenotype, disease defence but not ABA signalling." *Journal of Biosciences* 42 (3):501–508.
Goff, S A, D Ricke, T-H Lan, G Presting, R Wang, M Dunn, J Glazebrook, A Sessions, P Oeller, and H Varma. 2002. "A draft sequence of the rice genome (*Oryza sativa* L. ssp. japonica)." *Science* 296 (5565):92–100.
Greco, R, Pieter B F Ouwerkerk, C Sallaud, A Kohli, L Colombo, P Puigdomenech, E Guiderdoni, P Christou, J H C Hoge, and A Pereira. 2001. "Transposon insertional mutagenesis in rice." *Plant Physiology* 125 (3):1175–1177.
Green, E D. 2001. "Strategies for the systematic sequencing of complex genomes." *Nature Reviews Genetics* 2 (8):573.
Gutierrez, L, G Mongelard, K Floková, D I Păcurar, O Novák, P Staswick, M Kowalczyk, M Păcurar, H Demailly, and G Geiss. 2012. "Auxin controls *Arabidopsis* adventitious root initiation by regulating jasmonic acid homeostasis." *The Plant Cell*:tpc. 112.099119.
Hamer, L, T M DeZwaan, M V Montenegro-Chamorro, S A Frank, and J E Hamer. 2001. "Recent advances in large-scale transposon mutagenesis." *Current Opinion in Chemical Biology* 5 (1):67–73.
Han, B, Y Xue, J Li, X-W Deng, and Q Zhang. 2007. "Rice functional genomics research in China." *Philosophical Transactions of the Royal Society of London B: Biological Sciences* 362 (1482):1009–1021.
Harushima, Y, M Yano, A Shomura, M Sato, T Shimano, Y Kuboki, T Yamamoto, S Y Lin, B A Antonio, and A Parco. 1998. "A high-density

rice genetic linkage map with 2275 markers using a single F2 population." *Genetics* 148 (1):479–494.
Heng, H HQ, B Spyropoulos, and P B Moens. 1997. "FISH technology in chromosome and genome research." *Bioessays* 19 (1):75–84.
Hieter, P, and M Boguski. 1997. "Functional genomics: it's all how you read it." *Science* 278 (5338):601–602.
Hirochika, H, A Miyao, M Yamazaki, S Takeda, K Abe, R Hirochika, GK Agrawal, T Watanabe, K Sugimoto, and T Sasaki. 2001. "Retrotransposons of rice as a tool for the functional analysis of genes." In *Rice Genetics IV*, 279–292. World Scientific.
Hirochika, H, H Otsuki, M Yoshikawa, Y Otsuki, K Sugimoto, and S Takeda. 1996. "Autonomous transposition of the tobacco retrotransposon Tto1 in rice." *The Plant Cell* 8 (4):725–734.
Hossain, M R, G W Bassel, J Pritchard, G P Sharma, and B V Ford-Lloyd. 2016. "Trait specific expression profiling of salt stress responsive genes in diverse rice genotypes as determined by modified significance analysis of microarrays." *Frontiers in Plant Science* 7:567.
Jangam, A P, R R Pathak, and N Raghuram. 2016. "Microarray analysis of rice d1 (RGA1) mutant reveals the potential role of G-protein alpha subunit in regulating multiple abiotic stresses such as drought, salinity, heat, and cold." *Frontiers in Plant Science* 7:11.
Jiang, Y, Z Cai, W Xie, T Long, H Yu, and Q Zhang. 2012. "Rice functional genomics research: progress and implications for crop genetic improvement." *Biotechnology Advances* 30 (5):1059–1070.
Jin, D, Y Wang, Y Zhao, and M Chen. 2013. "MicroRNAs and their cross-talks in plant development." *Journal of Genetics and Genomics* 40 (4):161–170.
Kang, H-G, J-S Jeon, S Lee, and G An. 1998. "Identification of class B and class C floral organ identity genes from rice plants." *Plant Molecular Biology* 38 (6):1021–1029.
Kang, H, Y Wang, S Peng, Y Zhang, Y Xiao, D Wang, S Qu, Z Li, S Yan, and Z Wang. 2016. "Dissection of the genetic architecture of rice resistance to the blast fungus *Magnaporthe oryzae*." *Molecular Plant Pathology* 17 (6):959–972.
Kanno, A, and A Hirai. 1993. "A transcription map of the chloroplast genome from rice (*Oryza sativa*)." *Current Genetics* 23 (2):166–174.
Kennedy, D. 2002. The importance of rice. American Association for the Advancement of Science.
Kikuta, M, D Makihara, N Arita, A Miyazaki, and Y Yamamoto. 2017. "Growth and yield responses of upland NERICAs to variable water

management under field conditions." *Plant Production Science* 20 (1):36–46.

Kumar, A, S Dixit, and A Henry. 2013. "Marker-assisted introgression of major QTLs for grain yield under drought in rice." *Translational Genomics for Crop Breeding: Abiotic Stress, Yield and Quality* 2:47–64.

Kusano, M, Z Yang, Y Okazaki, R Nakabayashi, A Fukushima, and K Saito. 2015. "Using metabolomic approaches to explore chemical diversity in rice." *Molecular Plant* 8 (1):58–67.

Lawrenson, T, O Shorinola, N Stacey, C Li, L Østergaard, N Patron, C Uauy, and W Harwood. 2015. "Induction of targeted, heritable mutations in barley and *Brassica oleracea* using RNA-guided Cas9 nuclease." *Genome Biology* 16 (1):258.

Li, W, Z Zhu, M Chern, J Yin, C Yang, L Ran, M Cheng, M He, K Wang, and J Wang. 2017. "A natural allele of a transcription factor in rice confers broad-spectrum blast resistance." *Cell* 170 (1):114–126. e15.

Li, X, Q Qian, Z Fu, Y Wang, G Xiong, D Zeng, X Wang, X Liu, S Teng, and F Hiroshi. 2003. "Control of tillering in rice." *Nature* 422 (6932):618.

Li, Y, C Fan, Y Xing, Y Jiang, L Luo, L Sun, D Shao, C Xu, X Li, and J Xiao. 2011. "Natural variation in GS5 plays an important role in regulating grain size and yield in rice." *Nature Genetics* 43 (12):1266.

Liu, H, S Wang, X Yu, J Yu, X He, S Zhang, H Shou, and P Wu. 2005. "ARL1, a LOB-domain protein required for adventitious root formation in rice." *The Plant Journal* 43 (1):47–56.

Liu, J, J Shen, Y Xu, X Li, J Xiao, and L Xiong. 2016. "Ghd2, a CONSTANS-like gene, confers drought sensitivity through regulation of senescence in rice." *Journal of Experimental Botany* 67 (19):5785–5798.

Lu, Q, X Niu, M Zhang, C Wang, Q Xu, Y Feng, Y Yang, S Wang, X Yuan, and H Yu. 2018. "Genome-wide association study of seed dormancy and the genomic consequences of improvement footprints in rice (*Oryza sativa* L.)." *Frontiers in Plant Science* 8:2213.

Lynch, Jonathan P, Joseph G Chimungu, and Kathleen M Brown. 2014. "Root anatomical phenes associated with water acquisition from drying soil: targets for crop improvement." *Journal of Experimental Botany* 65 (21):6155–6166.

McCouch, S R, M H Wright, C-W Tung, L G Maron, K L McNally, M Fitzgerald, N Singh, G DeClerck, F Agosto-Perez, and P Korniliev. 2016. "Open access resources for genome-wide association mapping in rice." *Nature Communications* 7:10532.

Meng, X, H Yu, Y Zhang, F Zhuang, X Song, S Gao, C Gao, and J Li. 2017. "Construction of a genome-wide mutant library in rice using CRISPR/Cas9." *Molecular Plant* 10 (9):1238–1241.

Mohan, M, S Nair, A Bhagwat, T G Krishna, M Yano, C R Bhatia, and T Sasaki. 1997. "Genome mapping, molecular markers and marker-assisted selection in crop plants." *Molecular Breeding* 3 (2):87–103.

Monna, L, N Kitazawa, R Yoshino, J Suzuki, H Masuda, Y Maehara, M Tanji, M Sato, S Nasu, and Y Minobe. 2002. "Positional cloning of rice semidwarfing gene, sd-1: rice "green revolution gene" encodes a mutant enzyme involved in gibberellin synthesis."*DNA Research* 9 (1):11–17.

Moore, G, K M Devos, Z Wang, and M D Gale. 1995. "Cereal genome evolution: grasses, line up and form a circle." *Current Biology* 5 (7):737–739.

Nada, R M, and G M Abogadallah. 2018. "Root-favoured biomass allocation improves growth and yield of field-gown rice (*Oryza sativa* L.) plants only when the shoot sink is expandable." *Acta Physiologiae Plantarum* 40 (6):123.

Nallamilli, B R Reddy, J Zhang, H Mujahid, B M Malone, S M Bridges, and Z Peng. 2013. "Polycomb group gene OsFIE2 regulates rice (*Oryza sativa*) seed development and grain filling via a mechanism distinct from *Arabidopsis*." *PLoS Genetics* 9 (3):e1003322.

Ogawa, S, R Kawahara-Miki, K Miyamoto, H Yamane, H Nojiri, Y Tsujii, and K Okada. 2017. "OsMYC2 mediates numerous defence-related transcriptional changes via jasmonic acid signalling in rice." *Biochemical and Biophysical Research Communications* 486 (3):796–803.

Ogawa, S, K Miyamoto, K Nemoto, T Sawasaki, H Yamane, H Nojiri, and K Okada. 2017. "OsMYC2, an essential factor for JA-inductive sakuranetin production in rice, interacts with MYC2-like proteins that enhance its transactivation ability." *Scientific Reports* 7:40175.

Palmgren, M G, A K Edenbrandt, S E Vedel, M M Andersen, X Landes, J T Østerberg, J Falhof, L I Olsen, S B Christensen, and P Sandøe. 2015. "Are we ready for back-to-nature crop breeding?" *Trends in Plant Science* 20 (3):155–164.

Park, C-J, T Wei, R Sharma, and P C Ronald. 2017. "Overexpression of rice auxilin-like protein, XB21, induces necrotic lesions, up-regulates endocytosis-related genes, and confers enhanced resistance to *Xanthomonas oryzae* pv. *oryzae*." *Rice* 10 (1):27.

Pasam, R K, R Sharma, M Malosetti, F A van Eeuwijk, G Haseneyer, B Kilian, and A Graner. 2012. "Genome-wide association studies for agronomical traits in a world wide spring barley collection." *BMC Plant Biology* 12 (1):16.

Paterson, A H, J E Bowers, and B A Chapman. 2004. "Ancient polyploidization predating divergence of the cereals, and its consequences for comparative genomics." *Proceedings of the National Academy of Sciences* 101 (26):9903–9908.

Peng, X, Y Hu, X Tang, P Zhou, X Deng, H Wang, and Z Guo. 2012. "Constitutive expression of rice WRKY30 gene increases the endogenous jasmonic acid accumulation, PR gene expression and resistance to fungal pathogens in rice." *Planta* 236 (5):1485–1498.

Qi, P, Y-S Lin, X-J Song, J-B Shen, W Huang, J-X Shan, M-Z Zhu, L Jiang, J-P Gao, and H-X Lin. 2012. "The novel quantitative trait locus GL3. 1 controls rice grain size and yield by regulating Cyclin-T1; 3." *Cell Research* 22 (12):1666.

Rice Chromosomes, 11 12 Sequencing Consortia. 2005. "The sequence of rice chromosomes 11 and 12, rich in disease resistance genes and recent gene duplications." *BMC Biology* 3:1–18.

Saha, S, A B Sparks, C Rago, V Akmaev, C J Wang, B Vogelstein, K W Kinzler, and V E Velculescu. 2002. "Using the transcriptome to annotate the genome." *Nature Biotechnology* 20 (5):508.

Sallaud, C, D Meynard, J V Boxtel, C Gay, M Bes, J-P Brizard, P Larmande, D Ortega, M Raynal, and M Portefaix. 2003. "Highly efficient production and characterization of T-DNA plants for rice (*Oryza sativa* L.) functional genomics." *Theoretical and Applied Genetics* 106 (8):1396–1408.

Sasaki, A, M Ashikari, M Ueguchi-Tanaka, H Itoh, A Nishimura, D Swapan, K Ishiyama, T Saito, M Kobayashi, and G S Khush. 2002. "Green revolution: a mutant gibberellin-synthesis gene in rice." *Nature* 416 (6882):701.

Sasaki, T. 2001. "The progress in rice genomics." *Euphytica* 118 (2):103–111.

Sasaki, T. 2005. "The map-based sequence of the rice genome." *Nature* 436 (7052):793.

Sasaki, T, and B Burr. 2000. "International Rice Genome Sequencing Project: the effort to completely sequence the rice genome." *Current Opinion in Plant Biology* 3 (2):138–142.

Sasaki, T, T Matsumoto, K Yamamoto, K Sakata, T Baba, Y Katayose, J Wu, Y Niimura, Z Cheng, and Y Nagamura. 2002. "The genome sequence and structure of rice chromosome 1." *Nature* 420 (6913):312.

Schoof, H, and W M Karlowski. 2003. "Comparison of rice and *Arabidopsis* annotation." *Current Opinion in Plant Biology* 6 (2):106–112.

Shen, C, Z Que, Y Xia, N Tang, D Li, R He, and M Cao. 2017. "Knock out of the annexin gene OsAnn3 via CRISPR/Cas9-mediated genome editing decreased cold tolerance in rice." *Journal of Plant Biology* 60 (6):539–547.

Shimono, M, H Koga, A Akagi, N Hayashi, S Goto, M Sawada, T Kurihara, A Matsushita, S Sugano, and C-J Jiang. 2012. "Rice WRKY45 plays important roles in fungal and bacterial disease resistance." *Molecular Plant Pathology* 13 (1):83–94.

Si, L, J Chen, X Huang, H Gong, J Luo, Q Hou, T Zhou, T Lu, J Zhu, and Y Shangguan. 2015. "OsSPL13 controls grain size in cultivated rice." *Nature Genetics* 47 (3):447.

Sikirou, M, A Shittu, K A Konaté, A T Maji, A S Ngaujah, K A Sanni, S A Ogunbayo, I Akintayo, K Saito, and K N Dramé. 2018. "Screening African rice (*Oryza glaberrima*) for tolerance to abiotic stresses: I. Fe toxicity." *Field Crops Research* 220:3–9.

Singh, K, T Ishii, A Parco, N Huang, D S Brar, and G S Khush. 1996. "Centromere mapping and orientation of the molecular linkage map of rice (*Oryza sativa* L.)." *Proceedings of the National Academy of Sciences* 93 (12):6163–6168.

Song, W-Y, G-L Wang, L-L Chen, H-S Kim, L-Y Pi, T Holsten, J Gardner, B Wang, W-X Zhai, and L-H Zhu. 1995. "A receptor kinase-like protein encoded by the rice disease resistance gene, Xa21." *Science* 270 (5243):1804–1806.

Sreenivasulu, N, and T Schnurbusch. 2012. "A genetic playground for enhancing grain number in cereals." *Trends in Plant Science* 17 (2):91–101.

Sun, X, Y Cao, Z Yang, C Xu, X Li, S Wang, and Q Zhang. 2004. "Xa26, a gene conferring resistance to *Xanthomonas oryzae* pv. *oryzae* in rice, encodes an LRR receptor kinase-like protein." *The Plant Journal* 37 (4):517–527.

Sun, Y, X Zhang, C Wu, Y He, Y Ma, H Hou, X Guo, W Du, Y Zhao, and L Xia. 2016. "Engineering herbicide-resistant rice plants through CRISPR/Cas9-mediated homologous recombination of acetolactate synthase." *Molecular Plant* 9 (4):628-631.

Takatsuji, H, and N Hayashi. 2017. "Broad-spectrum blast resistance: harnessing a natural allele of a transcription factor in rice." *Molecular Plant* 10 (9):1144–1146.

Tan, Y F, Y Z Xing, J X Li, S B Yu, C G Xu, and Q Zhang. 2000. "Genetic bases of appearance quality of rice grains in Shanyou 63, an elite rice hybrid." *Theoretical and Applied Genetics* 101 (5–6):823–829.

Tanaka, W, Y Ohmori, T Ushijima, H Matsusaka, T Matsushita, T Kumamaru, S Kawano, and H-Y Hirano. 2015. "Axillary meristem formation in rice requires the WUSCHEL ortholog TILLERS ABSENT1." *The Plant Cell*:tpc. 15.00074.

Topp, C N, A S Iyer-Pascuzzi, J T Anderson, C-R Lee, P R Zurek, O Symonova, Y Zheng, A Bucksch, Y Mileyko, and T Galkovskyi. 2013. "3D phenotyping and quantitative trait locus mapping identify core regions of the rice genome controlling root architecture." *Proceedings of the National Academy of Sciences* 110 (18):E1695–E1704.

Turcotte, K, S Srinivasan, and T Bureau. 2001. "Survey of transposable elements from rice genomic sequences." *The Plant Journal* 25 (2):169–179.

Tyagi, A K, J P Khurana, P Khurana, S Raghuvanshi, A Gaur, A Kapur, V Gupta, D Kumar, V Ravi, and SVij. 2004. "Structural and functional analysis of rice genome." *Journal of Genetics* 83 (1):79–99.

Ueda, Y, F Frimpong, Y Qi, E Matthus, L Wu, S Höller, T Kraska, and M Frei. 2014. "Genetic dissection of ozone tolerance in rice (*Oryza sativa* L.) by a genome-wide association study." *Journal of Experimental Botany* 66 (1):293–306.

Uga, Y, K Sugimoto, S Ogawa, J Rane, M Ishitani, N Hara, Y Kitomi, Y Inukai, K Ono, and N Kanno. 2013. "Control of root system architecture by DEEPER ROOTING 1 increases rice yield under drought conditions." *Nature Genetics* 45 (9):1097.

Uji, Y, K Akimitsu, and K Gomi. 2017. "Identification of OsMYC2-regulated senescence-associated genes in rice." *Planta* 245 (6):1241–1246.

Uji, Y, S Taniguchi, D Tamaoki, H Shishido, K Akimitsu, and K Gomi. 2016. "Overexpression of OsMYC2 results in the up-regulation of early JA-rresponsive genes and bacterial blight resistance in Rice." *Plant and Cell Physiology* 57 (9):1814–1827.

Umehara, Y, A Inagaki, H Tanoue, Y Yasukochi, Y Nagamura, S Saji, Y Otsuki, T Fujimura, N Kurata, and Y Minobe. 1995. "Construction and characterization of a rice YAC library for physical mapping." *Molecular Breeding* 1 (1):79–89.

Urano, D, A Colaneri, and A M Jones. 2014. "Gα modulates salt-induced cellular senescence and cell division in rice and maize."*Journal of Experimental Botany* 65 (22):6553–6561.

Velculescu, V E, L Zhang, B Vogelstein, and K W Kinzler. 1995. "Serial analysis of gene expression."*Science* 270 (5235):484–487.

Vo, K TX, C-Y Kim, T V Hoang, S-K Lee, G Shirsekar, Y-S Seo, S-W Lee, G-L Wang, and J-S Jeon. 2018. "OsWRKY67 plays a positive role in basal and XA21-mediated resistance in rice."*Frontiers in Plant Science* 8:2220.

Wang, Nili, Tuan Long, Wen Yao, Lizhong Xiong, Qifa Zhang, and Changyin Wu. 2013. "Mutant resources for the functional analysis of the rice genome."*Molecular Plant* 6 (3):596–604.

Wang, S, J Wang, J Jiang, and Q Zhang. 2000. "Mapping of centromeric regions on the molecular linkage map of rice (*Oryza sativa* L.) using centromere-associated sequences."*Molecular and General Genetics MGG* 263 (1):165–172.

Wang, S, K Wu, Q Yuan, X Liu, Z Liu, X Lin, R Zeng, H Zhu, G Dong, and Q Qian. 2012. "Control of grain size, shape and quality by OsSPL16 in rice."*Nature Genetics* 44 (8):950.

Wang, W, G Li, J Zhao, H Chu, W Lin, D Zhang, Z Wang, and W Liang. 2014. "DWARF TILLER1, a WUSCHEL-related homeobox transcription factor, is required for tiller growth in rice."*PLoS Genetics* 10 (3):e1004154.

Wang, X-Y, X-L Shi, and B-L Hao. 2004. "The tRNA and rRNA genes in the *Oryza sativa* genome."*Yi Chuan Xue Bao* 31:871–877.

Wang, Y, L Cao, Y Zhang, C Cao, F Liu, F Huang, Y Qiu, R Li, and X Luo. 2015. "Map-based cloning and characterization of BPH29, a B3 domain-containing recessive gene conferring brown planthopper resistance in rice."*Journal of Experimental Botany* 66 (19):6035–6045.

Wang, Y, D Li, J Gao, X Li, R Zhang, X Jin, Z Hu, B Zheng, S Persson, and P Chen. 2017. "The 2′-O-methyladenosine nucleoside modification gene OsTRM13 positively regulates salt stress tolerance in rice."*Journal of Experimental Botany* 68 (7):1479–1491.

Wang, Y, C Pang, X Li, Z Hu, Z Lv, B Zheng, and P Chen. 2017. "Identification of tRNA nucleoside modification genes critical for stress response and development in rice and *Arabidopsis*."*BMC Plant Biology* 17 (1):261.

Wei, F-J, G Droc, E Guiderdoni, and C H Yue-ie. 2013. "International consortium of rice mutagenesis: resources and beyond."*Rice* 6 (1):39.

Wei, K, and H Chen. 2018. "Global identification, structural analysis and expression characterization of cytochrome P450 monooxygenase superfamily in rice."*BMC Genomics* 19 (1):35.

Wu, H, Y Liu, J He, Y Liu, L Jiang, L Liu, C Wang, X Cheng, and J Wan. 2014. "Fine mapping of brown planthopper (*Nilaparvata lugens* Stål) resistance gene Bph28 (t) in rice (*Oryza sativa* L.)."*Molecular Breeding* 33 (4):909–918.

Wu, J, T Maehara, T Shimokawa, S Yamamoto, C Harada, Y Takazaki, N Ono, Y Mukai, K Koike, and J Yazaki. 2002. "A comprehensive rice

transcript map containing 6591 expressed sequence tag sites."*The Plant Cell* 14 (3):525–535.
Wu, J, H Yamagata, M Hayashi-Tsugane, S Hijishita, M Fujisawa, M Shibata, Y Ito, M Nakamura, M Sakaguchi, and R Yoshihara. 2004. "Composition and structure of the centromeric region of rice chromosome 8."*The Plant Cell* 16 (4):967–976.
Wu, Tao, C Yang, B Ding, Z Feng, Q Wang, J He, J Tong, L Xiao, L Jiang, and J Wan. 2016. "Microarray-based gene expression analysis of strong seed dormancy in rice cv. N22 and less dormant mutant derivatives."*Plant Physiology and Biochemistry* 99:27–38.
Xie, M, S Zhang, and B Yu. 2015. "microRNA biogenesis, degradation and activity in plants."*Cellular and Molecular Life Sciences* 72 (1):87–99.
Xing, Y, Y Tan, J P Hua, X Sun, C Xu, and Q Zhang. 2002. "Characterization of the main effects, epistatic effects and their environmental interactions of QTLs on the genetic basis of yield traits in rice."*Theoretical and Applied Genetics* 105 (2–3):248–257.
Xu, F, J Fang, S Ou, S Gao, F Zhang, L Du, Y Xiao, H Wang, X Sun, and J Chu. 2015. "Variations in CYP 78 A 13 coding region influence grain size and yield in rice."*Plant, Cell and Environment* 38 (4):800–811.
Yamamoto, K, and T Sasaki. 1997. "Large-scale EST sequencing in rice." In *Oryza: From Molecule to Plant*, 135–144. Springer.
Yamanouchi, U, M Yano, H Lin, M Ashikari, and K Yamada. 2002. "A rice spotted leaf gene, Spl7, encodes a heat stress transcription factor protein."*Proceedings of the National Academy of Sciences* 99 (11):7530–7535.
Yang, C, D Li, D Mao, X Liu, C Ji, X Li, X Zhao, Z Cheng, C Chen, and L Zhu. 2013. "Overexpression of micro RNA 319 impacts leaf morphogenesis and leads to enhanced cold tolerance in rice (*Oryza sativa* L.)."*Plant, Cell and Environment* 36 (12):2207–2218.
Yang, Y, Y Li, and C Wu. 2013. "Genomic resources for functional analyses of the rice genome."*Current Opinion in Plant Biology* 16 (2):157–163.
Yang, Z, X Sun, S Wang, and Q Zhang. 2003. "Genetic and physical mapping of a new gene for bacterial blight resistance in rice."*Theoretical and Applied Genetics* 106 (8):1467–1472.
Yi, K, Z Wu, J Zhou, L Du, L Guo, Y Wu, and P Wu. 2005. "OsPTF1, a novel transcription factor involved in tolerance to phosphate starvation in rice."*Plant Physiology* 138 (4):2087–2096.
York, L M, E Nord, and J Lynch. 2013. "Integration of root phenes for soil resource acquisition."*Frontiers in Plant Science* 4:355.

Yoshimura, S, U Yamanouchi, Y Katayose, S Toki, Z-X Wang, I Kono, N Kurata, M Yano, N Iwata, and T Sasaki. 1998. "Expression of Xa1, a bacterial blight-resistance gene in rice, is induced by bacterial inoculation."*Proceedings of the National Academy of Sciences* 95 (4):1663–1668.

Yu, J, J Wang, W Lin, S Li, H Li, J Zhou, P Ni, W Dong, S Hu, and C Zeng. 2005. "The genomes of *Oryza sativa*: a history of duplications."*PLoS Biology* 3 (2):e38.

Yuan, Q, S Ouyang, A Wang, W Zhu, R Maiti, H Lin, J Hamilton, B Haas, R Sultana, and F Cheung. 2005. "The institute for genomic research Osa1 rice genome annotation database."*Plant Physiology* 138 (1):18–26.

Yuan, X, H Sun, Z Tang, H Tang, H Zhang, and J Huang. 2016. "A novel little membrane protein confers salt tolerance in rice (*Oryza sativa* L.)."*Plant Molecular Biology Reporter* 34 (2):524–532.

Zhang, B. 2015. "MicroRNA: a new target for improving plant tolerance to abiotic stress."*Journal of Experimental Botany* 66 (7):1749–1761.

Zhang, B, and Q Wang. 2015. "MicroRNA-based biotechnology for plant improvement."*Journal of Cellular Physiology* 230 (1):1–15.

Zhang, H-B, S Choi, S-S Woo, Z Li, and R A Wing. 1996. "Construction and characterization of two rice bacterial artificial chromosome libraries from the parents of a permanent recombinant inbred mapping population."*Molecular Breeding* 2 (1):11–24.

Zhang, L, Z Cheng, R Qin, Y Qiu, J-L Wang, X Cui, L Gu, X Zhang, X Guo, and D Wang. 2012. "Identification and characterization of an epi-allele of FIE1 reveals a regulatory linkage between two epigenetic marks in rice."*The Plant Cell*:tpc. 112.102269.

Zhang, M, J Ye, Q Xu, Y Feng, X Yuan, H Yu, Y Wang, and Y Yang. 2018. "Genome-wide association study of cold tolerance of Chinese indica rice varieties at the bud burst stage."*Plant Cell Reports* 37 (3):529–539.

Zhang, Y, Y Huang, L Zhang, Y Li, T Lu, Y Lu, Q Feng, Q Zhao, Z Cheng, and Y Xue. 2004. "Structural features of the rice chromosome 4 centromere."*Nucleic Acids Research* 32 (6):2023–2030.

Zhao, J, S Zhang, J Dong, T Yang, X Mao, Q Liu, X Wang, and B Liu. 2017. "A novel functional gene associated with cold tolerance at the seedling stage in rice."*Plant Biotechnology Journal* 15 (9):1141–1148.

Zhao, K, C-W Tung, G C Eizenga, M H Wright, M L Ali, A H Price, G J Norton, M R Islam, A Reynolds, and J Mezey. 2011. "Genome-wide association mapping reveals a rich genetic architecture of complex traits in *Oryza sativa*."*Nature Communications* 2:467.

Zheng, M, Y Wang, Y Wang, C Wang, Y Ren, J Lv, C Peng, T Wu, K Liu, and S Zhao. 2015. "Deformed floral organ1 (DFO1) regulates floral organ identity by epigenetically repressing the expression of OsMADS58 in rice (*Oryza sativa*)."*New Phytologist* 206 (4):1476–1490.

Zhou, H, B Liu, D P Weeks, M H Spalding, and B Yang. 2014. "Large chromosomal deletions and heritable small genetic changes induced by CRISPR/Cas9 in rice."*Nucleic Acids Research* 42 (17):10903–10914.

Zhu, T, P Budworth, W Chen, N Provart, H-S Chang, S Guimil, W Su, B Estes, G Zou, and X Wang. 2003. "Transcriptional control of nutrient partitioning during rice grain filling."*Plant Biotechnology Journal* 1 (1):59–70.

CHAPTER 18

GREEN NANOTECHNOLOGY

KHAZIMA MUAZIM, SAMEEN RUQIA IMADI, ALVINA GUL*

Atta-ur-Rahman School of Applied Biosciences, National University of Sciences and Technology, Islamabad, Pakistan
alvina_gul@yahoo.com

Abstract

Nanotechnology refers to the exploitation of nano-particles for solving the problems faced by humankind for centuries. Green nanotechnology deals with applications of nanotechnology in agriculture and agricultural research. Nanotechnology can be helpful in fields like energy saving, environmental issues, analytical studies, cosmetics, biology, pharmaceutics, construction work, pest control and food processing. There are different types of nano-particles available. The most commonly used ones are titanium and silver particles, which have been helpful in preventing waste production and in solar cell production. Green nanotechnology is also associated with reducing pollution, particularly water pollution. This chapter discusses the goals of green nanotechnology with reference to its applications in nano-fertilisers, nano-pesticides and nano-herbicides along with plant disease management.

Keywords: Nano-particles, Green Revolution, Crop Improvement, Controlled Delivery, Nano-tubes, Disease Management

Introduction

"There is plenty of room at the bottom". This revolutionary belief was propounded in 1959 by the legendary physicist Richard Feynman, and was followed by many scientific luminaries such as Eric Drexler who made this dream come true through their tireless efforts. This dream turned into reality in the form of "Nanotechnology". Nanotechnology is the face of modern technology, which has progressed by leaps and bounds over the last decade.

Nanotechnology, by definition, means "understanding and control of matter at the nano-scale, with dimensions approximately 1 to 100 nanometers, where unique phenomena enable novel application" (Thanos and Emerich 2008). Nanotechnology finds myriads of applications in various fields ranging from electronics and medicine to energy/environmental sustenance and the fabrication of biomaterials. One of the modern edges that this field of science has recently evolved is "Bio-nanotechnology", which is the intersection of biology and nanotechnology (Fakruddin, Hossain, and Afroz 2012), which is reshaping the field of biomedical science.

1. Nanotechnology and its Branches

Nanotechnology is an applied field of science. It has certainly brought scientists together from various fields, i.e. medicine, materials, physics and chemistry for the development of pyramids of products for consumer consumption. According to an inventory kept by an upcoming project, about 400 nano-technological products have been identified/produced for various commercial uses ranging from clothing and cosmetics to auto-motives and molecular computing. Nanotechnology seems to work in three major branches of science:

- Material science
- Biotechnology
- Microelectronics leading to the use of quantum mechanics effects and nano-robotics

Nanotechnology is playing a vital role in almost every field of science. There are two approaches that nanotechnology has adopted to carve its way to work: 1) a top-down approach; and 2) a bottom-up approach.

The top-down approach involves breaking down or whittling down bulk materials until we obtain their features at a nanometer scale. The bottom-up approach, by definition, means the "spontaneous self-association of molecules to build nanoscale assemblies which are widely used in nano-technological designs" (Colombo, Soto, and Gazit 2007). Bottom-up is more commonly used due to its flexibility, versatility and control over manipulation. A molecular self-assembly is an approach that is similar to the bottom-up approach whereby natural monomers form complex, multi-component three-dimensional (3D) structures. Natural building blocks, e.g.

nucleic acids, peptides and phospholipids, follow this phenomenon to form well-organized structures.

Here are the major areas where nanotechnology is applied:

- Energy applications
- Environmental applications
- Analytical applications
- Cosmetics
- Biological (therapeutic and diagnostic) applications
- Pharmaceutical applications
- Construction applications
- Military applications
- Applications in information and technology
- Agricultural productivity enhancement
- Water treatment and remediation
- Food processing and storage
- Air pollution and remediation
- Pest control

Nanotechnology is an intensively researched area today. This phenomenon is not only limited to laboratories but some of its products have already also reached the market. For example,

- Silver and titanium nanoparticles have been used in sunscreens/cosmetics (Dermatone SPF 20 and Loreal)
- Nanotex is a fabric that resists stains; it is also a wonder of nanotechnology (Eddie Bauer Ruston Fit)
- Scratch-resistant and glossy paints are made using nano-sized ceramic particles (Mercedes-Benz)
- Nanotechnology is also being used to build efficient computer hard drive read heads (Giant magnetoresistant material) which enables them to move at the speed of 80 miles an hour (IBM)

2. Green Technology: Clean Technology

Lux research had estimated that by 2014 nano-technological products would contribute about $2.6 trillion of manufactured goods, sharing about 10 million jobs (McCarter 2007), and the number of nano-products is bound to increase in the coming years, but their impact on human health, environment and safety are unknown. To circumvent unseen potential hazards from these particles, scientists and researchers are focusing on finding and optimizing methods where nano-particles/nano-materials can be fabricated using environmental- or planet-friendly reagents. Green nanotechnology and green chemistry are closely associated where these two emulate the natural processes for nanomaterial/nanoparticle synthesis. The nanoparticles produced through microbes or green synthesis include silver, gold, copper, iron, palladium, ruthenium, lead sulphate, cadmium sulphate, copper oxide, dioxycerium, iron oxide, titanium dioxide, and zinc oxide (Hussain et al. 2016).

Green nanotechnology serves two major functions:

- Helps to start a system initially with environmentally friendly reagents
- Helps to increase their environmental efficiency

Green nanotechnology relies on using LCA (Life Cycle risk Assessment) where the impact of materials/products on the environment is extrapolated. Green nanotechnology will enable the industry to function as an ecosystem where benign materials are used wisely, waste is recycled, and energy is used efficiently.

Following are the criteria/principles of green nanotechnology:

- Prevent waste: use chemical synthesis methods with the least toxic reagents leading to no waste to recycle or treat.
- Design less toxic chemical synthesis methods
- Use of renewable feedstock: Use raw materials and feedstocks that are renewable rather than depleting
- Maximize the atom economy: use the processes that incorporate most of the initial material. There should be a few atoms wasted, if any.
- Design materials or substances that degrade into innocuous materials after use. Minimize their chances of causing human or environmental toxicity
- Analyze the system in the real-time process to assess synthesis of toxic by-products

2.1. The Goals of Green Nanotechnology

Following are the main goals of green nanotechnology:

- Producing nanometer-scale materials with the least toxic synthesis method along with increasing the uptake of starting materials
- Fostering nano-technological materials that are benign to the environment; clean and green
- Using nano-technological products or materials to detect, manage and remediate environmental problems such as water pollution
- Replacing less environmentally-friendly products with green technological ones

2.2. Current Research in Green Nanotechnology

Green nanotechnology is the most significant scientific breakthrough of twenty-first-century science. With its advent, it has added to the change in lifestyles and jobs, as well as economies (Lu and Ozcan 2015). It is expected to facilitate greener and more sustainable economic growth. The six application areas of green nanotechnology include nano-enabled solar cells, energy storage, nano-generators, thermal energy, fuel catalysis, and water treatment (Shapira and Youtie 2015).

In the last decade, research has focused on the development of simple, cost-effective, clean, non-toxic and eco-friendly processes for the synthesis of nanoparticles. For achieving this, biosynthesis of nanoparticles was initiated, which involves the use of biochemical agents for the reduction of dissolved metal ions into nano-metals. Biological agents including plant material, fungi and bacteria have been used for this purpose (Ahmed and Ikram 2016). One of the examples of green nanotechnology is the nano-scale patterning of metal surfaces. In this technique, patterns are generated using bimolecular lithography DNA scaffolds. This technique maximizes material uptake, involves fewer steps and minimizes the use of water or other solvents than traditional top-down lithography (Schmidt 2007).

At the Georgia Institute of Technology, Atlanta, efforts are being made to fabricate electrical conductive adhesives which are lead-free, and thus environmentally friendly. This group has designed ECA using nano-silver fillers and molecular monolayers which are environmentally friendly and can be used as interconnects. Efforts are underway to produce quantum dots, a semiconducting material made from environmentally-friendly chemical synthesis methods rather than from cadmium selenide, a toxic compound. Quantum dots have numerous applications in sensing, biological or molecular imaging and electronic devices (Gardea-Torresdey et al. 2003).

The exploitation of biomaterials for the synthesis of nanoparticles is considered to be an interesting and important approach in green nanotechnology. Research has been conducted on algae, fungi, bacteria, and plants, and they are found to be important in the production of low cost, energy-efficient, non-toxic, and environmentally-friendly metal nanoparticles (Saif, Tahir, and Chen 2016). Studies prove that gall extract of *Pistacia integerrema* is a very good bioreductant for the synthesis of gold nanoparticles, which possesses antifungal and muscle relaxant properties leading to various biomedical and pharmaceutical applications (Islam et al. 2015). Among all the organisms, plants are predicted to be the best candidates for the large-scale green synthesis of silver nanoparticles. The silver nanoparticles produced in plants are more stable and have a faster rate of synthesis (Roy and Das 2015).

Silver nanoparticles synthesized by the reduction of silver nitrate using *Prunus japonica* leaves as a green reducing agent were observed to show potent antibacterial activity against *Escherichia coli, Proteus vulgaris, Staphylococcus aureus,* and *Bacillus cereus* to different extents (Saravanakumar et al. 2017). Leaf extracts of *Carissa spinarum* can act as a potential bio-resource for the synthesis of rapid, cheap, and effective silver nanoparticles having mosquitocidal effects, as observed through testing on the larvae of *Anopheles subpictus, Aedes albopictus,* and *Culex tritaeniorhynchus* (Govindarajan et al. 2016). It has been observed that silver nanoparticles prepared using *Lantana camara* leaf extract are a highly active antibacterial agent. The antibacterial activity of these nanoparticles increases with the decrement of particle size against all tested bacterial cultures (Ajitha, Reddy, and Reddy 2015).

Banana peel extract has been used for the synthesis of green metal nanoparticles. The average size of silver nanoparticles produced using banana peel extract was 23.7 nm as determined by dynamic light scattering (Ibrahim 2015). Green copper oxide nanoparticles ranging from 5 to 10 nm can be synthesized using *Gloriosa superba* plant extract. The nanoparticles possess significant activity against gram-negative *Klebsiella aerogenes, Pseudomonas desmolyticum, Escherichia coli,* and gram-positive *Staphylococcus aureus* (Naika et al. 2015). Another environmentally friendly and green method for the synthesis of copper oxide nanoparticles involves the use of leaf extract of *Ixoro coccinea.* The method produces nanoparticles with an average size ranging from 80 to 110 nm (Vishveshvar et al. 2018).

It has been discovered that algae can reduce metal ions, and hence be utilized for the synthesis of nanoparticles. Algae-mediated biosynthesis of nanoparticles is energy efficient, ecofriendly, expeditious, economical, and high yielding. Various classes of algae used for the production of green nanoparticles include *Cyanophyceae, Rhodophyceae, Chlorophyceae,* and *Phaeohyceae* (Sharma et al. 2016). The aqueous extract of alga *Pithophora oedogonia* was used for the reduction of silver nitrate to produce silver nanoparticles. It was observed that the synthesis process was completed within minutes, and the average size of the nanoparticles pertained to 34.03 nm. The nanoparticles also possessed a significant antibacterial activity against 7 tested pathogenic bacteria (Sinha et al. 2015).

The extract of marine alga *Colerpa racemosa* was also used for the synthesis of silver nanoparticles. After synthesis, the green nanoparticles were characterized using Fourier Transform Infrared Spectroscopy, X-ray Diffraction, and Transmission Electron Microscopy. The micrograph shows that the silver nanoparticles ranged from 5 to 25 nm and possessed antibacterial activity (Kathiraven et al. 2015).

Another green method for the synthesis of silver nanoparticles is by using carambola fruit extract, which not only helps in rapid reduction but also leads to stabilization of newly formed nanoparticles. It has been observed that the particles produced this way are pure and crystalline with a face-centered cubic structure (Gavade et al. 2015). The size of biologically synthesized nanoparticles ranges from 1 to 100 nm and is characterized through Scanning Electron Microscopy, UV spectroscopy, Transmission Electron Microscopy, X-ray diffraction, and Fourier Transform Infrared Spectroscopy. These green nanoparticles are potent antimicrobial agents and can be applied to engineering and medical applications (Chung et al. 2016).

The honey-mediated green synthesis of nanoparticles is a novel approach, which has been used to synthesize gold, silver, carbon, platinum, and palladium nanoparticles. Honey acts as both a reducing and stabilizing agent. The method uses room temperature and does not produce any toxic secondary product. It has been observed that honey-mediated green synthesis of nanoparticles is a simple, rapid, safe, cost-effective, reproducible, and biocompatible method (Balasooriya et al. 2017).

In recent years, interest has been shifting towards arginine, which is a unique, non-toxic, and biocompatible biomolecule, which can be used for the functionalization and stabilization of magnetic nanoparticles. With defined concentrations of arginine, different characteristics of synthesized nanoparticles can be controlled (Bagherpour et al. 2018). Nanofibres made from starch and proteins using electrospinning techniques also have the potential to be used in green technology. These nanofibres have a variety of applications in drug delivery, packaging material and filtration membranes. The rewards and risks of using green nanotechnology over traditional methods are shown in Table 1.

Table 1: Rewards and risks of using green nano-particles over traditional methods

Rewards	Risks
Used in a small amount	Bio-accumulation
Possess large surface area	Aggregation and difficulty in dispersion
Highly reactive	Toxicity of particle's by-product unknown
Friendly, biodegradable or recycled	Higher surface energy; stabilized using surfactants
Cost-effectiveness	Wave particle nature may lead to unknown/undiscovered features in a system under investigation
Requires less energy for production; energy efficient	Synergistic effect with other organic materials/metals may enhance their toxicity

2.3. Solar Cells

According to a survey conducted by Richard Smalley at Rice University of the next 10 major issues to be faced, energy was top of the list, whereas environmental safety occupied fourth place after water and food. Today, the global economy is dependent upon carbon (fossil fuels), thus the problem of climate/global warming is increasing. Under these circumstances, green nanotechnology will play an important role. Renewable energy sources such as solar, wind and geothermal can be alternative environmental sources in this regard.

Solar cells are also known as photovoltaic cells: these directly convert sunlight into electric energy. The working principle of these photovoltaic cells involves the following major steps:

- Absorption of light, generation of exciton (electron-hole pair)
- Diffusion of exciton to interface
- Separation of exciton (charges) at the interface due to electric field
- Separated charges reach opposite polls

Many modifications have been made to the basic structure of photovoltaic cells. Given below are a few widely investigated modifications, which will not only increase the efficiency of solar cells but also make them more environmentally-friendly.

- Organic photovoltaic cells
- Using quantum dots in solar cells
- Using nanofibres (ZnO electrospun) in photovoltaic cells

Organic photovoltaic cells were made using organic materials/polymers, e.g. Fullerenes, P3HT, PCBM and PCDTBT. Organic photovoltaics are preferable because they enable the idea of flexible solar cells since they can be made on any kind of substrate, unlike first-generation PVC using only silicon wafers. These OPV are made using the solution-processing technique, which is considered to be an important aspect of green technology. This type of PVC is highly influenced by its morphology, material used and the design of the device. About 3–9 % power efficiency has been achieved using OPV. The challenges further lie in optimizing their structure and interface for charge separation and collection (Brabec et al. 2010). The organics used do not have long-term stability. OPV has applications in the following areas.

- Organic memory devices
- Organic thin-film transistors

The adjustable bandgap allows the QD to be used as an energy-effective material in solar cells. QD is known to emit three electrons per photon in comparison with the classical one electron from a standard PVC panel. In addition, the absorption spectrum of these QD can also be tuned (Gao et al. 2011). Quantum-dot sensitized solar cells are the third generation of solar cells, and their theoretical efficiency has been calculated at around 44%. Electrospun TiO_2 and ZnO nanofibres provide OPV with better structural, electrical and environmentally friendly properties. These nanofibres have a high aspect ratio, thereby reducing the diffusion length for separated charges to reach the electrodes. This also increases the efficiency of solar cells and optimizes the surface area of OPV for absorption and exciton dissociation. The thickness of OPV is controlled by Electro-spinning time (Kim et al. 2013).

2.4. Water Treatment

Water is a necessity of life. As the human population is increasing, so is the competition for survival where abiotic factors play an important role, among which water is at the top of the list. Drinking water, wastewater or the contamination of water reservoirs is an important concern. Nanotechnology provides important means that are efficient, cost-effective and fast to treat/remediate the water bodies. Nanotechnology can be not only used for wastewater/drinking water remediation, but also for desalination of water and water management.

- Many nano-particles such as nano zero-valent iron particles can be used to remediate water from contamination (Shu et al. 2010).
- Nano-particles; zinc oxide and titanium oxide have also been used for similar functions.
- The use of nano zero-valent iron oxide along with ethylenediaminetetracetic acid (EDTA) and oxygen has also been used to remove xenobiotics present in water bodies. This method is an ideal example of green nanotechnology where it works under mild conditions by utilizing abundant non-toxic substances.
- Membranous structures consisting of nano-particles or nanopores are solutions stemming from the concept of nanotechnology and innovation.
- Bio palladium-based structures have been used for the removal of water contaminants (De Corte et al. 2012).
- The use of protein-based structures such as aquaporins is under investigation for their use in modelling systems that can be used as membranes for water purification.

Many tiny marine organisms such as algae and phytoplankton living on the water surface are not only a primary source of food but also help regulate climate balance by removing carbon from the atmosphere. TiO_2 is an effective disinfectant. These nano-particles are thought to be active under ultraviolet (UVR) radiation. These TiO_2 nano-particles are antibacterial coatings and wastewater disinfectants and have also been investigated as anti-cancer agents. These nano-particles change into atomic structures which make them highly reactive towards water and oxygen, which, in turn, leads them to form highly reactive oxygen species (ROS). These ROS have the power/ability to damage the cell membrane and DNA of bacteria and phytoplanktons including algae in wastewater. Recent studies have suggested that illuminating UV under artificial conditions may lead to marine toxicity; recent research was therefore conducted whereby TiO nanoparticles were illuminated under natural sunlight conditions penetrating 1m into pond water. In this study, four species of algae were used: this treatment seemed to be effective against three of these algal species. Significant quantities of ROS were produced. Although ROS can form naturally in marine waters, in the laboratory the rate of production of one type of ROS, OH, was six times higher in seawater containing TiO_2 nano-particles than that typically found in temperate coastal waters (Miller et al. 2012). A scanning electron microscope showed TiO_2 particles adhering to algal surfaces where their ROS activity not only attacked the membrane but also limited the process of photosynthesis.

3. Agriculture

In agriculture, green nanoparticles can be used to reduce the number of chemicals, reduce nutrient losses during fertilization, and enhance the yield of crops through pest and nutrient management. The nanofertilisers and nanopesticides are used to increase the productivity of crops (Prasad, Bhattacharyya, and Nguyen 2017).

3.1. Herbicides

Weeds or herbs are a great threat to agriculture because they greatly reduce crop yield. Herbicides are chemicals that are used to control, manipulate and eradicate the growth of undesirable plants and vegetation. These are commonly known as weedkillers and are used to kill unwanted plants.

3.1.1. Conventional Herbicides and Their Negative Impact on Human Health and Environment

Conventional herbicides are commercially available globally nowadays. Exposure to these herbicides is associated with acute toxicity and long-term exposure may even lead to carcinogenesis. These herbicides are also seen to affect the brain and can cause Parkinson's disease (Dewar et al. 2003). Studies have shown that herbicides are associated with the development of cancers like soft tissue sarcoma and non-Hodgkin's lymphoma. Studies have shown that the herbicides which are commercialized these days have negative effects on bird populations. (Blus and Henny 1997). The frog population has also been negatively affected by the use of herbicides.

Conventional herbicidal spraying of plants is seen to be directly related to de-masculinization in male frogs (Hayes et al. 2002).

3.1.2. NANOHERBICIDES: A Solution to Conventional Herbicides:

Nanotechnology has the potential to get rid of herbs in an environmentally friendly way because nano-herbicides do not leave any toxic or harmful residues in the environment. It is observed that less herbicide is used if the active ingredient in herbicide is combined with a smart delivery system (Pérez-de-Luque and Rubiales 2009). In this case, the delivery system is a nano-particle. Being nano-dimensional, these nano-herbicides get mixed with soil particles and prevent the growth of any unwanted weed. They are also manipulated to kill already grown unwanted herbs (Ali et al. 2014). Nano-herbicides are also shown not to be harmful to underground plant associations and growths like tubers and rhizomes (Satapanajaru et al. 2008). Nano-particles with a target-specific herbicide molecule are aimed at a specific receptor in the roots of herbs; it then enters the root system and inhibits the glycolysis of food reserves which results in starvation and finally death. Nano-herbicides are engineered to solve the problems caused by perennial weed management (Chinnamuthu and Boopathi 2009).

3.2. Fertilisers

Fertilisers are known to be organic or inorganic materials and chemicals that are added to the soil to increase the nutrient efficiency of soil (Erisman et al. 2008). These fertilisers are necessary for plant growth but they harm the soil and environment.

3.2.1. Limitations of Conventional Fertilisers

Conventional fertilisers are either organic or inorganic, and are used commercially for the improvement and proper growth of crops. These fertilisers have certain limitations and disadvantages. For example, inorganic fertilisers are associated with water pollution as they contain nitrates which enter underground waters and toxify them (Chaney 2012). Conventional fertilisers are also associated with soil contamination. It is observed that phosphate fertilisers have high quantities of fluorine in them which results in increased fluorine concentration in soil, which in turn decreases the quality of plants grown in such soil (Davis, Epp, and Riordan 2004). High aluminium content in inorganic fertilisers results in aluminium accumulation in soil and thus reduction in plant growth. If the aluminium concentration exceeds the threshold limit, this may cause acidification of soil (Thomas 2007).

The use of fertilisers is also seen to be associated with a decrease in the trace mineral content of the soil. Fertilisers are mostly composed of nitrogen, phosphorus and potassium, and over-use of these fertilisers decreases the soil availability of some of the important trace minerals, leading to crops with a low mineral content (Fan et al. 2008). A high level of fertilisers in the soil also damages symbiotic relationships among plant roots and mycorhizal fungi (Enwall, Philippot, and Hallin 2005).

3.2.2. Nanofertilisers

Nano-fertilisers are slow and controlled-release fertilisers, which means that they are compacted in a structure which releases them in an appropriately controlled amount. Conditional release of fertilisers has become important for promoting the development of environmentally friendly and sustainable agriculture (Chinnamuthu and Boopathi 2009). Encapsulation of a fertiliser with nano-particles to synchronize the release of active ingredients for crop uptake also prevents nutrient losses of soil and avoids the interaction of nutrients with the environment (DeRosa et al. 2010). Nano-porous zeolite surfactants are used to nano-structure the fertilisers. These zeolites are used to capture nitrogen and release it slowly in soil (Li 2003).

3.3. Pesticides

Pesticides are chemicals, used to attract, reduce and destroy pests and insects which attack plants. "Pesticide" is a vast term that comprises insecticides, fungicides and herbicides. Pesticides are one of the most common chemicals used or sprayed over cultivated plants (Gilden, Huffling, and Sattler 2010).

3.3.1. Chemical Pesticides and Their Disadvantages

Chemical pesticides are pesticides that are used conventionally and commercially on a large scale. These pesticides adversely affect human health and the environment. Some of the disastrous effects of pesticides are:

Acute Poisoning

Pesticides are connected to serious illnesses, which may include acute and chronic poisoning depending upon the length and degree of exposure. They are potent in inducing acute poisoning in agricultural labourers.

Pesticide poisoning can result in skin damage, skin allergy and irritation (Bassil et al. 2007).

Chronic Exposure

It has been reported in different studies that pesticide usage is associated with an increased incidence of cancer, which, in most cases, is non-Hodgkin's lymphomas and leukaemia (Sanborn et al. 2007). Pesticide poisoning is also seen to mimic hormones resulting in reproductive problems, neurological defects, birth defects, and even the death of foetuses (Jurewicz and Hanke 2008).

Pollution

Pesticides are observed to be a source of soil contamination and pollution (Miller 2004). They are also observed to be the major cause of water pollution.

Wild Life

Almost 95% of pesticide sprays are associated with environmental deterioration and the killing of non-targeted species. This also causes a serious threat to biodiversity leading to endangered species (Sims and Cupples 1999).

Pesticide resistance

Pesticide resistance starts when they are used excessively. The term relates to excessive use of pesticides due to which the species against which pesticides are targeted become resistant to chemicals, and instead of dying from these, become immune. Pesticide resistance is why many pesticides are now useless (Sims and Cupples 1999).

3.3.2. Nano-pesticides

It is time that the use of conventional pesticides was reduced to save the environment and reduce agricultural costs. This reduction in pesticide use can be achieved by increasing retention time and maintaining controlled delivery of pesticides (Sharon, Choudhary, and Kumar 2010). Nano-pesticides provide a solution, which is the controlled release of chemicals from the particles. One of the common pesticides which can be used is Halloysite, which is composed of nano-tubes. The use of nano-pesticides reduces the cost and the number of pesticides by increasing the release time of a chemical. They are also shown to have minimum effect on the environment (Allen 1994). For example, hollow porous silica nanoparticles can be used with validamycin as an effective pesticide. The pesticide is absorbed in the cuticular lipid of insects and results in insect death without any harmful environmental side-effects (Liu et al. 2006).

In another study, silver nanoparticles were synthesized using a cheap aqueous leaf extract of *Anisomeles indica.* It was observed that the nanoparticles pose significant mosquitocidal activity (Govindarajan et al. 2016). Different types of green nanoparticles, including metal nanoparticles, bimetallic nanoparticles, metal oxide nanoparticles, carbon nanotubes and halloysite nanotubes have been used for the detection, degradation and removal of pesticides (Rawtani et al. 2018).

Conclusion and Future Prospects

Agriculture in the twenty-first century faces many challenges relating to increasing food and fibre production; these challenges have increased because fewer agricultural workers are owing to increased urbanization. Variable climatic conditions also pose a threat to agriculture. In the face of all these threats, nanotechnology offers a solution to these agricultural challenges. Nanotechnology ensures environmental stability and agricultural sustainability. It is expected to increase crop yield in an environmentally friendly way soon. Adopting nanotechnology as a solution will pave the way to reducing world starvation by feeding the increasing population (Wheeler 2005).

At the same time, the general public's negative attitude to nanotechnology stands as a challenge to the creation of a better world. The public response to genetically modified food has already made clear their perception of nanotechnology. It is time to make the public aware of the positive and potential advantages of green nanotechnology. This technology, in the future, promises to relieve the world from plant disease and the high cost of different chemicals in agriculture (Sugunan and Dutta 2008).

References

Ahmed, S, and S Ikram. 2016. "Biosynthesis of gold nanoparticles: a green approach." *Journal of Photochemistry and Photobiology B: Biology* 161:141–153.

Ajitha, B, Y A K Reddy, and P S Reddy. 2015. "Green synthesis and characterization of silver nanoparticles using *Lantana camara* leaf extract." *Materials Science and Engineering: C* 49:373–381.

Ali, M Azam, I Rehman, A Iqbal, S Din, A Q Rao, A Latif, T R Samiullah, S Azam, and T Husnain. 2014. "Nanotechnology, a new frontier in Agriculture." *Advancements in Life Sciences* 1 (3):129–138.

Allen, R. 1994. "Agriculture during the industrial revolution." *The Economic History of Britain Since* 1700 (3):96–123.

Bagherpour, A R, F Kashanian, S A S Ebrahimi, and M Habibi-Rezaei. 2018. "L-arginine modified magnetic nanoparticles: green synthesis and characterization." *Nanotechnology* 29 (7):075706.

Balasooriya, E R, C D Jayasinghe, U A Jayawardena, R W D Ruwanthika, R M de Silva, and P V Udagama. 2017. "Honey mediated green synthesis of nanoparticles: new era of safe nanotechnology." *Journal of Nanomaterials* 2017:10.

Bassil, K L, C Vakil, M S, DC Cole, J S Kaur, and K J Kerr. 2007. "Cancer health effects of pesticides: systematic review." *Canadian Family Physician* 53 (10):1704–1711.

Blus, L J, and C J Henny. 1997. "Field studies on pesticides and birds: unexpected and unique relations." *Ecological Applications* 7 (4):1125–1132.

Brabec, C J, S Gowrisanker, J JM Halls, D Laird, S Jia, and S P Williams. 2010. "Polymer–fullerene bulk-heterojunction solar cells." *Advanced Materials* 22 (34):3839–3856.

Chaney, R L. 2012. "Food safety issues for mineral and organic fertilisers." In *Advances in agronomy*, 51–99. Elsevier.

Chinnamuthu, C R, and P M Boopathi. 2009. "Nanotechnology and agroecosystem." *Madras Agricultural Journal* 96 (1/6):17–31.

Chung, I-M, I Park, K Seung-Hyun, M Thiruvengadam, and G Rajakumar. 2016. "Plant-mediated synthesis of silver nanoparticles: their characteristic properties and therapeutic applications." *Nanoscale Research Letters* 11 (1):40.

Colombo, G, P Soto, and E Gazit. 2007. "Peptide self-assembly at the nanoscale: a challenging target for computational and experimental biotechnology." *TRDENS in Biotechnology* 25 (5):211–218.

Davis, D R, M D Epp, and H D Riordan. 2004. "Changes in USDA food composition data for 43 garden crops, 1950 to 1999." *Journal of the American College of Nutrition* 23 (6):669–682.

De Corte, S, T Hennebel, B De Gusseme, W Verstraete, and N Boon. 2012. "Bio-palladium: from metal recovery to catalytic applications." *Microbial Biotechnology* 5 (1):5–17.

DeRosa, M C, C Monreal, M Schnitzer, R Walsh, and Y Sultan. 2010. "Nanotechnology in fertilisers." *Nature Nanotechnology* 5 (2):91.

Dewar, A M, M J May, I P Woiwod, L A Haylock, G T Champion, B H Garner, R J N Sands, A Qi, and J D Pidgeon. 2003. "A novel approach to the use of genetically modified herbicide tolerant crops for environmental benefit." *Proceedings of the Royal Society of London B: Biological Sciences* 270 (1513):335–340.

Enwall, K, L Philippot, and S Hallin. 2005. "Activity and composition of the denitrifying bacterial community respond differently to long-term fertilization." *Applied and Environmental Microbiology* 71 (12):8335–8343.

Erisman, J W, M A Sutton, J Galloway, Z Klimont, and W Winiwarter. 2008. "How a century of ammonia synthesis changed the world." *Nature Geoscience* 1 (10):636.

Fakruddin, Md, Z Hossain, and H Afroz. 2012. "Prospects and applications of nanobiotechnology: a medical perspective." *Journal of Nanobiotechnology* 10 (1):1–8.

Fan, M-S, F-J Zhao, S J Fairweather-Tait, P R Poulton, S J Dunham, and S P McGrath. 2008. "Evidence of decreasing mineral density in wheat grain over the last 160 years." *Journal of Trace Elements in Medicine and Biology* 22 (4):315–324.

Gao, J, J M Luther, O E Semonin, R J Ellingson, A J Nozik, and M C Beard. 2011. "Quantum dot size dependent J– V characteristics in heterojunction ZnO/PbS quantum dot solar cells." *Nano Letters* 11 (3):1002–1008.

Gardea-Torresdey, J L, E Gomez, J R Peralta-Videa, J G Parsons, H Troiani, and M Jose-Yacaman. 2003. "Alfalfa sprouts: a natural source for the synthesis of silver nanoparticles." *Langmuir* 19 (4):1357–1361.

Gavade, S J M, G H Nikam, R S Dhabbe, S R Sabale, B V Tamhankar, and G N Mulik. 2015. "Green synthesis of silver nanoparticles by using carambola fruit extract and their antibacterial activity." *Advances in Natural Sciences: Nanoscience and Nanotechnology* 6 (4):045015.

Gilden, R C, K Huffling, and B Sattler. 2010. "Pesticides and health risks." *Journal of Obstetric, Gynecologic, and Neonatal Nursing* 39 (1):103–110.

Govindarajan, M, M Rajeswary, K Veerakumar, U Muthukumaran, SL Hoti, and G Benelli. 2016. "Green synthesis and characterization of silver nanoparticles fabricated using Anisomeles indica: mosquitocidal potential against malaria, dengue and Japanese encephalitis vectors." *Experimental Parasitology* 161:40–47.

Hayes, T B, A Collins, M Lee, M Mendoza, N Noriega, A A Stuart, and A Vonk. 2002. "Hermaphroditic, demasculinized frogs after exposure to the herbicide atrazine at low ecologically relevant doses." *Proceedings of the National Academy of Sciences* 99 (8):5476–5480.

Hussain, I, N B Singh, A Singh, H Singh, and S C Singh. 2016. "Green synthesis of nanoparticles and its potential application." *Biotechnology Letters* 38 (4):545–560.

Ibrahim, H MM. 2015. "Green synthesis and characterization of silver nanoparticles using banana peel extract and their antimicrobial activity against representative microorganisms." *Journal of Radiation Research and Applied Sciences* 8 (3):265–275.

Islam, N Ul, K Jalil, M Shahid, N Muhammad, and A Rauf. 2015. "Pistacia integerrima gall extract mediated green synthesis of gold nanoparticles and their biological activities." *Arabian Journal of Chemistry*.

Jurewicz, J, and W Hanke. 2008. "Prenatal and childhood exposure to pesticides and neurobehavioral development: review of epidemiological studies." *International Journal of Occupational Medicine and Environmental Health* 21 (2):121–132.

Kathiraven, T, A Sundaramanickam, N Shanmugam, and T Balasubramanian. 2015. "Green synthesis of silver nanoparticles using marine algae *Caulerpa racemosa* and their antibacterial activity against some human pathogens." *Applied Nanoscience* 5 (4):499–504.

Kim, A, Y Won, K Woo, C-H Kim, and J Moon. 2013. "Highly transparent low resistance ZnO/Ag nanowire/ZnO composite electrode for thin film solar cells." *ACS Nano* 7 (2):1081–1091.

Li, Zh. 2003. "Use of surfactant-modified zeolite as fertiliser carriers to control nitrate release." *Microporous and Mesoporous Materials* 61 (1–3):181–188.

Liu, F, L-X Wen, Z-Z Li, W Yu, H-Y Sun, and J-F Chen. 2006. "Porous hollow silica nanoparticles as controlled delivery system for water-soluble pesticide." *Materials Research Bulletin* 41 (12):2268–2275.

Lu, Y, and S Ozcan. 2015. "Green nanomaterials: On track for a sustainable future." *Nano Today* 10 (4):417–420.

McCarter, S. 2007. "Nanotechnology provides green path to environmentally sustainable economy." *EurekAlert*.

Miller, G T. 2004. Sustaining the Earth. California: Incorporated Pacific Grove.

Miller, R J, S Bennett, A A Keller, S Pease, and H S Lenihan. 2012. "TiO2 nanoparticles are phototoxic to marine phytoplankton." *PloS One* 7 (1):e30321.

Naika, H R, K Lingaraju, K Manjunath, D Kumar, G Nagaraju, D Suresh, and H Nagabhushana. 2015. "Green synthesis of CuO nanoparticles using *Gloriosa superba* L. extract and their antibacterial activity." *Journal of Taibah University for Science* 9 (1):7–12.

Pérez-de-Luque, A, and D Rubiales. 2009. "Nanotechnology for parasitic plant control." *Pest Management Science* 65 (5):540–545.

Prasad, R, A Bhattacharyya, and Q D Nguyen. 2017. "Nanotechnology in sustainable agriculture: recent developments, challenges, and perspectives." *Frontiers in Microbiology* 8:1014.

Rawtani, D, N Khatri, S Tyagi, and G Pandey. 2018. "Nanotechnology-based recent approaches for sensing and remediation of pesticides." *Journal of Environmental Management* 206:749–762.

Roy, S, and T Kumar Das. 2015. "Plant Mediated Green Synthesis of Silver Nanoparticles-A." *International Journal of Plant Biology and Research* 3 (3):1044.

Saif, S, A Tahir, and Y Chen. 2016. "Green synthesis of iron nanoparticles and their environmental applications and implications." *Nanomaterials* 6 (11):209.

Sanborn, M, K J Kerr, L H Sanin, D C Cole, K L Bassil, and C Vakil. 2007. "Non-cancer health effects of pesticides: systematic review and implications for family doctors." *Canadian Family Physician* 53 (10):1712–1720.

Saravanakumar, A, M M Peng, M Ganesh, J Jayaprakash, M Mohankumar, and H T Jang. 2017. "Low-cost and eco-friendly green synthesis of silver nanoparticles using *Prunus japonica* (Rosaceae) leaf extract and their antibacterial, antioxidant properties." *Artificial Cells, Nanomedicine, and Biotechnology* 45 (6):1165–1171.

Satapanajaru, T, P Anurakpongsatorn, P Pengthamkeerati, and H Boparai. 2008. "Remediation of atrazine-contaminated soil and water by nano zerovalent iron." *Water, Air, and Soil Pollution* 192 (1–4):349–359.

Schmidt, K. 2007. "Green nanotechnology: it's easier than you think."

Shapira, P, and J Youtie. 2015. "The economic contributions of nanotechnology to green and sustainable growth." In *Green Processes for Nanotechnology*, 409–434. Springer.

Sharma, A, S Sharma, K Sharma, S PK Chetri, A Vashishtha, P Singh, R Kumar, B Rathi, and V Agrawal. 2016. "Algae as crucial organisms in advancing nanotechnology: a systematic review." *Journal of Applied Phycology* 28 (3):1759–1774.

Sharon, M, A Kr Choudhary, and R Kumar. 2010. "Nanotechnology in agricultural diseases and food safety." *Journal of Phytology* 2 (4):83–92.

Shu, H-Y, M-C Chang, C-C Chen, and P-E Chen. 2010. "Using resin supported nano zero-valent iron particles for decoloration of Acid Blue 113 azo dye solution." *Journal of Hazardous Materials* 184 (1–3):499–505.

Sims, G K, and A M Cupples. 1999. "Factors controlling degradation of pesticides in soil." *Pesticide Science* 55 (5):598–601.

Sinha, S N, D Paul, N Halder, D Sengupta, and S K Patra. 2015. "Green synthesis of silver nanoparticles using fresh water green alga *Pithophora oedogonia* (Mont.) Wittrock and evaluation of their antibacterial activity." *Applied Nanoscience* 5 (6):703–709.

Sugunan, A, and J Dutta. 2008. "Pollution treatment, remediation and sensing." *Nanotechnology*.

Thanos, C G, and D Emerich. 2008. "On the use of hydrogels in cell encapsulation and tissue engineering systems." *Recent Patents on Drug Delivery and Formulation* 2 (1):19–24.

Thomas, D. 2007. "The mineral depletion of foods available to us as a nation (1940–2002)–a review of the 6th Edition of McCance and Widdowson." *Nutrition and Health* 19 (1–2):21–55.

Vishveshvar, K, M V A Krishnan, K Haribabu, and S Vishnuprasad. 2018. "Green synthesis of copper oxide nanoparticles using *Ixiro coccinea* plant leaves and its characterization." *BioNanoScience*:1–5.

Wheeler, S. 2005. "Factors influencing agricultural professionals' attitudes towards organic agriculture and biotechnology."

CHAPTER 19

POTENTIAL OF ORGANIC FERTILISERS FOR VEGETABLE PRODUCTION: AN ECO-AGRONOMIC PERSPECTIVE

SYED HAMMAD RAZA, FAHAD SHAFIQ*, ABIDA PARVEEN, MUHAMMAD IQBAL

Department of Botany, Government College University Faisalabad
fahadsheikh1800@gmail.com

Abstract

Organic fertilisers are a cheap source of nutrients. The addition of organic fertilisers alters the soil's physicochemical properties, including its water-holding capacity, soil hydraulic conductivity, porosity, bulk density, cation exchange capacity, and adsorption characteristics. In addition, the influence of organic matter in the modulation of soil microbial communities is substantial. Due to these properties, organic fertilisers are better alternatives to traditional fertilisers, which pose risks of groundwater contamination via nitrate leaching. Also, inorganic nutrients negatively alter the soil's physical characteristics, and negatively affect soil microbes. Insights into various types of organic fertiliser, including farmyard and poultry manure, compost, vermicompost, humic substances and biochar are presented in this chapter. Lastly, the interactive effects of these fertilisers on soil characteristics and vegetable production are discussed.

Keywords: biochar, organic agriculture, organic fertilisers, soil health, vegetables

Introduction to Fertilisers and Soil Fertility

The use of fertilisers on a massive scale on agricultural fields has become common practice ever since the Second Agricultural Revolution. Soil fertility influences crop growth and productivity, and the addition of nutrients to the soil enhances agricultural productivity. The nutrient status of soils is rarely sufficient to meet the growth requirements of plants, and farmers tend to compensate for this by the addition of synthetic and natural amendments that are rich in nitrogen, potassium and phosphorus (Ahmad et al. 2009). Unfortunately, the application of huge amounts of nutrients to the soil initiates nitrate leaching, groundwater contamination, and is linked with non-point source pollution (Kourakos et al. 2012, Eriksen et al. 2015). Moreover, the application of inorganic or synthetic fertilisers deteriorates the soil structure by altering its physical properties, field capacity, porosity, aggregate stability, bulk density, hydraulic conductivity and water infiltration rates (Haynes and Naidu 1998). Higher economic costs, lower soil retention times, and volatilisation are some other common problems also associated with the use of synthetic, inorganic fertilisers. Yet, massive amounts of fertilisers are added to the soil without prior knowledge of its fertility, nutrient status and fertiliser mineralisation rates (Balasubramanian et al. 1998).

Organic matter (OM) improves the soil's physical characteristics via humic substances and polysaccharides that bind and provide aggregate stability (Shukla et al. 2003, Zhang and Fang 2007, Singh Brar et al. 2015). Soil OM increases water-retention capacity and also promotes nitrogen mineralisation via biological transformations (Evanylo et al. 2008). Moreover, it increases soil hydraulic conductivity through improvement in soil porosity (Ndiaye et al. 2007) and field capacity (Pernes-Debuyser and Tessier 2004, Hati et al. 2007). Agricultural fields with lower OM values exhibit reduced fertility and productivity rates (Sharma and Subehia 2003). The reduction in soil OM is primarily due to excessive cultivation which negatively affects the soil's physical characteristics (Bhattacharyya et al. 2007). In this context, higher nitrogen leaching rates are linked with soil particle size, and OM contents with more coarsely textured soil exhibited lesser leaching rates and *vice versa* (Eriksen et al. 2015); and adverse environmental impacts of synthetic inorganic fertilisers should be reduced (Li et al. 2017). Therefore, the maintenance of soil OM fraction is a critical aspect in the improvement of soil quality and fertility. Environmentally-friendly agricultural practices will improve soil health and agricultural quality and productivity, whilst simultaneously reducing costs associated with inorganic fertilisers (Islam et al. 2017).

Interactive effects of organic fertilisers on vegetables

The positive influence of a variety of organic fertilisers on the growth and productivity of both crops and vegetables has been reported in many studies. For instance, the application of biochar improved rice yield (Vinh et al. 2014), whilst both biochar and perlite improved the yield of leafy vegetables (Awad et al. 2017). Similarly, the application of chicken manure increased tomato production (Li et al. 2017) and farmyard manure improved potato yield (Mama, Jeylan, and Aseffa 2016). The comparative effects of different organic fertilisers on multiple vegetable species have been presented in Table-1.

Organic farming

The term "organic farming" refers to the cultivation of crops and vegetables in the absence of chemicals or chemical fertilisers of synthetic origin. From an environmental perspective, organic farming improves the soil's physical and chemical properties, thereby restoring soil fertility (Chowdhury 2004, Ortiz-Escobar and Hue 2007). Restoration of soil fertility is an important aspect as repeated cultivation cycles deplete soil nutrients and deteriorate soil structure, ultimately reducing plant growth and productivity (Hughes and Venema 2005). Carbon-rich organic amendments could feed the soil's biological processes, releasing nutrients and improving soil physical characteristics (Lal 2006). Therefore, organic fertilisers have been regarded as environmentally-friendly, carbon-neutral alternatives (Evanylo et al. 2008, Oyewole, Opaluwa, and Omale 2012). They are also rich sources of beneficial soil microbes and provide macro- and micro-nutrients, vitamins and growth-promoting substances like indole acetic acid and gibberellic acid (Sreenivasa, Naik, and Bhat 2010). A brief account of various organic fertilisers is discussed below.

Farmyard manure

Livestock is a cheap source of organic fertiliser and the application of farmyard manure (FYM) to the agricultural fields is a traditional practice (Seefeldt 2013, Singh Brar et al. 2015). An average 1000pound cow produces 15 tonnes of FYM annually, which is a source of 200 pounds of nitrogen, 190 pounds of phosphorus (P_2O5) and about 250 pounds of potassium (K_2O) fertilisers (Ahmad et al. 2016). FYM also provides essential micronutrients, including iron, zinc, molybdenum, copper, boron, manganese, calcium, magnesium and sulphur (Stowell and Bickert 1995). The increase in soil OM content is dependent upon the decomposition rates of FYM (Haynes and Naidu 1998, Gregorich, Drury, and Baldock 2001).

The addition of FYM also improves soil porosity, resulting in better water-holding capacity (Eghball 2002, Ahmad et al. 2016). At present, the application of FYM is more commonly associated with irrigated agricultural practices where it has resulted in crop yield improvement through better soil physicochemical and biological characteristics (Flavel and Murphy 2006, Whalen and Chang 2002). The combined application of FYM with inorganic fertilisers also improved soil OM, microbial communities, soil physical properties and hydraulic conductivity (Bhattacharyya et al. 2007).

Chicken litter

Chicken litter is a heterogeneous blend of wasted feed, chicken faeces and feathers (Wilkinson et al. 2011). Over 15 million tonnes of chicken litter is produced annually in the United States, which is recycled as a low-cost organic fertiliser (Enticknap et al. 2006). Chicken manure also contains all the essential mineral nutrients for optimum plant growth (Ahmad et al. 2016). The most important among its nutrients is nitrogen, which is released into the soil after chicken litter application. Significant proportions of N have been reported in poultry manure due to the presence of higher levels of amino acids and proteins (Chen and Jiang 2014). Apart from the nutrient release-related aspects, its application improved the soil's physical properties, including soil porosity, organic matter contents, water holding capacity and nutrient status (Celik, Ortas, and Kilic 2004, Ouyang 2005).

Table-1 - Interactive effects of organic fertilizers on vegetables

Sr. No.	Vegetable	Organic fertiliser applied	Growth medium	Biological effects	References
1	Leafy vegetables (cabbage, dill, mallow, red lettuce and tatsoi)	Biochar Perlite	Hydroponics	– Improvements in root and shoot growth by 2 fold – Improvement in nutritional status – Biochar was more effective than perlite	(Awad et al. 2017)
2	*Basella alba* L. *Ipomoea aquatic Forssk.*	Biochar Compost Biochar and Compost	Soil	– Increase in nutrient uptake – Biochar application caused the increase in vegetable yields by 4.7–25.5%	(Vinh et al. 2014)
3	Tomato	Chicken manure	Soil	– Increase in yield and quality – The inappropriate application causes environmental pollution or an increase in nitrate contents – Maximum yield occurred in ROF treatment – ROF as best fertiliser to improve yield and quality with least environmental risk	(Li et al. 2017)

Sr. No.	Vegetable	Organic fertiliser applied	Growth medium	Biological effects	References
4	Potato	Farmyard manure and Nitrogen	Soil	– FYM and N increased growth attributes – Yield components increased – A combination of both is suitable to achieve increased yield	(Mama, Jeylan, and Aseffa 2016)
5	Potato	Manure and Fertiliser (Cow dung, Mustard oilcake as N, P_2O_5,K_2O	Soil	– Plant height increased – The number of tubers increased per hill – The yield of tubers per hill increased – Tuber size increased	(Hossain, Hakim, and Onguso 2003)
6	Tomato	Vermicompost, Compost, Mixed fertiliser, Inorganic fertiliser	Soil	– Fruit yield and diameter significantly increased – Number of fruits and plant height higher in mixed fertilisers	(Islam et al. 2017)
7	Tomato	Cattle manure poultry manure	Soil	– Improvement in yield	(Geboloğlu, Yanar, and Aydin 2011)
8	Carrot	Biochar	Soil	– Improvement in growth attributes of carrot in marginal soil	(Carpenter and Nair 2014)
9	Turnip	Minerals or	Soil	– Biomass increased maximally with	(El-Sherbeny et

Sr. No.	Vegetable	Organic fertiliser applied	Growth medium	Biological effects	References
		Organic Fertilizers (NPK, Humic acid, Compost tea)		the application of NPK and humic acid – Humic acid and compost tea increased carbohydrate contents – Mineral contents, total lipids and fatty acid contents increased in response to humic acid application.	al. 2012)
10	Cucumber	Humic plus Vegeamino	Sand	– Increase in dry biomass of plants – Improvements in chlorophyll contents – Beneficial effects on N and P uptake – Improved nutritional status – Absorption of nutrients increased – Both hummus and Vegeamino enhanced cucumber growth and yield	(Plant 2016)
11	Onion and Turnip	Chicken Manure Compost	Soil	– Compost improved the growth of seedlings – Chicken manure application enhanced both growth and yield – Chicken manure increased plant growth more in turnip	(Salamandane, Muetanene, and Cuiana 2016)

Sr. No.	Vegetable	Organic fertiliser applied	Growth medium	Biological effects	References
12	Tomato	Chicken manure FYM Agro-fish	Soil	– Chicken manure application increased plant height, root length, fresh and dry weights – Mixed manure increased fresh and dry weights more than that with split application – Agro-fish enhanced stem diameter, flower and fruit numbers – Chicken manure is regarded as the best among three types studied	Kalbani et al. 2016
13	Tomato	Urea	Soil	– Significantly increased growth, plant height, number of leaves, – Increased number of flowers, fruit clusters, fruit diameter	(Mondal and Al Mamun 2011)
14	Tomato	Biochar	Soil	– Improvement in growth attributes – Biochar mediated increase in the up-regulation of gibberellins was evident	(French and Iyer-Pascuzzi 2018)
15	Peppers	Vermicompost FYM Food and paper waste	Soil	– Vermicompost increased microbial biomass – An increase in the growth and yield attributes of pepper was evident	(Arancon et al. 2005)

Compost

The aerobic and anaerobic decomposition of OM results in the synthesis of compost that is a stable, final organic product (Talyan, Dahiya, and Sreekrishnan 2008). The addition of compost to the soil restores fertility by affecting soil pH and nutrient availability (Eghball 2002). In the absence of FYM, compost application to agricultural soil is an environmentally friendly alternative.

Over the past few years, the synthesis of compost has gained considerable attention, and the chemical conversion of animal and poultry waste into organic fertilisers, such as compost, emerged (Chen and Jiang 2014). The process of composting is a spontaneous oxidation process that results in the synthesis of compost (Bernal, Alburquerque, and Moral 2009). During this multistep process, microbes degrade complex organic molecules into relatively simple and stable organic compounds over a period of four to six weeks (Chen and Jiang 2014, Ryckeboer et al. 2003). Composting is associated with cheaper economic costs (Lim et al. 2015, Millner 2009), and the finished product is uniform compared with FYM; moreover, there is a significant reduction in pathogens (Bernal, Albuquerque, and Moral 2009). Compost is a nutrient-rich, organic fertiliser that, upon application to the soil, improves its physical properties along with its water retention capacity.

Vermicompost

A modified form of composting is vermicomposting, which involves the use of earthworms for the biodegradation of organic matter; a technique that was established a decade ago (Wu et al. 2014). Earthworms and associated symbiotic microbial communities result in the breakdown of organic matter into stable forms that can be used by plants (Dominguez and Edwards 2004, Garg and Gupta 2009). During the process of vermicomposting, an earthworm ingests OM by mouth and passes it through the gizzard where it is finely ground and passed into the gut (Palsania et al. 2008). In the gut of the earthworm, the release of nutrients by bioconversion occurs (Palsania et al. 2008). The end product is vermicompost, and it is very effective in improving the soil's physical characteristics, similar to those of compost (Singh et al. 2011). The addition of vermicompost improves soil moisture retention capacity, nutrient availability and soil microbial communities (Singh et al. 2011). Interestingly, vermicompost is richer in nutrients and carbon contents compared to compost, whilst other elements are higher in compost (Norbu

2002). Vermicompost also contains higher humus contents, which makes it more suitable for agricultural applications (Lim et al. 2015).

Humic substances

About 80% of the soil's organic carbon is in the form of humic substances, which is also a considerable source of sulphur and nitrogen (Canellas and Olivares 2014). Humus or humic substances (HS) represent the partially decayed assemblage of heterogeneous organic compounds which are resistant to microbial degradation and insoluble in water (Piccolo 2012). These heterogeneous compounds are irregularly associated with hydrophobic and hydrophilic phases within the soil (Piccolo 2002). The formation of HS takes place from plant and animal organic material after a combination of physical, biological and microbial transformation processes (Hernandez et al. 2015). The presence of humic substances in the soil represents the most stable form of soil carbon, and improves soil fertility, water retention capacity, porosity, and hence hydraulic conductivity.

The influence of HS on the production of vegetables and fruits has been reported in detail by Hernandez et al. (2015). An important aspect is that HS can act as biostimulants, and thereby modulate the growth of horticultural crops upon exogenous application (Hernandez et al. 2015). The application of HS promotes root growth in several plant species through modulation in primary and secondary metabolites (Canellas and Olivares 2014). Another study reported that HS mediated an increase in the enzyme activity of nitrogen assimilation and metabolism enzymes, viz. nitrate reductase, glutamate dehydrogenase and glutamine synthetase (Canellas et al. 2013, Hernandez et al. 2015). Application of HS also modulated activities of glycolysis and Krebs cycle enzymes (Nardi et al. 2007). The HS induced stimulation in the uptake of nitrates through the upregulation of H^+/NO_3^- symporter (Nardi et al. 2007), and also improved soil C status (Puglisi et al. 2008).

Biochar

Biochar is a porous organic material that can improve the soil's physical characteristics. The black carbon material can be prepared from organic materials via thermal processes or by the action of chemicals (Novak et al. 2009). It is chemically composed of aromatic carbon rings and produced by the thermal decomposition of organic matter in the absence, or very limited supply, of oxygen through pyrolysis (Atkinson,

Fitzgerald, and Hipps 2010, Karhu et al. 2011). If the resultant material is prepared for use in soil, then it will be termed biochar; otherwise, black carbon or charcoal (Barrow 2012). The addition of biochar to agricultural soils affected the soil pH and modulated soil nutrient retention through adsorption (Laghari et al. 2016, Novak et al. 2009).

Biochar-mediated improvements in the soil cation exchange capacity, and nutrient retention improved crop biomass and productivity (Schnell et al. 2012, Xu et al. 2012) as well as soil carbon contents (Lehmann et al. 2011). Other prominent effects of biochar on soil included improvement in particle size and bulk density, porosity and other physical properties, like soil structure and soil texture (Ding et al. 2016, Xu et al. 2012). Biochar also inhibited nitrogen volatilisation and improved soil N pool (McHenry 2011). It is relatively resistant to microbial decomposition, and its addition to the soil improves soil microbes by providing a niche for diverse types of microorganisms (Barrow 2012, Pietikäinen, Kiikkilä, and Fritze 2000).

Challenges and future prospects

Despite the greatly beneficial effects of organic fertilisers on the soil's physical and chemical characteristics, some of the limitations are associated with the application of these, which must be addressed properly. Of these, the most important one is the potential presence of pathogens in organic fertilisers, like chicken litter (Chen and Jiang 2014). In the composting of chicken litter, removing the pathogens from it prior to its application in the fields is generally suggested (Chen and Jiang 2014). Some physical and chemical treatments also target these pathogens in order to eliminate them from the animal and plant wastes of organic fertilisers (Chen and Jiang 2014, Kim et al. 2011, Wilkinson et al. 2011). Another important issue is the cost associated with the processing of organic wastes, such as cost efficiency and the selection of appropriate waste processing techniques (Kelleher et al. 2002). The process of composting also needs to be optimised as it results in considerable nitrogen loss during the process (Moore, Daniel, and Edwards 2000). Various amendments, like peat, straw and other organic wastes have been recommended to compensate for the loss. As a future prospect, organic farming is an environmentally-friendly approach to reducing ground-water contamination and promoting soil health.

References

Ahmad, A A, A Fares, F Abbas, and J L Deenik. 2009. "Nutrient concentrations within and below root zones from applied chicken manure in selected Hawaiian soils." *Journal of Environmental Science and Health Part B* 44 (8):828–843.

Ahmad, A A, T JK Radovich, H V Nguyen, J Uyeda, A Arakaki, JCadby, R Paull, J Sugano, and G Teves. 2016. "Use of organic fertilizers to enhance soil fertility, plant growth, and yield in a tropical environment." In *Organic Fertilizers-From Basic Concepts to Applied Outcomes*. InTech.

Arancon, N Q, C A Edwards, P Bierman, J D Metzger, and C Lucht. 2005. "Effects of vermicomposts produced from cattle manure, food waste and paper waste on the growth and yield of peppers in the field." *Pedobiologia* 49 (4):297–306.

Atkinson, C J, J D Fitzgerald, and N AHipps. 2010. "Potential mechanisms for achieving agricultural benefits from biochar application to temperate soils: a review." *Plant and Soil* 337 (1–2):1–18.

Awad, Y Mahmoud, S-E Lee, M Bedair M Ahmed, NT Vu, M Farooq, I S Kim, H Soo Kim, M Vithanage, A R A Usman, and M Al-Wabel. 2017. "Biochar, a potential hydroponic growth substrate, enhances the nutritional status and growth of leafy vegetables." *Journal of Cleaner Production* 156:581–588.

Balasubramanian, V, AC Morales, RT Cruz, and S Abdulrachman. 1998. "On-farm adaptation of knowledge-intensive nitrogen management technologies for rice systems." *Nutrient Cycling in Agroecosystems* 53 (1):59–69.

Barrow, CJ. 2012. "Biochar: potential for countering land degradation and for improving agriculture." *Applied Geography* 34:21–28.

Bernal, M P, JA Alburquerque, and R Moral. 2009. "Composting of animal manures and chemical criteria for compost maturity assessment. A review." *Bioresource Technology*100 (22):5444–5453.

Bhattacharyya, R, S Chandra, RD Singh, S Kundu, AK Srivastva, and HS Gupta. 2007. "Long-term farmyard manure application effects on properties of a silty clay loam soil under irrigated wheat–soybean rotation." *Soil and Tillage Research* 94 (2):386–396.

Canellas, L P, and F L Olivares. 2014. "Physiological responses to humic substances as plant growth promoter." *Chemical and Biological Technologies in Agriculture* 1 (1):3.

Canellas, L Pasqualoto, D M Balmori, L O Médici, N O Aguiar, E Campostrini, R CC Rosa, A R Façanha, and F L Olivares. 2013. "A combination of humic substances and *Herbaspirillumseropedicae*inoculation enhances the growth of maize *(Zea mays* L.)." *Plant and Soil* 366 (1–2):119–132.
Carpenter, B H, and ANair. 2014. "Effect of biochar on carrot production."
Celik, I, I Ortas, and S Kilic. 2004. "Effects of compost, mycorrhiza, manure and fertilizer on some physical properties of a Chromoxerert soil." *Soil and Tillage Research* 78 (1):59–67.
Chen, Z, and X Jiang. 2014. "Microbiological safety of chicken litter or chicken litter-based organic fertilizers: a review." *Agriculture* 4 (1):1–29.
Chowdhury, R. 2004. "Effects of chemical fertilizers on the surrounding environment and the alternative to the chemical fertilizers." *IES-ENVIS Newsletters* 7:4–5.
Ding, Y, Y Liu, S Liu, Z Li, X Tan, XHuang, G Zeng, LZhou, and B Zheng. 2016. "Biochar to improve soil fertility. A review." *Agronomy for Sustainable Development*36 (2):36.
Dominguez, J, and CA Edwards. 2004. "Vermicomposting organic wastes: a review." *Soil zoology for sustainable development in the 21st century, Cairo*:369–395.
Eghball, B. 2002. "Soil properties as influenced by phosphorus-and nitrogen-based manure and compost applications." *Agronomy Journal* 94 (1):128–135.
El-Sherbeny, S E, SF Hendawy, AA Youssef, NY Naguib, and MS Hussein. 2012. "Response of turnip (*Brassica rapa*) plants to minerals or organic fertilizers treatments." *Journal of Applied Sciences Research* 8 (2):628–634.
Enticknap, J J, H Nonogaki, A R Place, and R T Hill. 2006. "Microbial diversity associated with odor modification for production of fertilizers from chicken litter." *Applied and Environmental Microbiology* 72 (6):4105–4114.
Eriksen, J, M Askegaard, J Rasmussen, and K Søegaard. 2015. "Nitrate leaching and residual effect in dairy crop rotations with grass–clover leys as influenced by sward age, grazing, cutting and fertilizer regimes." *Agriculture, Ecosystems and Environment* 212:75–84.
Evanylo, G, C Sherony, J Spargo, D Starner, M Brosius, and K Haering. 2008. "Soil and water environmental effects of fertilizer-, manure-, and compost-based fertility practices in an organic vegetable cropping system." *Agriculture, Ecosystems and Environment* 127 (1–2):50–58.

Flavel, T C, and D V Murphy. 2006. "Carbon and nitrogen mineralization rates after application of organic amendments to soil." *Journal of Environmental Quality* 35 (1):183–193.

French, E, and A S Iyer-Pascuzzi. 2018. "A role for the gibberellin pathway in biochar-mediated growth promotion." *Scientific Reports* 8 (1):5389.

Garg, VK, and R Gupta. 2009. "Vermicomposting of agro-industrial processing waste." In *Biotechnology for Agro-Industrial Residues Utilisation*, edited by P.S Nigam and A Pandey, 431–456. Springer Science

Geboloğlu, N, Y Yanar, and M Aydin. 2011. "Effect of different organic fertilizers on yield and fruit quality of ındeterminate tomato (*Lycopersicon esculentum*)." *Scientific Research and Essays* 6 (17):3623–3628.

Gregorich, EG, CF Drury, and JA Baldock. 2001. "Changes in soil carbon under long-term maize in monoculture and legume-based rotation." *Canadian Journal of Soil Science* 81 (1):21–31.

Hati, K M, A Swarup, AK Dwivedi, AK Misra, and KK Bandyopadhyay. 2007. "Changes in soil physical properties and organic carbon status at the topsoil horizon of a vertisol of central India after 28 years of continuous cropping, fertilization and manuring." *Agriculture, Ecosystemsand Environment* 119 (1–2):127–134.

Haynes, R J, and R Naidu. 1998. "Influence of lime, fertilizer and manure applications on soil organic matter content and soil physical conditions: a review." *Nutrient Cycling in Agroecosystems* 51 (2):123–137.

Hernandez, OL, A Calderín, R Huelva, D Martínez-Balmori, F Guridi, N O Aguiar, F L Olivares, and L P Canellas. 2015. "Humic substances from vermicompost enhance urban lettuce production." *Agronomy for Sustainable Development* 35 (1):225–232.

Hossain, ABMS, MA Hakim, and J M Onguso. 2003. "Effect of manure and fertilizers on the growth and yield of potato." *Pakistan Journal of Biological Sciences* 6 (14):1243–1246.

Hughes, O, and JH Venema. 2005. "Farmer field schools facilitators' manual.-v. 1: Integrated soil, water and nutrient management in semi-arid Zimbabwe."

Islam, M A, S Islam, AAkter, M HRahman, and D Nandwani. 2017. "Effect of organic and inorganic fertilizers on soil properties and the growth, yield and quality of tomato in Mymensingh, Bangladesh." *Agriculture* 7 (3):18.

Karhu, K, T Mattila, I Bergström, and K Regina. 2011. “Biochar addition to agricultural soil increased CH4 uptake and water holding capacity–results from a short-term pilot field study.” *Agriculture, Ecosystems and Environment* 140 (1–2):309–313.

Kelleher, BP, James J Leahy, AM Henihan, TF O’dwyer, D Sutton, and MJ Leahy. 2002. “Advances in poultry litter disposal technology–a review.” *Bioresource Technology*83 (1):27–36.

Kim, J, JDiao, M W Shepherd, R Singh, S D Heringa, C Gong, and X Jiang. 2011. “Validating thermal inactivation of *Salmonella* spp. in fresh and aged chicken litter.” *Applied and Environmental Microbiology*:AEM. 06671-11.

Kourakos, G, F Klein, A Cortis, and T Harter. 2012. “A groundwater nonpoint source pollution modeling framework to evaluate long-term dynamics of pollutant exceedance probabilities in wells and other discharge locations.” *Water Resources Research* 48 (6):2805–2814.

Laghari, M, R Naidu, B Xiao, Z Hu, MSaffarMirjat, M Hu, M N Kandhro, Z Chen, D Guo, and QJogi. 2016. “Recent developments in biochar as an effective tool for agricultural soil management: a review.” *Journal of The Science of Food and Agriculture* 96 (15):4840–4849.

Lal, R. 2006. “Enhancing crop yields in the developing countries through restoration of the soil organic carbon pool in agricultural lands.” *Land Degradation and Development* 17 (2):197–209.

Lehmann, J, M C Rillig, J Thies, C A Masiello, W C Hockaday, and D Crowley. 2011. “Biochar effects on soil biota–a review.” *Soil Biology and Biochemistry* 43 (9):1812–1836.

Li, S, J Li, B Zhang, D Li, G Li, and Y Li. 2017. “Effect of different organic fertilizers application on growth and environmental risk of nitrate under a vegetable field.” *Scientific Reports* 7 (1):17020.

Lim, S L, T Y Wu, PN Lim, and K P Y Shak. 2015. “The use of vermicompost in organic farming: overview, effects on soil and economics.” *Journal of the Science of Food and Agriculture* 95 (6):1143–1156.

Mama, A, J Jeylan, and A W Aseffa. 2016. “Effects of different rates of organic and inorganinc fertilizer on growth and yield components of potato (*Solanum tuberosum* L.) in jimma are, south west Ethiopia.” *International Journal of Research* 4 (11): 115–121

McHenry, M P. 2011. “Soil organic carbon, biochar, and applicable research results for increasing farm productivity under Australian agricultural conditions.” *Communications in Soil Science and Plant Analysis* 42 (10):1187–1199.

Millner, P D. 2009. *Manure Management*. Edited by K Matthews, E Solomon and G Sapers. Vol. 1, *The Production Contamination Problem*. Beltsville, MD, USA: Elsevier.

Mondal, A B, and A Al Mamun. 2011. "Effect of foliar application of urea on the growth and yield of tomato." *Frontiers of Agriculture in China* 5 (3):372–374.

Moore, PA, TCz Daniel, and DR Edwards. 2000. "Reducing phosphorus runoff and inhibiting ammonia loss from poultry manure with aluminumsulfate 1." *Journal of Environmental Quality* 29 (1):37–49.

Nardi, S, A Muscolo, S Vaccaro, S Baiano, R Spaccini, and A Piccolo. 2007. "Relationship between molecular characteristics of soil humic fractions and glycolytic pathway and krebs cycle in maize seedlings." *Soil Biology and Biochemistry* 39 (12):3138–3146.

Ndiaye, B, J Molénat, V Hallaire, C Gascuel, and Y Hamon. 2007. "Effects of agricultural practices on hydraulic properties and water movement in soils in Brittany (France)." *Soil and Tillage Research* 93 (2):251–263.

Norbu, T. 2002. "Pretreatment of municipal solid waste by windrow composting and vermicomposting."A thesis for the degree of Master of ScienceAsian Institute of Technology, School of Environment Resource and Development, School of Environment Resource and Development, Asian Institute of Technology.

Novak, J M, I Lima, B Xing, J W Gaskin, C Steiner, KC Das, MAhmedna, D Rehrah, D W Watts, and W J Busscher. 2009. "Characterization of designer biochar produced at different temperatures and their effects on a loamy sand." *Annals of Environmental Science* 3 (1):2.

Ortiz-Escobar, M E, and N V Hue. 2007. *Current developments in organic farming*. Edited by S.G Pandalai. Vol. 2, *Recent Research Developments in Soil Science*. Kerala, India: Research Signpost.

Ouyang, Z. 2005. "Nitrogen transformation in maize soil after application of different organic manures." *Journal of Environmental Sciences* 17 (2):340–343.

Oyewole, CI, H Opaluwa, and R Omale. 2012. "Response of Tomato (*Lycopersicon esculentum*): Growth and yield, to rates of mineral and poultry manure application in the guinea savanna agro-ecological zone in Nigeria." *Journal of Biology, Agriculture and Healthcare* 2:44–56.

Palsania, J, R Sharma, JK Srivastava, and D Sharma. 2008. "Effect of moisture content variation over kinetic reaction rate during vermicomposting process." *Applied Ecology and Environmental Research* 6 (2):49–61.

Pernes-Debuyser, A, and D Tessier. 2004. "Soil physical properties affected by long-term fertilization." *European Journal of Soil Science* 55 (3):505–512.

Piccolo, A. 2002. "The supramolecular structure of humic substances: A novel understanding of humus chemistry and implications in soil science." *Advances in Agronomy* 75:57–134.

Piccolo, Alessandro. 2012. "The nature of soil organic matter and innovative soil managements to fight global changes and maintain agricultural productivity." In *Carbon Sequestration in Agricultural Soils*, edited by Piccolo A, 1–20. Springer.

Pietikäinen, J, O Kiikkilä, and H Fritze. 2000. "Charcoal as a habitat for microbes and its effect on the microbial community of the underlying humus." *Oikos* 89 (2):231–242.

Plant, Y O C. 2016. "Effect of adding different organic fertilizers in the absorption of some of nutrient elements from soil in properties of plant growth and yield of cucumber plant." *Global Journal of Agricultural Research* 4 (6):13–19.

Puglisi, E, G Fragoulis, A AM Del Re, R Spaccini, A Piccolo, G Gigliotti, D Said-Pullicino, and MTrevisan. 2008. "Carbon deposition in soil rhizosphere following amendments with compost and its soluble fractions, as evaluated by combined soil–plant rhizobox and reporter gene systems." *Chemosphere* 73 (8):1292–1299.

Ryckeboer, J, J Mergaert, J Coosemans, K Deprins, and J Swings. 2003. "Microbiological aspects of biowaste during composting in a monitored compost bin." *Journal of Applied Microbiology* 94 (1):127–137.

Salamandane, A R, B F Muetanene, and A AP Cuiana. 2016. "Performance evaluation of organic fertilizers for onion (*Allium cepa* L.) and turnip (*Brassica rapa* L.)." *Imperial Journal of Interdisciplinary Research* 2 (10):1137 – 1142.

Schnell, R W, D M Vietor, T L Provin, C L Munster, and S Capareda. 2012. "Capacity of biochar application to maintain energy crop productivity: Soil chemistry, sorghum growth, and runoff water quality effects." *Journal of Environmental Quality* 41 (4):1044–1051.

Seefeldt, S. 2013. "Animal manure as fertilizer." *University of Alaska Fairbanks, Cooperative Extension Service*.

Sharma, SP, and SK Subehia. 2003. "Effects of twenty-five years of fertilizer use on maize and wheat yields and quality of an acidic soil in the western Himalayas." *Experimental Agriculture* 39 (1):55–64.

Shukla, MK, R Lal, LB Owens, and P Unkefer. 2003. "Land use and management impacts on structure and infiltration characteristics of soils in the North Appalachian region of Ohio." *Soil Science* 168 (3):167–177.

Singh Brar, B, J Singh, G Singh, and G Kaur. 2015. "Effects of long term application of inorganic and organic fertilizers on soil organic carbon and physical properties in maize–wheat rotation." *Agronomy* 5 (2):220–238.

Singh, R P, P Singh, A SF Araujo, M H Ibrahim, and O Sulaiman. 2011. "Management of urban solid waste: Vermicomposting a sustainable option." *Resources, Conservation and Recycling* 55 (7):719–729.

Sreenivasa, MN, N Naik, and SN Bhat. 2010. "Beejamrutha: A source for beneficial bacteria." *Karnataka Journal of Agricultural Sciences* 22 (5).

Stowell, RR, and WG Bickert. 1995. "Storing and handling sand-laden dairy manure." *Extension Bulletin (Michigan State University. Cooperative Extension Service) (USA).*

Talyan, V, RP Dahiya, and TR Sreekrishnan. 2008. "State of municipal solid waste management in Delhi, the capital of India." *Waste Management* 28 (7):1276–1287.

Vinh, NC, NV Hien, MTL Anh, J Lehmann, and S Joseph. 2014. "Biochar treatment and its effects on rice and vegetable yields in mountainous areas of northern Vietnam." *International Journal of Agricultural and Soil Science* 2:5–13.

Whalen, J K, and C Chang. 2002. "Macroaggregate characteristics in cultivated soils after 25 annual manure applications." *Soil Science Society of America Journal* 66 (5):1637–1647.

Wilkinson, KG, E Tee, RB Tomkins, G Hepworth, and R Premier. 2011. "Effect of heating and aging of poultry litter on the persistence of enteric bacteria." *Poultry Science*90 (1):10–18.

Wu, T Y, S L Lim, P N Lim, and K P Y Shak. 2014. "Biotransformation of biodegradable solid wastes into organic fertilizers using composting or/and vermicomposting." 17th Conference on Process Integration, Modelling and Optimisation for Energy Saving and Pollution Reduction PRES.

Xu, G, Y L, J Sun, H Shao, and L Wei. 2012. "Recent advances in biochar applications in agricultural soils: benefits and environmental implications." *Clean–Soil, Air, Water* 40 (10):1093–1098.

Zhang, M-K, and L-P Fang. 2007. "Effect of tillage, fertilizer and green manure cropping on soil quality at an abandoned brick making site." *Soil and Tillage Research* 93 (1):87–93.

CHAPTER 20

LANDSCAPE-ECOLOGICAL ZONING OF AGRICULTURAL AREAS IN TURKESTAN OBLAST

GULNARA NYUSSUPOVA,[*1]
AYGUL TOKBERGENOVA,[1] MEHMET ARSLAN,[1]
KANAT ZULPYKHAROV[1]

[1] Al-Farabi Kazakh National University, Department of geography, land management and cadastre, 050040, 71 al-Farabi Ave., Almaty, Kazakhstan
[1]https://orcid.org/0000-0001-5294-2671
[2]https://orcid.org/0000-0002-0776-7242
[3] https://orcid.org/0000-0003-0222-4297
[4]https://orcid.org/0000-0002-0275-2463
*Corresponding author Gulnara.nyusupova@kaznu.kz

Abstract

A complex analysis of natural, man-made and environmental components was used within the areas of the Turkestan Oblast and highlighted the major environmental problems in rural areas.

The methods of landscape-ecological zoning used were: evaluation of the landscape's ecological state of the Turkestan Oblast within the administrative areas, based on the earlier series of thematic and specialised maps, the bindings of which were the "Physical Map" and the "Map of Anthropogenically Disturbed Landscapes". In addition, research was conducted on numerous published stocks of official statistics.

The results of the study were that landscape-ecological zoning of the Turkestan Oblast reflects the spatial structure of natural complexes, the degree of conversion and anthropogenic disturbance, and the level of environmental stress areas.

This article describes the developmental trends of the landscape structure of the agricultural lands of the Turkestan Oblast; it reports on the interaction of all components of the landscape inherent in this region, and their changes under the influence of anthropogenic impact. It discusses the forecast of the development of modern negative processes and presents a proposal for the environment/agricultural lands of the Turkestan Oblast, taking into account characteristics of the natural resource potential of the region.

The results can be used by local executive bodies for the industrial and agricultural development of the territory of the Turkestan Oblast.

Keywords: landscape, ecology, zoning, rural areas

1. Introduction

The system of reforming the agriculture of the Republic of Kazakhstan carried out in recent years, requires new approaches for assessing the impact of agricultural production on natural systems. In the framework of the organisation, and to create various management forms, including peasant farming, there are environmental problems that can be solved on the basis of landscape-ecological, socio-economic and ecological-demographic criteria (Kochurov & Rozanov, 1994). The diversity of natural conditions in the Turkestan Oblast, namely, the significant distribution of sandy and clayey plains in the North, East and South-East regions; flood plain complexes in the Central part; and mountain ranges in the Eastern and South-Eastern parts of the region has led to uneven industrial-agricultural development of the territory, of population distribution and of infrastructure (Akiyanova et al. 2014). The Turkestan Oblast is the main agricultural development district. The Central and South-Eastern parts of the region are dominated by areas of irrigated and rainfed agriculture. The long-term agricultural impact on the natural complexes of the South-Kazakhstan region has contributed to the qualitative and quantitative restructuring of the structure and the development of negative ecological processes, which are manifested in:

- the change and instability of the water regime; the level, quality and salinity of surface and groundwater.
- qualitative and quantitative changes in agricultural land (change of water-salt regime, fertility, and physical and chemical properties of soil).

- qualitative and quantitative transformations of biological diversity (change in species composition, productivity, percent cover, etc.).
- change in the structure of agricultural land and reducing diversity of zonal landscapes.
- dissemination of anthropogenic causes by negative processes (water erosion, deflation, degradation of soil and vegetative cover, salinisation, waterlogging and flooding, etc.).

Area of research: the agricultural landscapes of the Turkestan Oblast

2. Materials

Analysis of the landscape-ecological state of the territory of the Turkestan Oblast has allowed identification of the degree of tension following landscape-ecological zones, namely, as stable, satisfactory and hard (Karnieli et.al.2008).

Stable area (I) within the South-Kazakhstan region, which covers about 25% of the territory represented in the North-West, South-East and South-West regions and is characterised by a weak expression of anthropogenically-caused ecological landscape changes. At the termination of anthropogenic impact on natural complexes, they returned to their original state. Stable landscape-ecological conditions of agricultural land predominate within the Tulkubas (60%), and take place in the territories of Suzak (30%); Baydibek (30%); Kazygurt (20%); Otrar (30%); Tolebi (30%) administrative districts, and are grossly under-represented in the Arys (10%); the Shardara (10%) administrative districts (table 1).

A stable ecological landscape is most evident in the Aeolian landscapes of ridge-hilly plains composed of sands with ephemeroidea-components, psammolittoral with a predominance of rushuna, saxaul vegetation, sometimes with the participation of promoting national conservation plans on rough sands; and the landscape of relatively elevated plains (plateaus of Betpakdala), represented by weakly sloping, flat plains covered by clays, sands, sandstone and gravel with transketolase—chernobayevka, sagebrush-Karaulova, ephemerals—the thin vegetation on grey-brown loamy soils and gravelly solonetzes deserts.

The main degradative process for areas of the stable landscape-ecological condition is the degradation of the vegetative cover due to irrational herding of cattle and the fragmentary manifestation of the processes of deflation.

Satisfactory zone (II) within the Turkestan Oblast. More than 60% of the area is characterised by a moderate degree of anthropogenic transformation of the components of the landscape, and the region is most widespread in the North-East, South-East and Central parts, capturing all the administrative districts of the region. The zones of the state having satisfactory ecological landscapes predominate in Shardara (70% of its area); Sairam (90%); Ordabasy (95%); Saryagash (85%); Kazygurt (70%); and Baidibek (70%); Otrar (30%); Tolebi (30%) administrative districts (table 1). The territories within the area of satisfactory landscape-ecological status are characterised by processes of degradation of arable land, degradation of pasture vegetation, land salinisation, water and wind erosion, local contamination of surface waters and waters used for drinking purposes. Much less common are factors such as soil contamination and atmospheric air. Zones having satisfactory ecological states are most represented in the landscapes relative to the omitted plains formed by the alluvial plains' ancient alluvial folded stratified loam, sandy loam, sand with one-year saline-agricolo-shrub, sagebrush-golfinopoulou, melkozerova-couch grass, wormwood-one year saline vegetation on meadow-grey soils and saline solonetzes and solonchaks meadow; alluvial-proluvial with erosional plains dismemberment, stacked pebbles, loam, sandy loam with ephemer-ephemeroid, with sagebrush vegetation in tall grass meadows on the light grey soils of the South, combined with golfinopoulou vegetation on meadow-grey soils of the solonetz meadow and along riverbeds; aeolian hummocky plains, folded by sands with ephemeroidea-psammolittoral, eurotia, ephemeroidea-blocksaway vegetation (Tokbergenova at al. 2018). The landscape is relatively elevated plains within the area of the satisfactory ecological state of the reservoir and is confined to a sloping plain, complicated upgrades composed of clays, sands, sandstones and pebbles with Keyreu-Whiteland sagebrush, sometimes with significant participation of balica, sagebrush-black haloxylon; and vegetation on grey-brown light loamy saline soils in combination with the thin vegetation on the salt marshes.

Within the hummocky plains, a moderate ecological status is characterised by tectonically-denuded hummocks, composed of Neogene-Quaternary loess-like loam, gravel, conglomerates with ephemeroid-sagebrush-short grass, ephemeroid-grass vegetation and sierozems in the South.

The data of natural ecosystems in economic terms represent photophone (rainfed and irrigated) land and moderately modified pastures.

Tense zone (III) within the South-Kazakhstan region occupies about 15% of the area, confined mainly to the valley of the Syr Darya river, the areas adjacent to major industrial centres and oil fields, and burial grounds for the disposal of polluting and toxic substances. In addition, the hard zone comprises a rural area with a high density of population and highly developed irrigation development.

Within the Turkestan Oblast, the stress landscape-ecological condition is most pronounced in the Maktaaral administrative region (90% of the area), and the territory subordinated to the Maslikhat of the city of Shymkent (100%), and the city of Turkestan (50%). Zones of intense landscape and environmental status are also represented in Otrar (30%); Suzak (20%); Tolebi (20%) administrative districts, and is negligible in the Arys, Kazygurt, Ordabasy and Sairam districts. The areas are characterised by a high degree of anthropogenic transformation of natural landscapes. The nature of spatial spillover processes is predominantly linear and locally-areal. This zone is confined to dry plains.

Corrected consisting of stratified loam, sandy loam, sand with one-year thistle-agricolo-shrub, sagebrush-golfinopoulou, one-year thistle vegetation on meadow-grey soils and saline solonetzes and solonchaks meadow, ephemeral, ephemeral-wormwood, sedge-keyreu plants on takyr soils, ephemeral-mint-flavoured-sedge, one year-ephemeral-grass vegetation on the light grey soils of the southern light. Stressful environmental conditions are characterised as alluvial-proluvial plains, complicated beds of temporary streams, composed of loam, gravel loess-like loams with ephemeroidea-residential areas, ephemer-ephemeroid, melkozerova vegetation on the light grey soils of the North with the participation of meadow-grey soils, soccavo-feather grass-fescue with the participation of savanorola kropotova, ephemer-ephemeroid, sagebrush) in tall grass meadows with vegetation on sierozem. Landscapes of hummock plains on most of the tectonically modified-denuded areas are covered with hummocks of ephemeroid-short grasses and ephemeroid vegetation on sierozems in the South. The main reasons for the creation of the tense ecological situation in the above-mentioned areas are pollution of the Syrdarya river, groundwater contamination and the ambient air of rural areas close to developed fields and industrial complexes, salinisation and pollution of irrigated areas, deflation and dehumification, rainfed arable land, and degradation of pastures due to uneven grazing.

Table 1 –Landscape and ecological status (Consolidated analytical report, 2015)

Administrative district	Zone		
	Stable %	Satisfactory %	Tense %
Arys	10	80	10
Baydibek	30	70	-
Kazygurt	20	70	10
Maktaaral	-		100
Ordabasy	-	95	5
Otyrar	30	40	30
Sayram	-	70	30
Saryagash	-	85	15
Sozak	30	50	20
Tolebi	30	50	20
Tulkubas	60	40	-
Shardara	10	70	20
The territory subordinated to the Maslikhat of Shymkent			
The territory subordinated to the Maslikhat of Turkestan			50

Environmental analysis of the qualitative state of farmland in the Turkestan Oblast (Consolidated analytical report, 2015), showed that on 01.01.2015, no negative impacts were seen on 18% of the total agricultural land and on 82% of the farmland's impacts of adverse natural and anthropogenic processes are visible (figure 1).

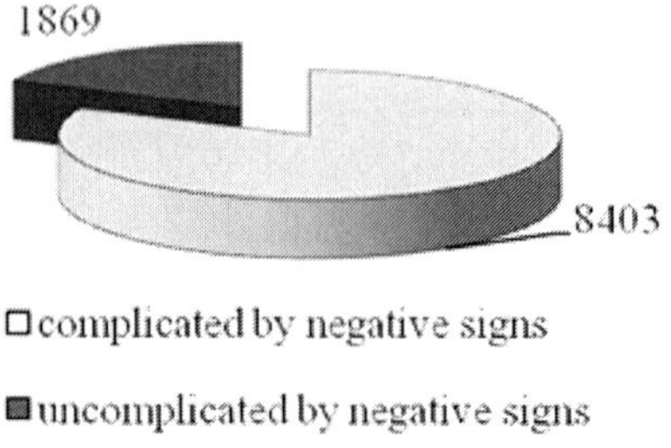

Figure 1 – Characteristics of agricultural land by the level of complexity with negative signs on 1 January 2015, in thousands of hectares (Consolidated analytical report, 2015).

In agricultural lands, complicated by negative signs, the largest area is occupied land, subject to deflationary, sodic processes, replaced and deflated (figure 2).

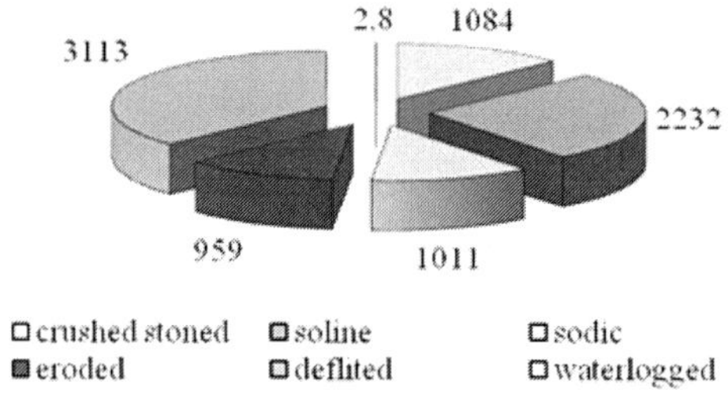

Figure 2 – Ecological condition complicated by negative characteristics of agricultural land, in thousands of hectares (Consolidated analytical report, 2015).

Method of research. The study of the problem of desertification and degradation of the lands of Kazakhstan was based on the following methods: comparative-geographical, cartographical, spatial analysis method, economic and statistical analysis, etc.

3. Results

Landscape-ecological zoning of rural areas of the Turkestan Oblast has allowed us to establish the types and processes of degradation, the degree of their manifestation and the territorial distribution within the administrative districts of the region. What follows is a closer look at the leading anthropogenic processes:

Degradation of vegetation from the standpoint of its economic use is manifested in the degradation of pastures, grasslands and forests.

Pastures in the Turkestan Oblast is 88.6% of the area of all agricultural lands. Of the total area of pastures, medium downed are 969.6 thousand hectares, much downed are 370.3 thousand hectares. Of them—modified—308.6 thousand hectares; weedy and poisonous plants—995.6 thousand hectares (Consolidated analytical report, 2015, Materials NPC Animal and Veterinary Sciences, 2015).

The ecological condition of rangelands, characterised by such indicators as the degree of their degradation, culture technical condition, and water invasion, depends on the evenness of pasture use (table 2).

Table 2 – Ecological condition of rangelands (Statistical Yearbook on agriculture in Turkestan oblast, 2015)

Administrative district	Area pastures thousand hectares	Watering pastures thousand hectares	Watering %		Degraded pastures, thousand hectares
			Engineering structures	Natural sources	
Baydybek	460.1	377.9	81.4	18.6	136.4
Kazygurt	207.7	198.2	78.2	21.8	10.70
Maktaaral	3.0	2.5	85.7	14.3	2.1
Ordabasy	146.1	97.1	82.7	17.3	124.7
Otyrar	1647.9	454.1	89.2	10.8	894.8
Sayram	44.9	44.3	84.4	15.6	29.7
Saryagash	650.4	293.3	87.2	12.8	138.4
Sozak	3503.2	2424.9	89.6	10.4	1206.5
Tolebi	133.1	124.9	76.3	23.7	86.1
Tulkybas	94.8	87.4	76.2	23.8	81.1
Shardara	1094.4	1031.1	75.7	24.3	1055.3

The cause of land degradation, in this case, is overgrazing. Environmental manifestations of the degradation of pastures in the Turkestan Oblast are the dominance of secondary groups, deflation or water erosion as a result of cattle failure. In about 75% of the area of degraded pastures presented, pastures were weedy and poisonous plants (figure 3). The most severely degraded pasture is in the Suzak, Otrarsky, Shardarinsky, Saryagashsky and Kazygurt districts.

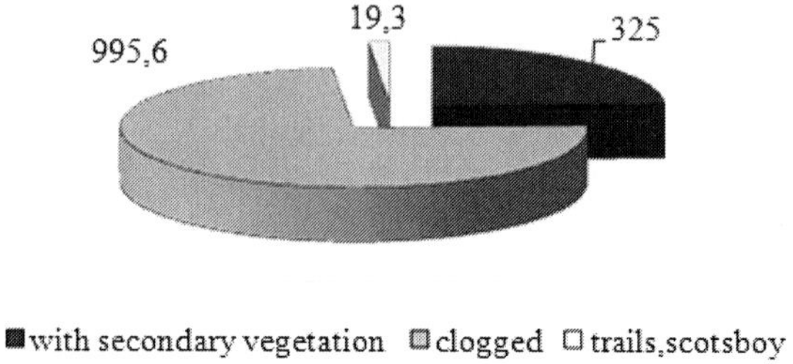

Figure 3 – Characteristics of degraded pastures by category on January 1, 2015, thousand hectares rangelands (Statistical Yearbook on agriculture in Turkestan oblast, 2015)

Analysis culture technical status of pasture ecosystems of South-Kazakhstan region indicates their satisfactory ecological condition. Clean pastures are 55.9% of the area of the latter (figure 4).

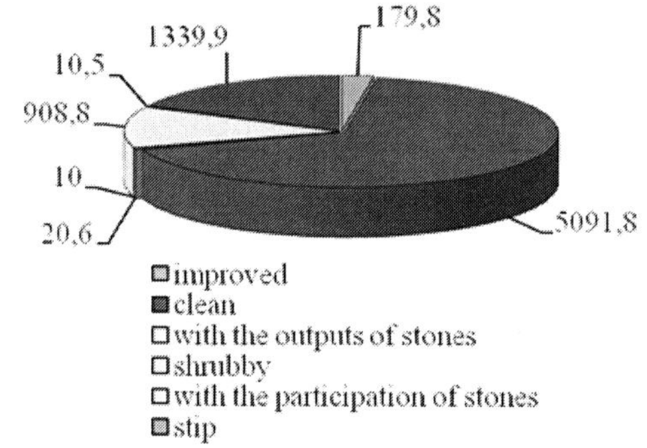

Figure 4 Culture technical condition of pastures for 1st January 2015, one thousand in hectares rangelands (Statistical Yearbook on agriculture in Turkestan oblast, 2015).

Degradation of rangelands in varying degrees is thus observed in all areas. However, the ecological condition of pastures in the Turkestan Oblast is characterised as satisfactory. Around villages and wells are areas of ecological intensity, accompanied by the processes of deflation and water erosion. Most susceptible to degradation are the grassland Aeolian plains, with great deflation.

The hayfields in the overall structure of the agricultural land in the Turkestan Oblast are just 142.7 thousand hectares or 1.4% of the area and are confined to the floodplains of the Syr-Darya, Shu, Arys, Bugun, Sarysu rivers and their tributaries, and the natural depressions and shores of lakes (Statistical Yearbook on agriculture in the Turkestan Oblast, 2015). Grasslands are very dynamic and depend primarily on hydrological conditions of river systems and the method of use of the hayfields. Background grasslands of the Turkestan Oblast were presented by reed beds, grass, cane, forb-grass communities. A general trend of reduced river flow due to regulation and diverse economic activity in lowland complexes led to a change in environmental conditions, reduction of the area of hayfields and change in their quality. Currently, the environmental manifestation of the degradation of grassland is the enhancement of salt marsh processes and the increase of halophytic species in the structure of grasslands, on the one hand, and the reduction of yield and quality of the land on the other. More than 88% of grasslands within the South-Kazakhstan Oblast belong to the category of pure (figure 5), indicating their satisfactory culture technical condition.

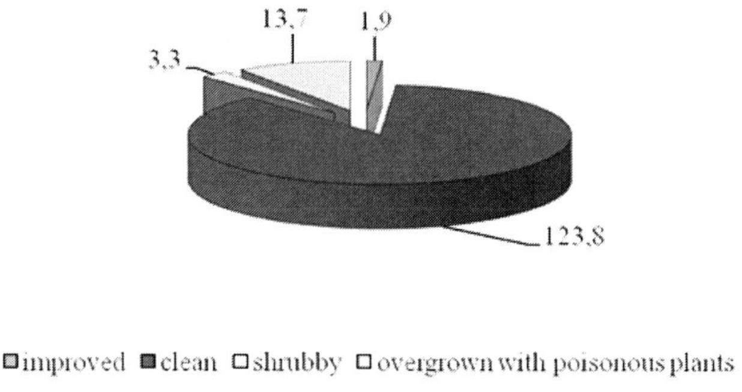

Figure 5 – The culture technical condition of hayfields on January 1, 2015, thousand hectares rangelands (Statistical Yearbook on agriculture in the Turkestan Oblast, 2015).

However, in the Suzak, Otrar, Shardara, Sairam and Ordabasy districts, the degradation of grasslands in the direction of strengthening the halophytic grass species and reducing productivity was noted.

Forest. In the composition of the forest Fund are lands covered with forests, as well as territory not covered with forest but granted for needs of forestry. The area of the forest Fund on the territory of the Turkestan Oblast on January 1, 2015, amounted to 3120.7 thousand hectares; of them to forest land belongs 453.9 thousand hectares (consolidated analytical report, 2015). Structural characteristics of the forest area of the region are shown in figure 6.

Forests on the territory of the Turkestan Oblast have a limited distribution, riparian, lowland desert and mountain forests confined to floodplains, sand and mountain ranges. Because of low forest cover, forests in the area are classified as forests of the 1st group. Over the last 5 years reforestation has been carried out in the Turkestan Oblast on an area of 2.5 thousand hectares, out of this 50 % of the forest area recovered comes from the year 2014 (Regions of Kazakhstan-Almaty: RK statistics Agency, 2004).

Tugai forests are formed by a loch and willow and Russian olive, confined to the floodplains of the Syr Darya, Sarysu, Shu, Bugun, etc., and are of great environmental importance in the area for the stabilisation of the water regime and channel processes for the improvement of the microclimate; they are most common in the Arys, Suzak, Shardara and Otrar districts.

Desolate forest in the Turkestan Oblast is confined to the sandy Kyzylkum massifs and Moyynqum (Suzak, Otrar and administrative Shardara districts) and are represented mainly by haloxylon thickets. The main causes of haloxylon degradation are deforestation for fuel and the reduction of groundwater, leading to the death of the latter.

Within the Kazygurt, Tolebi, Tyulkubas, Tolebi and, partly, Suzak administrative areas within the mountain system Karatau, the Kyrgyz range and the Western Tien-Shan, are high altitude common juniper woodlands with a shrubby part of under tier formed by honeysuckle, cotoneaster, spirea and deciduous and small-leaved forests formed by ash, apple, hawthorn, maple, and almond.

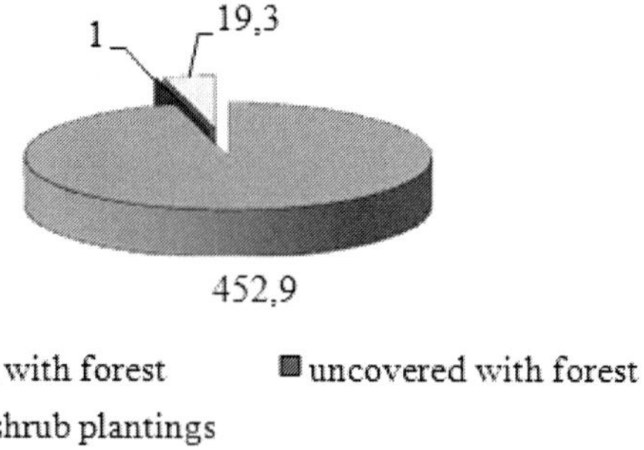

Figure 6 – Distribution of forest area, trees and shrubs plantings on 1 January 2015, hectares rangelands (Statistical Yearbook on agriculture in the Turkestan Oblast, 2015).

Forests play an important role in improving the environmental situation in the region. In river valleys, they perform water protection, protective, sanitary-hygienic and health functions, and improve the microclimate. The ecological condition of the forests of the South-Kazakhstan region as a whole is estimated as satisfactory. The main causes of forest degradation in the Turkestan Oblast are deforestation and fires.

Forest fires bring significant damage. Forest fire zonation within the Turkestan Oblast is allocated in the Ugam, Boraldayskom and desert areas. So, the number of fires on one million hectares in Tyulkubas, Tolebi, Kazygurt districts is from 4.4 to 19.6 (Arkhipov, 2004). The area of one fire ranges on average from 7.0 to 100.0 thousand hectares.

Forest degradation is noted in the Arys, Suzaksky, Shardarinsky, Otrarsky, Kazygurt, Tolebi, and Tyulkubas districts.

Degradation of arable land. In the general structure of farmland in the Turkestan Oblast, arable land occupies only 8.2% of the area of farmland (Statistical Yearbook on agriculture in the Turkestan Oblast, 2015) and established irrigated and rainfed arable land. Most large tracts of arable land are used under rainfed agriculture, concentrated in the Saryagash, Kazygurt, Tolebi, Ordabasy, and Baidibek districts.

Irrational use and long-term operation, without taking into account crop rotation, have led to a significant reduction in fertility due to the loss of humus and deterioration of soil-conservation properties of soil because of the development of erosive and deflationary processes. Currently, the total arable land in the region is complicated by the negative signs in more than 52% of the area, mainly in the Saryagash, Kazygurt, Tolebi, Ordabasy and Baidibek administrative districts (figure 7).

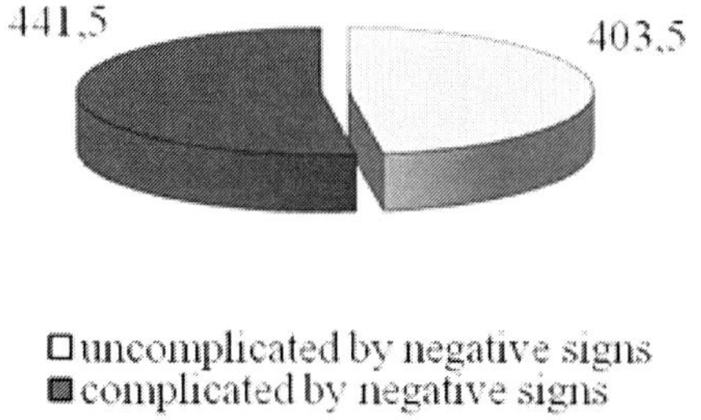

Figure 7 – Characteristics of rainfed arable land at the level of complication and negative signs on 1 January 2015, hectares rangelands (Statistical Yearbook on agriculture in the Turkestan Oblast, 2015).

The main causes of the degradation of rainfed arable land are salinisation and water erosion. Thus, among the disturbed arable land, about 53% of arable land is degraded because of the manifestation of the processes of water erosion (figure 8).

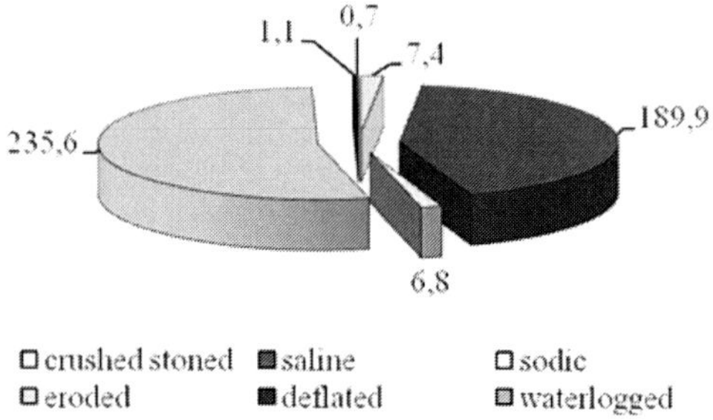

Figure 8 – Ecological condition complicated by negative symptoms of arable land on 1st January 2015, hectares rangelands (Statistical Yearbook on agriculture in the Turkestan Oblast, 2015).

Irrigated arable land in the Turkestan Oblast is represented slightly and takes only 442.3 thousand hectares or 2.2% of the area of arable land; and is confined to the river valleys and sloping old alluvial plains in the territory of the Maktaaral, Shardara, Suzak, Ordabasy administrative districts.

The ecological state of the irrigated areas is estimated as unsatisfactory. More than 65% of the area of irrigated lands is complicated by the adverse symptoms affecting their fertility, such as salinisation, waterlogging, erosion, etc. It should be noted that among the violated irrigated areas, about 90% are degraded because of salinisation and the development of solonetz processes (Vereshchaka & Dobbs, 1997) (figure 9).

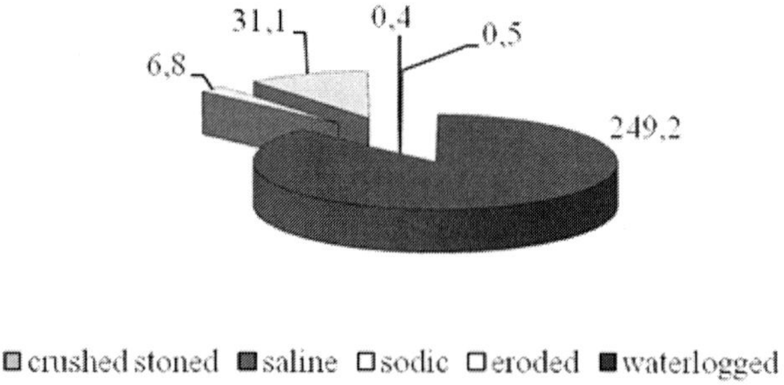

Figure 9 – Ecological condition complicated by negative symptoms of irrigated arable land on 1 January 2015, hectares rangelands (Statistical Yearbook on agriculture in the Turkestan Oblast, 2015).

The development of negative processes in irrigated lands is caused by the poor quality of water, its high salinity, unstable hydrological regime, deterioration of reclamation condition of soil and violation of technology of processing, and the low technical operation of irrigation and collector drainage networks. All of the above leads to the development of the processes of secondary salinisation, reduced soil fertility and crop yields (GLASOD, 1990).

4. Discussion

The most important indicators of landscape-ecological conditions are the indicators of the quality of agricultural land, which are not only a criterion for assessing the anthropogenic impact on natural complexes but also indicators of the manifestation of the degradation processes.

Integrated analysis of natural, anthropogenic and environmental components within the areas allowed us to identify the main environmental problems of rural territories of the Turkestan Oblast, which include:

- Scarcity and deterioration of quality of surface and groundwater, and silting up of reservoirs.
- The problem of rational use of water resources due to uncoordinated inter-state actions in matters of protection and regulation of transboundary watercourses. In the territories of Uzbekistan and Kyrgyzstan, the discharge of water in the Syr Darya river occurs in winter, which is extremely unfavourable for Kazakhstan.
- Water pollution of the river basin in the Badam and the Keles rivers by sewage of cities.
- Insufficient provision of rural population with quality drinking water.
- The distressed state of the drinking water supply in the Turkestan cities of Arys and Shardara in particular in the districts of Baidibek, Kazygurt and Tolebi.
- The lack and poor technical condition of treatment facilities and water supply facilities.
- The reduction of the fertility of arable lands as a result of secondary salinisation.
- Widespread around rural and urban informal settlements are illegal dumps, industrial waste and the lack of measures for their disposal.
- The strong influence on the environment of the Mirgalimsay mine in Kentau.

- The accumulation of large amounts of radioactive waste on the enterprises in the Central and Steppe mining areas and the lack of specially equipped places for the burial of pollutants.
- Within rural areas, accumulated pesticides, fertilisers and chemicals, the shelf life of which expired long ago. The latter pose a pollution threat to the soils and waters (Weipert &Walter &Wetzel, 1988). This problem remains acute due to the lack of on-site specialised areas for landfill storage and disposal of toxic waste.
- Degradation of vegetation, water and wind erosion on agricultural lands.
- Degradation of ecosystems in the Chushkakolskie lake system.

5. Conclusion

Analysis of the landscape structure in the Turkestan Oblast, according to the nature and direction of economic use of the territory, provided an opportunity to establish the landscape-ecological conditions of rural areas and their degree of disturbance.

Some reasons for the failures in the balanced use of the rural territories of the region are the extensive way of the development of agriculture, weak implementation of environmental protection measures, and weak control over their implementation.

For landscapes of the Turkestan Oblast we identified the following priority areas for environmental activities:

- measures to improve the state of rainfed and irrigated arable land.
- protection and rational use of pastures and hayfields.
- agroforestry events and activities for the protection of desert and Tugai forests.
- measures for the protection and rational use of water resources and reduction of water scarcity.
- measures for the prevention of ground-level pollution of atmospheric air.
- measures for the prevention and elimination of soil pollution.
- anti-deflation measures.
- anti-erosion measures.
- measures to combat salinisation.
- interventions to prevent swamping.
- measures for the prevention of negative consequences and prevention of natural phenomena.
- general activities for sustainable development of rural territories.

- activities to establish a network of protected areas.

Priority actions recommended are for denudation, accumulative alluvial, deluvial-proluvial plains and valley complexes with intense mining development, mining and processing complexes, rainfed and irrigated agriculture, and grazing of livestock. Moderately elevated transformed landscapes, denudation, and accumulative plains and alluvial plains, which are cultivated land and moderately modified pastures and within the area of satisfactory landscape-ecological status of implementation of environmental protection measures, are characterised as a secondary priority. Environmental protection measures are recommended for landscapes of plains formation, structural plateau, Aeolian, deluvial-proluvial plains, which are grazing lands and are characterised as weakly disturbed, and maintain a stable state of natural complexes.

On the territory of the Turkestan Oblast there are several specially protected natural areas: the Aksu-Zhabagly nature reserve; the nature reserve of Karatau; the Zhambyl reserve; the Ursk reserve; the Zadarinsky reserve; the Akdalinsky reserve; the Boraldai reserve; the Irsu-Daubabinsk reserve; the Sairam-Ugam reserve; and the Arys-Karatausky conservation area. The environmental assessment of rural areas must take into account the landscape-ecological condition of these, especially protected natural territories.

References

Akiyanova, F Z, A D Abitbayeva, K Baratovna R K Yegemberdiyeva, Temirbayeva. 2014. "Problems of desertification of the territory of Kazakhstan: status and forecast". *Life Sciences* 11 (10): 341–345.

Arkhipov V A. 2004. Forest Fire zoning, Schuchinsk

Consolidated analytical report on the status and use of lands of RK for 2015. The Committee for construction, housing and utilities and land resources management. Astana, 2015.

GLASOD – A Global Assessment of Human-induced Soil Degradation //UNEP and ISRIC. The Netherlands: Wageningen, 1990.

Karnieli, A, U Gilad, M Ponzet, T Svoray, R O Mirzadinov. 2008. "Fedorina Assessing land-cover change and degradation in the Central Asian deserts using satellite image processing and geostatistical methods". *Journal of Arid Environments* 10 (1016): 72

Kochurov, B I, L L Rozanov. 1994. Development of criteria and indicators for evaluation of ecological situation of the territory //the Problems of the environment and natural resources: an overview. inform. Vol. 5.

Materials NPC Animal and veterinary Sciences. Almaty, 2015.

Regions of Kazakhstan-Almaty: RK statistics Agency, 2004.

Statistical Yearbook on agriculture in Turkestan oblast in 2015. The Committee on statistics of the Republic of Kazakhstan, Shymkent, 2016.

Tokbergenova, A, G Nyussupova, M Arslan, L S Kiyassova. 2018. Causes and Impacts of Land Degradation and Desertification: Case Study from Kazakhstan, Vegetation of Central Asia and Environs. Springer pp. 291–302.

Vereshchaka, T V, A R Dobbs. 1997. (Syria) the Methodology of comprehensive mapping assessment of the ecological status of the territory according to integral indexes //Geodesy and cartography. No. 4.

Weipert, E, R Walter, T Wetzel. 1988. etc. Bioindication of pollution of terrestrial ecosystems: TRANS. with it. – M.: Mir

CHAPTER 21

ORGANELLEOMICS: MITOCHONDRIA AND CHLOROPLAST OF PLANTS

BUKHTAWAR FATIMA,[1] TAHIRA KHAN,[1] NAMRA HAQ,[1] RABIA AMIR,[1] ALVINA GUL,[1] VOLKAN ALTAY,[2] MUNIR OZTURK[3*]

[1]Atta-ur-Rahman School of Applied Biosciences, National University of Sciences and Technology, Islamabad, Pakistan
[2]Biology Department, Faculty of Science and Arts, Hatay Mustafa Kemal University, Hatay, Turkey
[3]Botany Department & Centre for Environmental Studies, Ege University, Izmir, Turkey
munirozturk@gmail.com

Abstract

The current review brings forth "organelleomics", an emerging research area, in view of the scientific efforts made in the last decade. Several theories have been propounded to elucidate the evolution of both mitochondria and chloroplasts: the endosymbiont theory of evolution; lateral gene transfer theory; and nucleosymbiosis or fusion theory, to name a few. In the molecular sciences, the mechanism of RNA splicing, and the different proteins involved in the process, i.e. MATURASES, CRM (Chloroplast Ribosome Maturation Protein), and PPR (Pentatricopeptide Repeat) have been explained in this regard. The techniques, such as in-gel digestion and 3D gel electrophoresis, have successfully thrown a light on the proteomics of both mitochondria and chloroplasts in response to abiotic stress, including the recently discovered protein family, recognised as WHIRLY. The proposed function of this family is to repair mtDNA and mtDNA recombination; nevertheless, their role as transcription factors is still being debated owing to their nonspecific binding nature. Mitochondrial proteins vary in terms of both concentration and expression in various different organs of plants. Regarding the genomic sciences, transformation experiments in chloroplasts have put forward the possibility of dark-operative photosynthetic plants. Furthermore, the effects of temperature and reactive oxygen species production have been explained concerning these organelles. In a nutshell, these studies ultimately lead us towards achieving the age-old goals of crop improvement and yield.

Keywords: evolution theories, proteomics, genomics, ROS effect, temperature effect

Introduction

Several theories have been propounded regarding the evolution of two vital organelles—mitochondria and chloroplast, one being that these two organelles evolved from prokaryotes via an event called endosymbiosis. According to this theory, these two organelles were independent, free-living bacteria; nonetheless, they were somehow ingested by the eukaryotes, and later on both the host and ingested bacteria developed a beneficial association, thus giving rise to an inseparable relationship (Lazcano and Peretó 2017). These ingested bodies later became semi-autonomous and self-replicating in advanced eukaryotes, namely, the mitochondria and the chloroplast of plants. The endosymbiont theory was proved during the 1970s by using RNA sequencing, known as RNA cataloguing (Archibald 2015). By means of RNA sequencing, an evolutionary relationship was detected between algal bacteria and F cyanobacteria, and also between mitochondria and proteobacteria (York 2017). These ingested bacteria progressively transformed and adjusted themselves as two principal components of the eukaryotic cell, i.e. mitochondria and chloroplasts, being typical of the plant kingdom. Another theory regarding the origin of these organelles suggests that they are the products of lateral gene transfer, which occurred between eukaryotes and eukaryotes (Koonin 2016). These gene transfer events occurred episodically, culminating in the formation of pan-genomes (gene content variation within species) in both the mitochondria and chloroplasts (Ku et al. 2015). Amongst the various theories being propounded to suggest the evolution of these two semi-autonomous organelles, one theory suggests that eukaryotes have possibly emerged as a consequence of the fusion of prokaryotic cells, in which one prokaryote turned into a nucleus and the other one turned into the mitochondrion and chloroplast (also called fusion or nucleosymbiosis model) (Boyle 2017).

Origin of Mitochondria

Recent molecular phylogenetic studies have revealed that enzymes underlying carbon metabolic pathways of plants unearth their evolutionary traces in prokaryotes (Pronina, Kupriyanova, and Igamberdiev 2017). Further studies have revealed that the majority of prokaryotic genes relocate to the host plant during endosymbiosis (Matassi 2017). The evolutionary account of mitochondrial genes encoded by the nucleus of the plant was studied by performing phylogenetic analyses on protein sequences using the maximum likelihood method. The results of the study revealed that most of the genes possessed a sequence similar to genes found in prokaryotes, particularly eubacteria (alphaproteobacteria) (Matassi 2017, Pronina, Kupriyanova, and Igamberdiev 2017), further suggesting and endorsing the notion that horizontal gene transfer between host, i.e. archeon and eubacteria, had probably happened in the event of endosymbiosis. Additionally, it can be concluded that the prokaryotic genome was reduced during the same event. (Koonin 2016). Enzymes responsible for carbon substrate metabolism and ATP production are associated with eubacterial genes, i.e. proteobacteria. The entirety of nuclear genes responsible for the tricarboxylic acid pathway (occurring inside mitochondria) found in host plants have evolved from proteobacteria (Schmitz et al. 2017). Similar conclusions can be drawn regarding genes encoding proteins responsible for the glyoxylate pathway (Koonin 2016).

Chloroplast origin

Chloroplasts trace their origin and emergence from prokaryotes, more precisely cyanobacteria, in an event called endosymbiosis (Sánchez-Baracaldo et al. 2017). This notion has been established using a molecular analysis technique employing rRNA of maize plant chloroplast, utilising Maxium and Gilbert's sequencing technique (Chotewutmontri and Barkan 2016). It has been reported that chloroplast membranes have various phosphate transporters intended to transport the photosynthetic products to the host plant. It has also been reported that host plants have also managed to develop transporters designed to supply food to chloroplasts found in them (Hirakawa et al. 2017). Additionally, researchers have reported genes governing other metabolic pathways in host plants, such as amino acid biosynthesis, carotenoid biosynthesis, non-Shemin tetrapyrrole biosynthesis, non-mevalonate isoprenoid biosynthesis, and nitrate and sulphate assimilation. These genes were initially encoded by chloroplasts (plastids); nevertheless, later relocation to the host plant was observed (Daniell et al. 2016). Genome sequencing has revealed that chloroplasts quintessentially possess an estimated one hundred genes, existing as operons. The chloroplast DNA has operons comparable to the bacterial genome, endorsing the notion that the chloroplast genome is a reduced form of the bacterial genome (Mozes-Koch et al. 2017).

An estimated 2000 nuclear genes of host plants have been reported to encode plastid proteins that are amongst the products produced during the event of endosymbiosis. These nuclear proteins possess transit peptides attached to them as a precursor that directs their transport to the outer membrane of chloroplasts (Meheust et al. 2016). The first protein studied in this context was RUBISCO. The small subunit of RUBISCO was studied with reference to a transit peptide is attached to in the form of N terminal extension, serving as its precursor. Proteolytic cleavage of the said extension takes place in the chloroplast (Nishimura, Kato, and Sakamoto 2017). ATP, being an energy currency unit of a cell, runs these transport processes and is essentially required for how the process was later recognised. Furthermore, the internal distribution of proteins within the chloroplast was displayed, where it was found that a GTP-regulated switch regulates the translocation of proteins through a receptor named Toc 75, which is a quintessential member of bacterial beta-barrel outer membrane proteins (O'Neil et al. 2017).

Subsequent research conducted on the plastocyanin protein occurring inside chloroplasts revealed that, following initial transport to the outer thylakoid membrane of the chloroplast, a peptide signal, comparable to bacterial secretion peptide, serves as an aid directing further transport to the inner thylakoid membrane. The import of protein to the inner membrane is, additionally, regulated by the ubiquitin-regulated chaperone system, which not only governs, but also regulates, protein balance within the respective organelle. The protein import to the inner membrane is regulated by the ubiquitin-regulated chaperone system to maintain protein balance within the organelle. This entire process of protein regulation and maintenance was termed "conservation sorting" (Richardson, Singhal, and Schnell 2017).

Molecular Mechanisms of Organelles

A variety of molecular mechanisms, such as DNA replication, RNA transcription, and protein translation, and RNA modification processes, such as RNA editing, RNA processing and RNA splicing occurring in organelles, are believed to be regulated by various splicing factors, as well as by a diverse group of specific RNA binding proteins, which are also encoded by the nucleus of a host plant (Chan et al. 2016).

In eukaryotes, RNA splicing is carried out by proteins encoded by intron, known as intron-encoded proteins (IEPs). These IEPs are embedded with highly coveted domains that underpin the splicing process. These coveted domains that govern the splicing process are endonuclease, reverse transcriptase and protein maturase (Elliott and Ladomery 2017). Nevertheless, in green plants the process of RNA splicing is unique, and differs from the one explained earlier, with reference to the other eukaryotes. Green plants do not possess these IEPs and, instead, have two principal maturase domains, which are encoded by the genomes of the said organelles. Furthermore, the factors responsible for splicing are encoded by the host plant to bring about the RNA splicing within an organelle (Lee and Kang 2016).

Generally speaking, in eukaryotes, introns are typically divided into two distinct forms based on the conserved primary and secondary regions present in them and on their characteristic splicing mechanisms (Majed and Karim 2017).

There are proteins that contribute towards the RNA splicing process and the following section will detail some of those proteins precisely.

Maturases: Maturases are encoded by host plants as well as by the organelle. In chloroplasts, these are encoded by an intron of the trnK gene and are known as MatK (encodes RNA binding protein); in mitochondria, these are encoded by the fourth intron of the nad1 gene (encodes NADH dehydrogenase) and these are referred to as MatR (Schmitz-Linneweber and Small 2008, Schmitz-Linneweber et al. 2006). Research studies conducted on *Arabidopsis* and the rice genome revealed that four nuclear genes are typically responsible for encoding maturases. Of these four maturases, three are targeted at mitochondria and the remaining one has dual locations (either in mitochondria or in chloroplasts). The start codon of these maturases is decisive as it serves as a factor responsible for governing their movement to either the mitochondria or chloroplast (Malik, Upadhyaya, and Khurana 2017). Reverse genetic approaches have emphatically described the potential tasks performed by two of these four nuclear-encoded maturases in the event of RNA splicing/removal of group II introns. It has been revealed through studies that maturase II contributes during splicing of nad1 intron (mitochondrial NADH dehydrogenase) (Cohen et al. 2014); nad7 intron (mitochondrial NADH dehydrogenase); and cox2 intron (encoding mitochondrial cytochrome oxidase, a respiratory enzyme in plants) (Sultan et al. 2016).

CRM (Chloroplast Ribosome Maturation Proteins): There are approximately sixteen CRM proteins that can be found in the *Arabidopsis* genome (Hotto et al. 2015) and nearly fourteen CRM proteins in the rice genome (Kusumi et al. 2011). The CRM proteins are classified into fourteen separate groups, which are further divided into four subfamilies. Out of these, nearly seven are directed at the chloroplast, two are directed at mitochondria and five are destined for two specific sites, either mitochondria and chloroplast or mitochondria and nucleus (Lee and Kang 2016). The CRM proteins were primarily identified in prokaryotes where they were involved in ribosomal maturation and were serving as a ribosome binding protein (Mutwil et al. 2017). In plants, they were reported as splicing factors found in chloroplasts. CRS 1 is the first characterised CRM gene reported to be found in chloroplasts where it was involved in the splicing of group II intron found in the chloroplast atpF (ATP) gene (Jenkins, Kulhanek, and Barkan 1997). The CRM proteins underpin splicing of group1 introns, group II introns and also the processing of 23srRNA (Liu et al. 2016).

PPR (Pentatricopeptide repeat): The Pentatricopeptide repeat (PPR) were primarily recognised in *Arabidopsis* as tandem repeats, consisting of 35 amino acid motifs, characteristically RNA binding proteins (Hsieh et al. 2015). This family is extraordinarily diverse in plants, comprising nearly 450 members compared with both mammals and prokaryotes where it comprises a mere seven members (Manna 2015). They contribute emphatically towards splicing, transcription, editing and RNA maturation (Lee and Kang 2016). It has been reported that four of these PPR take part in RNA splicing of introns in the chloroplast. One of these, termed PP4, was primarily reported in maize with reference to group II intron of rps 12 mRNA (Schmitz-Linneweber et al. 2006). Two of these PPR, i.e. TANG2 and Organelle Transcript Processing 439, are reported to take part in mitochondrial RNA splicing, i.e. in trans-splicing of nad1 (NADH dehydrogenase) (Guillaumot et al. 2017).

Proposed RNA Splicing Model in Organelles: Researchers have proposed a model to explain RNA splicing in these organelles. To begin with, the process of a primary transcript is produced in these organelles. The primary transcript then binds to the specific RNA binding protein, such as PPR. These proteins stabilise their conformation once attached to specific RNA sequences. This stabilised conformation thus facilitates the binding of different splicing factors to it (Schallenberg-Rüdinger and Knoop 2016).

Proteomics of Organelles

Researchers have conducted proteome analysis on different cellular organelles under unfavourable conditions, such as water deficiency, highly saline environments and extreme temperatures. These studies revealed that when plants are susceptible to such adverse environmental conditions they exhibit various protein interactions at various levels of cellular organisation (Zhu 2016). Studies conducted to understand the stress-sensing patterns in plants revealed that it is a complex science; initially the stress signal is sensed by the cell surface receptors of the plants, subsequently these signals are transduced into the plant cell. Amongst the various types of stress-elicited signals, the production of reactive oxygen species (ROS) is an emphatic response on the part of plants, which contributes greatly towards the triggering of a stress-related response in plants (Choudhury et al. 2017). These ROS are generally generated as a consequence of some sensitisation or elicitation, usually following an onslaught, either in the mitochondria or in the chloroplast of green plants (Kleine and Leister 2016).

Further studies revealed that the communication of these stress-related mediators between organelles, their synergy, and retrograde signalling under such perturbations essentially contributes towards generating stress-related responses in plants (Zhu 2016).

The stress-related proteins have been classified based on their functions into the following groups: osmoregulatory proteins, oxidative stress-related proteins, ion transporters, water transporters, molecular chaperons and proteins intended for degradation of other proteins (Dahal and Vanlerberghe 2017). Additionally, post-translational modification of proteins has been noticed, explicitly with reference to emerging abiotic stress. A study performed on the mitochondria of peas revealed that the majority of mitochondrial proteins possess a site for S-nitrosylation as such (Begara-Morales et al. 2016).

Structural proteomics of chloroplast: Approximately 2000–3000 chloroplast proteins are sourced from the nucleus, and their transport is dependent on the existence of N-terminal signal peptides (Baginsky 2016). Analysis of proteins present in a chloroplast and an etioplast (an immature form of a chloroplast which on light develops into a chloroplast) of barley has shown that about eight protein complexes exist in both chloroplasts and etioplasts. Out of these eight protein complexes, subunits, such as ATP synthetase, cytochrome B6 and NADPH dehydrogenase are abundant, whereas PS1 and PS11 are particularly found in the chloroplast (Shmakov et al. 2016).

Proteome-based analysis of chloroplasts of barley, wheat and *Arabidopsis* has also been carried out using gel electrophoresis. The following proteins in chloroplasts have been identified (Melonek, Oetke, and Krupinska 2016):

Proteomics of envelope membrane: Envelope membrane proteins contribute towards successful processing of many processes, such as synthesis of glycolipids, transport of fatty acids outside chloroplast, and transport of various metabolites and proteins (Nishimura, Kato, and

Sakamoto 2017). Research studies using the technique of in-gel digestion so far report that the envelope proteins of *Arabidopsis* comprise 644 proteins that can be located on membranes, including 616 protease proteins to maintain the functionality of membrane proteins (Kmiec et al. 2016, Rolland et al. 2003).

Proteomics of thylakoid membrane: The thylakoid membrane comprises various types of proteins that are essential during the light-harvesting process, as these are centres of light-harvesting (Moyet et al. 2018). SDS PAGE (Polyacrylamide gel electrophoresis) analysis revealed three major complexes: PSI, PSII, and light-harvesting chlorophyll/chlorophyll, with reference to the light-harvesting process associated with it (Rantala, Tikkanen, and Aro 2017).

Researchers studied thylakoid proteome of barley and found 45 thylakoid membrane proteins (Woolhouse and Batt 2016), of which 17 were found in photosystem II; 16 were found in photosystem I; 7 were found in cytochrome B6; and 5 were found in ATP synthetase (Woolhouse and Batt 2016).

Proteomics of stroma: Stroma is imperative as it plays a significant part in metabolic pathways, such as the CELVIN cycle; the presence of genetic material of chloroplasts further adds to its significance. The significance of stoma can be estimated from the fact that it is the home of approximately 590 proteins in *Arabidopsis thaliana* (Feike et al. 2016).

Furthermore, in the case of wheat, a total of 607 proteins are reported with reference to chloroplasts, of which 145 are recognised to be stromal proteins and the rest, 342, can be recognised as thylakoid membrane proteins; 163 can be recognised as lumen proteins; and 166 can be recognised as integral membrane proteins (Si et al. 2017).

Proteins of chloroplasts under abiotic stress: **A** diverse collection of proteins can be found in plant chloroplasts in response to various abiotic stress conditions. Some of these proteins have been characterised, and their repercussion on the physiology of plants in response to stress can be elucidated as follows:

B-carotene Hydroxylase: It is expressed in response to the paucity of water leading to stress in plants, affecting both the synthesis cycle of yellow carotenoid pigment C40H56O2 and abscisic acid synthesis (Pattanayak and Tripathy 2016).

Glycine betaine (cod A): It is expressed in various types of stress conditions, such as salinity stress, water paucity, oxidative stress, etc. It affects the plant by causing an increase in plant growth and in the repair of photodamage photosystem II (Mansour and Ali 2017).

Homogentisate phytyltransferase (HPT1): It can be instituted in response to light stress and effectively contributes towards the regulation of vitamin E (Sekine et al. 2016).

Betaine aldehyde dehydrogenase (BADH): It is expressed when plants are subjected to various stress conditions, such as salinity, or cold, oxidative stress. It contributes significantly towards photosynthesis enhancement and ROS elimination (Hore et al. 2017).

Plastidal protein synthesis elongation factor (zmeftu1): It is expressed in response to extremely high temperatures, where it causes a significant reduction in leaf protein accumulation, reduces thylakoid injury, and intensifies the photosynthetic rate (Qiu et al. 2018).

Chloroplast small heat shock protein (CPsHSP): It is expressed in response to extremely low temperatures, where it underpins a significant reduction in chlorophyll damage, and also thwarts electrolyte loss (Bernfur, Rutsdottir, and Emanuelsson 2017).

Aldehyde dehydrogenase (ALDH3I1): It is particularly expressed in response to salinity stress and paucity of water and significantly curbs oxidative burst and causes a reduction in lipids oxidation (Zhao, Missihoun, and Bartels 2017).

Proteomics of Mitochondria: Mitochondrial proteome studies have revealed that mitochondrial proteome is diverse as far as energy requirements and stress-related issues are concerned. Mitochondrial proteins, encoded by the nucleus of either tissue or organ, produce tissue or organ-specific mitochondrial proteins (Roger, Muñoz-Gómez, and Kamikawa 2017). Researchers conducted mutation studies into these proteins, which revealed that mutations in them cause them to lose their function at organ level. For instance, in one of the studies, it was revealed that these mutations actually resulted in delayed flowering and growth owing to the loss of PRR (pseudo response regulators proteins). Furthermore, alterations in root structures, as well as respiratory activity, were reported, subject to the mutation in fumarase (Yang 2017). The expression of various mitochondrial proteins encoded by the nucleus is regulated at the transcription level, which endows them with target specificity. Researchers have revealed some interesting facts regarding these proteins, such as the presence of a regularly occurring common motif site designated as II, and also that these proteins are associated with respiratory potential of mitochondria. Amongst these with respiratory potential, complex I (C1) and complex IV (CIV) show sustained expression in both controlled and stress-linked proteins, such as heat shock proteins; additionally, transcription in specific tissues during photorespiration is of significance (Gualberto and Newton 2017). The expression of particular proteins specific to different organs and tissues exclusively erects the unique mitochondrial proteome, casting light on the fact that mitochondrial proteome composition varies significantly in different tissues and organs (Rao et al. 2017).

Mitochondrial proteome in various organs was studied using 2D gel electrophoresis. Mostly, organs derived from cell culture root and shoot plant parts, known as vegetative organs, and stem, flower and silique, known as reproductive organs, were studied. It was revealed that constituents associated with respiratory potential are of uniform conformity with reference to vegetative organs; nevertheless, the respiratory potential varied in these organs, and was dependent on the particular type of substrate employed (Khan and Komatsu 2017). The stated research study has revealed the fact that metabolic activity varies in different tissues and organs, and the particular enzymes underlying the metabolic variation can be recognised as such. The mitochondrial proteome, with reference to reproductive organs, underlines metabolic pathways rather than being part of the respiratory pathway (Khan and Komatsu 2017). The expression of mitochondrial proteome varies in various tissues and organs as per their functions, such as in vegetative parts expression of isoform 1 of arginase, which was observed at an impressively high rate; a similar expression of isoform 2 was observed in reproductive parts (Astier, Gross, and Durner 2017).

tRNA import in mitochondria: The majority of mitochondrial proteins are the translation products of mitochondrial DNA; nevertheless, recent studies have revealed that tRNA, vital for the translation of mitochondrial proteins, do not suffice and thus a mechanism exists within a plant cell to transfer tRNA encoded by the nucleus (Chatterjee et al. 2018). Evolutionary studies regarding the origin of mitochondria have revealed that during the process of mitochondrial evolution from prokaryotes, some of its genome had been lost, thus giving a clue that it had lost the genes responsible for coding tRNA (Chen et al. 2017). Research studies suggest that discrete mechanisms can be found inside a plant cell to import tRNA to mitochondria; nevertheless, the mechanism pertaining to import machinery entails further research (Chatterjee et al. 2018). tRNA are large-sized, negatively-charged molecules, having a three-dimensional structure, with anticodon stem and D and T domains. *In vitro* studies have provided a clue that outer mitochondrial membrane proteins, such as translocase and voltage-governed anion channels are responsible for tRNA transport to mitochondria (Campo, Peixoto, and Martínez-Caballero 2017); nevertheless, the tRNA import is not the same in all plant species and varies greatly depending on the plant species. Further studies are deemed essential to understanding in detail the actual process underlying tRNA import. Research studies regarding the specificity of tRNA import into mitochondria across various plant species have revealed that mutations in anticodon regions do not affect the import process (Chatterjee et al. 2018); even so, mutations in the D and T regions are found to negatively affect the import of tRNA into mitochondria, thereby providing a clue that these two domains contribute towards regulating the tRNA import process. Additionally, tRNA aminoacylation is indispensable towards tRNA importation into mitochondria. (Kim et al. 2017). Three-dimensional structures of tRNA and aminoacylation are Conclusively found to be essential factors needed to import tRNA into mitochondria (Chatterjee et al. 2018, Kim et al. 2017); nonetheless, various other factors, depending upon the species, are found to be imperative in this regard and are an area of potential research (Campo, Peixoto, and Martínez-Caballero 2017).

Mitochondrial proteome under abiotic stress: The *Glycine*-rich RNA binding protein, 2(GRP2), can be found under various types of abiotic stress conditions encountered by plants: salinity, extremely low temperatures and osmotic stress are generally known in this regard. This endows the plant with a high seed germination rate, seedling growth and also the potential to cope with low-temperature stress (Laloum, Martín, and Duque 2018); whereas the uncoupling protein A (AtUCP1) contributes towards improving plant metabolic processes, such as photosynthesis, respiration, seed growth and leaf water content (Barreto et al. 2017).

Whirly Protein Family

The Whirly protein family is one that the scientific community has recently come across. It has been found that these proteins are plant DNA binding proteins. These proteins are present in both plant nuclear DNA and plant organellar DNA (Reichel et al. 2016). The first protein identified belonging to this family was St WHY1, which was found in potato tubers. There, it was found as the one which attaches to the promoter region of a gene PR-10 alpha gene, which is typically expressed during pathogen onslaught in potato tuber (Desveaux, Maréchal, and Brisson 2005). Further studies have revealed that these Whirly proteins are targeted at mitochondria and chloroplasts (plastids) as a transit peptide sequence, and they can be located in their amino acid sequence (Reichel et al. 2016). Additionally, in organelles contribute towards ds DNA repair, DNA recombination, and also in the export of nucleus-encoded peptides from organelles back to the nucleus (Ren et al. 2017).

In *Arabidopsis*, three members of the Whirly family can be found, i.e. WHY1, WHY2, WHY3, whereas in monocots, two members of it can be found, i.e. WHY1 and WHY2. Additionally, the one occurring in mitochondria of *Arabidopsis* can be found to influence the mt DNA replication and expression of genes encoded by mitochondria (Christensen 2017). Besides these, it has been revealed that both *Arabidopsis* WHY2 protein and Barley WHY1 protein bind to ss DNA probes and mitochondrial DNA regions in vitro. Thus, it can be concluded that these proteins contribute towards mt DNA repair, mt DNA recombination, and as a transcription factor (Kucharewicz et al. 2017). More recently, it has been revealed that a protein of the Whirly family undergoes phosphorylation with CIPK14. Additionally, it has been discovered that WHY1 phosphorylation is steering some functions occurring in the cell, such as serving as a senescence regulator in the nucleus, and as a transcription factor in the plastids and mitochondria (Ren et al. 2017).

Genomics of Organelles

Genomics of organelles is one of the most crucial subjects of research, particularly those extensively performing cellular organelles. Various transformation procedures have been performed on the chloroplast genome. These transformation procedures shed light on several distinguishing features of the chloroplast genome. One of the most fascinating of the various transformation procedures, known as "knock out", revealed that dual-transcription system functions in the chloroplast genome (Nevarez et al. 2017). In one such attempt, genes of transcription, generally the ones being secured from cyanobacteria, such as the rpo gene, were knocked out. This revealed the existence of an alternate polymerase enzyme, similar to the phage polymerase in the chloroplast genome, which actually brings forth the transcription of genes in the chloroplast genome (Fenstermacher et al. 2018). The chloroplast genome is amongst the choices for such transformation procedures, the reason being it is one with higher DNA replication potential and also has various sites that facilitate foreign gene integration (Nevarez et al. 2017).

Chloroplasts, one being plastids, makes it imperative to adumbrate them as well. Characteristically, a plastid genome comprises a circular zone of DNA, which is approximately 120 to 200kb in size, which effectively encodes RNA genes and 120 proteins (Zampini et al. 2017). A reduction in the number of proteins being encoded sheds light on the evolutionary noticeable transfer of a genome to the nucleus of the plant (Daniell et al. 2016). Typically, proteins are encoded by a specific region, termed the plastome, which possesses specific inverted repeats. These plastids are unique as they possess their very own machinery designed for transcription and translation. This unique feature can be attributed to the fact that they are actually cyanobacterial derivatives (Ponce-Toledo et al. 2017).

Recently, another transformation attempt has been made on tobacco leaf cells using another remarkable technique, known as biolistic transformation. The vector designed for the transformation process carries with it both the target gene and the homologous region to the plastid genome, thus transferring the target gene using homologous recombination. The target genes (operon) of focus include ChlB, ChlL and ChlN. These genes contribute towards reduction of p-childe to childe, which is vital during the dark-operated photosynthetic pathway occurring in the chloroplast. The two proteins encoded by these genes are labelled L-protein and NB proteins (Labiak and Karol 2017). The aforementioned transformation experiments have revealed that the transformation of these genes, either individually or collectively in the form of operon, increase the total chlorophyll content of the leaf cell, which, in turn, positively affects the overall photosynthetic activity of a plant under study. Researchers anticipate that chloroplast improvement through metabolic engineering can potentially assist us towards achieving goals of food security since the overall crop yield could be improved using this approach (Pandey, Kumar, and Tiwari 2017).

Microsatellites, being an essential part of both the chloroplast and mitochondrial genome, can be found in coding as well as non-coding regions of their genomes, occurring in the form of simple sequence repeats. The web repository, named Chloro Mito SSRDB, is a source where a wide collection of these can be found. The simple sequence repeats occurring in the coding region can be used as potential markers to identify the conservative/ uniform regions among species, providing evolutionary relationships among them, thus serving to further phylogenetic studies, and also helping analysis and evaluation of genetic resources of the potential subject (Ivanova et al. 2017).

Stress factors and their effects on organelles

Effect of temperature: It has been established over time and backed up by a great deal of scientific evidence that temperature beyond tolerance limit serves as a stress to plants. For example, it has recently been reported that the leaf of *Saintpaulia* sp., which is commonly known as iceberg, suffers more damage if it undergoes sudden temperature decline compared with a leaf undergoing a more gradual temperature decline. It was observed that the temperature plummet caused damage to the leaf's vacuolar membrane, discharging the vacuolar contents of it in the cytosol of palisade mesophyll cell, plummeting cytosolic pH, damaging the chloroplast of palisade mesophyll cells, and ultimately affecting the chlorophyll colour. The conclusion was thus drawn that temperature plummet ultimately results in chloroplast damage (Ruelland 2017).

Effect of oxidative stress: Generation of reactive oxygen species (ROS) by plant mitochondria serves as a determiner of stress to plants. Stress can be wide-ranging, including environmental onslaughts, chemical excess, and oxygen variation, etc. (Choudhury et al. 2017). ROS accumulation negatively influences plant development and respiratory or metabolic pathways, whereas in severe situations plant death can be the result (Noctor, Reichheld, and Foyer 2018). ROS are the reduction products of oxygen, production being principally brought about by terminal oxidases (Choudhury et al. 2017, Noctor, Reichheld, and Foyer 2018). Additionally, it has been revealed that components of the electron transport chain (ETC) also contribute towards ROS production; amongst the various factors which contribute towards its production, CII is the most prominent one. Subsequently, ROS accumulation leads towards altering the mitochondrial proteome, typically affecting ETC (Letts and Sazanov 2017).

Anti-ROS proteins: Reactive oxygen species is a generic response when plants are faced with any calamity: nevertheless, their production—or, more precisely, overproduction—termed oxidative burst, further exacerbates the situation, so, to cope with such a situation, plants have developed a very intricate strategy. In response to these ROS, plants produce specific agents in mitochondria: these include Mn-superoxide dismutase, ascorbate peroxidase, monodehydroascorbate reductase, and glutathione peroxidase (Sharma, Kiran, and Sopory 2017). The production of Glutathione (GST) and Heat shock proteins (HSP) during stress conditions has been a topic of extensive research (Jacob, Hirt, and Bendahmane 2017). It has been revealed that GST, along with aldehyde dehydrogenase, contributes towards the elimination of residual matter of lipid degradation; HSP contributes towards stress reduction when produced in response to associated stress in mitochondria (Jacob, Hirt, and Bendahmane 2017). Furthermore, the production of serine protease for the elimination of undesirable proteins, which are the consequence of plant vulnerability to some stress that elicited the generation of ROS, is valuable in this regard (Jacob, Hirt, and Bendahmane 2017).

ROS signalling pathway: The ROS-induced signalling pathway is one of the intricate signalling pathways found in plants. Studies on *Arabidopsis* have revealed that mitochondrial ROS signals production of H_2O_2 (Hydrogyen peroxide), and causes alteration in complex 1(CI) of the electron respiratory chain responsible for signalling the transcription of GSTs (Noctor, Reichheld, and Foyer 2018). Further studies revealed that GST1 and GST

III production increase noticeably due to CI mutation (Choudhury et al. 2017). Moreover, C1 mutation also caused increased production of HSPs, suggestive of a signalling pathway described earlier. Nonetheless, it was also revealed that ROS induced the production of ROS in the other cell. This type of ROS generation is known as ROS-induced ROS release (RIRR), which is indicative of NADPH oxidase production in one cell, inducing ROS production and discharge in an adjacent cell (Zandalinas and Mittler 2017). The pattern followed as a consequence of ROS can be adumbrated as ROS signals depolarisation of the mitochondrial inner membrane, which causes signalling peptides of various amino acid sequences to be transported from mitochondria to the cell nucleus, and then these signalling peptides lead towards the production of HSPs. This mechanism can be described based on the findings that signalling peptides produced by mitochondria undergo proteolytic cleavage, and these proteolytically- cleaved signalling peptides cause the production of HSPs (Demidchik 2017). The type of stress is the determiner of the response being induced; more precisely, stress signals are dependent on the nature of the stress-inducing them. As there are various types of stress that plants undergo, the stress signals being generated are also diverse, and their mechanism has not been fully understood until now. Currently, researchers are focusing on various signalling pathways activated by mt ROS signals (Sharma, Kiran, and Sopory 2017).

Conclusions

The current review throws light on the origin of the two main contributing organelles of cells. Amongst the various theories being presented over time regarding the origin of these organelles, the one suggesting their origin from prokaryotes has been given preference by several evolutionists. Additionally, these two exuberantly performing organelles have been described in the current review, which will help us further our knowledge in this regard.

Transformation experiments conducted on one of these organelles—the chloroplast—suggests that by over-expressing certain genes governing photosynthesis of green plants, other alternatives can be worked out, leading towards better prospects.

Proteome-based experiments on both these impressively performing organelles reveal that a good collection of their proteins is mainly sourced from their host genome. Researchers employed transgenic plants to conduct expression studies, specifically targeting the plant proteins that could ultimately help us with the impending threat of food shortage owing to population explosion. More precisely, the newly-developed varieties will be better able to adapt to an ever-changing environment that has been changing at a drastic pace at present. The discovery of the Whirly protein family is one of the greatest achievements of proteome-based studies, especially as these take part in mitochondrial DNA repair and mitochondrial DNA recombination.

Research studies conducted for understanding signalling pathways in mitochondria, elicited as a consequence of oxidative stress, are tremendous, as the pathways associated with ROS generation and follow-up are complex and multifaceted: work in this regard is a gruelling task for the researcher—nonetheless, comprehending these pathways will open up ways of developing crop varieties better able to cope up with oxidative stress.

Research studies conducted in order to broaden our horizons towards the effect of temperature and the associated plant's response is noteworthy. In the current review, a recent study of the leaf of *Saintpaulia* sp., commonly known as iceberg, has been analysed in detail. The possible measures towards its survival can further add towards achieving our goals of coping with stress associated with temperature in plants. Additionally, to conserve the vast collection of these genomes and to further their research globally, online web-based data services are available so that the researcher can benefit from different sets available in the repository. In short, organelleomics can thus help us further our goals of crop improvement and resolving issues of global food insecurity.

References

Archibald, J M. 2015. "Endosymbiosis and eukaryotic cell evolution." *Current Biology* 25 (19): R911–R921.

Astier, J, I Gross, and J Durner. 2017. "Nitric oxide production in plants: an update." *Journal of Experimental Botany* 69 (14):3401–3411.

Baginsky, S. 2016. "Protein phosphorylation in chloroplasts–a survey of phosphorylation targets." *Journal of Experimental Botany* 67 (13): 3873–3882.

Barreto, P, JECT Yassitepe, Z A Wilson, and P Arruda. 2017. "Mitochondrial uncoupling protein 1 overexpression increases yield in *Nicotiana tabacum* under drought stress by improving source and sink metabolism." *Frontiers in Plant Science* 8:1836.

Begara-Morales, J C, B Sánchez-Calvo, M Chaki, R Valderrama, C Mata-Pérez, M N Padilla, F J Corpas, and J B Barroso. 2016. "Antioxidant systems are regulated by nitric oxide-mediated post-translational modifications (NO-PTMs)." *Frontiers in Plant Science* 7:152.

Bernfur, K, G Rutsdottir, and C Emanuelsson. 2017. "The chloroplast-localized small heat shock protein Hsp21 associates with the thylakoid membranes in heat-stressed plants." *Protein Science* 26 (9):1773–1784.

Boyle, R. 2017. "Eukaryotic origins and the proterozoic earth system: A link between global scale glaciations and eukaryogenesis?" *Earth-Science Reviews* 174:22–38.

Campo, M L, P M Peixoto, and S Martínez-Caballero. 2017. "Revisiting trends on mitochondrial mega-channels for the import of proteins and nucleic acids." *Journal of Bioenergetics and Biomembranes* 49 (1):75–99.

Chan, K X, S Y Phua, P Crisp, R McQuinn, and B J Pogson. 2016. "Learning the languages of the chloroplast: retrograde signaling and beyond." *Annual Review of Plant Biology* 67:25–53.

Chatterjee, K, R T Nostramo, Y Wan, and A K Hopper. 2018. "tRNA dynamics between the nucleus, cytoplasm and mitochondrial surface: location, location, location." *Biochimica et Biophysica Acta (BBA)-Gene Regulatory Mechanisms* 1861 (4):373–386.

Chen, Z, N Zhao, S Li, C E Grover, H Nie, J F Wendel, and J Hua. 2017. "Plant mitochondrial genome evolution and cytoplasmic male sterility." *Critical Reviews in Plant Sciences* 36 (1):55–69.

Chotewutmontri, P, and A Barkan. 2016. "Dynamics of chloroplast translation during chloroplast differentiation in maize." *PLoS Genetics* 12 (7):e1006106.

Choudhury, F K, R M Rivero, E Blumwald, and R Mittler. 2017. "Reactive oxygen species, abiotic stress and stress combination." *The Plant Journal* 90 (5):856–867.

Christensen, A C. 2017. "Mitochondrial DNA repair and genome evolution." *Annual Plant Reviews, Plant Mitochondria* 50:11.

Cohen, S, M Zmudjak, C C des Francs-Small, S Malik, F Shaya, I Keren, E Belausov, Y Many, G G Brown, and I Small. 2014. "nMAT4, a maturase factor required for nad1 pre-mRNA processing and maturation, is essential for holocomplex I biogenesis in *Arabidopsis* mitochondria." *The Plant Journal* 78:253–268.

Dahal, K, and G C Vanlerberghe. 2017. "Alternative oxidase respiration maintains both mitochondrial and chloroplast function during drought." *New Phytologist* 213 (2):560–571.

Daniell, H, C-S Lin, M Yu, and W-J Chang. 2016. "Chloroplast genomes: diversity, evolution, and applications in genetic engineering." *Genome Biology* 17 (1):134.

Demidchik, V. 2017. "Reactive oxygen species and their role in plant oxidative stress." *Plant Stress Physiology*:64.

Desveaux, D, A Maréchal, and N Brisson. 2005. "Whirly transcription factors: defense gene regulation and beyond." *Trends in Plant Science* 10 (2):95–102.

Elliott, D, and M Ladomery. 2017. *Molecular Biology of RNA*. 2nd ed. Oxford UK: Oxford University Press.

Feike, D, D Seung, A Graf, S Bischof, T Ellick, M Coiro, S Soyk, S Eicke, T Mettler-Altmann, and K-J Lu. 2016. "The starch granule-associated protein Early Starvation1 (ESV1) is required for the control of starch degradation in *Arabidopsis thaliana* leaves." *The Plant Cell*:tpc. 00011.2016.

Fenstermacher, K J, V Achuthan, T D Schneider, and J J DeStefano. 2018. "An evolutionary/biochemical connection between promoter-and primer-dependent polymerases revealed by selective evolution of ligands by exponential enrichment (SELEX)." *Journal of Bacteriology*: JB. 00579-17.

Gualberto, J M, and K J Newton. 2017. "Plant mitochondrial genomes: dynamics and mechanisms of mutation." *Annual Review of Plant Biology* 68:225–252.

Guillaumot, D, M Lopez-Obando, K Baudry, A Avon, G Rigaill, A F de Longevialle, B Broche, M Takenaka, R Berthomé, and G De Jaeger. 2017. "Two interacting PPR proteins are major *Arabidopsis* editing factors in plastid and mitochondria." *Proceedings of the National Academy of Sciences* 114 (33):8877–8882.

Hirakawa, Y, T Nakayama, N Keilert, and K-I Ishida. 2017. "Evolution of UhpC-type hexose-phosphate transporters in dinoflagellates." *Phycological Research* 65 (2):166–170.

Hore, P, S Nahar, M Hasampura, and B Tanti. 2017. "Phylogenetic analysis and homology modelling of betaine aldehyde dehydrogenase-an aroma producing protein in rice (*Oryza Sativa* L.)." *International Journal of Multidisciplinary Approach & Studies* 4 (3).

Hotto, A M, B Castandet, L Gilet, A Higdon, C Condon, and D B Stern. 2015. "*Arabidopsis* Chloroplast Mini-Ribonuclease III Participates in rRNA Maturation and Intron Recycling." *The Plant Cell* 27 (3):724–740.

Hsieh, W Y, J C Liao, C Y Chang, T Harrison, C Boucher, and M H Hsieh. 2015. "The slow growth3 pentatricopeptide repeat protein is required for the splicing of mitochondrial NADH dehydrogenase subunit7 intron 2 in *Arabidopsis*." *Plant Physiology* 168 (2):490–501.

Ivanova, Z, G Sablok, E Daskalova, G Zahmanova, E Apostolova, G Yahubyan, and V Baev. 2017. "Chloroplast genome analysis of resurrection tertiary relict *Haberlea rhodopensis* highlights genes important for desiccation stress response." *Frontiers in Plant Science* 8:204.

Jacob, P, H Hirt, and A Bendahmane. 2017. "The heat-shock protein/chaperone network and multiple stress resistance." *Plant Biotechnology Journal* 15 (4):405–414.

Jenkins, B D, D J Kulhanek, and A Barkan. 1997. "Nuclear mutations that block group II RNA splicing in maize chloroplasts reveal several intron classes with distinct requirements for splicing factors." *The Plant Cell* 9 (3):283–296.

Khan, M N, and S Komatsu. 2017. "Organ-specific proteomic analysis of early vegetative stage soybean." *Protein and Peptide Letters* 24 (10):897–908.

Kim, K M, J H Noh, K Abdelmohsen, and M Gorospe. 2017. "Mitochondrial noncoding RNA transport." *BMB Reports* 50 (4):164.

Kleine, T, and D Leister. 2016. "Retrograde signaling: organelles go networking." *Biochimica Et Biophysica Acta (BBA)-Bioenergetics* 1857 (8):1313–1325.

Kmiec, B, P F Teixeira, M W Murcha, and E Glaser. 2016. "Divergent evolution of the M3A family of metallopeptidases in plants." *Physiologia Plantarum* 157 (3):380–388.

Koonin, E V. 2016. "Horizontal gene transfer: essentiality and evolvability in prokaryotes, and roles in evolutionary transitions." *F1000Research* 5.

Ku, C, S Nelson-Sathi, M Roettger, F L Sousa, P J Lockhart, D Bryant, E Hazkani-Covo, J O McInerney, G Landan, and W F Martin. 2015. "Endosymbiotic origin and differential loss of eukaryotic genes." *Nature* 524 (7566):427.

Kucharewicz, W, A Distelfeld, W Bilger, M Müller, S Munné-Bosch, G Hensel, and K Krupinska. 2017. "Acceleration of leaf senescence is slowed down in transgenic barley plants deficient in the DNA/RNA-binding protein WHIRLY1." *Journal of Experimental Botany* 68 (5):983–996.

Kusumi, K, C Sakata, T Nakamura, S Kawasaki, A Yoshimura, and K Iba. 2011. "A plastid protein NUS1 is essential for build-up of the genetic system for early chloroplast development under cold stress conditions." *The Plant Journal* 68 (6):1039–1050.

Labiak, P H, and K G Karol. 2017. "Plastome sequences of an ancient fern lineage reveal remarkable changes in gene content and architecture." *American Journal of Botany* 104 (7):1008–1018.

Laloum, T, G Martín, and P Duque. 2018. "Alternative splicing control of abiotic stress responses." *Trends in Plant Science* 23 (2):140–150.

Lazcano, A, and J Peretó. 2017. "On the origin of mitosing cells: a historical appraisal of Lynn Margulis endosymbiotic theory." *Journal of Theoretical Biology* 434:80–87.

Lee, K, and H Kang. 2016. "Emerging roles of RNA-binding proteins in plant growth, development, and stress responses." *Molecules and Cells* 39 (3):179.

Letts, J A, and L A Sazanov. 2017. "Clarifying the supercomplex: the higher-order organization of the mitochondrial electron transport chain." *Nature Structural & Molecular Biology* 24 (10):800.

Liu, C, H Zhu, Y Xing, J Tan, X Chen, J Zhang, H Peng, Q Xie, and Z Zhang. 2016. "Albino Leaf 2 is involved in the splicing of chloroplast group I and II introns in rice." *Journal of Experimental Botany* 67 (18):5339–5347.

Majed, S O, and A Y Karim. 2017. "Phylogenetic Relationship among Phosphomannose Isomerase (PMI), Phosphomannose Mutase (PMM) and GDP-D-Mannose Pyrophosphorylase (GMP) in Photosynthetic Eukaryotes." *ZANCO Journal of Pure and Applied Sciences* 28 (6):27–38.

Malik, S, K C Upadhyaya, and S M P Khurana. 2017. "Phylogenetic analysis of nuclear-encoded RNA maturases." *Evolutionary Bioinformatics* 13:1176934317710945.

Manna, S. 2015. "An overview of pentatricopeptide repeat proteins and their applications." *Biochimie* 113:93–99.

Mansour, M M F, and E F Ali. 2017. "Glycinebetaine in saline conditions: an assessment of the current state of knowledge." *Acta Physiologiae Plantarum* 39 (2):56.

Matassi, G. 2017. "Horizontal gene transfer drives the evolution of Rh50 permeases in prokaryotes." *BMC Evolutionary Biology* 17 (1):2.

Meheust, R, E Zelzion, D Bhattacharya, P Lopez, and E Bapteste. 2016. "Protein networks identify novel symbiogenetic genes resulting from plastid endosymbiosis." *Proceedings of the National Academy of Sciences of the United States of America* 113 (13):3579–3584.

Melonek, J, S Oetke, and K Krupinska. 2016. "Multifunctionality of plastid nucleoids as revealed by proteome analyses." *Biochimica et Biophysica Acta (BBA)-Proteins and Proteomics* 1864 (8):1016–1038.

Moyet, L, D Salvi, M Tomizioli, D Seigneurin-Berny, and N Rolland. 2018. "Preparation of membrane fractions (Envelope, Thylakoids, Grana, and Stroma Lamellae) from *Arabidopsis* chloroplasts for quantitative proteomic investigations and other studies." In *Plant Membrane Proteomics*, 117–136. Springer.

Mozes-Koch, R, E Tanne, A Brodezki, R Yehuda, O Gover, H D Rabinowitch, and I Sela^. 2017. "Expression of the entire polyhydroxybutyrate operon of *Ralstonia eutropha* in plants." *Journal of Biological Engineering* 11 (1):44.

Mutwil, M, J Marcin, R Zoschke, L Scharff, S M Jaime, C Ferrari, S Proost, N Omranian, M Musialak-Lange, and Z Nikoloski. 2017. "AtRsgA from *Arabidopsis thaliana* controls maturation of the small subunit of the chloroplast ribosome." *bioRxiv*:225052.

Nevarez, P A, Y Qiu, H Inoue, C Y Yoo, P N Benfey, D J Schnell, and M Chen. 2017. "Mechanism of dual-targeting of a transcriptional co-activator to plastids and the nucleus." *Plant Physiology*:pp. 00116.2017.

Nishimura, K, Y Kato, and W Sakamoto. 2017. "Essentials of proteolytic machineries in chloroplasts." *Molecular Plant* 10 (1):4–19.

Noctor, G, J-P Reichheld, and C H Foyer. 2018. "ROS-related redox regulation and signaling in plants." Seminars in Cell & Developmental Biology.

O'Neil, P K, L GL Richardson, Y D Paila, G Piszczek, S Chakravarthy, N Noinaj, and D Schnell. 2017. "The POTRA domains of Toc75 exhibit chaperone-like function to facilitate import into chloroplasts." *Proceedings of the National Academy of Sciences* 114 (24):E4868-E4876.

Pandey, S S, D Kumar, and B S Tiwari. 2017. "Chloroplast Metabolic Engineering for Sustainable Agriculture." In *Current Developments in Biotechnology and Bioengineering*, 149–162. Elsevier.

Pattanayak, G K, and B C Tripathy. 2016. "Modulation of biosynthesis of photosynthetic pigments and light-harvesting complex in wild-type and gun5 mutant of Arabidopsis thaliana during impaired chloroplast development." *Protoplasma* 253 (3):747–752.

Ponce-Toledo, R I, P Deschamps, P López-García, Y Zivanovic, K Benzerara, and D Moreira. 2017. "An early-branching freshwater cyanobacterium at the origin of plastids." *Current Biology* 27 (3):386–391.

Pronina, N A, E V Kupriyanova, and A U Igamberdiev. 2017. "Photosynthetic carbon metabolism and co 2-concentrating mechanism of Cyanobacteria." In *Modern Topics in the Phototrophic Prokaryotes*, 271–303. Springer.

Qiu, Z, S Kang, L He, J Zhao, S Zhang, J Hu, D Zeng, G Zhang, G Dong, and Z Gao. 2018. "The newly identified heat-stress sensitive albino 1 gene affects chloroplast development in rice." *Plant Science* 267:168–179.

Rantala, M, M Tikkanen, and E-m Aro. 2017. "Proteomic characterization of hierarchical megacomplex formation in *Arabidopsis* thylakoid membrane." *The Plant Journal* 92 (5):951–962.

Rao, R S P, F S, B Thal, H Eubel, J J Thelen, and I M Møller. 2017. "The proteome of higher plant mitochondria." *Mitochondrion* 33:22–37.

Reichel, M, Y Liao, M Rettel, C Ragan, M Evers, A-M Alleaume, R Horos, M W Hentze, T Preiss, and A A Millar. 2016. "In planta determination of the mRNA-binding proteome of *Arabidopsis* etiolated seedlings." *The Plant Cell*:tpc. 00562.2016.

Ren, Y, Y Li, Y Jiang, B Wu, and Y Miao. 2017. "Phosphorylation of WHIRLY1 by CIPK14 shifts its localization and dual functions in *Arabidopsis*." *Molecular Plant* 10 (5):749–763.

Richardson, L G L, R Singhal, and D J Schnell. 2017. "The integration of chloroplast protein targeting with plant developmental and stress responses." *BMC Biology* 15 (1):118.

Roger, A J, S A Muñoz-Gómez, and R Kamikawa. 2017. "The origin and diversification of mitochondria." *Current Biology* 27 (21):R1177-R1192.

Rolland, N, M Ferro, D Seigneurin-Berny, J Garin, R Douce, and J Joyard. 2003. "Proteomics of chloroplast envelope membranes." *Photosynthesis Research* 78 (3):205–230.

Ruelland, E. 2017. "Plant responses to chilling temperatures." *Plant stress physiology, 2nd edn. CABI, Wallingford*:97–137.

Sánchez-Baracaldo, P, J A Raven, D Pisani, and A H Knoll. 2017. "Early photosynthetic eukaryotes inhabited low-salinity habitats." *Proceedings of the National Academy of Sciences* 114 (37):E7737-E7745.

Schallenberg-Rüdinger, M, and V Knoop. 2016. "Coevolution of organelle RNA editing and nuclear specificity factors in early land plants." In *Advances in Botanical Research*, 37–93. Elsevier.

Schmitz-Linneweber, C, and I Small. 2008. "Pentatricopeptide repeat proteins: a socket set for organelle gene expression." *Trends in Plant Science* 13 (12):663–670.

Schmitz-Linneweber, C, R E Williams-Carrier, P M Williams-Voelker, T S Kroeger, A Vichas, and A Barkan. 2006. "A pentatricopeptide repeat protein facilitates the trans-splicing of the maize chloroplast rps12 pre-mRNA." *The Plant Cell* 18 (10):2650–2663.

Schmitz, J, N V Srikanth, M Hüdig, G Poschmann, M J Lercher, and V G Maurino. 2017. "The ancestors of diatoms evolved a unique mitochondrial dehydrogenase to oxidize photorespiratory glycolate." *Photosynthesis Research* 132 (2):183–196.

Sekine, D, K Murata, T Kimura, K Nakagawa, and T Miyazawa. 2016. "Identification of a Genetic Factor Required for High γ-Isoform Concentration in Rice Vitamin E." *Journal of Agricultural and Food Chemistry* 64 (49):9368–9373.

Sharma, S, U Kiran, and S K Sopory. 2017. "Developing Stress-Tolerant Plants by Manipulating Components Involved in Oxidative Stress." In *Plant Biotechnology: Principles and Applications*, 233–248. Springer.

Shmakov, N A, G V Vasiliev, N V Shatskaya, A V Doroshkov, E I Gordeeva, D A Afonnikov, and E K Khlestkina. 2016. "Identification of nuclear genes controlling chlorophyll synthesis in barley by RNA-seq." *BMC Plant Biology* 16 (3):245.

Si, T, X Wang, L Wu, C Zhao, L Zhang, M Huang, J Cai, Q Zhou, T Dai, and J-K Zhu. 2017. "Nitric oxide and hydrogen peroxide mediate wounding-induced freezing tolerance through modifications in photosystem and antioxidant system in wheat." *Frontiers in Plant Science* 8:1284.

Sultan, L D, D Mileshina, F Grewe, K Rolle, S Abudraham, P Głodowicz, AK Niazi, I Keren, S Shevtsov, and L Klipcan. 2016. "The Reverse Transcriptase/RNA Maturase Protein MatR Is Required for the Splicing of Various Group II Introns in Brassicaceae Mitochondria." *The Plant Cell* 28 (11):2805.

Woolhouse, H W, and T Batt. 2016. "The nature and regulation of senescence in plastids." *Perspectives in Experimental Biology* 2:163–175.

Yang, D. 2017. "Phytochrome control of plant growth and metabolism in *Arabidopsis thaliana*." University of Edinburgh.

York, A. 2017. "Archaeal evolution: Evolutionary insights from the Vikings." *Nature Reviews Microbiology* 15 (2):65.

Zampini, É, S Truche, Lepage, S Tremblay-Belzile, and N Brisson. 2017. "Plastid Genome Stability and Repair." *Somatic Genome Variation: In Animals, Plants, and Microorganisms*.

Zandalinas, S I, and R Mittler. 2017. "ROS-induced ROS release in plant and animal cells." *Free Radical Biology & Medicine*.

Zhao, J, T D Missihoun, and D Bartels. 2017. "The role of *Arabidopsis* aldehyde dehydrogenase genes in response to high temperature and stress combinations." *Journal of Experimental Botany* 68 (15):4295–4308.

Zhu, J-K. 2016. "Abiotic stress signaling and responses in plants." *Cell* 167 (2):313–324.

CHAPTER 22

GREEN MANURE AND ORGANIC AGRICULTURE

BENGU TURKYILMAZ UNAL

Biotechnology Department, Faculty of Arts and Sciences, Nigde Omer Halisdemir University, Nigde, Turkey.
bturkyilmaz@ohu.edu.tr, bengu35540@gmail.com

Abstract

After World War II, a "Green Revolution" movement emerged and heavily encouraged the use of pesticides and synthetic fertilisers in agricultural production. The increased use of these chemicals in agriculture has negatively affected the environment and human health. To mitigate this, the implementation of different methods of agriculture has become very important. Organic agriculture is known to improve the organic matter in the soil, together with the physical and biological properties of soils. Some types of fertilisers used in organic agriculture to improve soil quality and yield are animal fertilisers, green manure, vermicompost, bio fertilisers, and seaweed. The use of green manure in particular can help degraded soils to recover and improve crop yields at low cost and with easy application.

Keywords: ecological agriculture, legumes, nitrogen fixation, organic fertiliser, soil

Introduction

The Green Revolution is the period of change in agricultural production that targeted "high grain production" in global agriculture after 1960, in which agricultural production increased and traditional farming methods were excluded. High-yield animal and plant varieties were promoted by various green revolutionary methods that include mechanical tillage, synthetic fertilisers and biocides; and nowadays, transgenic plants are required to produce adequate food for the growing human population (Borlaug 2000, Trewavas 2002). Green revolutionary agriculture has been a major technological achievement and has had negative effects such as increased soil and water pollution, erosion, greenhouse gas emissions, increased insect resistance and loss of biodiversity (Beman, Arrigo, and Matson 2005, Heller and Keoleian 2003, Relyea 2005).

In recent years, the negative effects of modern farming practices on land and water quality have caused worldwide concern and new methods need to be developed for the purposes of research, and to regulate nutrient runoff, reduce leakage (Gillingham and Thorrold 2000, Jarvis, Scholefield, and Pain 1995, Ledgard et al. 2000, Sharpley et al. 1994) and improve soil fertility (Dick 1992, Liebig and Doran 1999).

The use of excessive fertiliser in agricultural areas is the major factor contributing to the increase of nitrate in natural waters. Increased nitrate levels in freshwater, along with increased phosphor levels, accelerate algal growth. Agricultural practices that reduce nitrate leakage from arable soils are needed (Ozturk et al. 2013a, Hakeem et al. 2011, Hakeem et al. 2016). In this context, fertiliser use and management is an important issue. Owing to their environmental friendliness, alternative methods such as organic and biodynamic agriculture have been encouraged to improve the quality of land and water compared to conventional practices (Conacher and Conacher 1998, Reganold 1995). Some scientists acknowledge that organic farming treatments reduce nitrate penetration (Kirchmann and Bergström 2001). Also, it has been proven that organic and biodynamic agriculture can increase soil organic matter and associated soil chemicals, and enhance the physical and biological properties of soils (Clark et al. 1998, Conacher and Conacher 1998, Fraser et al. 1988, Liebig and Doran 1999, Reganold et al. 1993, Wander et al. 2002).

In environmentally-friendly organic agriculture, green manure, animal manure, vermicompost, biofertiliser or a limited amount of mineral fertiliser applications are used to improve the quality of the soil. In this chapter, the importance of green manure applications in organic farming is discussed.

Improving soil quality and productivity

In recent years, excessive and careless use of agricultural land has resulted in degraded soil quality. Different amendments to regulate soil quality and productivity have been put into practice to increase beneficial soil micro-organisms, soil organic matter, total carbon, plant-available water-holding capacity, cation exchange capacity (CEC) and to reduce pathogen populations, ultimately improving soil quality (Bulluck III et al. 2002, Doran 1995).

Fertilisers are considered to have a significant influence on soil organic matter when applied with the same nutrient inputs since they are an ecological source of organic matter. This may not always happen because organic matter accumulation depends on the net integration of organic matter with the soil (Dick 1992, Paustian, Collins, and Paul 1997). In some cases, when organic materials are used together with mineral fertilisers, they may result in more plant recovery than the use of mineral fertilisers alone (Palm, Myers, and Nandwa 1997).

Traditional organic fertilisers, such as animal manure, are not preferred because they are not available in sufficient quantities, are inadequate for foodstuffs, and are difficult to process and apply (Palm, Myers, and Nandwa 1997). It has been documented that the qualitative and quantitative structure of the soil is improved by the application of organic materials such as manure compost, sewage sludge, municipal solid waste compost, and sludge compost (Giusquiani et al. 1995). Plant toxicity due to high salt content (Meek et al. 1974) and accumulation of trace metals are the most important problems in these applications (Donahue, Miller, and Shickluna 1983).

Since the use of imported synthetic fertilisers has exceeded the economic power of impoverished farmers, several projects have recently been developed for the use of herbaceous legumes, like *Cajanus* sp., *Crotalaria* sp., *Mucuna* sp., *Mimosa* sp., *Sesbania* sp. and *Tephrosia* sp. (Drechsel, Steiner, and Hagedorn 1996).

Organic Agriculture

Organic agriculture, in which growth regulators, synthetic fertilisers and pesticides are not used or can only be used to the extent permitted by law, is usually considered to be more environmentally friendly than traditional agriculture (Tuck et al. 2014, Bengtsson, Ahnström, and Weibull 2005). It can contribute to the global food supply while reducing the harmful effects of traditional agriculture on the environment (Badgley et al. 2007). Research has shown that the areas supporting organic agriculture are less susceptible to erosion (Reganold, Elliott, and Unger 1987), leaching of nutrients is low (Drinkwater et al. 1995), and mixing of pesticides and mineral fertilisers with water is also low (Mäder et al. 2002). It has also been stated that micro-organism activity and biomass are higher in organic agriculture areas (Drinkwater et al. 1995).

Several studies have been conducted on organic agriculture in the areas of biodiversity, ecological balance, human-animal health and food safety

(Gabriel et al. 2013, Reganold 2012, Winqvist, Ahnström, and Bengtsson 2012). It aims to protect consumer health by producing organic products using organic management practices (Argyropoulos et al. 2013). In recent years, the increase in the use of organic agriculture has led to an increase in the popularity of organic foods and economic gains (Klonsky and Tourte 1998).

The continuation and enhancement of soil fertility in the farm, when the elemental cycle takes place with the use of natural fertilisers; neglected or limited use of growth regulators, pesticides and synthetic fertilisers; and lower use of high energy-consuming foodstuffs are the main principles of organic farming (Fließbach et al. 2007).

Because of the processes of nutrient conversion in organic agriculture, soil quality is an important productivity factor (Bloem, Hopkins, and Benedetti 2007, Drinkwater, Wagoner, and Sarrantonio 1998, Mäder et al. 2002, Reganold et al. 1993).

The circulation of elements in organic farms is based on three basic principles: (a) sustainability of soil fertility by farm grazing; (b) no or limited use of mineral fertilisers and pesticides; and (c) lower use of high-energy consuming foodstuffs. In many countries, economic support is provided to farmers for the implementation of organic farming methods (Fließbach et al. 2007).

Deterioration in soil quality, expressed as high bulk density, low porosity, poor surface soil aggregation and slow infiltration that reduce agricultural productivity and increase pollution of water sources, can be improved by using environmentally beneficial compost in organic agriculture (Evanylo et al. 2008, Oquist, Strock, and Mulla 2007, Wander et al. 2002). Some types of fertiliser used in organic farming to improve soil quality and yield are shown in Fig.1.

Soils in organic farming areas have become more productive owing to the favourable effects of long-term organic treatments on soil properties (Bulluck III et al. 2002). Bulluck III and Ristaino (2002) found that alternative soil changes could improve the biological, physical and chemical properties of soil and improve plant yield. They stated that the alternative application could lead to better quality soil and fewer plant diseases.

The reports show that calcium, magnesium, manganese and potassium increase in the soils treated with organic treatments, but not in the soils treated with synthetic ones (Bulluck III et al. 2002). Similarly, Clark et al. (1998), found that in soil containing cover crops and manure, the concentrations of calcium, carbon, magnesium, phosphorus, and potassium increased, and the amount of soil carbon, phosphorus and potassium decreased after manure was no longer applied.

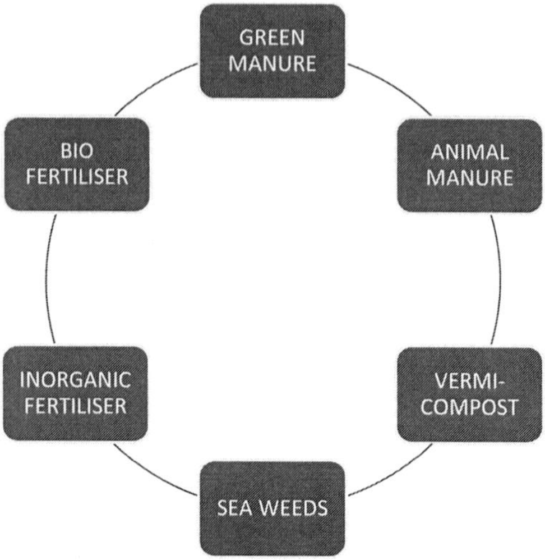

Fig. 1. Some types of fertiliser used in organic agriculture (Original)

In some studies, compost fertiliser was used to increase crop production and soil quality in organic farming. The added compost increased macro-nutrients (N, P, Mg, Na, Ca and K), micro-nutrients (Cu, Zn and Mn), and organic substances in the soil. In addition, a significant decrease in bulk density was apparent, while the significant increase in hydraulic conductivity and soil porosity was accompanied by changes to the soil's physical properties. In parallel, the productivity of *Brassica chinensis* and *Zea mays* also increased (Wong et al. 1999).

Biological nitrogen fixation (BNF) is one of the major sources of N in organic agriculture (Berry et al. 2002). While legumes take nitrogen by biological fixation, others provide soil mineral salts and/or organic wastes. In organic farms, the nitrogen in the soil is increased by the use of green manure such as legumes (Berntsen et al. 2006, Eriksen 2001).

Catch crops have positive effects on traditional and organic farming in different cropping systems (Kolbe and für Landwirtschaft 2004). Green manure catch crops have important functions such as increasing soil water-holding capacity, soil fertility and reducing erosion by covering the soil (Joyce et al. 2002). After harvesting the main crop, it can block the losses of N and other minerals remaining in the soil and transfer them to later crops (Stute and Posner 1995). In a long-term experiment by Hösch and Dersch (2003) in eastern Austria, growing crops with legumes and leguminous crops under annual precipitation of 900mm reduced the annual average nitrate infiltration. In organic agriculture, especially stockist farming, this can be a significant contributor to plant nutrition (Rinnofner et al. 2008). Another benefit of legume capture products is their ability to biodegrade nitrogen. Legumes use the existing nitrogen (N) in the soil and leave more inorganic N than non-soil crops (Thorup-Kristensen 2006).

Benefits of Organic Agriculture to Biodiversity

Global agricultural expansion is one of the greatest threats to global biodiversity (Tilman et al. 2001). Biodiversity, erosion, soil and water pollution, and threats to food safety and the quality of the land have increased the debate on the sustainability of intensive agricultural practices (Hole et al. 2005, Ozturk et al. 2010, Ozturk 2013b).

In the last quarter of the twentieth century, many dramatic reductions in both coverage and abundance were reported in agricultural land in Europe. Organic farming systems are now considered to be potential solutions for the continual loss of biodiversity and receive significant government support (Hole et al. 2005).

Organic farming methods lead to increased crop protection and biodiversity when compared to traditional methods. When comparing traditional and organic farming farms, the latter was found to have higher organic matter capacity and biodiversity, and lower nutrient losses (Tuomisto et al. 2012).

The application of herbicides reduces the number of weeds, which may have deleterious effects on insects and birds (Chiverton and Sotherton 1991). The use of insecticides is not only harmful to insects but will also reduce the number of beneficial insect species and predators that feed on them (Winston 1999).

Except for Weibull, Östman, and Granqvist (2003), all researchers, in their flora studies, have recorded more weed and species diversity in areas under organic management. The number of weed species in margins and grain areas in organic farming areas was more than twice that in conventional agricultural areas (Frieben and Köpke 1995) and the density of non-crop flora in conventional cereal fields is around 30% of organic farming areas (Hald 1999).

Vascular plants (Hyvönen and Salonen 2002), carabid beetles (Kromp 1989) and birds (Freemark and Kirk 2001) have also been reported to increase biological diversity in organic agricultural land. These studies suggest that organic farming methods generally support biodiversity (Benton, Vickery, and Wilson 2003) while loss of diversity is seen in modern agricultural practices (Ahnström 2003).

Despite the widespread belief that organic agriculture promotes biodiversity compared with conventional agriculture, this level of impact shows significant differences between different groups of organisms and types of land (Winqvist, Ahnström, and Bengtsson 2012).

Green Manuring

A rapid decline in soil fertility and high fertiliser costs are key factors limiting plant yield in many developing countries (Sangakkara and Stamp 2009). Green manure is a product used for soil remediation and as a food source for crops. While green manure production increases economic viability, it can reduce negative environmental effects related to agriculture. However, these approaches are complicated by the interactions between green manure, the environment, and how it is managed (Cherr, Scholberg, and McSorley 2006).

Increasing the physical structure of soil and soil fertility by adding large-branched green leafy plants grown in the field or leaves collected from outside to the soil is called green manuring. The goal of using green manure is to add an organic substance to the soil and to enrich the soil with mineral substances.

Advantages of Green Manuring:

- Improves soil structure
- Has a positive effect on the chemical and physical properties of soils
- Stimulates microbial activity by adding organic matter to the soil
- Helps protect the soil organic matter
- Accelerates decomposition of organic matter and breakdown of humus
- Causes the plant nutrients to be released in suitable forms for the use of plants
- Acts as a cover crop, facilitates penetration into rainwater, reduces erosion and runs away
- Food taken from deep layers by green manure plants returns the nutrients to the topsoil
- Increases the amount of some plant nutrients like P_2O_5, Ca, Mg and Fe
- Regulates air and water circulation in the ground
- Leguminous crops also fix the free nitrogen in the air

Green manure plants reduce soil erosion and increase the sustainability of agricultural systems by transferring nutrients to the main crops. This effectively inhibits the risk of nitrate infiltration. Biological nitrogen fixation by leguminous plants is an additional benefit of organic farming. However, such products may increase nitrate drain and reduce nitrogen uptake. When drought is observed, excessive water consumption may not be desirable (Rinnofner et al. 2008).

There are two types of green manuring:

1. Green fertilisation in-situ: growth as a pure product or intercrop with the main crop in the field and be embedded in the same area

a) Maincrop

b) Sequential cultivated crop

c) It can be applied as a naked fallow depending on the soil and climatic conditions of the zone

2. Green leaf fertilisation: application of green leaves and residues collected from bushes and trees (Fig.2.).

Green manuring is a suitable method for increasing soil fertility in the tropics (Joergensen 2002, Fageria 2007) and in organic agriculture for soil quality and microbial activity (Palm et al. 2001). Different species of leguminous plants and non-leguminous plants are applied as *ex situ* or *in situ* green manures (Sangakkara and Stamp 2009). Green manures are applied either after harvest as *in situ* or as *ex-situ* from external sources (Aulakh and Grant 2008). Kaewpradit et al. (2009) found that N fertilisation was promoted in rice after grinding residues were applied as green fertiliser. Some other studies have proved that *ex situ* application of green manure is useful in tropical crops (Gachengo et al. 2001).

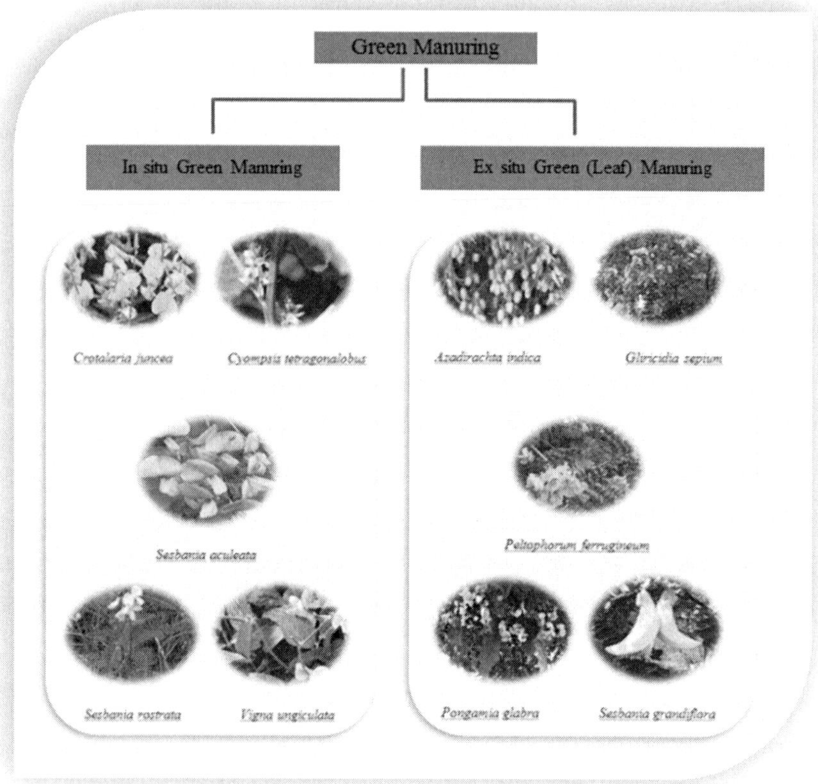

Fig.2. Green manuring types and some plant species belonging to these groups (Original)

The green manure crops should have

- Leaf and rapid growth in abundance at the beginning of the life cycle
- Abundant and succulent tops
- A deep and advanced root system
- Legumes with a good nodular growth
- The ability to grow even in weak and exhausted soil

Green manure and crop residues should be used with caution to recycle the nutrients on the farm. The management of soil organic substances, especially with short leys, is effective in providing good soil structure and biological activity. Carefully planned diverse rotations allow the application of cultural methods for weed control, while reducing pests and diseases (Watson et al. 2002).

Green manure and mineral fertiliser have similar effects on the long-term fertility of the soil compared with the application of no nutrients: the addition of nutrients in both forms, therefore, plays an important role in preserving soil quality. The Green manure has a greater influence on soil organic matter levels—and therefore soil biological activity—relative to mineral fertilisers (Edmeades 2003).

Legumes in Green Manuring

The use of soil-healing legumes in agriculture has long been debated. With increasing concerns about sustainability, it has become important to assess the role of soil-healing legumes in agricultural systems (Becker, Ladha, and Ali 1995).

Green manuring is commonly used synonymously with a cover-up (Richards, Wallace, and Turner 1996). However, green manures are the plants used for the next arable crop (Parsons 1984), or that are still immature and green (Atallah and Lopez-Real 1991), which increases their productivity when included in the soil.

Many green manure legumes are seen as a rich source of nitrogen obtained through biological N_2 fixation. With similar N utilization efficiency, green manure N has fewer loss mechanisms than synthetic N manure. For this reason, it can make a longer-lasting contribution to soil fertility (Becker, Ladha, and Ali 1995). In the long term, the use of green manure can help to recover degraded soils and increase crop yields (Eriksen 2001).

Green manure can lead to N_2O production during decomposition, nitrification and/or denitrification, the release of organic N in decomposition and formation of anaerobic fields in the soil, but N can reduce gas losses (Aulakh, Rennie, and Paul 1984, De Catanzaro and Beauchamp 1985). These emissions are generally higher than in cases where high N residues such as leguminous plants are included (De Catanzaro and Beauchamp 1985, McKenney et al. 1993).

Rinnofner et al. (2008) has found that the legume + legume mixture brought more product biomass to the plant and accumulated more plant nitrogen. In addition, biological nitrogen fixation reduces nitrate leaching by increasing N uptake from the soil (Rinnofner et al. 2008). A study on an organic farm in England compared dry matter and N accumulation, legume plants have grown during periods of 6 months to 2 years with non-leguminous ryegrass control. It was determined that red alfalfa is the most suitable species to be used as green manure (Stopes, Millington, and Woodward 1996).

The effects of growth and nitrogen fixation of legumes on soils with green manures and following rice crops have been investigated in many studies (Singh, Khind, and Singh 1991, Becker, Ladha, and Ali 1995). Green manure can be applied before rice or after rice (Garrity and Flinn 1988). Rice, which is 6-10 weeks old, is mixed with the soil during the land preparation for the new rice planting and used as before-rice green manure (Garrity and Flinn 1988). Post-rice green manures are added into the growing rice (Chungchu 1988) or sown after rice harvest (Lizhi 1988). Post-rice green manures were historically more abundant than pre-rice species.

Astragalus sinicus is the most widely used species in the cool tropics (Lizhi 1988) and *Calopogonium* spp., *Indigofera tinctoria* (Ali and Narciso 1995), *Clitoria ternatea, Crotalaria juncea, Desmanthus virgatus* and *Macroptilium atropurpureum* (Ladha et al. 1994) are the most widely used in the warm tropics, while *Sesbania cannabina* is the pre-rice green manure most commonly used in humid tropical regions of Asia and Africa, *Astragalus sinicus* is the most common post-rice green manure species for the cold tropics. Also, the *Titiania diversifolia* shrub has been used as green manure in rice fields in Asia and in corn and vegetable fields in Africa (Jama et al. 2000).

In organic farming, for exports in temperate regions, animal manure such as from cattle is generally preferred to meet the nutritional needs of plants. Feed grains such as red and white alfalfa (*Trifolium pratense* and *Trifolium repens)* grown in a mixture with grass in grazing or cutting-transport systems are the most commonly used (Høgh-Jensen and Schjørring 1997). As an alternative in recent years, crop rotation has started to be carried out in the autumn following the harvesting of the main crops, by harvesting catch crops (except for legumes) and green manure (legumes). Leguminous green manure keeps the atmospheric nitrogen which is present. Catch crops and green manures activate, pick up and re-mineralize other nutrients (Eriksen and Thorup-Kristensen 2002). It has been shown that the long-term use of green manure can greatly reduce the full-year fertiliser needs and increase soil fertility (Mueller and Thorup-Kristensen 2001).

Conclusion

In studies to increase productivity in crop production, it is necessary to pay attention to the effects of fertilisers on the environment and living things. Green fertilisers may be the preferred means for improving degraded soils and increasing crop yields owing to their low cost, environmental friendliness, biodiversity conservation and ease of use.

References

Ahnström, J. 2003. *Ekologiskt lantbruk och biologisk mångfald: en litteraturgenomgång*: Centrum för uthålligt lantbruk, Sveriges lantbruksuniv.

Ali, M, and J H Narciso. 1995. "Comparative evaluation and farmer's perceptions of green manure use in the rice-based farming systems." Proceedings of the International Rice Research Conference. The International Rice Research Institute, Los Baños, Philippines.

Argyropoulos, C, M A Tsiafouli, S P Sgardelis, and J D Pantis. 2013. "Organic farming without organic products." *Land Use Policy* 32:324–328.

Atallah, T, and J M Lopez-Real. 1991. "Potential of green manure species in recycling nitrogen, phosphorus and potassium." *Biological Agriculture and Horticulture* 8 (1):53–65.

Aulakh, M S, and C A Grant. 2008. *Integrated Nutrient Management for Sustainable Crop Production*. New York: Hawthorn Press.

Aulakh, MS, DA Rennie, and E A Paul. 1984. "The Influence of Plant Residues on Denitrification Rates in Conventional and Zero Tilled Soils 1." *Soil Science Society of America Journal* 48 (4):790–794.

Badgley, C, J Moghtader, E Quintero, E Zakem, M J Chappell, K Aviles-Vazquez, A Samulon, and I Perfecto. 2007. "Organic agriculture and the global food supply." *Renewable agriculture and food systems* 22 (2):86–108.

Becker, M, J K Ladha, and M Ali. 1995. "Green manure technology: Potential, usage, and limitations. A case study for lowland rice." In *Management of biological Nitrogen Fixation for the Development of More Productive and Sustainable Agricultural Systems*, 181–194. Springer.

Beman, J M, K R Arrigo, and P A Matson. 2005. "Agricultural runoff fuels large phytoplankton blooms in vulnerable areas of the ocean." *Nature* 434 (7030):211.

Bengtsson, J, J Ahnström, and A-C Weibull. 2005. "The effects of organic agriculture on biodiversity and abundance: a meta-analysis." *Journal of Applied Ecology* 42 (2):261–269.

Benton, T G, J A Vickery, and J D Wilson. 2003. "Farmland biodiversity: is habitat heterogeneity the key?" *Trends in Ecology and Evolution* 18 (4):182–188.

Berntsen, J, R Grant, J E Olesen, Ib S Kristensen, F P Vinther, J P Mølgaard, and B M Petersen. 2006. "Nitrogen cycling in organic farming systems with rotational grass–clover and arable crops." *Soil Use and Management* 22 (2):197–208.

Berry, P M, R Sylvester-Bradley, L Philipps, D J Hatch, S P Cuttle, F W Rayns, and P Gosling. 2002. "Is the productivity of organic farms restricted by the supply of available nitrogen?" *Soil Use and Management* 18:248–255.

Bloem, J, D W Hopkins, and A Benedetti. 2007. "Microbiological methods for assessing soil quality." *European Journal of Soil Science* 58 (5): 1214–1215.

Borlaug, N E. 2000. "Ending world hunger. The promise of biotechnology and the threat of antiscience zealotry." *Plant Physiology* 124 (2):487–490.

Bulluck III, L R M Brosius, G K Evanylo, and J B Ristaino. 2002. "Organic and synthetic fertility amendments influence soil microbial, physical and chemical properties on organic and conventional farms." *Applied Soil Ecology* 19 (2):147–160.

Bulluck III, L R, and J B Ristaino. 2002. "Effect of synthetic and organic soil fertility amendments on southern blight, soil microbial communities, and yield of processing tomatoes." *Phytopathology* 92 (2):181–189.

Cherr, C M, J M S Scholberg, and R McSorley. 2006. "Green Manure Approaches to Crop Production Florida Agric. Exp. Stn. Journal Series no. R–10702." *Agronomy Journal* 98 (2):302–319.

Chiverton, P A, and N W Sotherton. 1991. "The effects of beneficial arthropods of the exclusion of herbicides from cereal crop edges." *Journal of Applied Ecology* 1027–1039.

Chungchu, L. 1988. "Integrated use of green manure in ricefields in South China." *Sustainable Agriculture. Green manure in rice farming. International Rice Research Institute, PO Box* 933:319–331.

Clark, M S, W R Horwath, C Shennan, and K M Scow. 1998. "Changes in soil chemical properties resulting from organic and low-input farming practices." *Agronomy Journal* 90 (5):662–671.

Conacher, J, and A Conacher. 1998. "Organic farming and the environment, with particular reference to Australia: a review." *Biological Agriculture and Horticulture* 16 (2):145–171.

De Catanzaro, J B, and E G Beauchamp. 1985. "The effect of some carbon substrates on denitrification rates and carbon utilization in soil." *Biology and Fertility of Soils* 1 (4):183–187.

Dick, R P. 1992. "A review: long-term effects of agricultural systems on soil biochemical and microbial parameters." In *Biotic Diversity in Agroecosystems*, 25–36. Elsevier.

Donahue, R L, R W Miller, and J C Shickluna. 1983. *Soils. An introduction to soils and plant growth*: Prentice-Hall, Inc.

Doran, J. 1995. "Building soil quality." Proceedings of the 1995 Conservation Workshop on Opportunities and Challenges in Sustainable Agriculture. Red Deer, Alta., Canada, Alberta Conservation Tillage Society and Alberta Agriculture Conservation, Development Branch.

Drechsel, P, K G Steiner, and F Hagedorn. 1996. "A review on the potential of improved fallows and green manure in Rwanda." *Agroforestry Systems* 33 (2):109–136.

Drinkwater, L E, P Wagoner, and M Sarrantonio. 1998. "Legume-based cropping systems have reduced carbon and nitrogen losses." *Nature* 396 (6708):262.

Drinkwater, L E, D K Letourneau, F A H C Workneh, A H C Van Bruggen, and C Shennan. 1995. "Fundamental differences between conventional and organic tomato agroecosystems in California." *Ecological Applications* 5 (4):1098–1112.

Edmeades, D C. 2003. "The long-term effects of manures and fertilisers on soil productivity and quality: a review." *Nutrient Cycling in Agroecosystems* 66 (2):165–180.

Eriksen, J. 2001. "Nitrate leaching and growth of cereal crops following cultivation of contrasting temporary grasslands." *The Journal of Agricultural Science* 136 (3):271–281.

Eriksen, J, and K Thorup-Kristensen. 2002. "The effect of catch crops on sulphate leaching and availability of S in the succeeding crop." Book of abstracts.

Evanylo, G, C Sherony, J Spargo, D Starner, M Brosius, and K Haering. 2008. "Soil and water environmental effects of fertiliser-, manure-, and compost-based fertility practices in an organic vegetable cropping system." *Agriculture, Ecosystems and Environment* 127 (1–2):50–58.

Fageria, N K. 2007. "Green manuring in crop production." *Journal of Plant Nutrition* 30 (5):691–719.

Fließbach, A, H-R Oberholzer, L Gunst, and P Mäder. 2007. "Soil organic matter and biological soil quality indicators after 21 years of organic and conventional farming." *Agriculture, Ecosystems and Environment* 118 (1–4):273–284.

Fraser, D G, J W Doran, W W Sahs, and G W Lesoing. 1988. "Soil microbial populations and activities under conventional and organic management." *Journal of Environmental Quality* 17 (4):585–590.

Freemark, K E, and D A Kirk. 2001. "Birds on organic and conventional farms in Ontario: partitioning effects of habitat and practices on species composition and abundance." *Biological Conservation* 101 (3):337–350.

Frieben, B, and U Köpke. 1995. "Effects of farming systems on biodiversity." Organic Farming in Land Use Systems. Proceedings of the First ENOF (The European Network for Scientific Research Co-ordination in Organic Farming) Workshop.

Gabriel, D, S M Sait, W E Kunin, and T G Benton. 2013. "Food production vs. biodiversity: comparing organic and conventional agriculture." *Journal of Applied Ecology* 50 (2):355–364.

Gachengo, C N, C A Palm, B Jama, and C Othieno. 2001. "Tithonia and senna green manures and inorganic fertilisers as phosphorus sources for maize in Western Kenya." *Agroforestry Systems* 44 (1):21–35.

Garrity D P, J C Flinn. 1988. "Farm-level management systems for green manure crops in Asian rice environments". In: *Sustainable agriculture: green manure in rice fanning*. The International Rice Research Institute, Los Baiios, pp 111–129. IRRI, Philippines

Gillingham, A G, and B S Thorrold. 2000. "A review of New Zealand research measuring phosphorus in runoff from pasture." *Journal of Environmental Quality* 29 (1):88–96.

Giusquiani, P L, M Pagliai, G Gigliotti, D Businelli, and A Benetti. 1995. "Urban waste compost: effects on physical, chemical, and biochemical soil properties." *Journal of Environmental Quality* 24 (1):175–182.

Hakeem, K R, A Ahmad, M Iqbal, S Gucel, and M Ozturk. 2011. "Nitrogen-efficient rice cultivars can reduce nitrate pollution." *Environmental Science and Pollution Research* 18 (7):1184–1193.

Hakeem, K R, M Sabir, M Ozturk, M S Akhtar, and F H Ibrahim. 2016. "Nitrate and nitrogen oxides: sources, health effects and their remediation." In *Reviews of Environmental Contamination and Toxicology Volume 242*, 183–217. Springer.

Hald, A B. 1999. "Weed vegetation (wild flora) of long established organic versus conventional cereal fields in Denmark." *Annals of Applied Biology* 134 (3):307–314.

Heller, M C, and G A Keoleian. 2003. "Assessing the sustainability of the US food system: a life cycle perspective." *Agricultural systems* 76 (3):1007–1041.

Høgh-Jensen, H, and J K Schjørring. 1997. "Interactions between white clover and ryegrass under contrasting nitrogen availability: N 2 fixation, N fertiliser recovery, N transfer and water use efficiency." *Plant and soil* 197 (2):187–199.

Hole, D G, A J Perkins, J D Wilson, I H Alexander, P V Grice, and A D Evans. 2005. "Does organic farming benefit biodiversity?" *Biological Conservation* 122 (1):113–130.

Hösch, J, and G Dersch. 2003. *Einfluss von Bewirtschaftungsmassnahmen auf den zeitlichen Verlauf von N-Verlusten*: na.

Hyvönen, T, and J Salonen. 2002. "Weed species diversity and community composition in cropping practices at two intensity levels–a six-year experiment." *Plant Ecology* 159 (1):73–81.

Jama, B, C A Palm, R J Buresh, A Niang, C Gachengo, G Nziguheba, and B Amadalo. 2000. "Tithonia diversifolia as a green manure for soil fertility improvement in western Kenya: a review." *Agroforestry Systems* 49 (2):201–221.

Jarvis, S C, D Scholefield, and B Pain. 1995. "Nitrogen cycling in grazing systems." *Nitrogen fertilization in the environment* 38.

Joergensen, R G. 2002. "Challenges to organic farming and sustainable land use in the tropics and subtropics." *Journal of Agriculture and Rural Development in the Tropics and Subtropics* 103:105–105.

Joyce, B A, W W Wallender, J P Mitchell, L M Huyck, S R Temple, P N Brostrom, and T C Hsiao. 2002. "Infiltration and soil water storage under winter cover cropping in California's Sacramento Valley." *Transactions of the ASAE* 45 (2):315.

Kaewpradit, W, B Toomsan, G Cadish, P Vitayakoon, V Limpinuntana, P Saenjan, S Jogloy, and A Patanothai. 2009. "Mixing ground residues and rice straw to improve rice yield and N use efficiency." *Field Crops Research* 110:130–138.

Kirchmann, H, and L Bergström. 2001. "Do organic farming practices reduce nitrate leaching?" *Communications In Soil Science And Plant Analysis* 32 (7–8):997–1028.

Klonsky, K, and L Tourte. 1998. "Organic agricultural production in the United States: Debates and directions." *American Journal of Agricultural Economics* 80 (5):1119–1124.

Kolbe, H, and D S L für Landwirtschaft. 2004. *Zwischenfrüchte im Ökologischen Landbau: Fachmaterial*: Sächsische Landesanst. für Landwirtschaft.

Kromp, B. 1989. "Carabid beetle communities (Carabidae, Coleoptera) in biologically and conventionally farmed agroecosystems." In *Agricultural Ecology and Environment*, 241–251. Elsevier.

Ladha, J K, D K Kundu, M G A-V Coppenolle, M B Peoples, V R Carangal, and P J Dart. 1994. "Grain and forage legume effects on soil nitrogen dynamics and nitrogen availability to following rice crops." *Soil Science Society of America Journal (Submitted).*

Ledgard, S F, C A M De Klein, J R Crush, and B S Thorrold. 2000. "Dairy farming, nitrogen losses and nitrate-sensitive areas." Proceedings of the New Zealand Society of Animal Production.

Liebig, M A, and J W Doran. 1999. "Impact of organic production practices on soil quality indicators." *Journal of Environmental Quality* 28 (5):1601–1609.

Lizhi, C. 1988. "Green manure cultivation and use for rice in China." In: *Green manure in rice farming.* Proceedings of a symposium of sustainable agriculture 63–70.

Mäder, P, A Fliessbach, D Dubois, L Gunst, P Fried, and U Niggli. 2002. "Soil fertility and biodiversity in organic farming." *Science* 296 (5573):1694–1697.

McKenney, D J, S l W l Wang, C F Drury, and W I Findlay. 1993. "Denitrification and mineralization in soil amended with legume, grass, and corn residues." *Soil Science Society of America Journal* 57 (4):1013–1020.

Meek, B D, A J MacKenzie, T J Donovan, and W F Spencer. 1974. "The Effect of Large Applications of Manure on Movement of Nitrate and Carbon in an Irrigated Desert Soil 1." *Journal of Environmental Quality* 3 (3):253–258.

Mueller, T, and K Thorup-Kristensen. 2001. "N-fixation of selected green manure plants in an organic crop rotation." *Biological agriculture and horticulture* 18 (4):345–363.

Oquist, K A, J S Strock, and D J Mulla. 2007. "Influence of alternative and conventional farming practices on subsurface drainage and water quality." *Journal of Environmental Quality* 36 (4):1194–1204.

Ozturk, M, A Aksoy, S Gucel, and E Altundag. 2010. "Biodiversity & conservation—a case study from Europe & Turkey." Proceedings of 2nd International Conference. In: Bhatti R, Soomro SA (eds)" Biodiversity is our Life"(IC Biour-Life), Center for Biod. & Cons., Shah Abdul Latif University, Khairpur, Pakistan.

Ozturk, M, S Gucel, S Sakcali, and S Baslar. 2013a. "Nitrate and edible plants in the Mediterranean Region of Turkey: an overview." *Nitrate in leafy vegetables—toxicity and safety measures.* IK Intern. Publ. House Pvt. Ltd, New Delhi-Bangalore (India):17–51.

Ozturk, M., A Latiff, I Faridah-Hanum, K Hakeem. 2013b. "Ecosystem Biodiversity Prospects and Challenges." *Sains Malaysiana* 42 (10):1357–1538.

Palm, C A, C N Gachengo, R J Delve, G Cadish, and K E Giller. 2001. "Organic inputs for soil fertility management in tropical agroecosystems: application of an organic resource base." *Agriculture, Ecosystems and Environment.* 83:27–42.

Palm, C A, R J K Myers, and S M Nandwa. 1997. "Combined use of organic and inorganic nutrient sources for soil fertility maintenance and replenishment." *Replenishing soil fertility in Africa* (replenishing soil): 193–217.

Parsons, J W. 1984. "Green manuring." *Outlook on Agriculture* 13 (1):20–23.

Paustian, K, H P Collins, and E A Paul. 1997. *Management controls on soil carbon*: 1997a, CRC Press: Boca Raton, FL, USA. p.

Reganold, J P. 1995. "Soil quality and profitability of biodynamic and conventional farming systems: A review." *American Journal of Alternative Agriculture* 10 (1):36–45.

Reganold, J P. 2012. "The fruits of organic farming." *Nature* 485 (7397):176–177.

Reganold, J P, L F Elliott, and Y L Unger. 1987. "Long-term effects of organic and conventional farming on soil erosion." *Nature* 330 (6146):370.

Reganold, J P, A S Palmer, J C Lockhart, and A N Macgregor. 1993. "Soil quality and financial performance of biodynamic and conventional farms in New Zealand." *Science* 260 (5106):344–349.

Relyea, R A. 2005. "The impact of insecticides and herbicides on the biodiversity and productivity of aquatic communities." *Ecological Applications* 15 (2):618–627.

Richards, I R, P A Wallace, and I D S Turner. 1996. "A comparison of six cover crop types in terms of nitrogen uptake and effect on response to nitrogen by a subsequent spring barley crop." *The Journal of Agricultural Science* 127 (4):441–449.

Rinnofner, T, J K Friedel, R De Kruijff, G Pietsch, and B Freyer. 2008. "Effect of catch crops on N dynamics and following crops in organic farming." *Agronomy for Sustainable Development* 28 (4):551–558.

Sangakkara, U, and P Stamp. 2009. "Productivity and nitrogen use of maize as affected by in situ and ex situ green manuring in major and minor seasons of tropical Asia." *Acta Agronomica Hungarica* 57 (3):285–296.

Sharpley, A N, S C Chapra, Rl Wedepohl, J T Sims, T Cc Daniel, and K R Reddy. 1994. "Managing agricultural phosphorus for protection of surface waters: Issues and options." *Journal of Environmental Quality* 23 (3):437–451.

Singh, Y, C S Khind, and B Singh. 1991. "Efficient management of leguminous green manures in wetland rice." In *Advances in Agronomy*, 135–189. Elsevier.

Stopes, C, S Millington, and L Woodward. 1996. "Dry matter and nitrogen accumulation by three leguminous green manure species and the yield of a following wheat crop in an organic production system." *Agriculture, Ecosystems and Environment* 57 (2–3):189–196.

Stute, J K, and J L Posner. 1995. "Legume cover crops as a nitrogen source for corn in an oat-corn rotation." *Journal of Production Agriculture* 8 (3):385–390.

Thorup-Kristensen, K. 2006. "Root growth and nitrogen uptake of carrot, early cabbage, onion and lettuce following a range of green manures." *Soil use and management* 22 (1):29–38.

Tilman, D, J Fargione, B Wolff, C D'antonio, A Dobson, R Howarth, D Schindler, W H Schlesinger, D Simberloff, and D Swackhamer. 2001. "Forecasting agriculturally driven global environmental change." *Science* 292 (5515):281–284.

Trewavas, A. 2002. "Malthus foiled again and again." *Nature* 418 (6898):668.

Tuck, S L, C Winqvist, F Mota, J Ahnström, L A Turnbull, and J Bengtsson. 2014. "Land-use intensity and the effects of organic farming on biodiversity: a hierarchical meta-analysis." *Journal of Applied Ecology* 51 (3):746–755.

Tuomisto, H L, I D Hodge, P Riordan, and D White Macdonald. 2012. "Does organic farming reduce environmental impacts? – A meta-analysis of European research." *Journal of Environmental Management* 112:309–320.

Wander, M M, G L Walter, T M Nissen, G A Bollero, S S Andrews, and D A Cavanaugh-Grant. 2002. "Soil quality." *Agronomy Journal* 94 (1):23–32.

Watson, C A, D Atkinson, P Gosling, L R Jackson, and F W Rayns. 2002. "Managing soil fertility in organic farming systems." *Soil Use and Management* 18 (s1):239–247.

Weibull, A-C, Ö Östman, and Å Granqvist. 2003. "Species richness in agroecosystems: the effect of landscape, habitat and farm management." *Biodiversity and Conservation* 12 (7):1335–1355.

Winqvist, C, J Ahnström, and J Bengtsson. 2012. "Effects of organic farming on biodiversity and ecosystem services: taking landscape complexity into account." *Annals of the New York Academy of Sciences* 1249 (1):191–203.

Winston, M L. 1999. *Nature wars: People vs. pests*: Harvard University Press.

Wong, J W C, K K Ma, K M Fang, and C Cheung. 1999. "Utilization of a manure compost for organic farming in Hong Kong." *Bioresource Technology* 67 (1):43–46.